Application of Biomedical Engineering in Neuroscience

Sudip Paul
Editor

Application of Biomedical Engineering in Neuroscience

Springer

Editor
Sudip Paul
Department of Biomedical Engineering
North-Eastern Hill University
Shillong, Meghalaya, India

ISBN 978-981-13-7141-7 ISBN 978-981-13-7142-4 (eBook)
https://doi.org/10.1007/978-981-13-7142-4

This Springer imprint is published by the registered company Springer Nature Singapore Pte Ltd.
The registered company address is: 152 Beach Road, #21-01/04 Gateway East, Singapore 189721, Singapore

Preface

The book *Application of Biomedical Engineering in Neuroscience* creates a bridge between engineering and medicine. Usually, engineers develop novel tools and technologies, which are used by practitioners, doctors, clinicians, etc. In a country like India, we develop various prototypes/ideas/approaches, and then, with the help of industries/entrepreneurs/MSMES, these are used to develop and reconstruct tools for human applications. Obviously, ethical approval is needed so that doctors can use those products. There are several subbranches in biomedical engineering; one of the main areas of focus is the nervous system (the *brain*) and the science behind it, called "neuroscience." The brain forms a major part of the central nervous system and serves as the central processing unit of the body. Neuroscience is a multidisciplinary branch of biology that combines physiology, anatomy, molecular biology, developmental biology, mathematical modeling, and psychology to understand the fundamental and emergent properties of neurons and neural circuits. Neuroscientists generally focus on the functional aspects of neuroscience, but there is a gap between the theoretical approaches and practical applications. Keeping up with the developments of this new technology is challenging, and decisions about which tools to use and how best to use them can be difficult. This understanding between neuroscience and the biomedical science in the real world plays a vital role. Biomedical engineers develop certain circuits/technological innovations/designs/phenomena/ideas that are really helpful for neuroscientists. This book consists of nine different sections which provide sufficient information on biomedical engineering in neuroscience. It starts with an overview of human physiology, leading to the latest researched theories, concepts, tools, and techniques in an interconnected and well-organized manner to help readers better understand biomedical engineering and its latest applications in neuroscience disciplines.

Part I deals with human physiology, providing an in-depth overview of the various body parts and their functions in day-to-day life. This field includes physical and chemical composition of cells, tissues, and body organs. It creates an interrelationship between anatomy, body structure, and nutritional requirement.

Part II provides a basic understanding of the brain structure and how it works. It gives a brief description of the major regions and covers oxygen and blood supply to the brain. Our brain consists of numerous nerve cells working as basic building blocks to form a network in signal transportation. This network develops as we grow.

Part III covers various issues related to neurodegeneration and neuro-regeneration. Neurodegenerative disorders affect brain functionality, causing central nervous system disorders that might lead to nervous system dysfunction. They can cause Alzheimer's disease (AD), dementia, Parkinson's disease (PD), etc.

Part IV describes brain scanning for the detection of various brain-related disorders. These techniques have become an integral part of medical science. There are various factors that can affect the development of the brain, including genetic disorders, illness, malnutrition, blood clotting, etc. This may affect brain functionality, leading to life-threatening diseases.

Part V deals with the different types of biological signals and their significance in detection and treatment of various diseases. The signal origin/source can be at the molecular, cell, or system level. Such types of signals are employed in clinics and research laboratories. The common signals that are used are electroencephalogram (EEG), electrocardiogram (ECG), electromyogram (EMG), etc.

Part VI is divided into two chapters, covering various aspects of artificial intelligence (AI) and computer-aided diagnosis in medical science. AI has the potential to reduce manpower requirement in the healthcare field by replacing human deliverable care. Computer-aided diagnosis helps physicians make decisions in the treatment and management of various life-threatening diseases.

Part VII deals with nanomaterials in therapeutic application. Electrotherapy is beneficial in intervention and treatment of several diseases, but on the other hand, it also has negative impacts. Nanomaterials are potentially rich and are able to reduce the side effects of electro- and magnetotherapy.

Part VIII comprises three chapters relating to behavioral aspects like anxiety, stress, and emotion. All of these are caused due to neurological imbalance and dysfunction, leading to behavioral disorders. It has been seen that anxiety plays an important role in decreased quality of life with negative impact in the functionality of memory and cognition.

Part IX covers a very important field of biomedical engineering called neurological disability and rehabilitation. This differs from other fields of medicine in many ways. As we know, neurorehabilitation is a kind of teaching and learning process that aims to support the individual to cope up with family, friends, and society and to live a normal life.

As an editor, I hope that this book will be helpful from the engineering and medical points of view. I would like to see this book serve as a bridge between two different disciplines – a bridge that would help mankind with a wide range of possibilities in the healthcare industry in the near future.

Shillong, Meghalaya, India Sudip Paul

Acknowledgments

I would like to extend my gratitude to all the chapter authors for their sincere and timely support to make this book a grand success.

I am equally thankful to all the executive board members of Springer Nature for their kind approval and for granting me the permission as an editor of this book.

I would also like to extend my sincere thanks to Dr. Bhavik Sawhney, Associate Editor, Biomedicine, Springer Nature, and Ms. Vaishnavi Venkatesh, Production Editor (Books), Springer Nature, for their valuable suggestions and encouragement throughout the project.

It is with immense pleasure I express my thankfulness to my research scholars and colleagues for their support, love, and motivation in all my efforts during this project.

I am grateful to all the reviewers for their timely support and advice which improved the quality of the book.

I am speechless! I can barely find words to express all the wisdom, love, and support given to me by my beloved parents and wife and sweet son for their unconditional love, fidelity, endurance, and encouragement. I express my deep gratitude for their love, without which this book project would not have been completed.

There are so many others whom I may have inadvertently left out, but I sincerely thank all of them for their help.

Contents

About the Editor

Sudip Paul is currently an Assistant Professor in the Department of Biomedical Engineering, School of Technology, North-Eastern Hill University (NEHU), Shillong, India. He received his Ph.D. from the Indian Institute of Technology (Banaras Hindu University), Varanasi, with a specialization in Electrophysiology and Brain Signal Analysis, and subsequently completed postdoctoral research in Computational Neuroscience under the Biotechnology Overseas Associateship scheme for scientists working in the North Eastern States of India, supported by the Department of Biotechnology, Government of India. He has organized several workshops and conferences, most notably the 29th Annual Meeting of the Society for Neurochemistry, India, and the IRBO/APRC Associate School 2017. He has published more than 90 international journal and conference papers and has filed four patents.

A member of various societies and professional bodies, including the APSN, ISN, IBRO, SNCI, SfN, and IEEE, he received first prize in the Sushruta Innovation Award 2011, sponsored by the Department of Science and Technology, Government of India, and numerous other awards, including a World Federation of Neurology (WFN) travelling fellowship, Young Investigator Award, and IBRO and ISN Travel Awards. He also serves as an editorial board member for a number of international journals.

Part I

Introduction of Human Physiology

Overview of the Internal Physiological System of the Human Body

1

Karabi Ganguly

Abstract

Human physiology is described as the science of life as it correlates the study of the structure-function relation of all organs in in vivo system. Human physiology is the domain which states about the working of human system. It describes the biochemical and biophysical mechanisms from the molecular level including the cellular activity related to the behaviour of whole body. Physiology deals with the frameworks, the organs of the human body and their capacities to support the overall life processes within the body. Numerous frameworks and systems composed of several concerned organs perform a specific end task and bring about the homeostasis in a healthy living system. This chapter enriches us about the interrelated well-knitted cell-tissue-organ-system-organism functions individually to integrate all the systems to reflect ultimate vital activities of the human body.

Keywords

Cardiovascular system · Blood-vascular system · Digestive system · Respiratory system · Renal system · Musculoskeletal system · Nervous system

1.1 Blood Vascular System

Blood is a liquid connective tissue responsible for transport of nutrients, hormone and oxygen to the cells and removal of waste products from the cells. Heart is the central pumping organ that propels blood through the blood vessels throughout the body. Artery carries oxygenated blood (exception pulmonary artery) and veins carry deoxygenated blood [1].

K. Ganguly (✉)
JIS College of Engineering, Kalyani, West Bengal, India

S. Paul (ed.), *Application of Biomedical Engineering in Neuroscience*,
https://doi.org/10.1007/978-981-13-7142-4_1

Functions of Blood

- Haemoglobin of RBC carries oxygen as oxyhaemoglobin to various tissues of the body.
- Simplest soluble substances like glucose, amino acid and fatty acids are transported to tissues either as dissolved substances in blood or as bound substances to plasma proteins.
- Nitrogenous waste products and carbon dioxide are removed from the active tissues.
- White blood cells have immunological functions and play important role in detection of antigen and formation of antibody against it.
- Thrombocytes are involved in blood coagulation to stop bleeding from injured vessel.
- Hormones and enzymes are transported through blood.
- Blood is also having role in thermoregulation.

Constituents

Blood is composed of about 55% plasma which is a liquid matrix and contains 92% water by volume. Plasma along with water contains proteins, glucose, mineral ions, hormones, carbon dioxide and excretory products. Forty-five percent constituent of blood is corpuscles. There are mainly three types of blood cells or corpuscles or 'formed elements' which are red blood cells (RBCs or erythrocytes), white blood cells (WBCs or leucocytes) and platelets (thrombocytes). By volume, the red blood cells constitute about 45% of whole blood, the plasma and white cells about 0.7%. Iron containing protein haemoglobin is the major constituent of RBCs and is responsible for oxygen and carbon dioxide.

Human blood is bright red when its haemoglobin is oxygenated and dark red when it is deoxygenated. They have an adaptive immune system, based largely on white blood cells. White blood cells help to resist infections and parasites. Platelets are important in the clotting of blood. Circulatory blood is having optimum viscosity as haemoglobin is present in RBCs but not in plasma and cardiovascular system functions effectively.

Formed Elements

One microlitre of blood contains 4.7 to 6.1 million (male), 4.2 to 5.4 million (female) erythrocytes or red blood cells which contain haemoglobin and distribute oxygen. Mature red blood cells devoid of nucleus and organelles are found in mammals. The proportion of blood occupied by red blood cells is referred to as the haematocrit and is normally about 45% (Fig. 1.1).

One microlitre of blood contains 4000–11,000 leucocytes or white blood cells which are part of the body's immune system; they destroy and remove old or aberrant cells and cellular debris, as well as attack pathogens and foreign substances. 200,000–500,000 thrombocytes or platelets per microlitre of blood are having major role in blood clotting (coagulation). Fibrin forms the mesh during blood coagulation cascade over the damaged vessel [2].

Fig. 1.1 Formed elements of blood

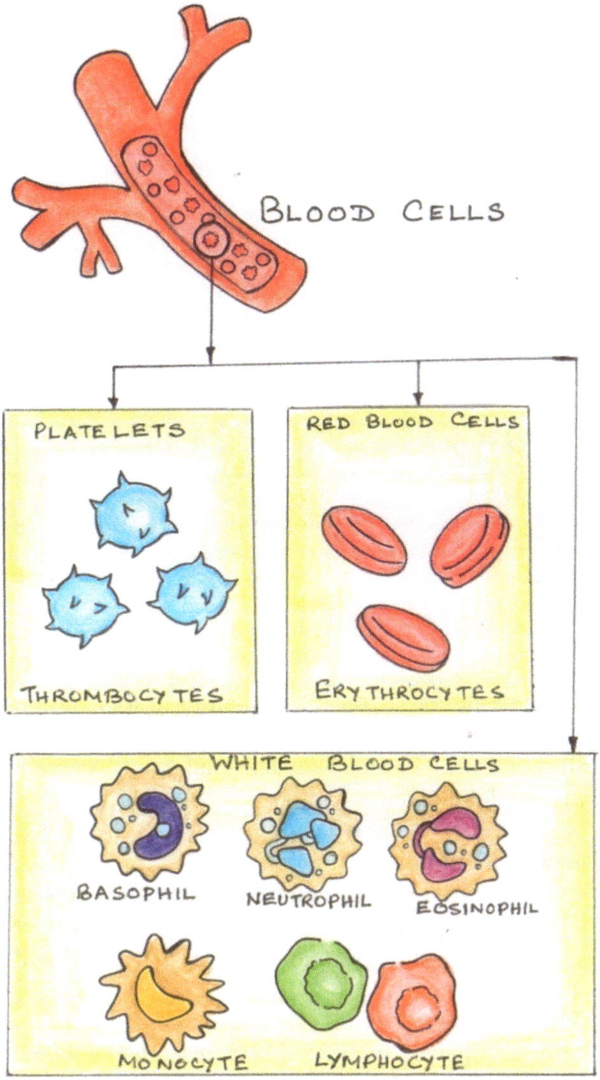

Plasma

About 55% of blood is plasma, a fluid that is the blood's liquid medium, which is straw yellow in colour. The blood plasma volume totals of 2.7–3.0 litres in an average human. It is essentially an aqueous solution containing 92% water, 8% plasma proteins and trace amounts of other materials. Plasma circulates dissolved nutrients, such as glucose, amino acids and fatty acids (dissolved in the blood or bound to plasma proteins), and removes waste products, such as carbon dioxide, urea and lactic acid [3].

Names and Sources of the Clotting or Coagulation Factors

Sl No.	Coagulation Factor	Common Name	Source
1	Factor I	Fibrinogen	Liver
2	Factor II	Prothrombin	Liver
3	Factor III	Tissue Factor Or Thromboplastin	Damaged Tissue Cells Release Tissue Thromboplastin. Platelets Release Platelet Thromboplastin.
4	Factor IV	Calcium Ions	Bone, And Absorption Through The Lining Of The Small Intestine
5	Factor V	Proaccelerin Or Labile Factor	Liver And Platelets
6	Factor VI	No Longer Used After It Was Discovered That This Chemical Is Factor Va	N/A
7	Factor VII	Proconvertin Or Stable Factor	Liver
8	Factor VIII	Anti-Hemophilic Factor	Platelets And The Lining Of Blood Vessels
9	Factor IX	Christmas Factor	Liver
10	Factor X	Stuart Prower Factor	Liver
11	Factor XI	Plasma Thromboplastin Antecedent	Liver
12	Factor XII	Hageman Factor	Liver
13	Factor XIII	Fibrin Stabilizing Factor	Liver

Fig. 1.2 Blood coagulation factors

Plasma proteins, blood coagulation factors and electrolytes are other important components of plasma. Plasma devoid of the clotting proteins (mainly fibrinogen) is called serum. Plasma contains albumin, globulin, fibrinogen, prothrombin, immunoglobulin, etc.

Blood Coagulation

Blood clotting or coagulation is a biological process that stops bleeding. Blood clotting is essential if there is bleeding due to surface injury and damage of blood vessels. Coagulation prevents haemorrhage and blood loss (Fig. 1.2).

Clotting Factors

Thirteen coagulation factors play important role in the biochemical cascade of coagulation. According to the order of their discovery, these factors are numbered. Along with these factors, vitamin K is essential for clot formation. Coagulation of blood is a complex multistep process. Liver synthesizes proteins which are sent into the bloodstream, circulate around the body and take essential part in the process. An external or internal injury that activates the proteins initiates the blood clotting process (Fig. 1.3).

Blood Group

Depending on the presence of inherited antigenic substances on the RBC surface, blood can be classified into different groups. Antigenic substances may be carbohydrate or protein or lipid. Blood type is an inherited character formed by the contribution of both parents.

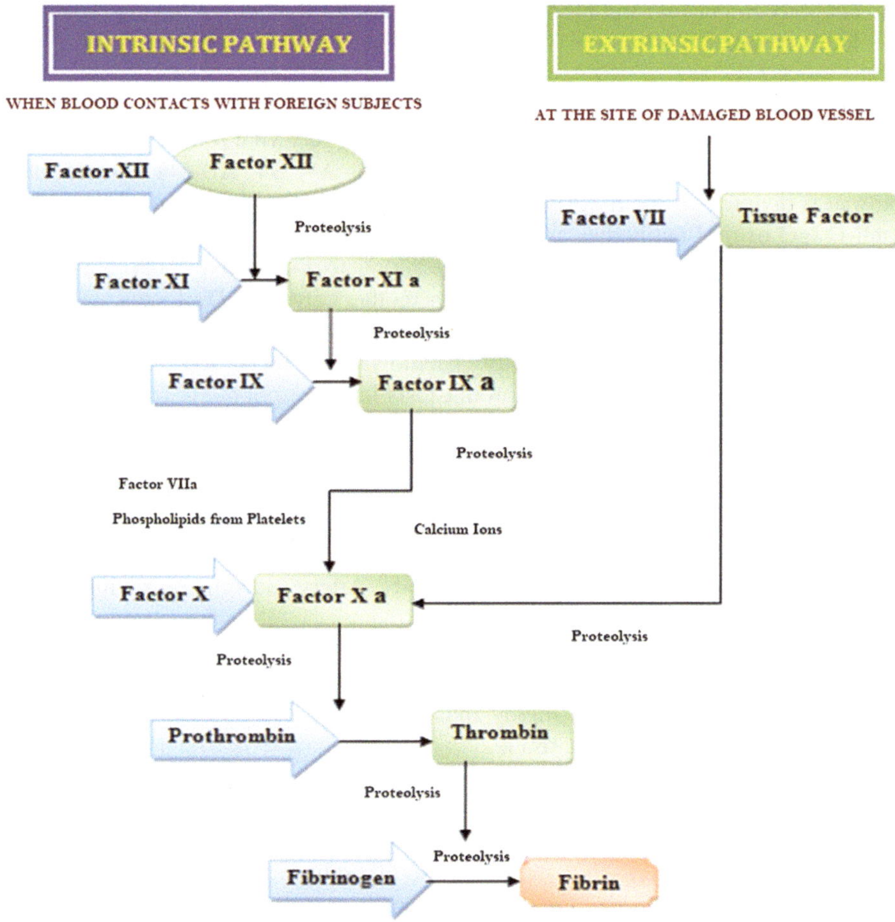

Fig. 1.3 Blood coagulation process

International Society of Blood Transfusion recognizes 35 different human blood group systems. A, B, AB and O are the four blood groups which are again divided into + or – depending on the Rhesus antigen status.

ABO system of blood grouping is done on the basis of two antigens, A and B, which are present on RBC surface. α and β are two antibodies present in blood plasma. Blood group A contains antigen A and antibody β; group B contains antigen B and antibody α; group AB contains both antigen A and B but no antibody; group O contains no antigen but both α and β antibodies.

Antibody present in recipient's plasma and antigen present in donor's RBC surface react with each other. So, blood group AB is the universal recipient having no antibody in plasma, and blood group O is the universal donor having no antigen on RBC surface.

Rh System

In human blood transfusion, Rhesus antigen plays a very vital role. In Rh antigen D, if present in blood, the blood group is denoted as A+, B+, AB+ or O+ and when absent is denoted as A-, B-, AB- or O-. D-negative individuals do not possess any anti-D IgG or IgM antibodies as Rh antigen D actually triggers the immune system response. Blood transfusion of D-positive blood to D-negative individual causes production of IgG against RhD. If an RhD- mother carries RhD+ foetus, antibodies against foetal RBCs are formed. These maternal antibodies if have the capability to cross the placental barrier can cause haemolysis of foetal RBCs. This disease condition of newborn is called erythroblastosis fetalis.

Blood Vessels Blood along with respiratory gases, nutrient and hormones is transported in various parts of the body through blood vessels which are muscular, elastic and flexible tubes. Blood vessels are of different types like arteries, arterioles, capillaries, venioles and veins. Network of these vessels is also called vascular tree which delivers blood to various part of the body. Artery and vein are having a three-layer membrane covering.

The innermost layer of the blood vessel is the tunica intima which is made of three sub-layers: epithelial cell layer, lamina propria and fenestrated membrane. Next to that is tunica media which is highly muscular layer. The outermost layer, tunica externa, is made of connective tissue.

Function of Valve

Valve prevents the backflow of blood as well as makes the flow unidirectional. The heart pumps blood directly into the arteries so blood flows in these vessels with high pressure with no backflow. Valves are absent in arteries. On the contrary, when crossing the capillaries, blood enters veins due to decrease in blood pressure, backflow is possible in veins, and valves present in veins prevent this backflow.

1.2 Cardiovascular System

Heart

The heart is a cone-shaped, hollow, muscular and central pumping organ of the body, having the size of a person's fist. The heart contracts and relaxes rhythmically to pump blood into the blood vessels throughout the life of an individual. It is placed under the sternum, above the diaphragm, in the middle of the thoracic cavity, flanked on each side by the lungs. It lies little to the left of the midline of thorax, with its pointed apex extending downwards, approximately at the level of the fifth intercostal space. It is found enclosed in a double sac of serous membrane, pericardium which is filled with pericardial fluid.

The heart wall is made of epicardium, myocardium and endocardium. Epicardium also called visceral pericardium, outermost layer of the heart wall, lubricates and protects the outside of the heart. Below epicardium lies thicker muscular layer of heart wall, myocardium. It makes up the majority thickness, mass and part of the heart that is responsible for pumping blood. Below the myocardium lies thin, smooth endocardium which prevents blood from sticking in the heart and in turn inhibits formation of blood clot.

Human heart is a four-chambered organ having two auricles or atria (left and right) and two ventricles (left and right). The right and left halves of the heart is separated by a wall or septum, named inter-auricular septum, in between auricles, and inter-ventricular septum in between ventricles. Auricles are relatively thin walled and act as reservoirs for the blood entering into the heart through the vena cava in the right atrium and pulmonary veins in the left atrium. Both contract at a time, filling the muscular ventricles which again contract forcing blood out of the heart through the aorta from the left ventricle and pulmonary artery from the right ventricle.

Each auricle opens into the ventricle of its own side through an auriculo-ventricular aperture. The right and left auriculo-ventricular apertures are guarded by the tricuspid valve and bicuspid (mitral) valve, respectively. These two valves prevent the backflow of blood from the ventricles to auricle. The flaps of tough tissue of auriculo-ventricular valves are held rigidly in place when under pressure by tiny cords (chordae tendineae) that extend to the inner muscular walls of the ventricles.

The left ventricle opens into the aorta and the right into the pulmonary artery. Both the openings are guarded by half-moon-shaped semilunar valves, named the pulmonary valve and aortic valve at the base of the pulmonary artery and aorta, respectively. Semilunar valves prevent backflow of blood from the pulmonary artery and aorta to the right and left ventricles.

The right-side chambers of the heart are small and maintain the pulmonary circulation, that is, circulation between the heart and lungs, whereas the left-side chambers of the heart are large and maintain the systemic circulation, that is, circulation between the heart and all the way to the extremities of the body [4].

Conducting Tissues of the Heart

The rhythmic contraction of the heart or heartbeat is myogenic in origin. Specialized impulse-conducting cells are present in the heart. These cells are quite different from the general cardiac muscle fibres. These cells constitute nodal tissues.

1. Sinu-auricular node (SAN): It has rich capillary blood supply. The SAN is present at the point where the superior vena cava empties into the right auricle. The SAN is also called the natural pacemaker because it is the region which initiates the heartbeat. The SAN originates 70–80 beats/min.
2. The impulse is transmitted from the SA node to AV node (atrioventricular node) by another special junctional tissue called inter-nodal bundles.

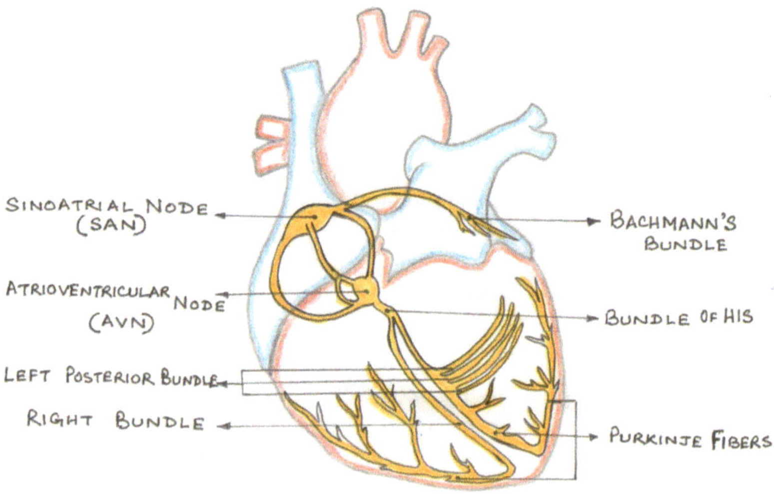

Fig. 1.4 Conduction system of the heart

3. The AV node (atrioventricular node) situates at the posterior part of the inter-auricular septum, near the coronary sinus. The AV node transmits the impulse originated in SAN to different parts of auricles and ventricles. It can create the impulse of 40–60 beats/min. If SAN fails to originate the beat, the AVN creates the impulse. So, it is called reserve pacemaker.
4. Bundle of HIS: It originates from the AVN and moves through the inter-ventricular septum. The bundle of HIS divides into two branches, right and left bundles, and moves up to the apex of the heart. The right and left bundles form fine branches which unite with the Purkinje fibres. It can create the impulse of 36 beats/min.
5. Purkinje fibres: The bundle of HIS gives out branches forming a network which spreads in the entire walls of the ventricles and spreads the excitation wave in the entire ventricles. It can generate an impulse like 30–35 beats/min (Fig. 1.4).

Coronary Systole and Diastole

Chambers of the heart remain either in systole state when cardiac muscle contracts to propel blood out of the chambers or in diastole state when cardiac muscle relaxes to allow the chambers to fill in blood. Pressure increases in major arteries during ventricular systole and decreases during ventricular diastole.

- During systole, the systolic pressure becomes high number (120 mm of Hg).
- During diastole, the diastolic pressure becomes low number (80 mm of Hg).

Cardiac Cycle

Atrial Diastole: The atrium is in diastole condition; the inferior and superior vena cava drain blood in the right atrium, and the pulmonary veins drain blood in the left atrium. The time taken is 0.7 s.

Atrial Systole: It takes 0.1 s. The atrium contracts draining blood into the respective ventricles. Due to higher atrial pressure, the first half of atrial systole is stronger than that of the last half.

Ventricular Systole: Just after the atrial systole, the ventricular systole begins. It takes a total of 0.3 s. At the onset of ventricular systole, atrioventricular valves close producing the first heart sound, LUBB. Isovolumetric phase starts just after the closure of valves. During this subphase, the pressure and volume remain constant, in spite of the contraction of the ventricles. The time taken is 0.05 s. The second subphase is rapid ejection phase when at first the outflow of blood is very rapid due to vigorous contraction. The pressure rises and semilunar valves open up; the blood gushes through the aorta and pulmonary artery. It takes 0.10 s. The third subphase is reduced ejection phase when the rate of blood flow slows down. It takes about 0.15 s.

Ventricular Diastole: It takes a total of 0.5 s. In the first subphase, protodiastolic phase, there is a brief interval between the beginning of the diastole and closing of semilunar valves. It takes around 0.04 s. At the end of this phase, semilunar valves close producing the second heart sound, DUBB. The second subphase is isovolumetric relaxation period where no alteration of volume takes place. This is the interval between the closing of semilunar valves and the opening of the atrioventricular valves. During this phase, ventricles relax as closed cavities and intraventricular pressure steeply falls. It takes 0.06 s. The third subphase is the first rapid filling phase where intraventricular pressure goes below that of the atria and the atrioventricular valves open. The atrial blood rushes into the ventricles producing a third heart sound. It takes 0.1 s. The fourth subphase is diastasis or slow filling phase where slow filling of blood occurs. Uniform filling is seen. In spite of this being the longest phase, here filling is minimum. It takes about 0.2 s. The fifth subphase is last rapid filling phase where ventricular diastole coincides with atrial systole. The atria squeeze blood with vigorous contraction, and blood rushes to the ventricles with force producing the fourth heart sound. It takes 0.1 s.

Blood Circulation Through Human Heart

The right atrium receives deoxygenated blood from the various parts of the body (except the lungs) through the two major vena cava, superior vena cava and inferior vena cava. Superior vena cava brings deoxygenated blood from head and upper parts of the body whereas Inferior vena cava returns deoxygenated blood from the lower parts of the body. From the right atrium, blood then passes through the tricuspid valve into the right ventricle. When the right ventricle contracts, blood is forced to propel through a set of semilunar valve into the pulmonary artery which in turn carries the deoxygenated blood to the lungs [5].

After the blood has passed through the lungs, it is brought back to the left atrium through the left and right pulmonary veins. From the left atrium through bicuspid valve, oxygenated blood passes to left ventricle. When the left ventricle contracts, oxygenated blood propels through a set of semilunar valve into the aorta which in turn carries the oxygenated blood to various parts of the body.

Though the heart is an organ filled with blood, its muscle layers are too thick to be nourished by this blood. The cells receive special nourishment through arteries called coronary arteries. There is an enlargement of the aorta at the point where it leaves the heart. This is called the aortic sinus. From here, the right and left coronary arteries branch off.

Cardiac Output

The amount of blood propelled out of the ventricles of the heart per minute is called cardiac output.

CO is having two important components, stroke volume and minute volume.

Stroke volume or systolic discharge refers to the output per ventricle per beat. For an adult, it is 70 ml. Minute volume refers to the output per ventricle per minute and for an adult it is 5–6 litres.

In other words, *cardiac output or minute volume* = *stroke volume* × *heart rate* (Figs. 1.5 and 1.6).

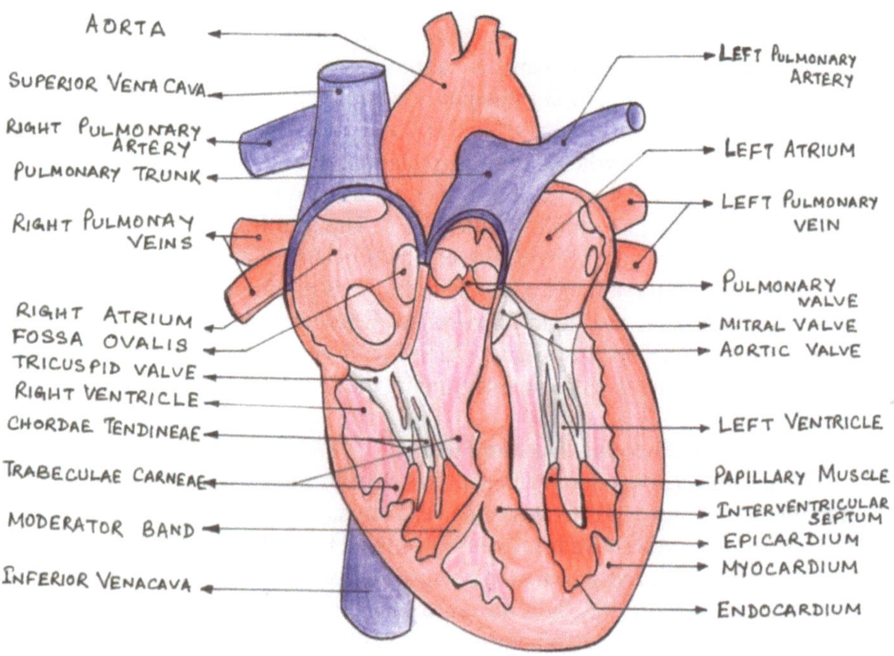

Fig. 1.5 Blood circulation through the heart

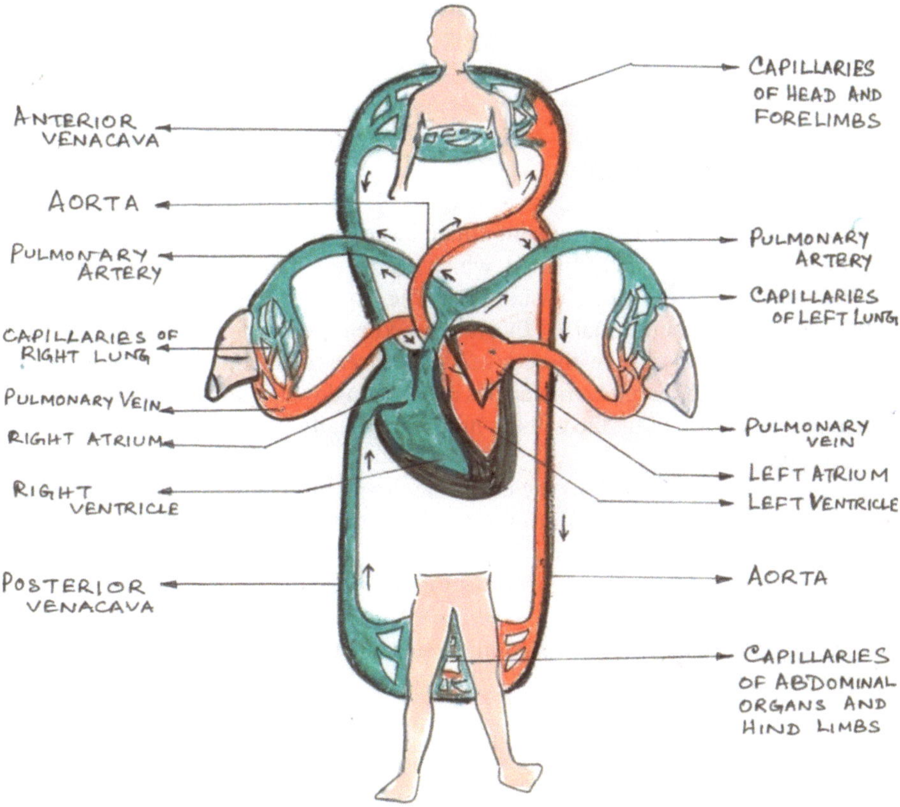

Fig. 1.6 Blood circulation through the human body

Electrocardiogram

The electrocardiogram (ECG) is the voltage-time graph of the electrical activities of the heart. Electrocardiograph is a non-invasive device that measures and monitors the cardiac activity from beat to beat and produces a distinctive waveform in response to the electrical changes taking place within the heart [6].

The first part of the wave, called the P wave, is a small positive (upwards) wave of ECG produced due to the atrial depolarization which is reflected by the increase in voltage of 0.1mV during atrial systole. Atrial repolarization merges with the ventricular complex and cannot be recorded in normal ECG. The next part of the ECG wave is the QRS complex which features a small drop in voltage (Q) a large voltage peak (R) and another small drop in voltage (S). So, Q wave is a small negative wave followed by large positive R wave, next followed by small negative S wave. QRS complex is produced due to the ventricular depolarization during ventricular systole. The final part of the ECG wave is the last positive wave, T wave, a small peak that follows the QRS complex. Variations in the waveform and distance between the waves of the ECG denote physiological abnormality and disease conditions like heart block, congenital cardiac abnormality, myocardial infarction, ischaemia, etc.

1.3 Digestive System

The digestive system is one of the important systems of the body, through which the energy is supplied externally as food, which is converted into the required chemicals for the nutrition of the cells, tissues, etc. The digestive system is also called the alimentary or gastrointestinal (GT) system. When food is ingested, the digestive system plays a vital role in the conversion process of complex sugars to simple sugars (glucose), and large fat molecules are broken down to fatty acids and glycerol, which can be absorbed by the cells as nutrients. Finally, the unwanted materials are egested through the anus.

Digestion can be defined as the complete process of changing the chemical and physical composition of food in order to facilitate assimilation of the nourishing ingredients of food by cells of the body.

Organs of the digestive system are as follows:

1. Oral cavity/buccal cavity – lips, tongue, teeth, salivary glands
2. Oesophagus
3. Stomach
4. Small intestine – duodenum, jejunum, ileum
5. Large intestine – caecum, colon
6. Rectum
7. Anus

1. *Oral cavity*: The gastrointestinal tract (GI) is a continuous tubular passageway that begins with the oral cavity or mouth.
   (a) Lips: It forms the opening for the oral cavity. It is a muscular vascular and motile organ.
   (b) Tongue: It is a muscular motile organ. The main functions of the tongue are to move the food around during mastication (chewing), deglutition (swallowing) and determination of tastes.
   (c) Teeth: Teeth are situated in the oral cavity. Teeth cut and tear the food into smaller pieces, crush and grind the food into finer particles.
   (d) Salivary gland: There are three pairs of salivary glands situated in the oral cavity. The parotid gland in the posterior part of the oral cavity, submaxillary gland under the maxillary bone and sublingual gland below the tongue on either side of the mouth secrete saliva into the oral cavity. Saliva contains digestive enzyme called ptyalin which begins the digestion of carbohydrates. It also contains a thick, lubricant called mucin.

2. *Oesophagus*: It is a tube-like structure which connects the buccal cavity and the stomach. The wave-like contractions, peristalsis, in the oesophagus propels the food to the stomach.
3. *Stomach*: The stomach is an elastic sac-like structure located in the abdominal cavity, below the diaphragm; the interior wall of the stomach is composed of mucous membrane and contains the glands that secrete HCl and gastric juice. Gastric juice is a clear and slightly acidic fluid, containing HCl, enzymes, minerals and salts.
   1. To begin the chemical digestion, pepsin reduces proteins to peptides and polypeptides. A further stage of protein digestion takes place in the small intestine.
   2. Rennin converts the soluble milk proteins caseinogens to the insoluble form, casein, which is then reduced to peptones by pepsin.
   3. Gastric lipase is an enzyme produced in small amounts, and it begins the digestion of fats. Once the food is mixed with gastric juices and HCl, it forms chime. When it is sufficiently digested, the chime leaves the stomach through the end part of it and enters into the small intestine [7].

Functions of the Stomach

1. Physical digestion.
2. Changing solid food into semisolid.
3. Chemical digestion takes place here (Fig. 1.7).

4. *Small intestine*: The small intestine is the continuation of the GI tract and is a connection between the stomach and large intestine. It is divided into three parts: (i) duodenum, (ii) jejunum and (iii) ileum.

The functions of the small intestine are completion of digestion of food and to absorb the essential soluble nutrients of digestion.

The pancreas secretes digestive enzymes like amylase for digestion of carbohydrates, trypsin and chymotrypsin for digestion of proteins and lipase for digestion of fats.

The gall bladder secretes bile into the duodenum through the bile duct. Bile is essential for the emulsification of fats into fat droplets.

In the walls of the entire small intestine, there are tiny fingerlike projections called villi. Through the tiny capillaries in the villi, the nutrients such as glucose, fatty acids and amino acids are absorbed into the bloodstream. Functions of small intestine are completion of physical and chemical digestion, release of digestive juice which contains enzymes, providing a large surface area for absorption of digested food, passage of waste materials to the large intestine by peristalsis and protection by screening bacteria.

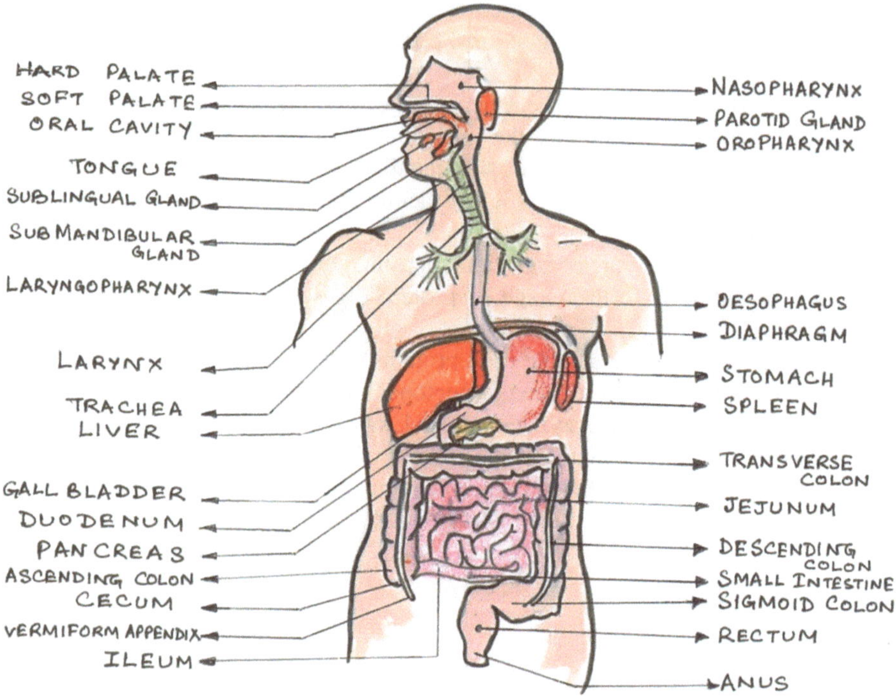

Fig. 1.7 Digestive system

5. *Large intestine*: The next part of the GI tract after the small intestine is the large intestine. It is composed of caecum and colon. The caecum or the first part of the large intestine is connected to the ileum. The vermiform appendix suspends from the caecum.

The colon is divided into ascending colon, transverse colon, descending colon and sigmoid colon.

Functions of the large intestine are absorption of water from the faecal matter, secretion of mucous and passage of the faeces into the rectum by peristalsis.

6. *Rectum*: The rectum extends from the sigmoid colon and ends in the lower opening of the GI tract called the anus. It serves as a storage area for the faeces. The rectum passes the faeces into the anus.
7. *The anus* is the end part of the GI tract that expels the faecal matter out of the body, and the process is known as defecation (Fig. 1.8) [8].

MAIN SITES OF DIGESTION IN HUMAN

1. MOUTH	When food is chewed, saliva starts digesting carbohydrates.
2. ESOPHAGUS	Peristaltic movement pushes the food down into the stomach.
3. STOMACH	Everything is blended with digestive juices. HCL kills bacteria, Enzymes breakdown the proteins.
4. LIVER	Green liquid called bile is secreted to break down the fats.
5. PANCREAS	Many kinds of digestive enzymes are synthesized here.
6. SMALL INTESTINE	Food is mixed with bile & pancreatic juices.
7. LARGE INTESTINE	Indigestible food and water are processed, stored & dispersed.
8. ANUS	Solid waste passes from the rectum in order to leave the body

Fig. 1.8 Digestion site

Accessory Organs of Digestion

Liver: It is the largest gland of the body. The vital functions of the liver are:

(a) Produces bile, which is used in the small intestine to emulsify fats into fat droplets and absorb fats after digestion.
(b) Removes glucose (sugar) from blood which is converted and stored as glycogen.
(c) Stores vitamins such as B12, A, D, E and K.
(d) Removes poisons or toxic substances from blood.
(e) Destroys old erythrocytes and releases bilirubin.
(f) Produces various blood proteins, such as prothrombin and fibrinogen which helps in clotting of blood.

Gall bladder: It is a small pear-shaped organ situated below the liver and acts as a reservoir for the bile synthesized in the liver and to increase the concentration of bile.

Pancreas: It is a leaflike structure situated behind the stomach between the loop of the duodenum and the spleen. It acts as both endocrine and exocrine gland. In the digestive system, it provides digestive juices that pass through the pancreatic duct.

Function of pancreas: Exocrine pancreatic juice contains three enzymes:

(a) Trypsin acts on protein.
(b) Amylase acts on maltose.
(c) Lipase acts on fats (Figs. 1.9, 1.10 and 1.11) [9, 10].

CARBOHYDRATE DIGESTION

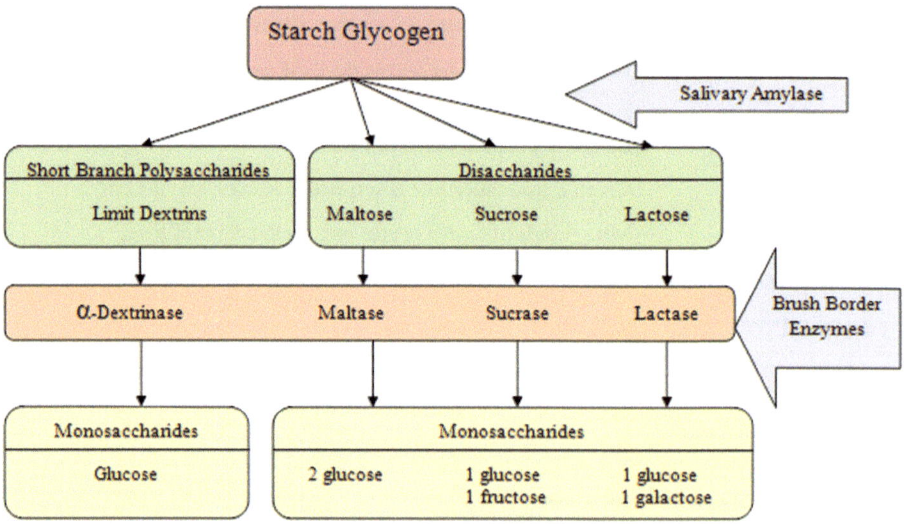

Fig. 1.9 Digestion of carbohydrate

PROTEIN DIGESTION

PROTEOLYTIC ENZYMES AND THEIR ACTIONS		
ENZYMES SECRETED	SECRETED IN	ACTION
Pepsin	Stomach	Converts complex proteins to small peptides
Trypsin	Pancreas	• Specifically acts on peptide bonds contributed by basic amino acids like arg, lys, his • Activates trypsinogen to trypsin • Procarboxypeptidase to carboxypeptidase, proelastase to elastase and proaminopeptidase to aminopeptidase
Chymotrypsin		Specifically acts on peptide bonds contributed by aromatic amino acids like phe, tyr, trp
Carboxypeptidase		Carboxy terminal amino acids
Elastase		
Amino Peptidase	Small Intestine	Amino terminal amino acids
Dipeptidase		Acts on dipeptides and release free amino acids

Fig. 1.10 Digestion of protein

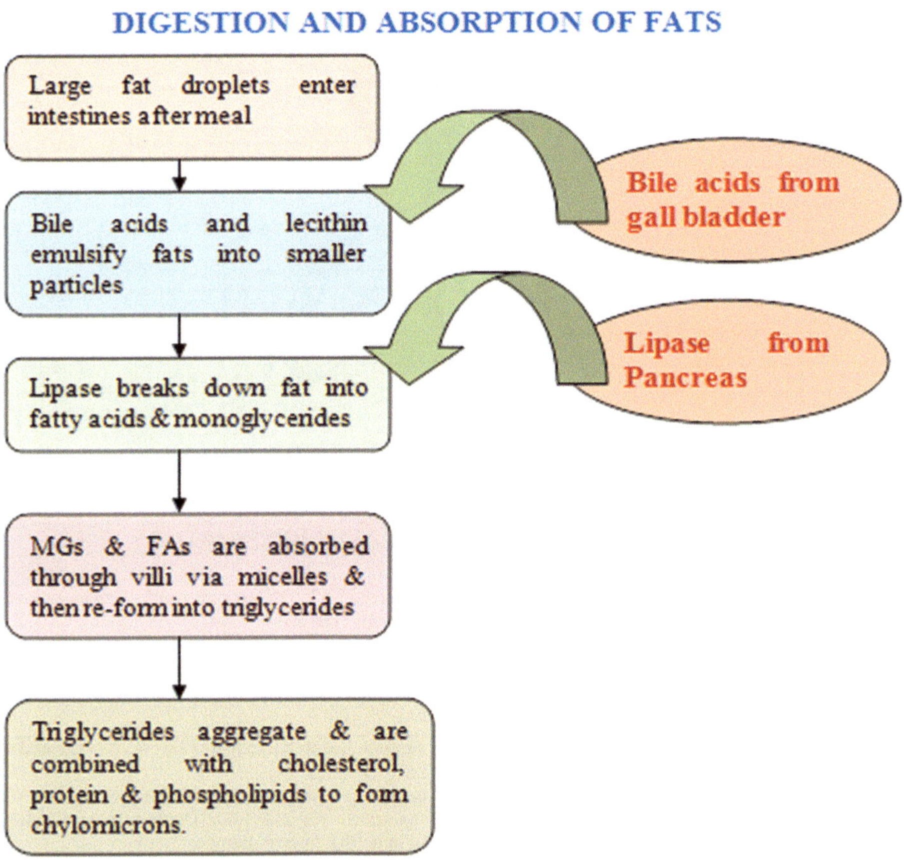

Fig. 1.11 Digestion and absorption of fats

1.4 Respiratory System

The organ system that aids the body in the exchange of gases between the air and blood is the respiratory system. It comprises of the respiratory tract, the pulmonary vessels, the lungs, the alveoli and breathing muscles. The alveoli and the alveolar ducts are responsible for the gaseous exchange in the body, and the rest of the organs provides a pathway of distribution of air. The organs of the respiratory system filter, warm and humidify the air that we breathe in and also play a role in speech and the sense of smell, help the body to maintain homeostasis or balance among the many elements of the body's internal environment (Figs. 1.12 and 1.13) [11].

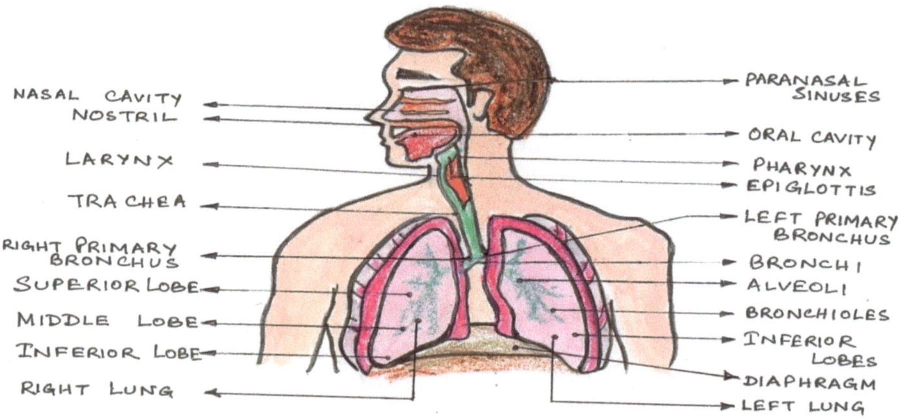

Fig. 1.12 Organs of the respiratory system

UPPER RESPIRATORY TRACT	LOWER RESPIRATORY TRACT
Composed of Nose, Pharynx and Larynx etc.	Composed of Trachea, Lungs and all segments of Bronchial Tree etc.
Organs of upper respiratory tract are located outside the chest cavity.	Organs of upper respiratory tract are located inside the chest cavity.
Nasal Cavity: Inside the nose the sticky mucous membrane lining the nasal cavity traps dust particles & tiny hairs called cilia help move them to the nose to be sneezed or blown out. **Sinuses:** Air filled spaces alongside the nose help makes the skull lighter. **Pharynx:** Both food and air pass through the pharynx before reaching their appropriate destination. **Larynx:** It is essential to human speech.	**Trachea:** Located just below larynx, it is the main airway to lungs. **Lungs:** One of the largest organs of the body. Responsible for providing Oxygen to capillaries and exhaling carbon dioxide. **Bronchi:** It branch from trachea into each lung and create the network of intricate passage that supply lungs with air. **Diaphragm:** It is the main respiratory muscle that contracts and relaxes to allow air into lungs.

Fig. 1.13 Subcomponents of the respiratory system

Structure of the Human Lungs

The lungs are placed in the thoracic cavity, above the diaphragm, on either side of the sternum within the rib cage. It is spongy due to the presence of alveoli and pink in colour as it is rich in blood capillaries for the exchange of respiratory gases. The two lungs are cone shaped, tapering at the apex and broad at the base. The right lung has three lobes and the left lung has two lobes. The left lung is slightly smaller to accommodate the heart in position. Each lung is covered by two-layered membrane called pleura – the inner visceral pleura and outer parietal pleura. The pleural cavity between these two layers is filled with pleural fluid. This arrangement favours free movement of expanding and contracting the lungs (Fig. 1.14).

RIGHT LUNG	LEFT LUNG
It has more lobes and segments than left.	It has less lobes and segments than the right lung.
It is divided into three lobes-upper, middle and lower by two fissures-one oblique and one horizontal	It is divided into two lobes-upper and lower by oblique fissure. It doesn't have a middle lobe.
The upper horizontal fissure separates the upper from middle lobe. It begins in lower oblique fissure near posterior border of lung and running horizontally forward cuts the anterior border on a level with sternal end of fourth costal cartilage.	It has a homologous -feature, a projection of upper lobe termed as "lingula" or little tongue. The lingula on the left serves as an anatomic parallel to the right middle lobe with both areas being predisposed to similar infections and anatomic complications.
The lower oblique fissure separates the lower from middle & upper lobes. It is closely aligned with oblique fissure in left lung.	There are two broncho-pulmonary segments of lingual-superior & inferior.

Fig. 1.14 Section of the lungs

Mechanism of Breathing

The entire process of respiration takes place in two phases:

1. Inspiration or inhalation
2. Expiration or exhalation

Exchange of Gases

After passing through the bronchioles, oxygenated air finally reaches the tiny air sacs (alveoli) by passing through. Each alveolus is thin walled and is surrounded by a network of very fine blood capillaries. The oxygen present in the air passes through the walls of the alveoli into the blood present in capillaries. This blood is returned to the heart so that it can be supplied to all the tissues of the body. In the tissues, oxygen is used for oxidation of food and is converted into carbon dioxide. From the tissues, the blood absorbs carbon dioxide and carries to the alveoli of lungs for expiration (Fig. 1.15) [12].

1.5 Renal System

The organ of the renal system is the kidney that is a bean-shaped organ located at the rear of the abdominal cavity in the retroperitoneal space in both sides of the vertebral column in the body. Nephrons are the structural and functional unit of the kidney. It performs various functions like filtering of blood, releasing and retaining of water, removal of wastes, acid-base balance, etc.

The two major structures of the kidney are the renal cortex and renal medulla. The renal cortex that surrounds a portion of the medulla is called the renal pyramid (of Malpighi). In between the renal pyramids, there are projections of cortex known

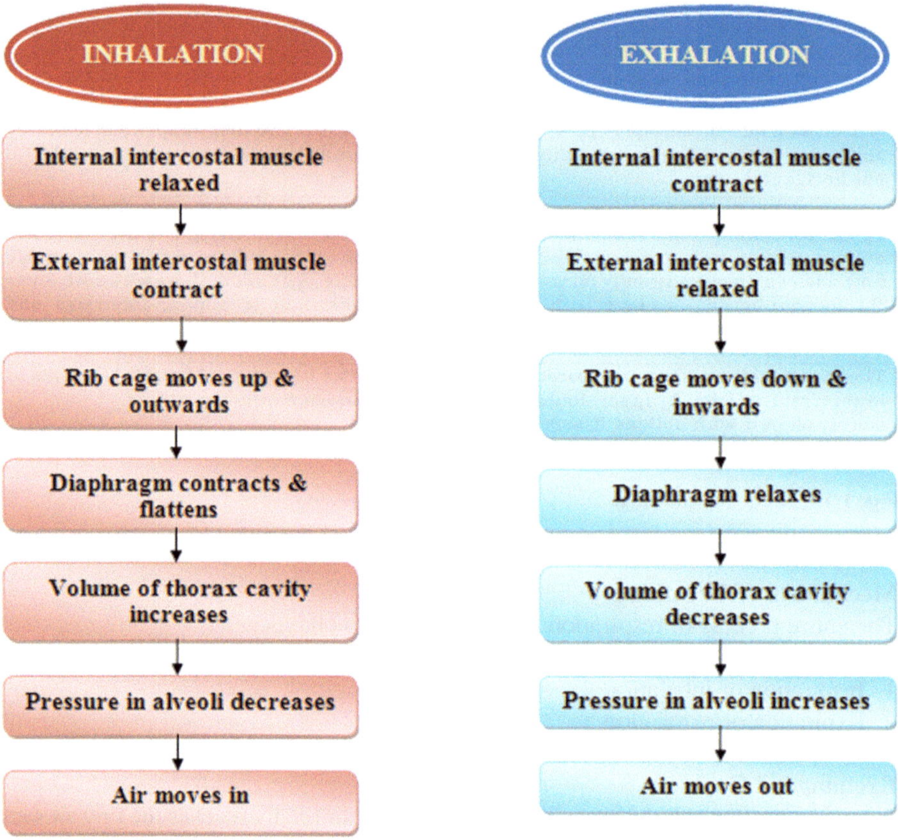

Fig. 1.15 Mechanism of breathing

as the renal columns or Bertin columns. The initial filtering portion of a nephron is the renal corpuscle which is located in the cortex. This is followed by a renal tubule that passes from the cortex deep into the medullary pyramids. Each kidney feeds urine into the bladder by means of a tube known as the ureter. Urine is a byproduct of the osmoregulatory function of kidneys, which filter blood, reabsorb water and nutrients and secrete wastes (Fig. 1.16).

Nephron

Nephron, the cardinal unit of kidney, regulates the blood volume, blood pressure, plasma osmolarity and volume of water and soluble substances in blood. By countercurrent multiplier system, nephron through filtration, reabsorption and secretion ultimately forms hypertonic, slightly acidic urine [13].

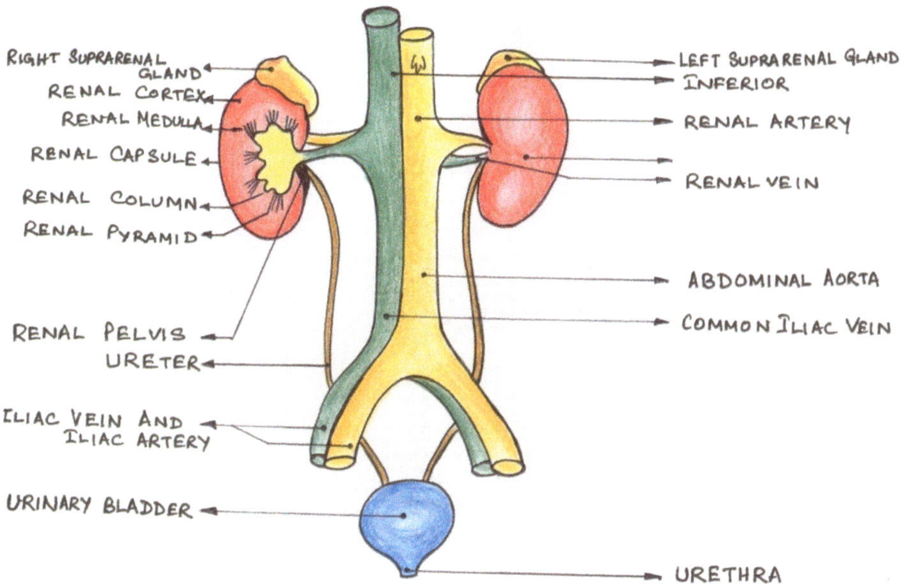

Fig. 1.16 Renal system

Different Parts of the Nephron

The Glomerulus

It is a tuft of capillaries that is surrounded by the Bowman's capsule. Afferent arteriole supplies the blood to glomerulus. The juxtaglomerular apparatus (JGA) is a group of specialized cells that are located around the afferent arteriole where it enters the renal corpuscle.

Proximal Convoluted Tubule (PCT)

This convoluted tubule lies next to the Bowman's capsule. The inner lining of this PCT consists of a single layer of brush-bordered cubical epithelial cells. Over 99% of the filtered load of sodium present in the glomerular filtrate is reabsorbed, and two thirds of the reabsorption occurs in the PCT by uniport (Na+ reabsorbed partly by carrier-mediated passive transport or by active transport where Na+ is subjected to strong electrochemical attraction) or antiport (Na+-H+ exchange) or symport (Na+ reabsorbed along with glucose, amino acid, etc.) mechanism (Fig. 1.17) [14].

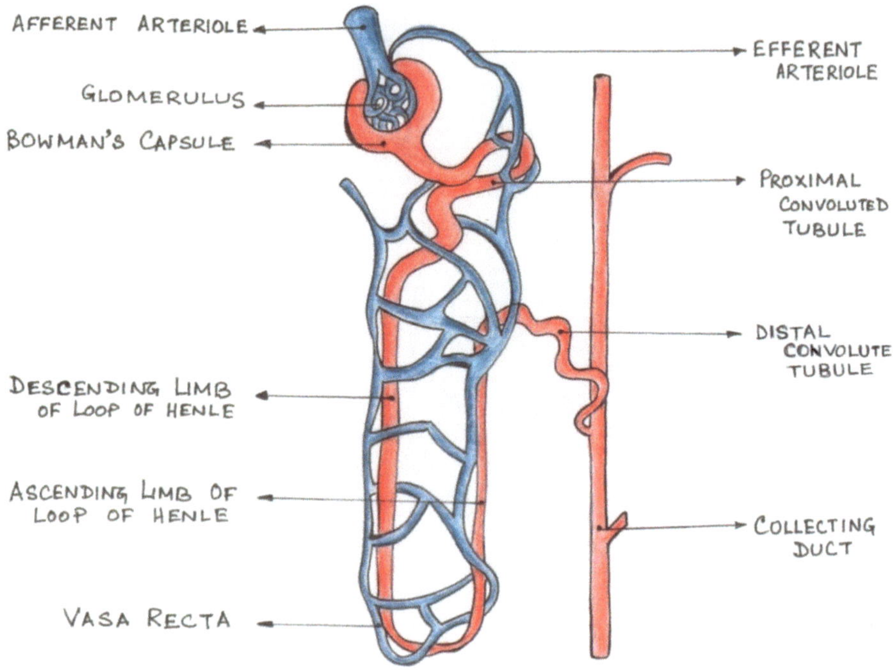

Fig. 1.17 Different parts of the nephron

Loop of Henle

It lies next to the PCT and has U-shaped appearance. It consists of a descending limb and ascending limb. The first part of the descending limb consists of cubical epithelial cells. The remaining part of the descending limb and the first part of the ascending limb consist of flattened epithelial cells. The loop of Henle is the narrowest part of the renal tubule, and it extends inwards within the medulla. The descending limb is permeable to water but not to electrolytes, whereas the ascending limb is permeable to electrolytes but not to water. So, water reabsorption takes place in the descending limb and electrolytes are reabsorbed in the ascending limb.

Distal Convoluted Tubule (DCT)

It is the part of the renal tubule which connects the collecting duct and loop of Henle. The first part of the DCT remains close to modified cells of afferent arteriole and is called the juxtaglomerular cells. Special cells like the macula densa, juxtaglomerular cells and Lacis cells form the juxtaglomerular apparatus which secretes rennin. The DCT is permeable to water but not to electrolytes so water reabsorption takes place here. ADH influences this water reabsorption [15].

Collecting Duct

This is the last part of nephron and consists of cubical epithelial cells. Water reabsorption here is also influenced by ADH. Collecting tubules of different nephron unite to form the duct of Bellini which opens through the pore in the renal pyramid, and next from there, the hypertonic slightly acidic excretory fluid is brought into the ureter and from there it leaves the kidney [16].

Functions of the Kidneys

Excretion of wastes → Wastes like urea; uric acid produced by metabolism is excreted into the urine by the kidney.
Reabsorption of vital nutrients → Vital nutrients like amino acids get reabsorbed in PCT by the sodium-dependent transporter; glucose is also reabsorbed in PCT by Na+/glucose cotransporter.
Acid-base homeostasis → It is the maintenance of pH around a stable value which is performed by the kidney by reabsorbing and regenerating bicarbonate from urine and excreting hydrogen ions and anions into urine.
Osmolality regulation → The level of water and salt in the body is maintained by the kidney. ADH is secreted with increase in osmolality which in turn results in water reabsorption by the kidney and increase in urine concentration.
Blood pressure regulation → Long-term regulation of BP mainly depends on the kidney. Changes in renin (chemical messenger) alter the hormones angiotensin II and aldosterone. Both the hormone increases the absorption of sodium chloride by the kidney, thus raising blood pressure [17, 18].

1.6 Musculoskeletal System

Muscular System
There are mainly three types of muscles – voluntary (as functions under our will), skeletal (attached to skeleton) or striated (have striations or light and dark band) muscle; cardiac (attached to heart) muscle; and involuntary (does not function under our will) or smooth muscle. A voluntary muscle has an outermost connective tissue covering called epimysium within which muscle is divided into several bundles called fasciculum. Each fasciculum is again bounded by a connective tissue covering called perimysium. The fasciculum consists of numerous muscle fibres each of which is surrounded by another connective tissue envelop called endomysium. Inner to endomysium lies the structureless membrane of the muscle fibre, sarcolemma. The cytoplasm of the muscle fibre is called the sarcoplasm within which numerous myofibrils are present.

Myofibrils have alternate light (I-band) and dark (A-band). Dark band is composed of myosin filament (contractile protein) and light band is composed of actin protein. Myosin filaments remain 45 nm apart. They extend from one end of A-band to the other. There is a less refractile area and so a less dark area in the middle of the A-band known as H-band. Central part of the H-band contains a darker zone called M-band [22, 23].

At rest the central portion of myosin is not overlapped by actin filament. Two more types of filament or contractile proteins are tropomyosin and troponin.

Tropomyosin is a coiled strand with two α helices and it fits in the groove of actin. Troponin is attached to tropomyosin and is distributed at intervals of about 40 nm along the filament forming reactive sites for contraction process. Troponin is of three types –tropomyosin-binding troponin (TpT), calcium-binding troponin (TpC) which bears four sites for calcium ion binding and troponin with binding sites for F-actin (TpI).

Each myosin filament is made of golf club-shaped molecule consisting of two parts – heavy meromyosin (HMM) and light meromyosin (LMM). HMM consists of large globular head and a short tail. Globular head has two important sites: one for binding the actin filament and the other for ATPase activity where ATP is hydrolysed.

Actin exists in two forms – G-actin (globular actin) and F-actin (fibrous actin). F-actin is formed by polymerization of G-actin to form a two-stranded helical chain [24].

Muscle Contraction Physiology

1. An action potential in a motor neuron causes acetylcholine to release in the synaptic cleft.
2. Acetylcholine binds with receptors on the cell membrane on the muscle fibre, opening Ca2+ -Na+ channels. Usually referred to as calcium channels.
3. Calcium is released from the terminal cisternae into the muscle fibre.
4. Calcium binds to troponin
5. Troponin shifts tropomyosin, which was blocking the active site on the actin.
6. Myosin heads attach to actin by breaking down ATP to ADP and a phosphate via myosin-ATPase
7. The myosin head forms a 'cross-bridge' on the active site of the actin filament.
8. The cross-bridge pulls actin, which slides over the myosin – known as the 'power stroke'.
9. The release of ADP completes the cross-bridge movement, and ATP attaches to myosin, breaking the actin-myosin cross-bridge.
10. Every time ATP is split into ADP + P, the myosin head 'cocks' into place to form another cross-bridge with actin.

This entire process shortens the sarcomere, which is the functional unit of a muscle cell.

Excitation-Contraction Coupling

Action potential initiates skeletal muscle contraction. Calcium ions play an important role in the process. Thus, what is needed is a way to link muscle excitation (the depolarization of the action potential) to Ca^{++} release from the sarcoplasmic reticulum. This link is known as *excitation-contraction coupling*. Invagination of the sarcolemma forms the T-tubule which moves inwards the muscle fibre and forms triad structure with two cisternae on either side. Action potential travels through T-tubule and depolarizes cisternal membrane. The lumen of the T-tubule is continuous with the extracellular fluid, and the membrane depolarization during an action potential occurs across the T-tubule membrane. On either side of the T-tubule are swellings of the sarcoplasmic reticulum called cisternae (*lateral sacs*). The lateral sacs are in close apposition to the T-tubule that forms triad structure. In fact, in very high magnification electron micrographs, one can see regularly spaced densities, which were dubbed *feet* when they were first discovered by electron microscopy about 40 years ago. Subsequent molecular studies have identified the feet as being part of the *calcium channel protein in the membrane of the sarcoplasmic reticulum (SR* Ca^{++} *channel)*. Depolarization in the T-tubule membrane opens a Ca^{++} channel in the sarcoplasmic reticulum membrane.

Muscle becomes fatigue and the capacity of muscle to generate and exert force declines. It can be a result of vigorous exercise, but abnormal fatigue may be caused by barriers to or interference with the different stages of muscle contraction. There are two main causes of muscle fatigue: the limitations of a nerve's ability to generate a sustained signal (neural fatigue) and the reduced ability of the muscle fibre to contract (metabolic fatigue).

Electrical impulses (nerve impulses) come from the brain and flow through T-tubule, cause depolarization of cisternal membrane and in turn trigger calcium ions release from cisternae of sarcoplasmic reticulum. Repeated stimulation may cause muscle fatigue either due to reduced ability of muscle to generate force or due to nerve (Fig. 1.18).

Rigor mortis (Latin: *rigor* 'stiffness', *mortis* 'of death'): It is a prominent sign of death caused due to a chemical change in the muscle in turn stiffening of limbs of corpse. Rigor mortis sets within 4 h post-mortem.

Isotonic and Isometric Muscle Contraction

Muscular system is very important as it can produce movement and provide protection and support for organs in the body. The unique, characteristic feature of muscle cell is the relative abundance and organization of actin and myosin filaments within the cells. These filaments are specialized for contraction. There are three muscle types present in vertebrates, namely, smooth muscles, skeletal muscles and cardiac muscles. The contraction of cardiac and smooth muscles is, generally, involuntary, while the skeletal muscle is under voluntary control. Depending on the pattern of tension production, muscle contraction can be classified as isotonic contraction and isometric contraction.

Skeletal System

Skeletal system makes the framework of the human body. Bone is the major constituent, but joints, cartilages, tendons (fibrous tissues that bind the bones and muscles together at joints) and ligaments (fibrous tissues that bind the bones together at joints) also have significant roles to play. Bone is light weight, hard material which can withstand and resist tension and forces applied on it. Bone is a connective tissue with calcium salt deposited on solid matrix of it. Bone is hard, strong due to its solid matrix but its flexibility and great tensile strength are due to the presence of organic parts (collagen fibres) [25, 26].

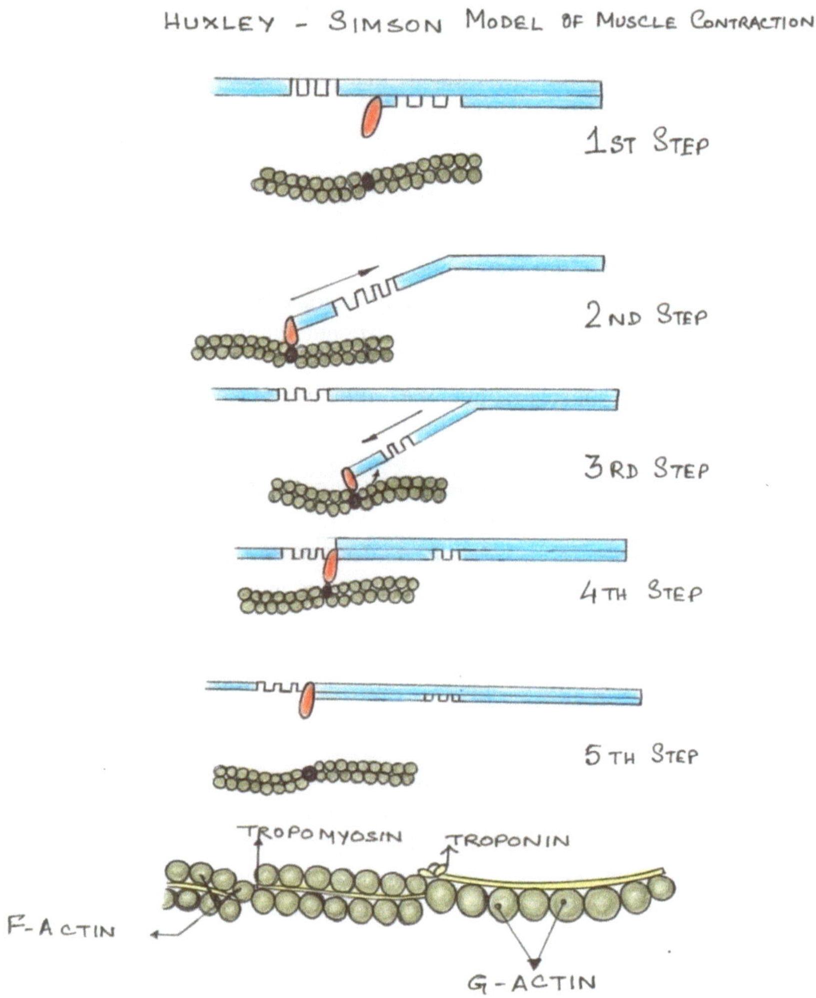

Fig. 1.18 Muscle contraction

The skeleton is subdivided into two divisions: the axial skeleton, the bones that form the longitudinal axis of the body, and the appendicular skeleton, the bones of the limbs and girdles (Fig. 1.19) [27].

Types of Bones

The bones of the skeleton are divided into various categories depending on their shape.

Long bone: This type of bone has long shaft with two extremities. They are found in the limbs. The bones of the arm, forearm, thigh and legs are typical examples. The

Mechanical Functions of bones:	• **Protection** • **Shape** • **Movement**
Synthetic Functions of Bones:	• **Synthesis of blood cells**
Metabolic Functions of Bones:	• **Mineral Storage.** • **Fat storage** • **Role in acid-base balance**

Fig. 1.19 Functions of bone

shaft is a cylinder of compact tissue which contains yellow bone marrow. The extremities have thin outer covering of compact tissue and inner network of cancellous or spongy bone which contains red bone marrow.

Short bone: This type of bone has no shaft but consists of small mass of spongy bones surrounded by compact bone. Carpals (bones of wrist) and tarsal (ankle) are short bones.

Flat bone: This type of bone provides a broad surface for muscular attachment and protects the internal organs. It is made of cancellous bone sandwiched by two compact bones. Skull, scapula and sternum are typical examples of flat bones.

Irregular bone: Bones of the face and vertebra are irregular bones which have peculiar shape and hence the name.

Sesamoid bone: It is developed in the tendon of the muscles and is found in close association with joint. The patella is the example.

Structure of Bones

Bone has a complex structure. It is having nerves, blood vessels and bone cells like osteoblast, osteoclast and tiny cavities called lacunae in the solid matrix of bone tissue. Blood in blood vessels supply nutrients to skeletal tissue and also provide the route for metabolic waste removal. Lacunae are arranged in concentric circles called lamellae around the central (Haversian) canal. The central canal and matrix ring together form the unit called osteon or Haversian system. Central canals run lengthwise through the bony matrix carrying the blood vessels and nerves to all areas of the bones. From the central canal, tiny tubular structures called canaliculi radiate outwards to reach all the lacunae. Bone cells are connected to the nutrient supply source through the hard bone matrix by the transportation system made by canaliculi. This extensive network of canal facilitates supply of sufficient nourishment and quick bone injuries heal. Perforating (Volkmann's) canals remain at the right angle to the shaft and form connecting pathway from outside to the interior of bone.

Joints

Joints are the junction of two or more number of bones. These are also termed as articulations which hold the bones firmly and securely and also give the rigid skeleton flexibility and mobility. Hyoid bone of the neck is the exception. The boric-binding function of joints is just as important as their role in providing mobility. The immovable joints of the skull, for instance, form a snug enclosure for our vital brain [19].

Joints are classified in two ways – functionally and structurally. Functional classification depends on the degree of joint movement. On this basis, synarthroses are immovable joints, amphiarthroses are slightly movable joints, and diarthroses are freely movable joints. Axial skeleton is made of either immovable or slightly movable joints to provide compact, rigid attachment and protection of the internal organs. The limbs have freely movable joints which provide mobility to these body parts.

Structurally, there are *fibrous, cartilaginous and synovial joints* based on whether fibrous tissue, cartilage or a joint cavity separates the bony regions at the joint.

Synovial Joint

Synovial joints are those in which the articulating bone ends are separated by a joint cavity containing synovial fluid. They account for all joints of the limb. All synovial joints have a distinguishing feature.

(a) Articular cartilage – Articular (Hyaline) cartilage covers the ends of the bones forming the joint.
(b) Fibrous articular capsule – The joint surfaces are enclosed by a capsule of fibrous connective tissue, and the capsule is lined with a smooth synovial membrane.
(c) Joint cavity – The articular capsule encloses a cavity called the joint cavity which contains lubricating synovial fluid.
(d) Reinforcing ligaments – The fibrous capsule is usually reinforced with ligaments.

Associated structures of synovial joints are bursae and tendon sheath. Bursae are sac full of synovial fluid and are lined with synovial membrane and reduce friction between associated structures and joints during joint activity. Tendon sheath is essentially extended bursa that swaddles completely a tendon subjected to friction (Fig. 1.20).

Classification of Synovial Joint

The type of movement of the bones at a joint is facilitated by the shape of the articulating bone surfaces. Based on such shapes, our synovial joints can be classified as plane, hinge, pivot, condyloid, saddle and ball-and-socket joints.

- In a *plane joint*, the articular surfaces are essentially flat, and only short slipping or gliding movements are allowed. The movements of plane joints are non-axial, that is, gliding does not involve rotation around any axis. The intercarpal joints of the wrist are the best examples of plane joints.
- In a *hinge joint*, the columnar end of one bone fits into a trough-shaped surface on another bone. Like a mechanical hinge, angular movement is allowed in just one plane. Examples are the elbow joint, ankle joint and the joints between the phalanges of the fingers. Hinge joints are classified as uniaxial; they allow movement around one axis only [20].

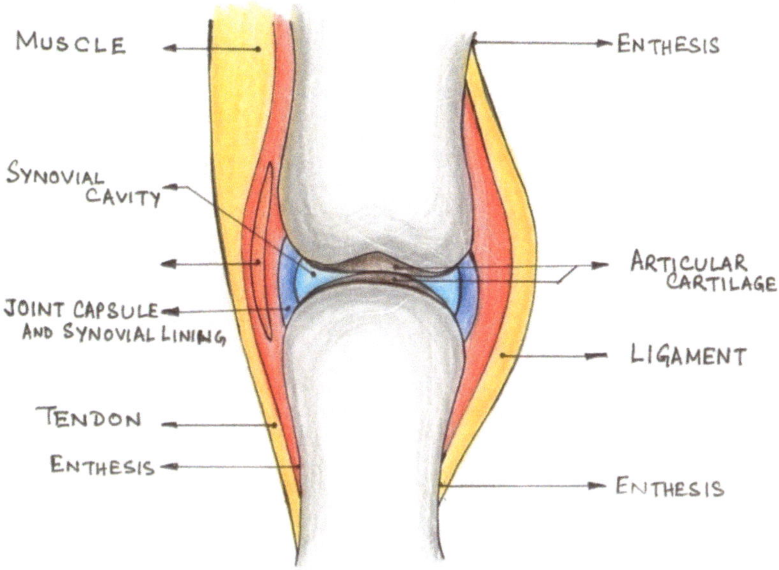

Fig. 1.20 Structure of synovial joint

- In a *pivot joint*, the rounded end of one bone fits into a sleeve or ring of bone (and possibly ligaments). Pivot joints are uniaxial joints where bone can only rotate around its long axis. The proximal radioulnar joint and the joint between the atlas and the dens of the axis are examples.
- In a *condyloid joint* ('knucklelike'), the egg-shaped articular surface of one bone fits into an oval concavity in another. Both of these articular surfaces are oval. Condyloid joints allow the moving bone to travel (1) from side to side and (2) back and forth, but the bone cannot rotate around its long axis. Movement occurs around two axes; hence, these joints are biaxial (bi = two), as in knuckle (metacarpophalangeal) joints.
- The *saddle joints* are having both convex and concave areas, so these joints allow essentially the same movements like condyloid joints. The best examples of saddle joints are the carpometacarpal joints in the thumb, and the movements of these joints are clearly demonstrated by twiddling your thumbs.
- In a *ball-and-socket joint*, the globular head of one bone fits into a circular socket in another. These multiaxial joints allow movement in all axes, including rotation, and are the most freely moving synovial joints. The shoulder and hip are examples [21].

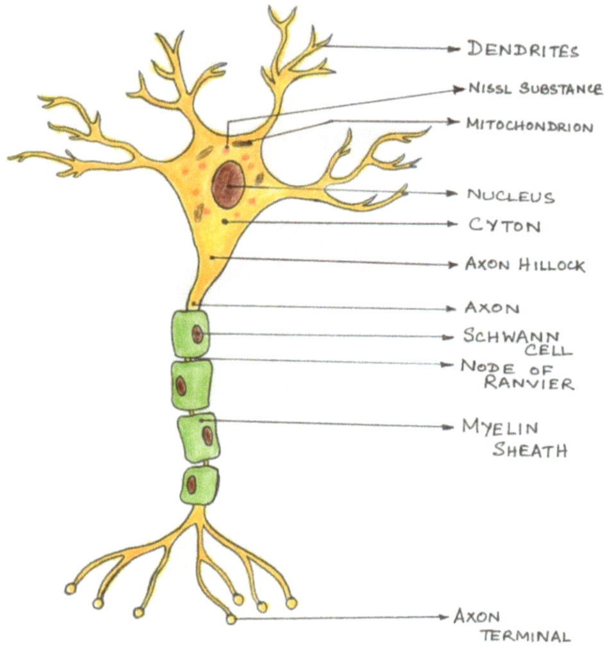

Fig. 1.21 Structure of neuron

1.7 Nervous System

Neuron

The nerve cells or neurons that are the primary components of the brain, spinal cord and autonomic ganglia of PNS transmit and process the messages by the means of different electrical and chemical signals through the synapses. Interconnection among the neurons forms neural networks.

The structure of a neuron is composed of the cell body called soma, thin branches arising from the cell body called dendrites and a cellular extension called axon. Multiple dendrites and only one axon can arise from the cell body (Fig. 1.21).

Generally, the signals in the body travel from one neuron (axon) to the other (dendrite). In some exceptional cases, this connection can occur between one axon and other or between one dendrite and other.

There are specialized neurons like as follows: the neurons that respond to stimuli like touch, sound or light and all other stimuli affecting the cells of the sensory organs are called sensory neurons; the neurons that receive signals from the brain and spinal cord to cause muscle contractions and affect glandular outputs are the motor neurons; and the neurons that connect neurons to other neurons within the same region of the brain or spinal cord in neural networks are known as interneurons [28, 29].

Classification of Neuron

According to Structure

Neurons are classified into apolar (without any process), unipolar (when both axon and dendron arise from the same pole), bipolar (when both axon and dendron arise from the opposite pole) and multipolar (when axon arises from the first pole and dendron arises from some other pole).

According to Function

1. Afferent neurons are those that transmit the information from tissues and organs into the central nervous system. They are also called sensory neurons.
2. Efferent neurons are those that transmit signals from the central nervous system to the effector cells. They are also called motor neurons.
3. Interneurons are those that connect neurons within specific regions of the central nervous system.

Action Potential

Synaptic transmission or neurotransmission is the process through which the nerve impulse travels from one neuron to the next crossing through the synapse or junction area.

In resting condition, due to accumulation of impermeable Na+ ions outside the membrane, the outer of it becomes positively charged, while inner part of the nerve fibre becomes negatively charged due to greater concentration of anions which include permeable Cl− ions and impermeable organic anions. When any point of the outer part of the membrane and inner part is connected by electrodes, a deflection will be observed in the galvanometer, which shows the existence of the potential known as resting membrane potential. The magnitude of it is around −70 mV.

This resting membrane potential is disturbed when a threshold or adequate stimulus is applied to a nerve fibre and modified potential is set up, which is known as impulse of action potential.

The action potential differs from resting membrane potential not only in its magnitude but also in its nature. That means it is propagatory in nature while resting membrane does not propagate.

Development of Action Potential

When a nerve fibre is stimulated at a particular point, permeability for Na+ ion (PNa +) increases at that point. As a result, Na+ ion enters from outside inwards. Normally Na+ ion tends to enter the nerve fibre by the influence of two forces – concentration gradient (higher concentration of Na + ion outside the membrane as compared to the inside). Na+ ion tries to pass from high concentration to low concentration area and by electrical gradient (inner part of the membrane being negatively charged while the outer part being positively charged, cations try to enter through the membrane of the nerve fibre). But this is opposed by the impermeability of the Na+ ion due to the smaller pore size of the membrane than the size of the Na+ ion at resting condition [30, 31].

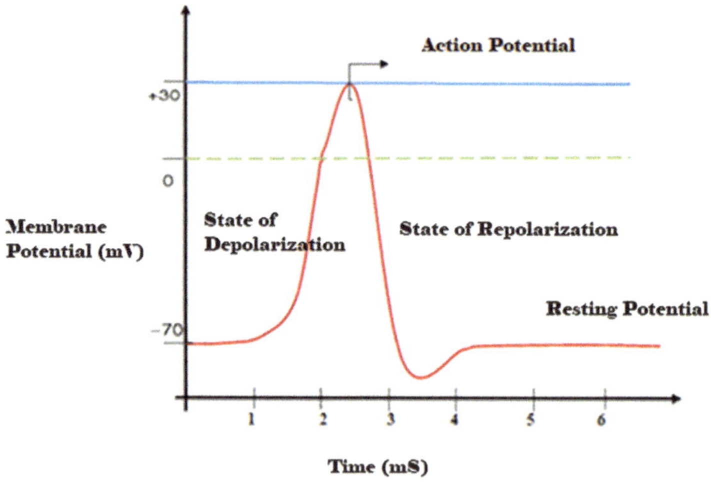

Fig. 1.22 Typical action potential

As the nerve fibre is stimulated, permeability for the Na+ ion increases and the Na + ion rushes in causing depolarization of the membrane at the point of stimulation. At the point of stimulation, the interior atmosphere of the membrane becomes positively charged and the exterior becomes negatively charged.

Entry of Na+ ion causes further increase of PNa+ resulting in further entry of Na+ ions. Very soon PNa+ at this region of stimulation becomes increased by several hundredfold. Thereafter, this massive entry of Na+ ion begins to fall due to rapid decrease of PNa+. At this phase, the permeability of membrane to K+ ion, that is, PK +, begins to rise resulting in efflux of K+ ions due to its concentration gradient. But PK+ soon returns to normal and efflux of K+ ions stops (Fig. 1.22).

The area of the nerve where the stimulus is being applied becomes depolarized as already said due to entry of Na+ ions. But very soon as it returns to the resting state again, it is said to be repolarized. During depolarization, the potential difference at the point of depolarization is mainly due to Na+ ions' entry into the cell. When a stimulus is applied to a nerve fibre, its polarity is reversed from −70 mV and it proceeds towards 0. In doing so, when the membrane potential becomes −60 mV, the negativity begins to diminish very rapidly. This point is called the firing level. After this, the curve goes to 0 and crosses it, i.e. overshoots it and reaches to a peak of +30 mV or more. Thereafter, it begins to decline. This part of the curve is called negative after potential.

Action potential, once developed, propagates throughout the membrane without decrementing until it is stopped. The stimulated area, i.e. the area of depolarization, acts as a battery and stimulates the adjoining polarized area, as a result of which, the latter area becomes depolarized and acts in turn as a battery to the next polarized area, which thus becomes depolarized. In this way, the action potential propagates [32].

Charge-carrying ions like sodium, potassium, calcium and chloride are responsible for the generation and propagation of the signals. Stimuli like pressure, stretch and chemical transmitters can activate a neuron by opening the specific ion channels which leads to ionic flow changing membrane potential.

The function of voltage-dependent ion channels is altered by the changes in the cross-membrane voltage. If there is a large voltage change, then action potential (AP) is generated which obeys all-or-none law. This AP travels in a rapid manner along the length of the axons and activates synaptic connections with other cells when it arrives. Thicker neurons can transmit the impulses more rapidly as compared to the thin neurons. There is a sheath of myelin around the axon. In the central nervous system (CNS), a glial cell forms this sheath, and in the peripheral nervous system (PNS), Schwann cells forms this sheath. It, because of these myelinated sheath action potentials, travels faster than the unmyelinated axons, minimizing metabolic expenses. In the peripheral nerves, the myelin sheath runs along the axon but becomes discontinuous at nodes of Ranvier.

Human Brain Structure and Functions

The most sensitive, delicate and important part of the human body is the brain. It is composed of neurons and nerve endings. The brain is protected by a bony shell, skull. This organ plays a vital role in coordination and regulation of all physical and mental activities of the body. It is the regulatory centre for emotion, judgement, reasoning, memory and consciousness.

Inner to skull lies the three-layered membrane of the brain called the meninges. The layers are dura mater, dense and tough lining; arachnoid mater, fine middle layer; and pia mater, inner layer covering the brain and spinal cord. Subarachnoid space is the gap between the arachnoid mater and pia mater. The meninges secretes cerebrospinal fluid and protects delicate nerve tissues.

The brain is divided into three parts – forebrain (cerebrum), midbrain (pons Varolii and medulla oblongata) and hindbrain (cerebellum).

Cerebrum

This is the largest part of the brain, divided into the right and left hemispheres which are connected by corpus callosum. Peripheral part of brain is covered with grey matter, and central part is made of white matter. Each hemisphere is divided into frontal, temporal, parietal and occipital lobes. The cerebrum has elevated convolutions called gyri and depressed fissures called sulci. The frontal lobe regulates motor activities, temporal lobe has olfactory and auditory areas, parietal lobe has sensory and gustatory areas, and occipital lobe is the site of visual sensations.

The main functions of the cerebrum are emotion, judgement, reasoning, association, memory and consciousness. It is also the site of registration and control of motor and sensory activities. Ventricles are spaces or canals in the cerebrum. These are filled with cerebrospinal fluid (CSF) which keeps the brain and spinal cord moist and prevents these delicate parts from shock and acts as a cushion and as flows throughout the brain and spinal cord; it maintains uniform pressure around these structures [33, 34].

Cerebellum

It is located beneath the posterior part of the cerebrum. The cerebellum maintains body balance and posture and coordinates voluntary movement and muscle tone. It receives the information to perform this work from the motor area of the cerebral cortex, sensory receptors and visceral receptors.

Pons

It lies between the midbrain and medulla oblongata and acts as bridge. It is the part of the brain which connects the cerebellum and cerebrum through nerve fibres. Trigeminal, abducent, facial and vestibule cochlear nerves arise from it.

Medulla Oblongata

It is the junction between the brain and spinal cord and is located at the base of the brain. The medulla oblongata is the site where the nerve tracts cross over. This part controls the heart rate, respiratory rate, swallowing, depth of breathing and vomiting.

Thalamus and Hypothalamus

The thalamus is a large mass of grey matter, situates below the cerebrum and acts as sensory relay station where the nerve fibres from the spinal cord and brainstem move inwards and nerve fibres move onwards to the cerebral cortex.

The hypothalamus is composed of a number of nuclei and situates below the thalamus at the base of the brain. It connects the posterior lobe of the pituitary gland (hypophysis) through neural connections and the anterior lobe of that gland through vascular connections. The hypothalamus controls the body temperature, cardiac rhythms, thirst, appetite, anterior pituitary secretion and synthesis of vasopressin.

Spinal Cord Structure and Functions

The spinal cord is an important part of the central nervous system. It extends from the medulla oblongata to the second lumbar vertebra within the vertebral column which protects the spinal cord against injury. Inner to the vertebral column is the three-layer covering of spinal cord, the meninges. It ends as the cauda equina, a fan of nerve fibres. It is composed of eight cervical segments, twelve thoracic segments, five lumbar segments, five sacral segments and one coccygeal segment. Thirty-one pairs of spinal nerves are attached to the spinal cord through an anterior root and a posterior root and divide into an anterior and a posterior ramus. The anterior root consists of motor fibres, whereas the posterior root consists of sensory fibres and a posterior root ganglion. The anterior ramus supplies the skin and muscle of the front and sides of the trunk and the limbs. The posterior ramus supplies the skin and muscle of the back of the trunk.

The spinal cord conducts nerve impulses in and out at the same level, up and down to other levels of the cord, to and from the brain. It is the major site of reflex actions which are involuntary responses to stimuli involving the brain, spinal cord and peripheral nerves [35, 36].

1.8 Chapter Summary

Human physiology is the scientific study of the structure-function relation of various living systems of the body. This chapter enlightens how the structures of the body function together to maintain life. The form and the function are closely associated with maintain homeostasis of the in vivo physiological systems. Overview of the internal physiological system of the human body depicts structure-function correlation of various organs of the cardiovascular system, blood-vascular system, digestive system, respiratory system, renal system, musculoskeletal system and nervous system and its holistic role in the functioning of the human being.

References

1. Hartenstein V, Mandal L (2006) The blood/vascular system in a phylogenetic perspective. BioEssays 28:1203–1210
2. Jacob M, Chappell D, Becker BF (2016) Regulation of blood flow and volume exchange across the microcirculation. Crit Care 20(1):319. https://doi.org/10.1186/s13054-016-1485-0
3. Monahan-Earley R, Dvorak AM, Aird WC (2013) Evolutionary origins of the blood vascular system and endothelium. J Thromb Haemost 11:46–66. https://doi.org/10.1111/jth.12253
4. Anderson RH, Webb S, Brown NA (1999) Clinical anatomy of the atrial septum with reference to its developmental components. Clin Anat 12:362–374
5. Keith A (1918) Harveian lecture on the functional anatomy of the heart. Br Med J 1:361–363. https://doi.org/10.1136/bmj.1.2987.361
6. Gross L, Kugel MA (1931) Topographic anatomy and histology of the valves in the human heart. Am J Pathol 7(5):445–474.7
7. Borgstrom B, Dahlqvist A, Lundh G, Sjovall J (1957 Oct) Studies of intestinal digestion and absorption in the human. J Clin Invest 36(10):1521–1536
8. Granger DN, Kvietys PR, Perry MA (1982 June) Role of exchange vessels in the regulation of intestinal oxygenation. 22. Am J Phys 242(6):G570–G574
9. Elsom KA, Chornock FW, Dickey FG (1942 Nov) Intubation studies of the human small intestine. Xxiii. A method of determining digestive activity in any portion of the gastrointestinal tract, with some measurements of protein digestion in the stomach and small intestine. J Clin Invest 21(6):795–800
10. Adibi SA (1969 May) The influence of molecular structure of neutral amino acids on their absorption kinetics in the jejunum and ileum of human intestine in vivo. Gastroenterology 56 (5):903–913
11. Wright GW (1961) Structure and function of respiratory tract in relation to infection. Bacteriol Rev 25(3):219–227
12. Ballenger JJ (1960) Experimental effect of cigarette smoke on human respiratory cilia. N Engl J Med 263:832–835
13. Mora-Fernández C, Domínguez-Pimentel V, de Fuentes MM, Górriz JL, Martínez-Castelao A, Navarro-González JF (2014) Diabetic kidney disease: from physiology to therapeutics. J Physiol 592:3997–4012
14. Elbadawi A (1996) Functional anatomy of the organs of micturition. Urol Clin North Am 23:177–210
15. Dantzler WH, Layton AT, Layton HE, Pannabecker TL (2013) Urine-concentrating mechanism in the inner medulla: function of the thin limbs of the loops of Henle. Clin J Am Soc Nephrol 9 (10):1781–1789. https://doi.org/10.2215/CJN.08750812
16. Taylor KA, Robertson JD (1984) Analysis of the three-dimensional structure of the urinary bladder epithelial cell membranes. J Ultrastruct Res 87:23–30

17. Lexell J, Henriksson-Larsén K, Winblad B, Sjöström M (1983) Distribution of different fiber types in human skeletal muscles: effects of aging studied in whole muscle cross sections. Muscle Nerve 6:588–595
18. Brown SHM, Ward SR, Cook MS, Lieber RL (2011) Architectural analysis of human abdominal wall muscles: implications for mechanical function. Spine 36:355–362
19. Kuo AD, Zajac FE (1993) Human standing posture: Multi-joint movement strategies based on biomechanical constraints. Prog Brain Res 97:349–358
20. Murphy AC, Muldoon SF, Baker D, Lastowka A, Bennett B, Yang M, Bassett DS (2018) Structure, function, and control of the human musculoskeletal network. PLoS Biol 16:e2002811
21. Langen UH, Pitulescu ME, Kim JM, Enriquez-Gasca R, Sivaraj KK, Kusumbe AP, Singh A, Di Russo J, Bixel MG, Zhou B, Sorokin L, Vaquerizas JM, Adams RH (2017) Cell-matrix signals specify bone endothelial cells during developmental osteogenesis. Nat Cell Biol 19:189–201
22. Sommer JR, Waugh RA (1976) The ultrastructure of the mammalian cardiac muscle cell – with special emphasis on the tubular membrane systems. A review. Am J Pathol 82(1):192–232
23. Johnson EA, Sommer JR (1967 Apr) A strand of cardiac muscle. Its ultrastructure and the electrophysiological implications of its geometry. J Cell Biol 33(1):103–129
24. Buller AJ, Kean CJ, Ranatunga KW, Smith JM (1984) Temperature dependence of isometric contractions of cat fast and slow skeletal muscles. J Physiol 355:25–31
25. Edwards RH, Harris RC, Hultman E, Kaijser L, Koh D, Nordesjö LO (1972) Effect of temperature on muscle energy metabolism and endurance during successive isometric contractions, sustained to fatigue, of the quadriceps muscle in man. J Physiol 220(2):335–352
26. Merton PA (1954) Voluntary strength and fatigue. J Physiol 123(3):553–564
27. Shadrin IY, Khodabukus A, Bursac N (2016) Striated muscle function, regeneration, and repair. Cell Mol Life Sci 73(22):4175–4202. https://doi.org/10.1007/s00018-016-2285-z
28. Hallett M, Rothwell J (2011) Milestones in clinical neurophysiology. Mov Disord 26 (6):958–967. https://doi.org/10.1002/mds.23572
29. Kress GJ, Mennerick S (2008) Action potential initiation and propagation: upstream influences on neurotransmission. Neuroscience 158(1):211–222. https://doi.org/10.1016/j.neuroscience.2008.03.021
30. Lodish H, Berk A, Zipursky SL et al (2000) Molecular cell biology, 4th edn. W. H. Freeman, New York. Section 21.1, Overview of Neuron Structure and Function. Available from: https://www.ncbi.nlm.nih.gov/books/NBK21535/
31. Neishabouri A, Faisal AA (2014) Saltatory conduction in unmyelinated axons: clustering of Na (+) channels on lipid rafts enables micro-saltatory conduction in C-fibers. Front Neuroanat 8:109. https://doi.org/10.3389/fnana.2014.00109
32. Eames RA, Gamble HJ (1970 May) Schwann cell relationships in normal human cutaneous nerves. J Anat 106(Pt 3):417–435
33. Messé A, Rudrauf D, Benali H, Marrelec G (2014) Relating structure and function in the human brain: relative contributions of anatomy, stationary dynamics, and non-stationarities. PLoS Comput Biol 10(3):e1003530. https://doi.org/10.1371/journal.pcbi.1003530
34. Cerf M, Thiruvengadam N, Mormann F et al (2010) On-line, voluntary control of human temporal lobe neurons. Nature 467(7319):1104–1108. https://doi.org/10.1038/nature09510
35. Brown AG (1981) Organization in the spinal cord. The anatomy and physiology of identified neurons. Springer, New York
36. Schoenen J (1982a) The dendritic organization of the human spinal cord: the dorsal horn. Neuroscience 7:2057–2087

Part II

The Brain

Animal Models of Ischemic Stroke

2

Harpreet Kaur, Deepaneeta Sarmah, Kiran Kalia, Anupom Borah, Kunjan R. Dave, Dileep R. Yavagal, and Pallab Bhattacharya

Abstract

Stroke is the second leading cause of death worldwide. Up to 80% of strokes are ischemic and take place due to occlusion of major cerebral arteries or its branches. The pathophysiology of stroke is multifaceted, involving excitotoxicity and activation of inflammatory pathways leading to disturbances in ion channels, oxidative damage, and apoptosis. Thrombolytics are the only FDA-approved drug for ischemic stroke. In order to study the pathophysiology, development of a reliable and reproducible model of ischemic stroke is of great importance. The ideal animal model is the one which can mimic the features of the pathology. This chapter summarizes the models of ischemic stroke with its advantages and limitations.

Keywords

Middle cerebral artery occlusion · Craniectomy · Photothrombosis · Endothelin-1 · Embolic stroke model

H. Kaur · D. Sarmah · K. Kalia · P. Bhattacharya (✉)
Department of Pharmacology and Toxicology, National Institute of Pharmaceutical Education and Research-Ahmedabad (NIPER-A), Gandhinagar, Gujarat, India

A. Borah
Department of Life Science and Bioinformatics, Assam University, Silchar, Assam, India

K. R. Dave
Department of Neurology, University of Miami Miller School of Medicine, Miami, FL, USA

D. R. Yavagal
Department of Neurology and Neurosurgery, University of Miami Miller School of Medicine, Miami, FL, USA

S. Paul (ed.), *Application of Biomedical Engineering in Neuroscience*,
https://doi.org/10.1007/978-981-13-7142-4_2

2.1 Introduction

Stroke is an etiologically heterogeneous clinical state often arising in majority of cases by thromboembolic occlusion of a major cerebral artery or its subdivisions [1]. This uncontrollable obstruction hampers the supply of oxygen resulting in energy failure, free radical formation, a burst of glutamate release, intracellular calcium accumulation, and the breakdown of transmembrane ionic gradients and subsequently stimulating numerous proinflammatory factors [2]. These series of events, termed as ischemic cascade, produce permanent tissue damage in the brain [1, 3]. The extent of resulting brain injury depends on many factors such as cause, localization, severity, and duration of collateral blood flow. At present, tissue plasminogen activator (rt-PA) is the only FDA-approved therapy for acute ischemic stroke [4]. Apart from thrombolytics, mechanical thrombectomy has also shown benefit in ischemic stroke patients. However, the therapeutic window of 4.5 hours with rt-PA makes it accessible to only 5% of stroke cases [5]. In this context, developing reliable and better therapeutic options is a considerably urgent need of the time [1].

Furthermore, clinical development of newer interventions is being halted by the requirement of large-scale human studies, approachability of subject immediately after tissue insult, and variability among subjects [6]. Rodent experimental models of ischemia have been established, aimed at exploring different underlying mechanisms, and these could possibly mimic the close neuropathological consequences of human ischemic stroke [6]. Most of our understanding related to cerebral ischemia pathophysiology is an outcome from animal experimentation which includes defining area of ischemic penumbra, identification of spreading cortical depolarization, local cerebral blood flow mapping, excitotoxicity, reperfusion injury, inflammation, blood-brain barrier injury, post-stroke neurogenesis, cell apoptosis, and phenomena of ischemic preconditioning [7]. Another important aspect of the animal model is its contribution to the evolution of novel imaging techniques like diffusion-weighted magnetic resonance imaging (DWI) techniques, which could detect the full extent of hyperacute brain infarction [8]. However, with the exception of rt-PA, none of the hundreds of neuroprotective compounds have proven to be efficacious on the basis of phase 3 trial and thereby fail to translate into effective stroke therapies [6]. To advance the development of acute and restorative stroke therapies, the Stroke Therapy Academic Industry Roundtable (STAIR) came into existence which issues a recommendation for strategies to maximize the use of intravenous thrombolytics and more refined clinical studies based on preclinical stroke research [1, 9]. Strict adherence to these recommendation has failed to fill the gap between the relevance of the animal model to human conditions and data interpretation in the process of clinical translation [6]. This chapter summarizes the knowledge on current available models of stroke and deliberates the associated benefits and limitation of each, as well as factors which influence the outcome of the model.

Briefly, cerebral ischemic model are divided mainly into two categories: focal and global models of cerebral ischemia [1]. Models for focal cerebral ischemia include the transient middle cerebral artery occlusion (MCAO) and permanent MCAO, embolic MCAO, distal MCAO, photochemically induced MCAO, and neonatal hypoxic-ischemic brain damage model [10]. Global models for cerebral ischemia include cardiac arrest-induced forebrain ischemia, bilateral carotid artery ligation (2-VO), and four-vessel occlusion (4-VO). However, if ample blood supply to ischemic tissue is restored within a short period of ischemia, selective death of neurons may occur, whereas the other regions are preserved, for instance, glial and vascular cells, which is commonly referred to as selective neuronal death [10]. The present chapter will focus on experimental models of focal ischemic stroke.

2.2 Types of Focal Ischemic Models of Stroke

2.2.1 Intraluminal Suture MCAO Model

Middle cerebral artery (MCA), the largest cerebral artery, is mainly affected, and its blockage leads to ischemia [1]. MCAO accounts for 70% of total infarct leading to neuronal death [11]. Therefore, we require techniques that occlude this artery in animals and mimic human ischemic stroke conditions [12]. Intraluminal suture MCAO model is the most regularly used experimental model of ischemic stroke as reported by Howells et al. and in approximately 40% of rodent experiments for testing neuroprotective agents [12]. In 1986, Koizumi et al. introduced this model, and soon it gained popularity in exploring the progression of edema with reperfusion in rats [6].

Adult rats or mice are most suitable rodents to be used in an MCAO model with intraperitoneal anesthesia, i.e., ketamine (80–100 mg/kg) and xylazine (5–10 mg/kg), or isoflurane inhalation anesthesia. The procedure requires femoral artery cannulation to monitor the blood gas parameters and mean arterial pressure throughout the surgery. Body temperature is regulated using a heating pad at 37 ± 0.3 °C. A midline neck incision is made, and common carotid artery (CCA), external carotid artery (ECA), and internal carotid artery (ICA) are isolated followed by insertion of a nylon monofilament or suture (heat blunted or silicon coated) into the ICA via ECA and advancing the filament until it interrupts the blood supply to the MCA [1, 6, 13]. Laser Doppler flowmetry (LDF) allows for continuous real-time assessment of changes in the cerebral blood flow (CBF). The reduction in CBF to less than 10–15% and the return to 70% of baseline after reperfusion represent successful occlusion of MCA. The inserted filament can be removed after a desired period of occlusion time to model clinical reperfusion. The occlusion period can vary from 60, 90, or 120 min, or a permanent MCAO can be done in animals.

Intraluminal suture MCAO model aids transient or permanent MCAO with reperfusion. Liu et al. reported that to maintain the anatomic integrity of cerebral arteries after reperfusion, the better choice for transient MCAO is by the ECA approach [14]. Till date, several modifications have been made in the intraluminal filament model since its introduction, to improve the recovery of animals [15]. Modifications, such as concerning kind, coating, and length of the filament in order to improve reproducibility, decrease mortality and hemorrhage when using heat-blunted filaments instead of silicon-coated filaments [16, 17].

Advantages

1. The model mimics majority of the aspects of human ischemic stroke.
2. The duration of ischemia and reperfusion time period can be controlled.
3. The model is well suitable for large infarct volumes and has high reproducibility.
4. The model does not require craniectomy and is therefore less invasive and reduces damage to the cranial structure.
5. With the help of laser Doppler flowmetry, real-time cerebral blood flow can be measured, and placement of suture can too be confirmed.

Disadvantages

1. Mortality rates are high due to chances of subarachnoid hemorrhage while performing MCAO procedure.
2. High variation in infarct volume.
3. In MCAO model factors such as variability in body and brain temperature, anesthesia, physical properties of suture, the timing of occlusion, the weight of animals, the strain of animals, and cerebral vasculature contribute to the variation in infarct volume [10, 18, 19].

2.2.2 Craniectomy Model

Craniectomy is another model for ischemic stroke which involves direct surgical exposure of a middle cerebral artery (MCA) via craniectomy [1]. The model was first introduced in Sprague-Dawley rats by Robinson et al. by ligating the distal MCA to induce cortical ischemia [20]. Till date, two common techniques have been performed: The first involves isolation of temporalis muscle and parotid gland and removing the skull superimposing the MCA. MCA occlusion is then performed by electrocoagulation which results in a permanent occlusion or alternatively by the use of ligatures, microaneurysm clip, or photochemical MCAO [21–24]. The second technique involves three-vessel occlusion (3VO). 3VO model involves the additional occlusion of the two CCAs, which reduces the collateral blood flow and thus consolidates the ischemic damage [7]. However, the size of the infarct depends on the occlusion of both MCA and CCAs which can be either permanently or transiently occluded [24, 25]. Ischemia induced by craniectomy mostly affects parietal, frontal, and rostral occipital cortices and exterior part of striatum [22, 26]. Craniectomy avoids the hypothalamic, thalamic, hippocampal, and midbrain damage and produces less infarct in contrast to the MCAO model [1].

Advantages

1. High reproducibility in infarct size and functional deficits.
2. The main advantage of this model is low mortality rate with visual confirmation of MCAO.

Disadvantages

1. The major limitation is it may induce cortical spreading depressions and inflammation through rupturing of vessels by electrocoagulation or drilling.
2. This technique also influences intracranial pressure, alterations in brain temperature, and BBB function due to the opening of the dura. Therefore, it requires efficient surgical skills to be performed [23].

2.2.3 Photothrombotic Model

The photothrombotic stroke model, also known as photochemical stroke model, was introduced in 1985 [17]. This model has been primarily used in rats and requires a photoactive dye, i.e., rose bengal or erythrosine [20]. The dye can be injected via intraperitoneal or intravenous route followed by illuminating the skull by a beam of laser in specific cortical location for few minutes at a specific wavelength [17, 27, 28]. Illumination by laser activates the dye leading to the generation of reactive oxygen species that ultimately damages endothelial membrane and causes platelet activation followed by aggregation and finally formation of thrombi and rapid progression of ischemic cell death in cortical and subcortical regions [27–29]. Variability is less with this model as there is only selective occlusion to the pial vessels surrounding the illuminated zone. A slightly modified version of the photothrombotic model is there in which the induction of proximal occlusion of MCA with rose bengal generates cortical and subcortical lesions [30]. Qian et al. performed magnetic resonance imaging (MRI) and revealed a sizeable volume of penumbra [30]. With the help of this model, which displays assessable penumbra, it is easy to evaluate the impact of reperfusion with thrombolysis and/or neuroprotectants in a model [30].

Advantages

1. It is minimally invasive and has less mortality and high reproducibility.
2. The model is widely used and well-characterized, as it produces a similar thrombus which is observed clinically.

Disadvantages

1. Damage to capillary endothelium is more due to its end-arterial occlusive nature resulting in edema formation and more cytotoxicity.
2. Local collateral flow/reperfusion occurs during model generation leading to smaller or no ischemic penumbra [31].
3. The pathophysiological mechanisms are different from a human ischemic stroke where cytotoxic edema is a crucial sign [32].

2.2.4 Endothelin-1 Model

Endothelin-1 (ET-1) is a potent and long-acting vasoconstrictor peptide suitable for inducing focal ischemia [1, 17]. Initially developed in rats, topical administration to the exposed surface of MCA resulted in constriction of vessels with gradual reperfusion [33]. Later on, the model was successively modified for intracerebral (stereotactic) injection onto tissues near to MCA [1] which results in a dose-dependent ischemic lesion with marginal ischemic edema. Application of ET-1 to MCA significantly decreases 70–93% of blood flow resulting in ischemia [6, 33]. With the help of this model, any region of the brain is targetable [20]. Discrete and targeted anatomical lesions can be induced by targeting specific neuroanatomical areas, such as white matter tracts, to produce specific behavioral deficit [34].

Advantages

1. ET-1 model is minimally invasive with a low mortality rate.
2. The procedure for ET-1 is quite reliable and easy to perform by directly targeting MCA without damaging other facial muscles [17].
3. The possibility of targeting the brain's deep and superficial region with ET-1 model [17].
4. ET-1 model can be performed in conscious animals with the help of a guide cannula implanted prior in advance. This minimizes any confounding effects of anesthesia.

Disadvantage

1. High variability in infarct size is the disadvantage due to variability in the response of the blood vessels to ET-1.

2.2.5 Embolic Stroke Model

Embolic stroke models have been developed which closely mimic human ischemic stroke and are appropriate for the study of thrombolytic agents [17]. The model relies on the generation of an embolus/clot for occlusion of the MCA [17]. An embolus is an artificial clot pre-formed from the autologous blood outside the body [35] and inserted into the ICA. On the basis of the size of an embolus, it can be divided into micro- (20–50 μm) and macrosphere (100–400 μm). The microsphere model uses materials such as dextran, superparamagnetic iron oxide, titanium dioxide (TiO_2), and ceramic [36] implantation into the MCA or ICA via ECA with the help of a fine catheter. This model produces disparity in infarct size as it is multifocal due to its micro size and extent of infarct increases up to 24 h after injection [37]. In the case of the macrosphere model, the diameter of the spheres implanted in the ICA is 100–400 μm. This model produces reproducible occlusion in the MCA, leading to focal ischemia similar to that of the intraluminal suture MCAO model [38].

Advantage

1. Embolic stroke model closely mimics human ischemic stroke and is suitable for preclinical investigation of thrombolytic drugs.

Disadvantages

1. High variability in lesion size and high mortality [39].
2. Microsphere clots can break up and result in multifocal ischemic lesions depending on where the fragments become lodged.

2.3 Conclusion

The number and diversity of experimental focal ischemia model have its own advantages and well-recognized limitations. Still, there is a lack of appropriate animal model which could possibly mimic the complex and heterogeneous nature of human ischemic stroke in the closest sense. This is probably due to a shortcoming in analyzing the multitude of modifying factors such as presence or absence of functioning collateral systems, duration and severity of ischemia along with acceptable blood pressure, age, etiology, sex, and localization of the infarct. Preclinical studies are valuable in analyzing the features of stroke which resemble closely to a human stroke. This issue is illustrated by the still persistent failure of successful translation of effective experimental neuroprotective strategies into patients suffering from this devastating disease. Setting high standards for experimental studies will improve their relevance to human stroke and will help in the development of novel therapies for stroke. Genetically modified animals have led to a new avenue of approach for understanding pathophysiology of stroke. All in all, this developing field holds great possibility toward improving stroke care, despite its limitations (Table 2.1).

Table 2.1 Rodent models of ischemic stroke: advantages and disadvantages

Animal model	Advantages	Disadvantages
Intraluminal suture MCAO	Imitates human ischemic stroke	Hemorrhage susceptibility with certain suture types
	Good reproducibility in neurological deficits and infarct size	Not suitable for study of thrombolytics
	Reperfusion highly controllable	Hyper-/hypothermia
Photothrombosis	Minimally invasive with less mortality	Causes early vasogenic edema
	High reproducibility of the infarct	Not suitable for exploring neuroprotective agents
	Well-defined localization of ischemic lesion	
Craniectomy	Good reproducibility in neurological deficits and infarct size	High invasiveness and consecutive complications
	High long-term survival rates	Surgical skill of high degree required
	Visual confirmation of successful model	
Endothelin-1	Low invasiveness	Duration of ischemia not controllable
	Low mortality	Interpretation of results complicated due to induction of astrocytosis and axonal sprouting
	High specificity	
Embolic stroke	Most close in replicating the pathogenesis of human stroke	Low reproducibility of infarcts
	Suitable for studying thrombolytic agents	High variability in size of lesion Spontaneous recanalization

References

1. Fluri F, Schuhmann MK, Kleinschnitz C (2015) Animal models of ischemic stroke and their application in clinical research. Drug Des Dev Ther 9:3445
2. Lipton P (1999) Ischemic cell death in brain neurons. Physiol Rev 79(4):1431–1568
3. Fisher M (2004) The ischemic penumbra: identification, evolution and treatment concepts. Cerebrovasc Dis 17(Suppl 1):1–6
4. Adams HP Jr et al (2007) Guidelines for the early management of adults with ischemic stroke: a guideline from the American Heart Association/American Stroke Association Stroke Council, Clinical Cardiology Council, Cardiovascular Radiology and Intervention Council, and the Atherosclerotic Peripheral Vascular Disease and Quality of Care Outcomes in Research Interdisciplinary Working Groups: the American Academy of Neurology affirms the value of this guideline as an educational tool for neurologists. Circulation 115(20):e478–e534
5. Fonarow GC et al (2011) Timeliness of tissue-type plasminogen activator therapy in acute ischemic stroke: patient characteristics, hospital factors, and outcomes associated with door-to-needle times within 60 minutes. Circulation 123(7):750–758

6. Durukan A, Tatlisumak T (2008) Animal models of ischemic stroke. Handb Clin Neurol 92:43–66
7. McAuley M (1995) Rodent models of focal ischemia. Cerebrovasc Brain Metab Rev 7(2):153–180
8. Sevick R et al (1990) Diffusion-weighted MR imaging and T2-weighted MR imaging in acute cerebral ischaemia: comparison and correlation with histopathology. In: Brain edema VIII. Springer, New York, pp 210–212
9. Fisher M et al (2009) Update of the stroke therapy academic industry roundtable preclinical recommendations. Stroke 40(6):2244–2250
10. Wang L, Qin C, Yang G-Y (2017) Animal models for ischemic stroke. In: Translational research in stroke. Springer, New York, pp 357–379
11. Bogousslavsky J, Van Melle G, Regli F (1988) The Lausanne Stroke Registry: analysis of 1,000 consecutive patients with first stroke. Stroke 19(9):1083–1092
12. Howells DW et al (2010) Different strokes for different folks: the rich diversity of animal models of focal cerebral ischemia. J Cereb Blood Flow Metab 30(8):1412–1431
13. Longa EZ et al (1989) Reversible middle cerebral artery occlusion without craniectomy in rats. Stroke 20(1):84–91
14. Liu S et al (2009) Rodent stroke model guidelines for preclinical stroke trials. J Exp Stroke Transl Med 2(2):2
15. Trueman RC et al (2011) A critical re-examination of the intraluminal filament MCAO model: impact of external carotid artery transection. Transl Stroke Res 2(4):651–661
16. Tsuchiya D et al (2003) Effect of suture size and carotid clip application upon blood flow and infarct volume after permanent and temporary middle cerebral artery occlusion in mice. Brain Res 970(1–2):131–139
17. McCabe C et al (2018) Animal models of ischaemic stroke and characterisation of the ischaemic penumbra. Neuropharmacology 134:169–177
18. Guan Y et al (2012) Effect of suture properties on stability of middle cerebral artery occlusion evaluated by synchrotron radiation angiography. Stroke 43(3):888–891
19. Yuan F et al (2012) Optimizing suture middle cerebral artery occlusion model in C57BL/6 mice circumvents posterior communicating artery dysplasia. J Neurotrauma 29(7):1499–1505
20. Sommer CJ (2017) Ischemic stroke: experimental models and reality. Acta Neuropathol 133(2):245–261
21. Shigeno T et al (1985) Recirculation model following MCA occlusion in rats: cerebral blood flow, cerebrovascular permeability, and brain edema. J Neurosurg 63(2):272–277
22. Buchan AM, Xue D, Slivka A (1992) A new model of temporary focal neocortical ischemia in the rat. Stroke 23(2):273–279
23. Sugimori H et al (2004) Krypton laser-induced photothrombotic distal middle cerebral artery occlusion without craniectomy in mice. Brain Res Protocol 13(3):189–196
24. Tamura A et al (1981) Focal cerebral ischaemia in the rat: 1. Description of technique and early neuropathological consequences following middle cerebral artery occlusion. J Cereb Blood Flow Metab 1(1):53–60
25. Yanamoto H et al (2003) Evaluation of MCAO stroke models in normotensive rats: standardized neocortical infarction by the 3VO technique. Exp Neurol 182(2):261–274
26. Göb E et al (2015) Blocking of plasma kallikrein ameliorates stroke by reducing thromboinflammation. Ann Neurol 77(5):784–803
27. Watson BD et al (1985) Induction of reproducible brain infarction by photochemically initiated thrombosis. Ann Neurol 17(5):497–504
28. Kleinschnitz C et al (2008) Blocking of platelets or intrinsic coagulation pathway–driven thrombosis does not prevent cerebral infarctions induced by photothrombosis. Stroke 39(4):1262–1268
29. Dietrich WD et al (1986) Photochemically induced cortical infarction in the rat. 1. Time course of hemodynamic consequences. J Cereb Blood Flow Metab 6(2):184–194

30. Qian C et al (2016) Precise characterization of the penumbra revealed by MRI: a modified photothrombotic stroke model study. PLoS One 11(4):e0153756
31. Lee VM et al (1996) Evolution of photochemically induced focal cerebral ischemia in the rat. Magnetic resonance imaging and histology. Stroke 27(11):2110–2118; discussion 2118-9
32. Provenzale JM et al (2003) Assessment of the patient with hyperacute stroke: imaging and therapy. Radiology 229(2):347–359
33. Macrae IM et al (1993) Endothelin-1-induced reductions in cerebral blood flow: dose dependency, time course, and neuropathological consequences. J Cereb Blood Flow Metab 13(2):276–284
34. Lecrux C et al (2008) Effects of magnesium treatment in a model of internal capsule lesion in spontaneously hypertensive rats. Stroke 39(2):448–454
35. Zhang L et al (2015) Focal embolic cerebral ischemia in the rat. Nat Protoc 10(4):539
36. Hossmann K-A (2008) Cerebral ischemia: models, methods and outcomes. Neuropharmacology 55(3):257–270
37. Mayzel-Oreg O et al (2004) Microsphere-induced embolic stroke: an MRI study. Magn Reson Med 51(6):1232–1238
38. Gerriets T et al (2003) The macrosphere model: evaluation of a new stroke model for permanent middle cerebral artery occlusion in rats. J Neurosci Methods 122(2):201–211
39. Macrae I (2011) Preclinical stroke research–advantages and disadvantages of the most common rodent models of focal ischaemia. Br J Pharmacol 164(4):1062–1078

Retrain the Brain Through Noninvasive Medically Acclaimed Instruments

3

Meena Gupta and Dinesh Bhatia

Abstract

Brain training with the use of transcranial magnetic stimulation and neurofeedback is approved by the Food and Drug Administration, USA. There are various neurological and psychological illnesses where this type of brain stimulation helps in improving learning and improving quality of life. This chapter includes present uses of brain stimulation training in various brain diseases. The procedure of brain retraining is related to operant conditioning in which amplitude, frequency, and coherence of electrical activity influence the brain electrical activity of the human being.

Keywords

Brain training · Neurofeedback · Neuroplasticity · Transcranial magnetic stimulation

3.1 Introduction

Transcranial magnetic stimulation (TMS) and neurofeedback (NFB) both are medically acclaimed instruments which are approved for therapeutic as well as diagnostic purposes by the Food and Drug Administration (FDA), USA, for different medical and research usages. Clinicians and health practitioners are using these instruments in neurological as well as psychological diseases such as stroke, dementia, illusion, hallucination, depression, schizophrenia, and cerebral palsy as a brain training as well as in retraining brain functioning. These instruments have great use in therapeutic, diagnostic, and research application in medical sciences. The procedure of brain retraining is related to operant conditioning in which amplitude, frequency, and

M. Gupta (✉) · D. Bhatia
Department of Biomedical Engineering, North-Eastern Hill University, Shillong, Meghalaya, India

S. Paul (ed.), *Application of Biomedical Engineering in Neuroscience*,
https://doi.org/10.1007/978-981-13-7142-4_3

coherence of electrical activity influence the brain electrical activity of human being. In NFB and TMS approches, the brain frequencies reduced which showing excess, and others which are deficit are increased. In various recent researches on brain activities and brain functioning, we realized the brain has the ability to adapt and show changes which is called neuroplasticity. It also has the ability to reorganize its function to some extent such as self-regulation, homeostasis, and stability which occurs due to sudden insult to the regulatory system.

3.2 Transcranial Magnetic Stimulation

The brain is a complex vital organ, situated inside the cranium. Due to various neurological problems or diseases, it starts degenerating and loses its normal functioning. The brain has the ability to change or learn if external stimuli are provided. Plasticity of the brain allows coping with negative emotions and learning useful behavior. TMS and neurofeedback are the new approaches of treatment available nowadays that directly influence brain physiology. Retraining or reeducating the brain helps to acquire functional enhancement which is lost due to neurological or psychological disease. In TMS, the short magnetic pulses targeted over the scalp are used to excite the motor cortex area. TMS was first introduced in 1985 by Barket as a safe and noninvasive method for stimulation of the cerebral cortex [1]. It doesn't require any oral medication or sedation for its action; it is directly used on scalp of the human brain for cortical stimulation. It has no side effect as compared to oral antidepressant medication which is associated with many side effects such as weight gain, nausea, constipation, and dryness in mouth.

However, it has certain limitations to be not used in some conditions such as any metallic implant or pacemaker inserted inside the human body. This technique of treatment is approved by the Food and Drug Administration (FDA) and brain way deep TMS system for treatment of major depressive disorder. Stimulation of motor cortex by TMS with single pulse generates contralateral muscle-evoked potential (MEP). MEP helps to provide assessment technique for mapping of motor cortex in functional aspect. It has the capacity to generate rapid, repetitive pulsed magnetic waves. It passed through the scalp, penetrating the cortex of the brain where it introduced the magnetic current. This magnetic current helps to activate the neurotransmitters such as norepinephrine, dopamine, and serotonin.

TMS as a tool provides new insights in brain behavior study due to its ability to directly alter cortical activity [2]. TMS device emits brief pulses of current through magnetic stimulation coil, and the current flows through this coil in less than milliseconds duration and provides rapidly changing magnetic field around the coil. TMS is a noninvasive instrument for magnetic stimulation of neural tissues such as neurons including the cerebral cortex, spinal roots, cranial nerves, and spinal nerves [3]. The TMS is applied as single pulses in different brain areas and repetitive pulses with various frequencies in neural cell through the scalp. Single pulses depolarize neurons and help to measure the evoked potential effects of neural cells. It provides novel insights to diagnose and manage various neural circuits underlying neurological and psychiatric disorders. It's a type of neurological tool.

Indications of TMS Therapy

- For the purpose of diagnosis.
- For the purpose of treatment.
- Medication management.
- Psychological treatment.
- Research purpose.

Contraindication for Usage of TMS

- Fever.
- Any metallic implant.
- Pacemaker.
- Seizures.
- Chronic pain.
- Pregnant women.
- Malignant tumor.
- Carcinogenetic abases.
- Severe mental retardation.
- Dystonia.
- Senile age.

3.3 Principle of TMS

Anthony Barket from the University of Sheffield, UK, first introduced TMS in 1985 [1]. It's a painless [4] method for activating the motor cortex of human brain and understanding the integrity of central motor pathways. From the time of its introduction in the medical field, its use in many clinical neurophysiology, neurology, neurosciences, and psychiatry has already spread worldwide. Its applications are famous in research application, but nowadays its craze is increasing in clinical domain. In 1838, Michael Faraday discovered electromagnetic induction. According to him, a pulse of currents passing through the coil which is placed over the head of person has the capacity to strengthen and show changes in magnetic pulses which is generated through the TMS and penetrate the scalp and skull to reach the brain with negligible attenuation. These pulses produce ionic current in the brain. Motor cortex is the site of stimulation of nerve fibers at which sufficient current depolarizes the neuron [5]. Depolarization of neurons by TMS depends on activating function which helps transmembrane current to flow and deviate it to the electrical field of nerve [3]. However, in case of bent nerve fiber, the magnetic current will continue in a straight line and pass out of the fiber around the brain membrane; thus, the spatial derivative of the electric field along the nerve is critical, again causing a bend to be a

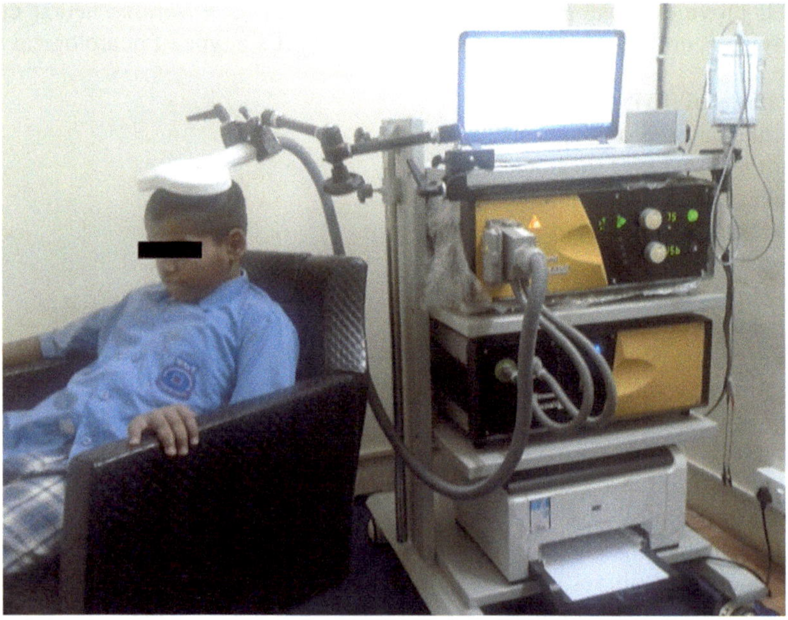

Fig. 3.1 Child undergoing TMS therapy [9]

preferential point of stimulation. These properties of TMS make it different from trancranial electrical stimulation (TES). This directly affects the pyramidal cells and indirectly affects the axons [6, 7]. In some neurological disorders, cortical excitability decreases due to damage or degeneration; in those cases, the TMS helps to increase the excitability of neurons. While operating the TMS, the operator controls the intensity of current moving in the coil, and generated magnetic field depends on the shape of coil. Two magnetic coil shapes are generally used, namely, circular shape and figure of eight shape [8]. For searching the evoked potential, magnetic coil is placed on the scalp of the patient and single pulse transmitted in the motor cortex area of the brain by coil. When the motor evoked potential find it fixed for the repetitive transcranial magnetic stimulation for the purpose of brain stimulation in term of excitatory or inhibitory responses (Fig. 3.1).

3.4 Diagnostic and Therapeutic Application of TMS

TMS stimulation on human scale introduces excitability of the motor cortex, functional integrity of intracortical neuronal structure, corticospinal pathway, and colossal fibers as well as function of nerve roots and peripheral motor pathway. Some TMS findings also help show its useful effect for an early diagnosis and prognostic prediction [10].

3.5 Motor Threshold

When TMS is applied on the cerebral cortex at appropriate intensity, it shows brain cells' excitatory effect, and it can be recorded from the opposite site of the extremity muscles. This stimulation is called motor-evoked potential or MEP. According to various studies on TMS, motor threshold is defined as the lowest intensity needed to elicit MEP.

3.6 Central Motor Conducting Time

Central motor conducting time is the difference between MEP induced by motor cortex stimulation and spinal stimulation-evoked potential which is estimated by subtracting the latency of motor potential induced by stimulation of the spinal motor root from that of the response to motor cortex stimulation.

While placing the magnetic coil on the back of the neck or in lumbosacral spine, the magnetic pulse stimulates spinal roots but not spinal descending pathways [11]. In neural foramina, induced electrical field is increases but in spinal canal its decreases [12].

3.6.1 Motor-Evoked Potential

Single or repetitive pulse stimulation of motor cortex results in the peripheral or spinal muscles to produce a neuroelectrical signal that is called motor-evoked potential. Clinically, it is used as one of the diagnostic tools in detecting many neurodevelopmental or neurodegenerative diseases such as dementia, autism, and cerebral palsy.

3.6.2 Silent Period

During voluntary muscle contraction, a single pulse is applied on motor cortex contralateral to the target muscle, to capture the electromyographic activity for few hundred milliseconds time duration after the MEP. This period of electromyographic suppression is called silent period. In corticospinal track dysfunction, finding silent period is very difficult. Most of the silent period occurs due to inhibitory mechanism at the motor cortex. There are some examples where silent period decreases movement disorder, amyotrophic lateral sclerosis, and intracortical inhabitation [13, 14].

3.6.3 Transcallosal Conduction

Transcallosal conduction is delay in the patient having lesion in the corpus callosum. This transcallosal conduction technique helps in gaining functional information with the help of MRI scanning that detects the interhemipheric lesion in patients who have multiple sclerosis.

3.6.4 Paired Pulse TMS

Paired pulse TMS is used to examine the intracortical inhibitory and facilitatory mechanisms. In this technique, inhibitory and facilitatory mechanisms were first introduced for the study of motor cortex [15].

3.7 TMS for Neurosurgery

TMS techniques are also used in preoperative assessment of specific brain area and for intraoperative monitoring of corticospinal tracts functions for optimized surgical processors. In presurgical process, it minimized the postsurgical deficits. It works similar to an MRI scan to investigate the lesion area. It's as compared with functional MRI, it gives information to the neurosurgeon about any lesson in the brain and it also show which part of brain can be show function enhancement after neurosurgery.

3.8 Therapeutic Uses of TMS in Various Conditions

Therapeutic interventions of TMS exist not only for motor cortex area but for other regions of the brain as well. It has a long-lasting effect on the area outside the motor cortex region. It shows measurable behavioral changes that affect visual, auditory, prefrontal, parietal cortex, and cerebellum regions of the brain [16–18]. This finding increases the probability of therapeutic uses of TMS in normalized pathologically to either decrease or increase cortical excitability.

Transcranial Magnetic Stimulation is a Proven Therapy Cure for the Following Conditions

- Severe neuropathic pain.
- Variation in muscle tone.
- Seizures.
- Headaches.
- Parkinson's disease.
- Tinnitus.
- Dysphasia.
- Headache.
- Paretic hand after stroke.

- Spinal cord injury.
- Fibromyalgia.
- Alzheimer's.

3.8.1 Therapeutic Effect of TMS in Alzheimer's Disease

Alzheimer's is a neurodegenerative disease which shows progressive neuronal loss with altered synaptic plasticity in neurotransmitter level. This disease is associated with sensory or motor functional loss and altered sensations. Pharmacological treatment of Alzheimer's such as acetyl cholinesterase inhibitors and N-methyl-D-aspartate receptor antagonist limits the effectiveness and restricts the disease progress. TMS as a diagnostic as well as therapeutic tool holds good potential for psychiatric patients.

3.8.2 Therapeutic Effect of TMS on Depression

Depression is one of the severe mental disorders with a variety of symptoms such as behavioral, cognitive, and effected somatic functions. Some typical symptoms of severe depression are frequent mood swings, low self-esteem, and poor social interactions leading to isolation from surroundings. Some other factor concerns are changes in eating or sleeping habits, and it could extend leading to tendency for repeated suicidal attempts. For the rehabilitation of depression, TMS is used as the purpose of therapeutic modality that reduce the risk of clinical symptoms seen in the depression patients. With various research works on TMS for rehabilitation of depression suggested, it is an effective way of reliving mood swing and major depressive disorder with the use of TMS. TMS is a promising novel therapy which provides great opportunity in the treatment of depression or major depressive disorder. In TMS, a magnetic coil will be placed on the patient's head, which helps to generate magnetic pulses, and due to the production of magnetic field, it undergoes to the scale and penetrates the membranes of the neurons as stated by Terao et al. in 2002. The mechanism of TMS is based on Faraday law of electromagnetic induction. According to this law, the magnetic pulse, situated near the conductors, will be transformed into an electric current which helps in depolarizing underlying neurons and nerve cells by these magnetic fields.

3.8.3 Therapeutic Effect of TMS on Spinal Cord Injury

In spinal cord-injured patients, neuropathic pain is the most common complication that results in poor impact on quality of life. TMS as a therapeutic tool has a potential to relive this chronic pain and suffering [19]. According to Defrin et al., years showed effects of rTMS for reliving the neuropathic pain after a spinal cord injury. They suggested that rTMS on the targeted area of leg, which implies giving magnetic stimulation to the central zone of the brain with chronic neuropathic pain, shows good impact to subside the pain and provide relief.

3.8.4 Therapeutic Effect of TMS on Stroke

The conventional process of neurorehabilitation focuses on the direct electric stimulation of peripheral nerves which stimulates and trains the muscles. The recovery of stroke patients shows improvement within a year. TMS is a new intervention approach as a rehabilitation and therapeutic purpose; it directly increases by stimulating the nervous system through the scalp [20]. It increases the rate of progress by directly stimulating the denervated muscles of the paretic extremities. It increases the neuroplasticity of the damaged brain area in lesser time than other therapies.

3.8.5 TMS and Its Applications in Children

In TMS, the induction of focal current modulates the brain function of the targeted cortex. It is used widely as a diagnostic tool in adults, but in the children, its application to date has been limited, even though TMS offers unique opportunities to gain insights into the neurophysiology of a child's brain [21]. It also helps in providing the opportunity to objectively assess the integrity of corticospinal tracts in various childhood disorders such as cerebral palsy and autism. TMS have good potential to evaluate maturity of the corticospinal pathways in normal children and as a diagnostic tool to evaluate the motor developmental delay. The TMS gives opportunity to assess integrity and action of corticospinal track in children's and is a safer method than electrical stimulation of the brain [22, 23]. TMS is one of the complementary methods for studying central motor pathways and corticocortical excitability, mapping cortical brain function, and studying the true relationship between focal brain activity and behavior in the children [24].

3.8.6 General Application of TMS in Children

- *TMS as a brain mapping tool*: Single pulse stimulation of TMS at different regions of the brain registers the effect of brain neural inhibitory or excitatory action. When TMS is employed, it maps the brain regions and helps in the identification of clinical pathologies.
- *TMS as a probe of neural network*: TMS as a probe is used to assess the significance of the functional elements of neural network by performing some functional activities. Combination of this study with neuroimaging helps to evaluate and explain the functional connectivity between cortical areas of different lobs of the brain [25].

- *TMS as a measure of cortical excitability*: There are some parameters which measure the cortical excitability and help to identify the underlying pathology. These parameters are following the motor threshold (MT), paired-pulse curve (PPC), cortical silent period (CSP), and input-output curve. Cortical excitability with TMS provides insights for neurotransmitter modulation of the various underlying pathologies such as cognitive impairment, cortical networking weakness, reorganization of cortical networks during brain development, maturation, rehabilitation, and learning impairment.
- *TMS as a modulator of brain function*: Cortical excitability depends on the intensity and frequency of stimulation. In brain, cortical excitability can be modulated with the help of rTMS [26] that increase the probability of exploring therapeutic use of TMS.

References

1. Barker AT, Jalinous R, Freeston IL (1985) Non-invasive magnetic stimulation of human motor cortex. Lancet 1(1):1106–1107
2. Luber B (2009) Non-invasive brain stimulation in the detection of deception: scientific challenge and ethical consequences. Behav Sci Law 27(1):191–208
3. Kobayashi M, Pascual-Leone A (2003) Transcranial magnetic stimulation in neurology. Lancet Neurol 2(1):145
4. Merton PA, Morton HB (1980) Stimulation of the cerebral cortex in the intact human subject. Nature 285:227
5. Barket AT (1987) The history and basic principles of magnetic stimulation. Lancet 11(1):1106–1107
6. Kernell D, Chien-Ping W (1967) Post-synaptic effects of cortical stimulation on forelimb motoneurones in the baboon. J Physiol 191(1):673–690
7. Day BL, Dressler D et al (1989) Electric and magnetic stimulation of human motorcortex: surface EMG and single motor unit responses. J Physiol 412(1):449–473
8. Thickbroom GW, Mastaglia FT (2002) Mapping studies. In: Handbook of transcranial magnetic stimulation. Arnold, London, pp 127–140
9. Gupta M, Rajak BL, Bhatia D, Mukherjee A (2016) Effect of r-TMS over standard therapy in decreasing muscle tone of spastic cerebral palsy patients. J Med Eng Technol 40(4):210–216
10. Masahito K (2003) Transcranial magnetic stimulation in neurology. Lancet Neurol 2(3):145–156
11. Kobayashi M, Ohira T, Nakamura A, Gotoh K, Toya S (1997) Bony foramina facilitate magnetic stimulation: an experimental cat sciatic nerve model. Electroencephalogr Clin Neurophysiol 105(1):79–85
12. Maccabee PJ (2002) Basic physiology of peripheral and spinal cord magnetic stimulation. In: Handbook of transcranial magnetic stimulation. Arnold, London, pp 78–84
13. Rona S, Berardelli A, Vacca L, Inghilleri M (1998) Alterations of motor cortical inhibition in patients with dystonia. Mov Disord 13(1):118–124
14. Hallett M (1998) Physiology of dystonia. Adv Neurol 78(1):11–18
15. Oliveri M, Caltagirone C, Filippi MM (2000) Paired transcranial magnetic stimulation protocols reveal a pattern of inhibition and facilitation in the human parietal cortex. J Physiol 529(1):461–468
16. Kosslyn SM, Pascual-Leone A, Felician O (1999) The role of area 17 in visual imagery: convergent evidence from PET and rTMS. Science 284(1):167–170

17. Mottaghy FM, Gangitano M, Sparing R, Krause BJ, Pascual-Leone A (2002) Segregation of areas related to visual working memory in the prefrontal cortex revealed by rTMS. Cereb Cortex 12(1):369–375
18. Hilgetag CC, Theoret H, Pascual-Leone A (2001) Enhanced visual spatial attention ipsilateral to rTMS-induced 'virtual lesions' of human parietal cortex. Nat Neurosci 4(1):953–957
19. Defrin R, Grunhaus L, Zamir D, Zeilig G (2007) The effect of a series of repetitive transcranial magnetic stimulations of the motor cortex on central pain after spinal cord injury. Arch Phys Med Rehabil 88(1):1574–1580
20. Hankey GJ, Pomeroy VM, King LM, Pollock A, Baily-Hallam A, Langhorne P (2006) Electrostimulation for promoting recovery of movement or functional ability after stroke. Stroke 37:2441–2452
21. Kuang-Lin L, Pascual-Leone A (2002) Transcranial magnetic stimulation and its applications in children. Chang Gung Med J 25:7
22. Barker AT, Freeston IL, Jalinous R, Jarrett JA (1987) Magnetic stimulation of the human brain and peripheral nervous system: an introduction and results of an initial clinical evaluation. Neurosurgery 20(1):100–109
23. Barker AT, Jalinous R (1985) Non-invasive magnetic stimulation of human motor cortex. Lancet 11(1):1106–1107
24. Pascual-Leone A, Bartres-Faz D, Keenan JP (1999) Transcranial magnetic stimulation: studying the brain-behaviour relationship by induction of "virtual lesions". Phil Trans R Soc Lond B 1999:1229–1238
25. Beckers G, Zeki S (1995) The consequences of inactivating areas V1 and V5 on visual motion perception. Brain 118(1):49–60
26. Lubar JF (1997) Neocortical dynamics: implications for understanding the role of neurofeedback and related techniques for the enhancement of attention. Appl Psychophysiol Biofeedback 22(2):111–126

Part III

Introduction to Neurodegenerative and Regenerative Disorders

Scientific Basis of Ayurvedic Medicine: In Hunt for a Cure to Alzheimer's Disease

4

Atanu Bhattacharjee

Abstract

Alzheimer's disease (AD) is a degenerative brain disorder that manifests as a progressive deterioration of memory and mental function – a state of mind commonly referred as "dementia" – and causes changes in personality and behavior. AD was first described in 1906 by the German neuropsychiatrist Prof. Alois Alzheimer. Major pathological hallmarks of AD are characterized by deposition of abnormal amyloid β-protein plaques and formation of neurofibrillary tangles of tau protein and decline in cholinergic neurotransmission in the brain. Amid a decade ago, huge development in Alzheimer's pervasiveness has lighted the significance of more examines in the inquiry of new prescription. One of the major clinical advances in the treatment of AD has been the utilization of acetylcholinesterase (AChE) and butyrylcholinesterase (BChE) inhibitors to mitigate ACh level in the cerebrum albeit cholinergic compounds with nicotinic and muscarinic agonist properties additionally have pulled in some intrigue. At present, there are extremely constrained meds accessible to treat AD, and a large portion of the treatment is accessible just to defer the movement of side effects and symptomatic alleviation for a brief timeframe. Medicinal plants have a lot of undiscovered store of regular medications and potential wellspring of common AChE inhibitors. The basic assorted variety of their phytoconstituents makes them a significant wellspring of novel lead mixes for the mission of medications to treat AD. In this way, orderly ethnopharmacological screening of these plants may give valuable leads in the disclosure of new medications for AD treatment. With this foundation, an efficient survey is set up to deliver a refreshed learning in different phytoextracts and their subordinates alongside their conceivable activity on cholinergic sensory system to ease AD treatment. Electronic database was utilized for looking through the data identified with studies performed in plants amid decades ago.

A. Bhattacharjee (✉)
Faculty of Pharmaceutical Science, Assam Down Town University, Guwahati, Assam, India

S. Paul (ed.), *Application of Biomedical Engineering in Neuroscience*,
https://doi.org/10.1007/978-981-13-7142-4_4

Keywords
Alzheimer's disease · Acetylcholine · Cholinesterase inhibition · Medicinal plant

4.1 Introduction

Ayurveda (traditional Indian system of medicine) emphasizes about the root cause of any diseases. According to Ayurveda there are three different bio-energies (Dosha), viz., Vata (all types of movement), Pitta (digestive and metabolic activities), and Kapha (growth, union, stability, and composition), seven different body tissues (Dhatu), and three wastes (Mala) which are very significant for maintaining healthy life of human beings. Among all Dhatus, the Majja Dhatu (nervous tissue or bone marrow) is responsible for communication, intelligence, and movement. When there is an imbalance between Dosha and Majja Dhatu, neurological disorder occurs. Predominance of VATA which is dry and cold in nature, when combined with Majja Dhatu in the cerebral cortex, results in cortical shrinkage, shrinkage in hippocampus, and formation of amyloid β plaques, and gradually leads to Alzheimer's disease [1]. Alzheimer's disease (AD) is a dynamic neurodegenerative turmoil portrayed by low dimension of synapse acetylcholine (ACh) in the cerebrum. It is an irreversible age-related kind of dementia that continuously disintegrates the brain and burglarizes the individual memory and mental capacities and causes changes in personality and behavior [2, 3]. Promotion of AD was first depicted in 1906 by German neuropsychiatrist Prof. Alois Alzheimer. It is a complex, multifactoral, progressive disorder associated with diminished level of Ach in the brain [4]. Normally, ACh is stored in nerve terminal vesicles and released upon depolarization of nerve terminal, thereby binds with the receptor in synapses. However, in patients with AD, the ACh which is released has a very short half-life due to the presence of large amounts of the enzymes, viz., acetylcholinesterase (AChE) and butyrylcholinesterase (BChE), which are both present in the brain and are detected among neurofibrillary tangles and neuritic plaques. These enzymes hydrolyze the ester bond of ACh molecule to choline and acetate, leading to loss of stimulatory activity [5, 6]. Further, two major pathological hallmarks of AD are characterized by deposition of abnormal amyloid β-protein plaques and formation of neurofibrillary tangles of tau protein resulting in diminished impulse transmission from the end of one neuron to the beginning of another neuron [7–9].

Alzheimer's disease (AD) is a dynamic neurodegenerative turmoil described by low level of acetylcholine (ACh) in the synapse of the cerebrum. It is an irreversible age-related kind of dementia that slowly breaks up the brain and burglarizes the individual memory and mental capacities and causes changes in persoality and behavior [2, 3]. Promotion was first portrayed in 1906 by the German neuropsychiatrist Prof. Alois Alzheimer [10]. There are extensive economical and social and emotional burden attributed with the caring of patients with this malady. Truth be told, in cutting-edge mechanical way of life, where future is long, this illness is a noteworthy reason for grimness, and it forces serious strains on the social welfare frameworks. It is reported that in the USA alone, more than five million individuals are influenced by AD, whereas in India, the disease has just been developed as one of the deadliest dynamic danger of humankind after diabetes, cancer, and cardiac sickness [11].

One of the major clinical advances in the treatment of AD has been the usage of acetylcholinesterase (AChE) and butyrylcholinesterase (BChE) inhibitors to expand ACh level in the cerebrum albeit cholinergic operators with nicotinic and muscarinic agonist properties also have pulled in some interest [12, 13]. BChE, fundamentally connected with glial cells and explicit neuronal pathways, severs ACh in a similar manner to AChE to end its physiological activity. Such investigations, measurably slower decrease in the intellectual execution of dementia patients having explicit BChE polymorphisms that normally lower BChE movement, have focused on BChE as another way to deal with the progress of AD [1, 14]. At present, there are extremely obliged medications available to treat AD, and most of the available treatment is to delay signs and symptom for a brief time period. Therapeutic plants address a great deal of unfamiliar store of trademark remedies and potential well-spring of basic AChE inhibitors. The fundamental grouped assortment of their phytoconstituents makes them an imperative wellspring of novel lead blends for the mission of drugs to treat AD. Thusly, orderly ethnopharmacological screening of these plants may give significant leads in the revelation of new drugs for AD treatment.

Ways to deal with improved cholinergic capacity in AD have included incitement of cholinergic receptors or prolonging the accessibility of ACh discharged into the neuronal synaptic cleft by utilization of operators which reestablish the dimension of acetylcholine through hindrance of both AChE and BChE [15, 16]. Recent studies indicated that cholinesterase inhibitors not only hold a key role to enhance cholinergic transmission in the brain but also might play a modulatory role on Aβ plaque deposition and neurotoxic fibrils formation via suppressing amyloid precursor protein (APP) expression [17, 18]. Therefore, AChE and BChE inhibitors have become remarkable alternatives in the treatment of AD. Existing anticholinesterase drugs, viz., tacrine, donepezil, physostigmine, galantamine, and heptylphysostigmine, for the treatment of dementia are reported to have several dangerous adverse effects such as hepatotoxicity, short duration of biological action, low bioavailability, adverse cholinergic side effects in the periphery, and a narrow therapeutic window [19–21].

Restorative plants and their concentrates are accepting an essential part in the present treatment of mental issue either as standard or fundamental medication with explicit reference to their ethnopharmacological points of view [22]. The history of medication revelation has demonstrated that plants contain dynamic intensifies that have turned out to be new sources to examine for the pharmaceutical business. Plant constituents may not just act synergistically with different constituents from a similar plant yet may likewise improve the action of mixes or check dangerous impacts of mixes from other plant species. In traditional practices, various plants have been utilized to treat intellectual issue, including neurodegenerative sicknesses and diverse neuropharmacological issue [23, 24]. Ayurveda (traditional Indian system of medicine) classified these restorative plants as "medharasayanas" (Sanskrit: "medha" implies intellect/cognition and "rasayana" implies revival). Medharasayanas include four remedial plants to upgrade memory and discernment by Prabhava (specific action), viz., Mandukaparni (*Centella asiatica*; Family: Umbelliferae), Yastimadhu (*Glycyrrhiza glabra*; Family: Leguminosae), Guduchi (*Tinospora cordifolia*; Family: Menispermaceae), and Shankhapushpi (*Convolvulus pluricaulis*; Family: Convolvulaceae) [13, 25]. Other medicinal plants like Ashwagandha (*Withania somnifera*; Family: Solanaceae), Brahmi (*Bacopa monniera*; Family: Scophulariaceae), Jyothishmati (*Celastrus paniculatus*; Family: Celastraceae), Kushmanda (*Benincasa hispida*; Family: Cucurbitaceae), Vacha (*Acorus calamus*; Family: Araceae), and Jatamamsi (*Nardostachys jatamansi*; Family: Valerianaceae) are very much archived in Ayurvedic literatures as nootropics [26–28]. Indian Haladi (*Curcuma longa*; Family: Zingiberaceae), a rich source of curcumin, showed potent immunomodulatory and anticholinesterase activity and thereby delays neurodegeneration process [13, 29]. Yokukansan, a TCM remedy, is a revitalizer with no adverse effects [30]. Galantamine, an alkaloid from snowdrop, is used as herbal supplement in the USA to treat AD [31]. Till date the adverse drug reaction (ADR) of available synthetic drugs is huge, and hence shift of paradigm toward natural therapy has been the subject of current focus. With this scientific background, the chapter has been prepared to bring an up-to-date information about the role of alternative medicine, to counter AD.

4.2 Etiology and Pathogenesis

In patients with AD, the neurons become disabled. Early stage of AD interferes with the neuron's ability to produce the energy for efficient coordination between them, a process known as metabolism. They also mislay the ability to repair them which ultimately leads to death of neuron cells. Exactly what interferes with the normal functioning of neuron is still a mystery, but with recent progression of medical sciences, three pathological hallmarks are being postulated.

4.2.1 β-Amyloid Plaques

Tucked in the spaces between neurons, thick and sticky deposits of plaques are made up of a protein called β-amyloid. These plaques are also surrounded by other amino acid fragments, neuro-remnants, and immune cells, viz., microglia which in turn digests the damaged neurons and causes neuro-inflammation. β-Amyloid, a spontaneously aggregating peptide of 39–43 amino acids, is a snipped fragment of larger protein called amyloid precursor protein (APP). APP is an essential component of neuron growth which rests in part inside and outside the cell. β-Amyloid is derived from sequential proteolytic cleavage of the amyloid precursor protein (APP) by β- and γ-secretases [32]. Initial cleavage by β-secretase (BACE; β-site of APP cleaving enzyme), a membrane-anchored aspartic protease, generates a soluble N-terminal fragment and a membrane-associated C-terminal fragment. The C-terminal fragment then undergoes proteolysis by γ-secretase to give the Aβ peptide which is progressively attached with each other and converted into insoluble plaques [33, 34]. According to the "β-amyloid cascade," deposition of β-amyloid plaques triggers the neurotoxic cascade and promotes neuro-inflammation which ultimately resulted in neurodegeneration [17].

4.2.2 Neurofibrillary Tangles

One of the pathological hallmarks of AD is the formation of intracellular neurofibrillary tangles which consists of hyperphosphorylated tau protein [35]. Tau is an axonal protein which binds to microtubules and promotes their assembly and stability. Phosphorylation of tau protein is regulated by the balance between multiple kinases (e.g., glycogen synthase kinase-3 (GSK3) and cyclin-dependent kinase-5 (CDK5)) and phosphatases (e.g., PP-1 and PP-2A). Hyperphosphorylation of tau protein leads to destruction of microtubule-associated proteins and thereby prevents microtubule assembly and impairs axonal transport resulting in neuronal death [6, 36, 37].

4.2.3 Loss of Cholinergic Neurons

The cholinergic hypothesis claims that decrease in cognitive function in dementia is predominantly related to a decrease in cholinergic neurotransmission. AD leads to selective degeneration of cholinergic neurons of basal forebrain and results in the decline in the level of cholinergic markers, viz., ChAT and Ach, and increase in AChE [38, 39] (Fig. 4.1).

Moreover, it has been postulated that ACh acts as protein kinase inhibitor and thereby reduces tau phosphorylation. This inhibits intracellular neurofibrillary tangle formation and delays the progression of AD [35].

Formation of extracellular β-amyloid plaques with drastic neuronal and synaptic reduction in cholinergic system of the brain is considered as a major pathological hallmark of AD. Therefore, restoration of the central cholinergic function plays a significant role to improve cognitive impairment in patients with AD. There are three principal approaches by which the cholinergic deficit can be addressed: (i) nicotinic receptor stimulation, (ii) muscarinic receptor stimulation, and (iii) cholinesterase (AChE and BChE) inhibition [14, 41, 42].

4.3 Ayurvedic Perspective of AD

Although AD is not mentioned in Ayurveda, its symptoms like loss of memory and aging had been described in Ayurveda as Smṛtināsha and Jarāvasthā, respectively. Memory loss is mentioned among the prodromal symptom of jarā [28, 43].

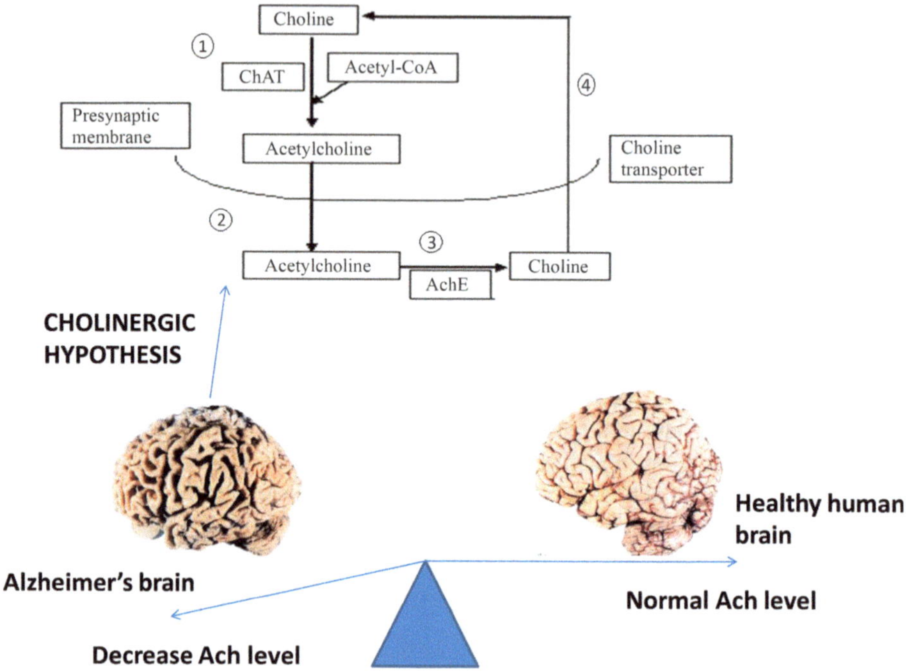

Fig. 4.1 Possible therapeutic approaches in modulation of cholinergic transmission (1). Increase in ChAT activity, (2) inhibition of Ach release, (3) inhibition of AchE, (4) promotion of choline transport. (Adapted and modified from [40])

AD can be named as "Smrti Nasha" in Ayurveda and is brought about by the exhaustion of dhatus (tissues) and weakness of Satva Guna (the quality of mind which speaks to immaculateness and awareness) by rajas (the characteristic of mind which generally speaks to vitality and dynamism) and tamas (the trait of psyche speaking to lack of involvement and obliviousness). At the point when the Manovaha srotas (channels conveying cyclic driving forces mainly in charge of memory) are tormented with Vata dosha, the individual is influenced, and there is breaking down of the psychological exercises [28].

The sign and manifestations of dementia can be comprehended regarding Ayurvedic ideas. To comprehend the etiopathogenesis of dementia as per Ayurveda, assessment of general Manasa (psycology) and Buddhi (intellect) is very essential. Therefore, it is worthwhile to find the correlation between Mana, Buddhi, Medha, Dhriti, and Smriti [44]. During the time spent learning, intellectual capacity that decides the nature, benefits, and bad marks of an object of information is Buddhi. Buddhi and Mana are connected with Karya Karana Sambandh as during the time spent advancement, Buddhi is the principal substance (Tatva), which is in charge of further improvement of Indriyas and Manas [13].

Conventional drug therapy is based on various theories to locate a viable treatment and relief for AD manifestations. These therapies include amyloid cascade and cholinergic hypothesis. Certain other important aspects like inflammatory, oxidative cascade and hormonal pathways also contribute a vital role in AD [45]. Some other lifestyle-associated possible cause may include lack of routine movement, intake of fast foods with high glycemic index and saturated fats, and less intake of plant-based nutrients, antioxidant substances [e.g., phytoestrogens], and seafood rich in ω-3 fatty acids. These may lead to insulin resistance, increased oxidative stress, and homocysteine and estrogen deficiency [46, 47].

In a planned investigation of hazard factors, expanding age, absence of instruction, and the apolipoprotein E epsilon4 allele were observed to be altogether connected with expanded danger of AD [48]. Then again, utilization of nonsteroidal anti-inflammatory drugs and wine and routine daily activity were observed to be altogether connected with diminished danger of AD [49–51]. Some confirmation proposes that oxidative harm of mind tissue might be engaged with pathogenesis, which goes before the event of side effects and the arrangement of amyloid-containing plaques and neurofibrillary tangles [52]. Hereditary variables may likewise be included; however the dilemma associated with the genetic factors responsible for AD has not been clarified in full yet. The disclosure of pathogenic transformations in the β-amyloid peptide (β-APP) and the presenilin qualities provides a sound findings to the hereditary theory on the grounds that amyloid β generation and statement add to the causes of AD [53]. Certain scientific investigations reported that modifications in lysosomal framework may play a significant role in the progression of neurodegeneration associated with AD. Lysosomal abnormality perhaps deals with the generation of free radicals in the cerebrum, and increased oxidative stress prompts steady loss of metabolic movement resulting in diminished life expectancy of a person [54].

4.4 Clinical Examination and Diagnosis

AD can be considered as advanced form of dementia which in Ayurveda is termed as "Smrti Nasha" (loss of memory). This occurs as a result of depletion of dhatus (tissues) and impairment of Satva Guna (the attribute of mind which represents purity and perception) by rajas (a state of mind representing energy and vigor) and tamas (the attribute of mind representing tameness and unawareness). Affliction of Manovaha srotas (retention of sequential events of learning responsible for memory) with Vata dosha may lead to deterioration of mental status of the person suffering from AD. Ayurvedic system of medicine had developed certain dietary and therapeutic measures to delay aging and rejuvenate whole functional dynamics of the body organs which is known as "Rasayana chikitsa" (rejuvenation therapy).

AD is a life-changing disease for both the patient and caregiver. At present, treatments are available to provide a symptomatic relief from the disease for a short period of time if diagnosed in early stages. According to Ayurveda, the physician must elaborate concept about the progression and treatment of AD. In the early stages, Rookshana and Amapachana medicines are prescribed along with Shadangam kashaya and Saddharanam churna. Rookshana and Amapachana help Srotas prepared for Snehana and Shodhana procedure. Often, Langhana (lightening therapy) is introduced followed by Brihana (nourishing) therapy for detoxification process. Further, Snehana therapy deals with Pranaavruta Samana and Chatushprakara snehana (four types of unctuous substances, i.e., ghee, oil, fat, and bone marrow). After Snehana (oilation), Swedana and Shodhana should be performed in the form of Vasti and Virechana, respectively.

4.5 Management

Ayurvedic system of medicine utilizes remedy from natural sources (extracts of plant, animal, and mineral) for the treatment of various mental disorders, including impaired memory condition. There are two methodologies of treatment: first one is general or aggregate prosperity of health, including mental activities (memory amusements, perusing, card recreations) and reflection; second methodology specifically targets intellect (medha) and memory (smriti) [55, 56]. The patient ought to be given appropriate guiding and mental help, for example, Satvavajaya Chikitsa is considered as one of the best exercise of Manovikara (mental unsettling influences). This will be useful to deal with the conduct indications of patient with Alzheimer's disease like agitation, meandering, tension, outrage, and sorrow. Lifestyle changes like healthy eating routine, fitness conduct, and following perfect set of principles as cited in Ayurveda are ideal to counteract neurodegenerative sicknesses, viz., AD specifically [57]. Some important medications routinely used to treat cognitive dysfunctions are depicted below.

4.5.1 Medhya Rasayana Drugs

Amid Vedic time the training framework was conferred by oral guidance, and the understudies held and reviewed this information by rehashed recitation. In this way, insight was viewed as identical with memory, and medications that diminished time of remembering the exercises were produced from regular sources. Several Ayurvedic medicines have been exploited for the treatment and management of acute and chronic neurological diseases associated with cognitive dysfunctions. These medicines from natural sources are believed to modulate neuroendocrine-immune systems via antioxidants and anti-inflammatory properties and thereby enhance memory and rejuvenate cognitive functions. Some of the traditionally reputed medhya rasayana drugs are mentioned below:

(i) Decoction of Mandhukparni (*Centella asiatica*).
(ii) Jastimadhu (*Glycyrrhiza glabra*) decoction with milk.
(iii) Decoction of Guduchi (*Tinospora cordifolia*).
(iv) Shankhpushpi (*Convolvulus pluricaulis*) leaves in paste form.

A significant integer of medicinal plants of Indian origin is described with brain tonic and memory-boosting impacts. Generally, these plants are prescribed in combination as it is believed to provide synergistic effects and provide substantial relief of the symptoms. Sometimes, the formulations are incorporated with mineral-derived drugs [58, 59]. A list of medicinal plants used as brain tonics are listed below (Table 4.1).

In conventional practice, the main medications at present affirmed for the treatment of AD are cholinomimetics. In fact, tacrine was the first clinically established synthetic AchE inhibitor introduced to improve Ach level at the nerve ending. Later, second-generation inhibitors are introduced, but resulting side effects and cost factor made their utilization limited. Although tacrine provides positive influence on few proportions of memory execution, the size of change was extremely unassuming. Further, long-term uses have severe side effects including stomachache, sickness, spewing, and looseness of bowels, which might be huge and economic constraining [8]. Significant research is being led in different territories, for example, utilizing particular mitigating drugs (specific cyclooxgenase-2 [COX-2] inhibitors) to decrease fiery movement in the mind and cancer prevention agents [49, 50]. Other current therapeutic interventions are antiamyloid procedures (e.g., vaccination, collection inhibitors, β-secretase inhibitors), modification of metal chelators (e.g., clioquinol), lipid-bringing down operators, anti-hypertensive, vitamins, and synapse receptors [48]. Phytoextracts on locally available plant materials with AchE inhibitory properties have demonstrated some promise. Ayurveda has numerous definitions, and herbo-mineral drugs talked about underneath are utilized to enhance intellectual capacities. The exact biochemical system of activity of these herbs isn't clear. It is likely that cell reinforcement and mitigating properties of these herbs might be in charge of their gainful impact in the treatment of AD.

4.6 Scientific Basis

4.6.1 Nootropic Plants

Ayurvedic medicinal plants recommended for AD treatment are mentioned below.

4.6.1.1 Pharmacological Studies

***Bacopa monniera* (Brahmi)**

The Central Drug Research Institute (CDRI), Lucknow, and other premiere scientific organizations of the Govt. of India, has conducted systematic study regarding the nootropic effects of *B. monnieri*, and based on the outcome, a polyherbal preparation "Mentat" is successfully marketed. It has been observed that ethanolic extract of *B. monnieri* facilitated acquisition, consolidation, and retention of memory in vivo with passive avoidance paradigm and maze tests.

Saponins and triterpenoidal type of phytoconstituents bacosaponins like bacopasides III to V, bacosides A and B, and bacosaponins A, B, and C, jujubogenin, bisdesmosides, bacosaponin D, E and F are found as major compounds. Minor phytoconstituents include lupeol, betulic acid, polyphenols, and thiol

Table 4.1 Indian medicinal plant with nootropic effect

Sl. No.	Common name	Scientific name	Family
1	Vacha	*Acorus calamus*	Acoraceae
2	Bada Ilaichi	*Amomum subulatum*	Zingiberaceae
3	Shatawar	*Asparagus racemosus*	Asparagaceae
4	Brahmi	*Bacopa monniera*	Plantaginaceae
5	Dhak, Palash	*Butea frondosa*	Fabaceae
6	Sankha Holi	*Canscora decussata*	Gentianaceae
7	Dalchini	*Cinnamomum zeylanicum*	Lauraaceae
8	Guggulu	*Commiphora wightii*	Burseraceae
9	Shankhpushpi	*Convolvulus pluricaulis*	Convolvulaceae
10	Haldi	*Curcuma longa*	Zingiberaceae
11	Sedge	*Cyperus rotundus*	Cyperaceae
12	Jadwar Shireen	*Delphinium denudatum*	Ranunculaceae
13	Choti Elaichi	*Elettaria cardamomum*	Zingiberaceae
14	Amla	*Emblica officinalis*	Phyllanthaceae
15	Vayu Vidang	*Embelia ribes*	Phyllanthaceae
16	Laung	*Eugenia caryophyllus*	Myrtaceae
17	Bari Saunf	*Foeniculum vulgare*	Umbelliferae
18	Ginko	*Ginkgo biloba*	Ginkoceae
19	Hypericum	*Hypericum perforatum*	Hypericaceae
20	Vidari Kand	*Ipomea paniculata*	Convolvulaceae
21	Lavander	*Lavandula officinalis*	Lamiaceae
22	Maca	*Lepidium meyenii*	Brassicaceae
23	Jatamansi	*Nardostachys jatamansi*	Caprifoliaceae
24	Nishoth	*Operculum tarpethum*	Convolvulacea
25	Uood Saleeb	*Paeonia emodi*	Paeoniaceae
26	Keora	*Pandanus odoratissimus*	Pandanaceae
27	Saunf	*Pimpinella anisum*	Apiaceae
28	Renuka	*Piper aurantiacum*	Piperaceae
29	Peepal, Pippali	*Piper longum*	Piperaceae
30	Kava	*Piper methysticum*	Piperaceae
31	Heal-all	*Prunella vulgaris*	Lamiaceae
32	Badam	*Prunus amygdalis*	Rosaceae
33	Gulab	*Rosa damascene*	Rosaceae
34	Kuth	*Sassurea lappa*	Compositae
35	Bahera	*Terminalia belerica*	Umbelliferae
36	Harar, Hareetaki	*Terminalia chebula*	Combretaceae
37	Giloe	*Tinospora cardifolia*	Menispermaceae
38	Tagar	*Valerian wallachia*	Caprifoliaceae
39	Khas	*Vetiverica zizinoides*	Caprifoliaceae
40	Ashwagandha	*Withania somnifera*	Solanaceae
41	Adrakh	*Zingiber officinale*	Zingiberaceae

Adapted and modified from Divya and Lakshmi [60]

compounds that confer strong antioxidant activity. In hippocampus, *B. monnieri* augmented protein kinase activity in hippocampus region of the brain and inhibited cholinergic degeneration in vivo. Further, standardized extract of *B. monnieri* reversed cognitive deficits induced by colchicines and ibotenic acid and also inverted (a) diminution of acetylcholine, (b) reduction in choline acetyltransferase activity, and (c) decrease in muscarinic cholinergic receptor binding in the frontal cortex and hippocampus. The extracts protected neurons from β-amyloid toxicity by inhibiting acetylcholinesterase activity [61, 62].

Centella asiatica (Mandookparni)
C. asiatica (Family: Umbelliferae) aqueous extract showed improved learning and memory potential in vivo. Its extract significantly alleviated oxidative stress by decreasing malonaldehyde (MDA) level and increasing glutathione level. Further it also showed AchE inhibitory activity which combinedly may be associated with cognition-enhancing efficacy [26].

Convolvulus pluricaulis Chois (Shankhpushpi)
C. pluricaulis (Family: Convolvulaceae) significantly increased ACh, catechol amine, and 5-hydroxytryptamine (5-HT) level in the brain [63]. In vitro assay showed its anticholinesterase activity of *Convolvulus pluricaulis* extract at IC_{50} 234 ± 38 μg/ml [64].

Eugenia caryophyllus spl. (Laung)
The aqueous extract of *E. caryophyllus* (Family: Myrtaceae) showed significant AchE activity which may be due to the presence of water-soluble phytoconstituents present in the extract. Further, another research report with Chavanprash containing *E. caryophyllus* showed significant anticholinesterase and antioxidant potential in vivo against scopolamine-induced amnesia [65].

Glycyrrhiza glabra Linn. (Yastimadhu)
The alcoholic extract of *G. glabra* and glabridin, an isolated bioactive component of *G. glabra*, showed both AchE and BuChE inhibitory activity in dose-dependent manner. Further, liquiritin and isoliquirin, isolated flavonoid constituents of the plant, showed significant antioxidant profile against scopolamine-induced amnesia in vivo [66].

Lawsonia inermis Linn. (Mehndi)
Acetone-soluble fraction (100 mg/kg b.w.) of petroleum ether extract of *L. inermis* leaves showed significant ($p \leq 0.001$) *nootropic* effect in vivo in a dose-dependent manner. Further, it affected 5-hydroxytryptamine- and noradrenaline-mediated behavior in experimental animals [67].

Nardostachys jatamansi DC. (Jatamansi)
Jatamamsi is a safe and highly reputed Ayurvedic medicinal plant used as sedative. The sedative activity is due to the presence of valeranone, a sesquiterpenes and

coumarins. Other terpenoids include spirojatamol, nardostachysin, jatamols A and B, and calarenol.

In vivo experiment with alcoholic extract suggested significant improvement in learning and memory paradigm against amnesia induced by diazepam and scopolamine. It also reversed amnesia due to the natural aging of mice which depicts that the phytoconstituents can be valuable natural alternative to delay age-associated dementia [68].

Pongamia pinnata (Karanj)

Petroleum ether extract of the seed of *P. pinnata* was tested for nootropic activity in an experimental model of Alzheimer's disease (caused by ibotenic acid-induced lesioning of nucleus basalis magnocellularis). It reversed both the cognitive deficits and the reduction in cholinergic markers after 2 weeks of treatment. Reversal of perturbed cholinergic function was considered as the possible mechanism [69].

Tinospora cordifolia F.Vill (Guduchi)

n-butanolic fraction of ethanolic extract of *T. cordifolia* stem which contains saponin showed significant nootropic reduction in cholinergic markers in vivo in dose-dependent manner. Further, extract also showed anti-inflammatory activity against lipopolysaccharide-induced neuro-inflammation. Ethanolic extract of this plant diminished ACh content in rat whole brain but increased it in the cortex. These may be correlated with its nootropic potentiality as anti-Alzheimer's drug [70].

Withania somnifera Dunal (Ashwagandha)

Ashwagandha, a solanaceous drug, is utilized widely in Ayurveda as a nerve tonic and "adaptogen" and encourages the body to adjust to stress. It acts via cell reinforcement and free radical scavenging action to boost the immune system. A total alkaloid concentrate of Ashwagandha root showed CNS depressant action. Liquid concentrates of this herb have been found to increment cholinergic action, incorporating increments in the acetylcholine content and choline acetyltransferase movement in rodents, and this may mostly clarify the insight-upgrading and memory-improving impacts. Additionally, ongoing reports recommended methanol concentrate of Ashwagandha which caused neurite outgrowth in a portion and time-subordinate way in human neuroblastoma cells. The dimensions of two dendritic markers, viz., MAP 2 and PSD-95, were observed to be extraordinarily expanded in cells treated with Ashwagandha, proposing that it invigorates dendrite development [27, 71, 72].

4.6.1.2 Clinical Studies

The data on clinical examinations on Indian therapeutic plants possessing nootropic is displayed underneath.

4.6.1.3 *Bacopa monniera* Linn. (Brahmi)

A clinical trial with 110 human volunteers for 9 months of Suksma (herbal formulation containing *B. monnieri* (dose: 200–300 mg)) exhibited significant increase in intelligence quotient (IQ) scores [73].

4.6.1.4 *Centella asiatica* Linn. (Mandookparni)

C. asiatica (Family: Umbelliferae) aqueous extract showed improved learning and memory potential in vivo. Its extract significantly alleviated oxidative stress by decreasing malonaldehyde (MDA) level and increasing glutathione level. Further it also showed AchE inhibitory activity which combinedly may be associated with cognition-enhancing efficacy (Anil K et al., 2009). In recent randomized clinical trials, 28 participants (> 61 years of age) were administered with *C. asiatica* extracts (250, 500, and 750 mg/daily; p.o.) for 2 months. After 2 months, patients with 750 mg dose extract showed significant improvement in cognitive function (as assessed by event-related potential and the computerized assessment battery test) and mood (using Bond-Lader visual analogue) [74, 75].

4.7 Nootropic Polyherbal and Mineral Preparations.

4.7.1 Indian Noni

Noni, a polyherbal formulation of *Morinda citrifolia* extract and *Garcinia cambogia*, was tested in vivo for its nootropic potential against scopolamine-induced amnesia. The formulation (dose: twice/day for 14 days) showed significant ($p < 0.01$) decline in transfer latency against scopolamine-induced group. Further, it significantly inhibited acetylcholinesterase enzyme in rat brain homogenate and showed antioxidant property. These results suggested that Noni formulation might offer a useful therapeutic choice in either the prevention or the treatment of AD [76].

4.7.2 Memorin (Phytopharma)

Memorin is a dietary supplement designed to enhance brain functions. The formulation consists of aqueous juice of seven well-known Ayurvedic medicinal plants, viz., *C. asiatica* (150 mg), *C. alsonoids* (150 mg), *G. glabra* (100 mg), *A. calamus* (100 mg), *R. serpentina* (30 mg), *W. somnifera* (100 mg), and *S. lappa* (50 mg). Pretreatment with the formulation (200 mg/kg daily once for 21 days) attenuated ECS-induced amnesia in vivo [15].

4.7.3 Mentat (Himalaya Healthcare)

Mentat, a polyherbal formulation containing extracts of *C. asiatica*, *B. monnieri*, and *W. somnifera*, is marketed by Himalaya Pvt. Ltd. *B. monnieri* is well known for its nootropic effect. In vivo experimentation revealed that the herb enhances memory and learning process and also calms restlessness in several mental disorders. *C. asiatica* possesses antiepileptic properties and is commonly used as an adjuvant to epileptic drugs. It balances amino acid levels, which is beneficial in treating depression. It also prevents cognitive impairment. Ashvagandha is used as a mood stabilizer in clinical conditions of anxiety and depression. Withanolides, the chemical constituents present in winter cherry, possess rejuvenating properties. The herb also reduces oxidative stress, which can cause mental fatigue [77, 78].

4.7.4 Shankhpushpi Syrup (Baidyanath)

Shankhpushpi Syrup is an Ayurvedic remedy for memory and intellect. It is beneficial in mental weakness, forgetfulness, memory loss, low retention power, etc. The formulation consists of *C. alsinoids* (200 mg), brahmi (50 mg), sugar (8000 mg), and citric acid (6 mg). Shankhpushpi is traditionally being used as brain tonic. Brahmi calms the mind and reduces anxiety, mental stress, work-related stress, and depression. The formulation showed excellent nootropic activity in vivo in normal and scopolamine-induced amnesia models. It (dose>40 ml/kg p.o.) also showed significant anticholinesterase activity but showed moderate effects on other neurotransmitter contents of whole brain tissue [79].

4.7.5 Trasina (Dey's Pharmaceuticals)

Trasina is a polyherbal capsule containing *O. sanctum* (190 mg) and *W. somnifera* (80 mg). Trasina (one cap/day for 21 days) showed beneficial effect in mental activity in two in vivo models, viz., intracerebroventricular injection of colchicine (15 mg/rat) and ibotenic acid-induced neurotoxicity (10 mg/rat). The drug caused significant reduction in time spent on maze apparatus transfer and augmented cholinergic markers and M1 receptor binding of rat brain. Further, it also showed significant anticholinesterase activity in dose-dependent manner [77, 78].

4.7.6 Saraswatarishta (Baidyanath)

Saraswatarishta is a polyherbal Ayurvedic liquid formulation containing juice of *C. asiatica* (960 mg), *A. racemosus* (240 mg), *P. tuberose* (240 mg), *T. chebula* (240 mg), *F. vulgare* (240 mg), *Z. officinale* (240 mg), *V. zizanioides* (240 mg), *W. fruticosa* (240 mg), *V. negundo* (240 mg), and *O. turpethum* (240 mg) and honey. Saraswatarishta contains 5–10% of self-generated alcohol in it. This self-generated alcohol and the water present in the product act as a media to deliver water- and alcohol-soluble active herbal components to the body. A clinical study in 25 patients with the formulation (20 ml two times/day for 1 year) showed prominent antiepileptic activity. The drug showed nootropic effects against various experimental models in vivo via anticholinesterase activity [77, 78].

4.7.7 Vidyarthi Amrit (Maharishi Ayurveda)

Vidyarthi amrit is a polyherbal Ayurvedic liquid formulation containing juice of *C. alsinoids*, *C. asiatica*, *V. wallichii*, *E. ribes*, *A. racemosus*, *W. somnifera*, *E. cardamomum*, *P. anisum*, and *A. calamus*. The formulation exhibited significant anticholinesterase and antiglutamate activity and increase in L-aspartate and GABA levels of whole brain tissue in vivo which may be associated with nootropic potentiality of the formulation [77, 78].

4.7.8 Dimag Paushtik Rasayan (Baidyanath)

Dimag Paushtik Rasayan is an Ayurvedic Rasayana formulation composed of *C. alsinoids*, *W. somnifera*, *C. asiatica*, *B. monnieri*, Makaradhwaj (sulfide of mercury), and *N. jatamansi*. This drug exhibited nootropic effects in in vivo experimental models. The memory-enhancing effects of the formulation were attributed to cholinergic mechanisms [77, 78].

4.7.9 Geriforte (Himalaya)

Geriforte is an Ayurvedic formulation. Each Geriforte tablet contains Chyawanprash concentrate (100 mg), Himsra (*C. spinosa*) (13.8 mg), Kasani (*C. intybus*) (13.8 mg), Daruharidra (*B. aristata*) (10 mg), Vasaka (*A. vasica*) (10 mg), Kakamachi (*S. nigrum*) (6.4 mg), Arjuna (*T. arjuna*) (6.4 mg), Biranjasipha (*A. millefolium*) (3.2 mg), Kasamarda (*C. occidentalis*) (3.2 mg),

Jhavuka (*T. gallica*) (3.2 mg), Ashvagandha (*W. somnifera*) (30 mg), Shatavari (*A. racemosus*) (20 mg), Yashtimadhu (*G. glabra*) (20 mg), Mandukaparni (*C. asiatica*) (20 mg), Shilajeet (purified) (20 mg), Haritaki (*T. chebula*) (15 mg), Makardhwaj (10 mg), Musali (*A. adscendens*) 910 mg), Udakiryaka (*C. digyna*) (10 mg), Kapikachchu (*M. pruriens*) (10 mg), Jatiphalam (*M. fragrans*) (10 mg), Pippali (*P. longum*) (10 mg), Jaatipatree [*M. fragrans* (mace)] (10 mg), Bhringaraja (*E. alba*) (10 mg), Vriddadaru (*A. speciosa*) (10 mg), Abhrak bhasma (10 mg), Jasad bhasma (10 mg), Kumkuma (*C. sativus*) (7 mg), Mandur bhasma (5 mg), Lavanga (*S. aromaticum*) (5 mg), Ela (*E. cardamomum*) (5 mg), Yawani (*C. copticum*) (5 mg), Haridra (*C. longa*) (5 mg), Jyothishmati (*C. paniculatus*) (5 mg), and Loha bhasma (5 mg). The natural ingredients in Geriforte work synergistically to prevent free radical-induced oxidative damage to various organs. The ingredients are natural rejuvenators and cardioprotective agents. As an immunomodulator, Geriforte stimulates the immune system to respond against disease-causing microorganisms. Geriforte is an adaptogenic which effectively combats stress and fatigue. It also increases stamina and improves overall performance. Results demonstrated an expanding feeling of prosperity, physical effectiveness, and enhancement in mental capacities. The treatment diminished uneasiness and aided in better nitrogen balance. The medication has all the earmarks of being a normal blend of natural parts for capturing fast beginning of mental and physical handicap in matured people [28, 80].

4.7.10 Mineral Preparations

In Ayurveda, minerals are used as therapeutic agents in their calcinated forms. Ayurvedic preparations like Bhasmas of metal are therapeutically effective and safe for internal use. Few such mineral preparations associated with learning and memory are mentioned below.

4.7.10.1 Siddh Makardhwaja (Mercury)

Makardhwaja I, /mn. s is traditionally administered along with milk. The preparation helped in overall progress of overall status of experimental animals. It revealed growth-promoting, memory-improving, and carbohydrate-sparing properties without affecting other central nervous system (CNS) parameters in rodents [81].

4.7.10.2 Swarna Bhasma (Gold)

Swarna Bhasma showed potent antioxidant and free radical scavenging activity. The preparation (25 mg/kg orally for 7 to 10 days) exhibited a nootropic effect in vivo compared with *P. ginseng* tea (350 mg/kg orally for 10 days). The preparation showed significant anticholinesterase activity which may be associated with beneficiary effect on cognitive function [82, 83].

4.8 Summary and Conclusion

At present, restricted solutions are accessible to treat AD, and most of them are used either to postpone the progressions or symptomatic alleviation for a shorter period of time. Yet, the subsequent adverse reactions with economic factors related to these medications have made their utility constrained. In this manner, it is beneficial to shift the paradigm from synthetic to natural medicine as the latter is considered to be more safe and economical by WHO [84]. In the field of AD therapy, several interventions on the possible role of natural products have already been supported by the experimental outcomes. Till date, the greatest successes have resulted from plant-based AchE inhibitors and antioxidant discovery programs, which have provided substantial progress in the treatment of AD [85]. Other approaches include antiamyloid agents, β-secretase inhibitors, anti-inflammatory drugs, nicotinic and muscarinic receptor agonists, and tau protein inhibitors [86]. Currently, the investigation of natural products that can act as functional foods is of utmost interest to delay the progression of and provide symptomatic relief from AD and promote other associated health benefits. A significant shift from unidirectional to multidirectional drug therapy is being observed for the management of AD. However, cellular-level assay to find out mechanism is indeed very much essential to establish their therapeutic claim as anti-Alzheimer's drug. In a recent study, lifestyle modifications like stress-free life, vacation, spending quality and happy moments with family members, etc. are found to be beneficial to reduce risk of dementia. Moreover, yoga and meditation also reduce dementia and AD progression [10, 87]. If the research is progressed in a successful manner, the drugs derived from natural sources could be a good substitute for the conventional medicine, as they are easily and economically available these days.

At present, confined arrangements are available to treat AD, and most of them are utilized to put off the development of either signs or symptomatic mitigation for a brief time allotment. However, the ensuing adverse responses with financial variables identified with these drugs have made their utility obliged. As such, it is gainful to move the worldview from synthetic to natural drug as the latter is viewed as progressively sheltered and efficient by WHO [84]. In the field of AD treatment, a few mediations on the conceivable job of common items have just been bolstered by the exploratory results. Till date, the best triumphs resulted from plant-based AchE inhibitors and antioxidant discovery programs, which have given significant advancement in the treatment of AD [85]. Different other methodologies incorporate antiamyloid agents, β-secretase inhibitors, anti-inflammatory drugs, nicotinic and muscarinic receptor agonists, and tau protein inhibitors [86]. Currently, the investigation of natural products that can act as functional foods is of utmost interest to delay the progression of and provide symptomatic relief from AD and promote other associated health benefits. A huge move from a solitary focus to a multi-target sedate methodology, particularly for chronic and complex illness disorders, is being seen for the administration of AD. Notwithstanding, cell-level examine to discover system is for sure especially fundamental to set up their restorative case as hostile to Alzheimer's medication. In an ongoing report, ways of life alteration like peaceful life, excursion, spending quality and glad minutes with relatives, and so on are observed to be useful to reduce danger of dementia. On the off chance that the examination is advanced in an effective way, the medications got from natural sources could be a decent substitute for conventional medicine, as they are effectively and monetarily accessible nowadays.

Acknowledgment The author is grateful to Assam Down Town University, Assam, for providing necessary support for the preparation of the manuscript.

References

1. Adewusi EA, Steenkamp V (2011) *In vitro* screening for acetylcholinesterase inhibition and antioxidant activity of medicinal plants from southern Africa. Asian Pac J Trop Med 5 (2):829–835
2. Amod PK, Laurie AK, Girish JK (2005) Herbal complement inhibitors in the treatment of neuroinflammation. Ann N Y Acad Sci 1056:413–429
3. Colucci L, Bosco M, Ziello AR et al (2012) Effectiveness of nootropic drugs with cholinergic activity in treatment of cognitive deficit: a review. J Exp Pharmacol 4:163–172
4. Dwivedi KK, Singh RH (1978) A clinical study of medhya rasayana therapy in the management of convulsive disorder. J Res Ayurveda Siddha 13:97
5. Anderson RN (2002) Deaths: leading causes for 2000. Natl Vital Stat Rep 50:1
6. Zhang F, Jiang L (2015) Neuro-inflammation in Alzheimer's disease. Neuropsychiatr Dis Treat 11:243–256
7. Doraiswamy PM (2002) Non-cholinergic strategies for treating and preventing Alzheimer's disease. CNS Drugs 16:24
8. Gupta BM, Bala A (2013) Alzheimer's disease research in India: a scientometric analysis of publications output during 2002–11. Res Neurol Int J 2013:204542. https://doi.org/10.5171/2013.204542
9. Lemstra AW, Eikelenboom P, Van Gool WA (2003) The cholinergic deficiency syndrome and its therapeutic implications. Gerontology 49:55
10. Atanu B, Shastry CS, Saha S (2015) Nootropic activity of *Crataeva nurvala* Buch-ham against scopolamine induced cognitive impairment. EXCLI J 14:335–345
11. Berrino F (2002) Western diet and Alzheimer's disease. Epidemiol Prev 26:107
12. Orhan I, Kartal M, Naz Q, Ejaz A, Yilmaz G, Kan Y et al (2007) Antioxidant and anticholinesterase evaluation of selected Turkish Salvia species. Food Chem 103:1247–1254
13. Rammohan VR, Olivier D, Varghese J, Dale EB (2012) Ayurvedic medicinal plants for Alzheimer's disease: a review. Alzheimers Res Ther 4(22):1–12
14. Vinutha B, Prashanth D, Salma K, Sreeja SL, Pratiti D, Padmaja R et al (2007) Screening of selected Indian medicinal plants for acetylcholinesterase inhibitory activity. J Ethnopharmacol 109:359–363
15. Dahanukar SA, Kulkarni RA, Rege NN (2000) Pharmacology of medicinal plants and natural products. Indian J Pharmacol 32:81–88
16. Harvey AL (2008) Natural products in drug discovery. Drug Discov Today 13:894–901
17. Mahesh G, Tasneem S (2014) Biomarker controversies and diagnostic difficulties in Alzheimer's disease. Am J Phytomed Clin Ther 2(4):463–468
18. Rinne JO, Kaasinen V, Jarvenpaa T, Någren K, Roivainen A, Yu M et al (2003) Brain acetylcholinesterase activity in mild cognitive impairment and early Alzheimer's disease. Neurol Neurosurg Psychiatry 74:113–115
19. Chatellier G, Lacomblez L (1990) Tacrine (teterahydroaminoacridine; THA) and lecithine in senile dementia of the Alzheimer type: a multicenter trial. Br J Med 300:495
20. Clegg BJ, Nicholson T, McIntyre L, De Broe S, Gerard K, Waugh N (2002) Clinical and cost effectiveness of donepezil, rivastigmine, and galantamine for Alzheimer's disease: a systematic review. Int J Technol Assess Health Care 18:497
21. Spiegel R (2002) Rivastigmine: a review of its clinical effectiveness. Rev Neurol 35:859
22. Philomena G (2011) Concerns regarding the safety and toxicity of medicinal plants – an overview. J Appl Pharm Sci 1(6):40–44
23. Perry EK, Pickering AT, Wang WW, Houghton PJ, Perry NS (1999) Medicinal plants and Alzheimer's disease: from ethnobotany to phytotherapy. J Pharm Pharmacol 51:527–533
24. Vaidya AB (1997) The status and scope of Indian medicinal plants acting on the central nervous system. Indian J Pharmacol 29(S340):228–234
25. Manyam BV (1999) Dementia in Ayurveda. J Altern Complement Med 5:81–88

26. Anil K, Samrita D, Atish P (2009) Neuroprotective effects of *Centella asiatica* against intracerebroventricular colchicine-induced cognitive impairment and oxidative stress. Int J Alzheimer's Disease 9:1–8
27. Mishra LC, Singh BB, Dagenais S (2000) Scientific basis for the therapeutic use of *Withania somnifera* (ashwagandha): a review. Altern Med Rev 5:334–346
28. Singh RH, Sinha BN, Sarkar FH, Udupa KN (1979) Comparative biochemical studies on the effect of four Medhya Rasayana drugs described by Caraka on some central neurotransmitters in normal and stressed rats. J Indian Med Yoga Homeopat 14:7–14
29. Ahmed T, Gilani AH (2009) Inhibitory effect of curcuminoids on acetylcholinesterase activity and attenuation of scopolamine-induced amnesia may explain medicinal use of turmeric in Alzheimer's disease. Pharmacol Biochem Behav 91:554–559
30. Howes MJ, Houghton PJ (2003) Plants used in Chinese and Indian traditional medicine for improvement of memory and cognitive function. Pharmacol Biochem Behav 75:513–527
31. Rhee IK, Meent MV, Ingkaninan K, Verpoorte R (2001) Screening for acetylcholinesterase inhibitors from Amaryllidaceae using silica gel thin-layer chromatography in combination with bioactivity staining. J Chromatogr 915:217–223
32. Roberts SB (2002) Gamma-secretase inhibitors and Alzheimer's disease. Adv Drug Deliv Rev 54:1579–1582
33. Ghosh AK, Osswald HL (2014) BACE1 (β-secretase) inhibitors for the treatment of Alzheimer's disease. Chem Soc Rev 43(19):6765–6813. https://doi.org/10.1039/c3cs60460h
34. Kar S, Slowikowski SPM, Westaway D, Mount HTJ (2004) Interactions between β-amyloid and central cholinergic neurons: implications for Alzheimer's disease. J Psychiatry Neurosci 29 (6):427–441
35. Prerna U, Vikas S, Mushtaq A (2010) Therapy of Alzheimer's disease: an update. Afr J Pharm Pharmacol 4(6):408–421
36. Gelb DJ (2000) Measurement of progression in Alzheimer's disease: a clinician perspective. Stat Med 19:1390–1398
37. Vladimir V, Sophia R (2018) Results of Beta secretase-inhibitor clinical trials support amyloid precursor protein-independent generation of Beta amyloid in sporadic Alzheimer's disease. Med Sci (Basel) 6(2):45–58
38. Jemima J, Bhattacharjee P, Singhal RS (2011) Melatonin – a review on the lesser known potential nutraceutical. Int J Pharm Sci Res 2(8):1975–1987
39. Talic S, Dragicevic I, Corajevic L, Martinovic BA (2014) Acetylcholinesterase and butyrylcholinesterase inhibitory activity of extracts from medicinal plants. Bull Chem Technol Bosnia Herzegovania 43:11–14
40. Jian-zhi W, Ze-fen W (2006) Role of melatonin in Alzheimer-like neurodegeneration. Acta Pharmacol Sin 27(1):41–49
41. Jivad N, Rabiei Z (2014) A review study on medicinal plants used in the treatment of learning and memory impairments. Asian Pac J Trop Biomed 4:780–789
42. Puchchakayala G, Akina S, Thati M (2012) Neuroprotective effects of meloxicam and selegiline in scopolamine-induced cognitive impairment and oxidative stress. Int J Alzheimers Dis 12:1–8
43. Abhang RY (1987) Medhya rasayana: past, present and future. Deerghayu Int 4:8–18
44. Akhlaq AF, Tahira F, Anil M, Jolin Hwee-Jing O, Wei-Yi O (2018) Ayurvedic medicine for the treatment of dementia: mechanistic aspects. Evid Based Complement Alternat Med 40(2):1–11
45. Bush AI (2003) The metallobiology of Alzheimer's disease. Trends Neurosci 26(4):207–214
46. Fernando GP (2008) Brain foods: the effects of nutrients on brain function. Nat Rev Neurosci 9 (7):568–578
47. Huang W, Zhang X, Chen W (2016) Role of oxidative stress in Alzheimer's disease. Biomed Rep 4(5):519–522
48. Huang Y, Mucke L (2012) Alzheimer mechanisms and therapeutic strategies. Cell 148 (6):1204–1222

49. Aisen PS (2002) Evaluation of selective COX-2 inhibitors for the treatment of Alzheimer's disease. J Pain Symptom Manag 23:S35–S42
50. Aisen PS (2002) Anti-inflammatory agents in Alzheimer's disease. Curr Neurol Neurosci Rep 2:405–452
51. Nillert W, Pannangrong W, Welbat JU (2017) Neuroprotective effects of aged garlic extract on cognitive dysfunction and neuroinflammation induced by β-amyloid in rats. Nutrients 9:24–36
52. Lindsay J, Laurin D, Verreault R, Hebert R, Helliwell B, Hill GB, McDowell I (2002) Risk factors for Alzheimer's disease: a prospective analysis from the Canadian Study of Health and Aging. Am J Epidemiol 156:445–452
53. Schneider LS (1996) New therapeutic approaches to Alzheimer's disease. Psychiatry 57:30–36
54. Bahr BA, Bendiske J (2002) The neuropathogenic contributions of lysosomal dysfunction. J Neurochem 83:481–492
55. Jagdeep SD, Prasad DN, Avinash CT, Rajiv G (2009) Role of traditional medicine in neuropsychopharmacology. Asian J Pharm Clin Res 2(2):72–76
56. Patel VS, Jivani NP, Patel SB (2014) Medicinal plants with potential nootropic activity: a review. Res J Pharm Bio Chem Sci 5(1):1–11
57. Frawley D (1989) Ayurvedic healing. Motilal Banarsidass Publishers, Delhi
58. Khare CP (2007) Indian medicinal plants an illustrated dictionary. Springer, New Delhi
59. Satyavati GV, Gupta AK, Tandon N, Seth SD (1987) Medicinal plants of India, vol 2. Indian Council of Medical Research, New Delhi
60. Divya SV, Lakshmi CM (2007) Alzheimer's disease. Scientific basis for Ayurvedic medicines. CRC Publishers, New York
61. Ahirwar S, Tembhre M, Gour S, Namdeo A (2012) Anticholinesterase efficacy of *Bacopa monnieri* against the Brain Regions of Rat – a novel approach to therapy for Alzheimer's disease. Asian J Exp Sci 26(1):65–70
62. Gohil KJ, Patel JA (2010) A review on *Bacopa monniera:* current research and future prospects. Int J Green Pharm 2:1–9
63. Parul A, Bhawna S, Amreen F, Sanjay KJ (2014) An update on Ayurvedic herb *Convolvulus pluricaulis* Choisy. Asian Pac J Trop Biomed 4(3):245–252
64. Maya M, Sarada S (2014) *In vitro* screening for anti-cholinesterase and antioxidant activity of methanolic extracts of Ayurvedic medicinal plants used for cognitive disorders. PLoS One 9(1): e86804
65. Akinrimisi ED, Ainwannde AI (1975) Effect of aqueous extract of *Eugenia caryolphyllus* on brain acetylcholinesterase in rats. West Afr J Pharmacol Drug Res 2:127–133
66. Muralidharan P, Balamurugan G, Venu B (2009) Cerebroprotective effect of *Glycyrrhiza glabra* Linn. root extract on hypoxic rats. Bangladesh J Pharmacol 4:60–64
67. Iyer MR, Pal SC, Kasture VS, Kasture SB (1998) Effect of *Lawsonia inermis* on memory and behaviour mediated via monoamine neurotransmitters. Indian J Pharmacol 30:181–185
68. Habibur R, Muralidharan P, Anand M (2011) Inhibition of AChE and antioxidant activities are probable mechanism of Nardostacys jatamansi DC in sleep deprived Alzheimer's mice model. Int J PharmTech Res 3(3):1807–1816
69. Prachi S, Lakshmayya L, Vinod SB (2017) Anti-Alzheimer activity of isolated karanjin from *Pongamia pinnata* (L.) pierre and embelin from Embelia ribes Burm.f. Int Q J Res Ayurveda 38 (1):76–81
70. Prakash R, Sandhya E, Ramya N, Dhivya R, Priyadarshini M, Sakthi PB (2017) Neuroprotective activity of ethanolic extract of Tinospora cordifolia on LPS induced neuroinflammation. Transl Biomed 8(4):135–142
71. Bhattacharya SK, Kumar A (1995) Effect of glycowithanolides from Withania somnifera on an animal model of Alzheimers disease and perturbed central cholinergic markers of cognition in rats. Phytother Res 9:110–116
72. Schiloers R, Liermann A, Bhattacharya SK, Kumar A, Ghoshal S, Rich V (1997) Systemic administration of defined extracts from Withania somnifera (Indian ginseng) and shilajit

differentially affects cholinergic but not glutaminergic and gabaergic markers in rat brain. Neurochem Int 30:181
73. Carlo C, William LG, Michael L, Dale K, Kerry B, Barry O (2008) Effects of a standardized *Bacopa monnieri* extract on cognitive performance, anxiety, and depression in the elderly: a randomized, double-blind, placebo-controlled trial. J Altern Complement Med 14(6):707–713
74. Kashmira JG, Jagruti P, Anuradha KG (2010) Pharmacological review on *Centella asiatica*: a potential herbal cure-all. Indian J Pharm Sci 72(5):546–556
75. Soumyanath S, Zhong Y, Henson E et al (2012) *Centella asiatica* extract improves behavioral deficits in a mouse model of Alzheimer's disease: investigation of a possible mechanism of action. Int J Alzheimers Dis 12:381–394
76. Uma G, Uma Maheswari S (2014) Neuroprotective effects of polyherbal formulation (Indian Noni) on scopolamine-induced memory impairment in mice. Int J Pharm Pharm Sci 6 (1):354–357
77. Handa SS, Bhargava VK (1997) Effect of BR 16-A (Mentat) on cognitive deficits in aluminum treated and aged rats. Indian J Pharmacol 28:258–264
78. Kulkarni SK, Verma A (1994) BR 16-A (Mentat): a herbal preparation improves learning and memory performance in mice. Probe 3:222, 28
79. Sharma PC, Yelne MB, Dennis TJ (2005) Database on medicinal plants used in Ayurveda and Sidha, 1st edn. New Delhi, CCRAS, Dept. of AYUSH, Ministry of Health and Family Welfare, Govt. of India
80. Boral GC, Gautam B, Anjan B, Das NN, Nandi PS (1989) Geriforte in anxiety neurosis. Indian J Psychiatry 31(3):258–260
81. Vohora SB, Kim HS, Shah SA, Khanna T, Dandiya PC (1995) CNS and adaptogenic effects of Siddh Makardhwaja: an Ayurvedic mercury preparation, in Trace Elements in Nutrition and Health. Jamia Hamdard and Wiley Eastern Ltd., New Delhi
82. Bajaj S, Vohora SB (2000) Anticataleptic, anti-anxiety and anti-depressant activity of gold preparation used in Indian systems of medicine. Indian J Pharmacol 32:339
83. Bajaj S, Ahamad I, Fatima M, Raisuddin S, Vohora SB (1999) Immunomadulatory activity of a Unani gold preparation used in Indian system of medicine. Immunopharmacol Immunotoxicol 21:151
84. Perry NSL, Houghton PJ, Theolad AE, Jenner P, Perry EK (2000) *In vitro* inhibition of human erythrocyte acetylcholinesterase by *Salvia lavandulaefolia* essential oil and constituent terpenes. J Pharm Pharmacol 52:895–902
85. Perry NSL, Houghton PJ, Sampson J, Theolad AE, Hart S, Lis-balchin M (2001) *In vitro* activities of *Salvia lavandulaefolia* (Spanish Sage) relevant to treatment of Alzheimer's disease. J Pharm Pharmacol 53:1347–1356
86. Ingkaninan K, Phengpa P, Yuenyongsawad S, Khorana N (2006) Acetylcholinesterase inhibitors from *Stephania venosa* tuber. J Pharm Pharmacol 58:695–700
87. Mukherjee PK, Kumar V, Mal M, Houghton PJ (2007) Acetylcholinesterase inhibitors from plants. Phytomedicine 14:289–300

Memory Dysfunction Correlates with the Dysregulated Dopaminergic System in the Ventral Tegmental Area in Alzheimer's Disease

5

Fawaz Alasmari, Naif O. Al-Harbi, Mohammed M. Alanazi, Abdullah F. Alasmari, and Youssef Sari

Abstract

Alzheimer's disease (AD) is one of the neurodegenerative diseases associated with neuroinflammation. Tau neurofibrillary tangles and amyloid beta (Aβ) plaques can activate microglia and then elevate the levels of neuroinflammatory mediators in AD models. The elevation of cytokines levels can lead to increased Aβ production, which is one of the causes of the pathogenesis of AD. Although it is noteworthy that AD is associated with deficit in cholinergic system, it also demonstrated that AD is associated with dopaminergic neurodegeneration in the ventral tegmental area (VTA). The VTA sends dopaminergic inputs into the hippocampus and regulates the memory and learning functions. The depletion of dopaminergic neurons in the VTA in AD models might lead to memory impairments and cognition deficit. We suggest here that that neurodegeneration in the dopamine neurons is involved in the development of dysregulated behaviors in AD animal models. In this chapter, we illustrate the role of AD-associated neuroinflammation in dopaminergic neurodegeneration in the VTA.

Keywords

Alzheimer's disease · Neuroinflammation · Memory dysfunction · Ventral tegmental area · Dopaminergic neurodegeneration

F. Alasmari (✉) · N. O. Al-Harbi · M. M. Alanazi · A. F. Alasmari
Department of Pharmacology and Toxicology, College of Pharmacy, King Saud University, Riyadh, Kingdom of Saudi Arabia
e-mail: ffalasmari@ksu.edu.sa

Y. Sari
Department of Pharmacology and Experimental Therapeutics, College of Pharmacy and Pharmaceutical Sciences, University of Toledo, Toledo, OH, USA

S. Paul (ed.), *Application of Biomedical Engineering in Neuroscience*,
https://doi.org/10.1007/978-981-13-7142-4_5

5.1 Introduction

The prevalence of Alzheimer's disease (AD) is growing rapidly in the United States (US) [1], and this disease has been associated with long-term pathophysiological effects [2, 3]. It is estimated that from 2010 to 2050, the projected number of AD patients aged 65 years and older could triple, while the number of AD patients aged 85 years or older could quadruple in the United States [4]. It is important to note that AD can lead to serious negative consequences, and the number of deaths resulted from AD in the United States could be more than double from 2010 to 2040 [5].

Tau neurofibrillary tangles formation, which has been observed in AD animals and patients, is one of the hallmarks associated with the pathology of AD [6–9], and these tangles have been found to induce neurodegeneration through various mechanisms [10, 11]. The elevation of neuroinflammatory cytokines levels may be a key factor in the induction of neurodegeneration, thus leading to AD pathogenesis. Several studies have demonstrated that the levels of neuroinflammatory cytokines such as interleukins (ILs) increase in the brain of AD patients/animal models [12–14]. In addition, other cytokines, including tumor necrosis factor-α (TNF-α) and interferon-γ (IFN-γ) concentrations, are increased in the circulatory system in AD patients [15–18]. Our previous opinion and review articles have discussed in detail the positive correlation between increased production of neuroinflammatory cytokines and the development of neurodegeneration in AD models [19, 20]. In this chapter, we highlight the role of tau neurofibrillary tangles-induced neuroinflammation in the development of AD.

In addition to tau neurofibrillary tangles, the presence of amyloid beta (Aβ) plaques is another biomarker for AD progression [21]. It is important to note that Aβ can increase the release of neuroinflammatory cytokines, including ILs and TNF-α, from microglia (for review see Ref. [19]). These cytokines can increase amyloid precursor protein (APP) expression [22], and decrease Aβ uptake [23], consequently leading to Aβ plaques formation in the brain. Proinflammatory mediators downregulate the expression and functions of the Aβ transporter protein such as ATP-binding cassette subfamily B member 1 in the brain (for review see Ref. [20]). Importantly, studies have indicated that Aβ plaques or oligomers can affect neurotransmitter neurons, such as dopaminergic and glutamatergic neurons [24, 25]. In this chapter, we discussed the effects of AD progression on the ventral tegmental area (VTA) dopaminergic system. We also discussed the effects of neuroinflammation induced by Aβ plaques on dopamine neurons in the VTA.

Dopamine is released from the VTA into other brain regions, including the nucleus accumbens (NAc) and medial prefrontal cortex (mPFC), and regulates memory and learning as well as the reward pathways of drug dependence [26, 27]. Drug abuse, ethanol and/or nicotine, exposure has been reported to increase the release of dopamine from the VTA to the NAc [28, 29], an effect that might be involved in dysregulated learning behaviors in animal models [30]. Importantly, memory and learning behaviors have been found to be regulated by VTA-hippocampus (HIP) dopaminergic circuit suggesting the critical role of this dopaminergic pathway [25, 31]. These findings suggest that stimulating the dopaminergic neurons in the mesocorticolimbic system, including VTA, might restore the dysregulation of memory and learning functions induced by neuropsychiatric disorders.

5.2 Alzheimer's Disease and Neurodegeneration

AD is a neurodegenerative disease associated with manifestations, including neuroinflammation and neurotoxicity [32, 33]. These may cause neurodegeneration [34, 35] that is characterized by a marked loss in neurons and impairments of memory and learning [36]. Development of Aβ plaques has been found to be caused by accumulation of Aβ oligomers, which have direct neurodegenerative effects [37]. Aβ oligomers also cause neuroinflammation by activation of microglia, which are the driving source of neuroinflammatory cytokines (for review see Refs. [19, 20]). The levels of the neuroinflammatory cytokines are positively correlated with accumulated Aβ, including Aβ oligomers and plaques [38]. In addition, Aβ oligomers can induce synaptic impairments through the formation of pore-like shapes on synaptic channels [39]. We suggest here that AD-associated neuroinflammatory responses may cause dopaminergic neurodegeneration in the VTA and other brain regions, thus leading to memory impairments.

Another mechanism underlying AD-associated neurodegeneration is the presence of tau neurofibrillary tangles, which are the consequences of aggregation of tau protein. This has been associated with the development of neuroinflammation via activation of microglia [40]. This microglial activation increased the release of proinflammatory mediators levels, leading to marked neurodegenerative effects [41, 42]. In addition, studies have found that the formation of aberrant phosphorylated tau and glycosylation of tau contribute to the pathology of neurodegeneration [43–45]. For instance, phosphorylated tau in the form of a paired helical filament or neurofibrillary tangles has been demonstrated to lead to neurodegeneration in animals, probably because of destabilization of microtubule assembly due to the interaction of neurofibrillary tangles with tubulin in the neurons [46, 47]. The molecular mechanisms involving neurodegeneration in AD animal models and, particularly, the roles of Aβ and tau protein in neurodegeneration are illustrated in this chapter.

5.2.1 Role of Amyloid Beta in Neurodegeneration in Alzheimer's Disease

Cleavage of APP at β and γ sites causes the formation of Aβ. This metabolic process of APP is found to increase significantly in AD animal models, and Aβ production and deposition also increase in AD patients (for review see Ref. [48]). Importantly, a positive association between the degree of neurodegeneration and the levels of Aβ has been found in AD animal models. A recent study has demonstrated the role of neuroinflammatory cytokines in Aβ accumulation in the brain, and this results in increased of Aβ production, which may lead to the progression of AD [49, 50]. Another study further indicates that carnosine, an antioxidant compound, treatments could prevent Aβ-induced neuroinflammation and oxidative stress. This study found that this compound was able to decrease the production of neuroinflammatory mediators such as IL-1β using BV-2 microglial cells [51].

Studies discussed extensively the involvement of neuroinflammation in the progression of AD. Increases in the production of neuroinflammatory cytokines from activated microglia might lead to neurodegeneration (for review see Ref. [52]). Importantly, a prior study has suggested that neuroinflammatory responses are found in association with upregulated APP in male mice [53], which may further increase the formation of Aβ oligomers and plaques. For example, inflammatory cytokines exposure was demonstrated to increase APP protein and mRNA expression. Additionally, the expression levels of APP, a metabolic enzyme, and β-secretase enzymes increase following exposure to an inflammatory cytokine, IL-18, using human-like neurons [54]. These effects induced by inflammatory mediators cause neurodegeneration, thus leading to AD pathogenesis.

In addition, Aβ oligomers can induce neurodegeneration through the formation of pore-like shapes on the synaptic structure and cause synaptic dysfunction, which further progresses AD [39]. A marked disruption in synaptic function can be induced through the binding of Aβ oligomers to the synapses [55]. Aβ oligomers impair synaptic plasticity, and the effect might be associated with the dysregulated balance of excitatory/inhibitory systems in presynaptic sites. Notably, a significant loss in the excitatory synapses has been observed in the areas of Aβ oligomers and senile plaques [56]. This study mentioned that a small subset of excitatory synapses is associated with deposition of Aβ oligomers in the cortex in APP/PS1 transgenic mice.

Impairment in synaptic activity is involved in the development of neurodegeneration. Studies have demonstrated that overstimulation of N-methyl-D-aspartate (NMDA) receptor can lead to synaptic loss and triggers signaling pathways involving phosphorylation of tau and activation of caspases [37]. The synaptic activity is also sensitive to Aβ oligomers in AD patients. Interestingly, it has been found that Aβ oligomers cause impairments in the synaptic function in hippocampal neurons, and this effect is abolished after stimulation of dopamine 1/5 receptors [57], indicating that dopamine neurons are highly sensitive to the Aβ oligomers.

5.2.2 Role of Tau Tangles in Neurodegeneration in Alzheimer's Disease

A link has been observed between the presence of tau tangles (neurofibrillary tangles) and neurodegeneration in AD models [58]. Tau mutation has been found in AD patients and can cause accumulation of tau proteins in the form of filaments, and these mutated tau proteins cannot bind and stabilize microtubules [59]. It is important to note that aggregated tau filaments can lead to neuroinflammation via activation of astrocytes and microglia [60], which may lead to an increase in the release of neuroinflammatory cytokines. This effect can progress AD through neurodegeneration mechanisms [19]. Thus, dysregulation of tau proteins can lead to negative consequences in AD pathogenesis.

In addition, the caspase cascade pathway is involved in the generation of tau fibrillary tangles in neurodegenerative diseases, including AD [61]. Neuronal apoptosis may be associated with abnormal tau proteins phosphorylation [62]. Therefore, further studies are warranted to determine the possible mechanisms underlying the link between activation of caspase cascade pathway and the formation of neurofibrillary tau tangles. Additionally, degeneration of fibrillary tau tangles can be abolished by inhibiting glycogen synthase kinase-3β or cyclin-dependent protein kinase 5 in AD [63], and modulation of calmodulin-dependent protein kinase II or protein phosphatase-2A could also inhibit neurofibrillary degeneration [63]. These observations indicate these pathways are interacted and involved in the activity of tau tangles in AD models. Accordingly, targeting these pathways is a possible strategy to reduce the effects of tau tangles in the progression of AD. For instance, glycogen synthase kinase-3β inhibitor can decrease the phosphorylation of tau tangles in an AD in vitro model [64]; calmodulin-dependent protein kinase II has been demonstrated to be involved in tau tangles-mediated neurodegeneration in an AD model [65].

5.2.3 Role of Signaling Pathways in Neurodegeneration in Alzheimer's Disease

In addition to tau tangles and amyloid pathways, other systems have been demonstrated to be implicated in neurodegeneration in AD models. Corticotropin-releasing hormone is involved in neurodegeneration in transgenic models of tau and Aβ-related pathologies [66], and stress hormones and the signaling pathways play critical roles in the progression and development of neurodegeneration in AD models. Importantly, stress and glucocorticoids induce neuronal dysfunction and atrophy by activating tau hyperphosphorylation leading to degeneration on the synaptic proteins (for review see Ref. [67]). Additionally, chronic stress has been found to enhance neuronal apoptosis through increasing caspase-3-positive neurons in the cerebral cortex of rats [68]. In addition, chronic stress induces Aβ accumulation in an AD mouse model [69], and this effect is abolished following exposure to corticotropin-releasing factor-1 (CRF-1), indicating that Aβ-mediated neurodegeneration is partially mediated by the CRF pathway. Moreover, it has been found that increased urinary cortisol level is a marker of AD pathogenesis [70]. Indeed, increases in cortisol levels are associated with reduced gray matter volumes and memory deficit in humans [71]. These observations suggest the hypothesis that stress hormone signaling is involved in the progression of neurodegenerative diseases.

Furthermore, it has been suggested that the calcium channels are also considered as targets to attenuate AD progression [72]. Delayed hippocampal neuronal death has been demonstrated to be induced by an imbalance of calcium levels; an effect was found in glutamate-treated neurons [73]. This study reported that a NMDA receptor antagonist could abolish this neuronal cell death. Excessive activation of NMDA receptor by glutamate and excitotoxicity is found in AD models (for review see Ref. [74]). Importantly, calcium channel blockers, including L-type and T-type blockers, induce neuroprotective effects in both cortical and hippocampal neurons in mice [75]. The increase in calcium influx is induced by stimulation of NMDA receptor in postsynaptic neurons. Notably, an NMDA receptor antagonist, memantine, has been approved for the treatment of AD symptoms [76]. Importantly, the production of Aβ oligomers can increase in calcium influx through stimulation of NMDA receptor, which can lead to neuronal death in rat cortical neurons [24]. This effect might be due to the induction of mitochondrial dysfunction, which is caused by increased calcium influx rate [24, 77, 78]. Therefore, modulation of the NMDA receptor-calcium pathway might have potential therapeutic effects on attenuation of the pathogenesis of AD.

Other pathways, such as GABAergic and serotonergic pathways, may also have crucial roles in the development of AD [79, 80]. This indicates that AD pathophysiology is complex and multifactorial biomarkers are interacted to promote the progression of AD. Therefore, further investigation is required to study the novel molecular signaling pathways that can be targeted for the treatment of AD symptoms.

5.3 Role of the Dopaminergic System in the Ventral Tegmental Area in Neurodegeneration

The mesocorticolimbic brain regions regulate the release of neurotransmitters from and into other brain regions [27]. Innervation of these neurotransmitters into those brain areas is implicated in the development of dysregulated neurobehavioral and neurobiological effects [31, 81]. In next sections, we review the correlation between VTA dopaminergic neurodegeneration and the presence of AD symptoms in AD models.

5.3.1 Dopaminergic Neurodegeneration in the Ventral Tegmental Area in Alzheimer's Disease

The VTA dopaminergic neurons are lost in AD models [25], an effect associated with reduced dopamine release from the cortex and VTA [25, 82]. This may suggest that histopathological alterations associated with AD may lead to degeneration of dopaminergic neurons in several brain areas, including the VTA. Similarly, a recent study showed VTA dopaminergic neuronal loss in AD mouse model at the early stage of formation of Aβ plaques [25]. In addition, a recent clinical study also demonstrated a significant deficit in dopaminergic neurons in the VTA at the early stage of AD [83]. These observations provide critical insights into the correlation between dopaminergic neurodegeneration and the development of AD. In addition, a marked degeneration of dopaminergic neurons has also been found in the lateral VTA in postmortem tissues of subjects with both Parkinson and dementia diseases [84]. In fact, the VTA projects dopaminergic inputs into the NAc, HIP, mPFC, and amygdala, and these pathways are suggested to be implicated in neuropsychiatric disorders (for review see Ref. [27]). Importantly, degeneration of the VTA dopaminergic neurons was associated with memory deficit in AD animal models [25].

5.3.2 Degeneration of Ventral Tegmental Area Dopamine Neurons and Neurobehavioral Symptoms

It is suggested in several studies that development of deficit in memory and learning behaviors is caused in part by impaired dopaminergic neurons activity in the VTA [83, 85–87]. Decrease in dopamine release from the cortex of 3xTg-AD mice is associated with memory dysfunction in AD models [82]. This study reported that activating dopamine system by a dopamine reuptake blocker could abolish this dysfunction. Moreover, the VTA-HIP dopaminergic pathway has essential role in cognitive abilities, including memory and learning behaviors [31]. Therefore, it is possible that dopamine release from the VTA into the HIP, a brain region responsible mainly for learning and memory, is deficit in AD, and decreased dopamine release leads to impaired memory functions.

A study found that adult female and male Sprague Dawley rats exposed to chronic stress showed lower VTA dopaminergic neuronal activity than age-matched controls [87], and the decrease in the activity of dopaminergic neurons is accompanied by an increase in immobility duration. Treatment with ketamine restored both impaired immobility duration and reduction in the activity of the VTA dopaminergic neurons using male and female mice exposed to chronic mild stress [87]. This suggests that dysregulated behaviors might be associated with reduced activity of the dopaminergic neurons in the VTA in neuropsychiatric models. In addition, the activation of a specific population of VTA dopaminergic neurons correlates with positive memory activity in rats, as determined by a motor function test [86]. In fact, accumulating studies have demonstrated that a marked deficit in the dopamine release is associated with impaired memory and cognition behaviors in AD models/patients [25, 82]. These findings indicate that neurodegeneration in the dopamine neurons might lead to the formation of dysregulated behaviors in AD clinical and preclinical models.

The VTA-NAc dopaminergic pathway has a crucial role in the reward, memory, and learning process [27]. Impairments in the memory and reward system have been found in a Tg2576 transgenic AD mouse model, which has VTA dopaminergic neurodegeneration at the pre-plaque stage [25]. Dysfunction of the dopaminergic projection from the VTA into the HIP and NAc might consequently lead to impairment of the hippocampal neuronal plasticity. In addition, treatment with L-DOPA and selegiline (monoamine oxidase-B inhibitors) at a normal dose could successfully restore the dysfunction of reward and memory behaviors, and the beneficial effect is associated with improved hippocampal dendritic spine and plasticity [25]. Moreover, the VTA dopaminergic circuit is involved in the generation of motivation, reward, and other drug-seeking behaviors [85]. Further, reduced dopaminergic activity in the VTA might be a critical factor linked to the early stage of AD in humans [83]. Therefore, degeneration of dopaminergic neurons in the VTA might affect dopamine outflow from the VTA and NAc, leading to pathogenesis/progression of neurological diseases, including drug dependence and AD.

5.4 The Dopaminergic System: A Pharmacotherapeutic Target for Attenuation of Alzheimer's Disease Symptoms

Previous studies have found that stimulation of the dopaminergic system in the mesocorticolimbic areas could provide beneficial therapeutic effects on AD symptoms in the preclinical stage [88–91]. For instance, treatment with levodopa, a precursor of dopamine, could improve spatial and recognition memories, determined by Barnes maze and object recognition assays, in a transgenic AD animal model [88]. Short latency afferent inhibition (SLAI) is widely used to study the function of the cholinergic system in the cerebral cortex [92, 93]. In humans, treatment with levodopa at a single dose can normalize AD-decreased SLAI [91]. These results suggest that manipulation of dopaminergic transmission by levodopa attenuates AD symptoms at least in part by modulating the cholinergic system.

Other studies suggest that dopamine receptor agonists can improve AD symptoms in part by modulating signaling pathways [89, 90]. A dopamine agonist, rotigotine, can restore cholinergic function and elevate the excitability of the cortical area in AD clinical models [90]. Further, treatment with rotigotine for 4 weeks restores cortical plasticity of long-term potentiation (LTP) assessed by intermittent theta burst stimulation protocols compared with that with placebo or rivastigmine in humans [89], and this effect on LTP is associated with increased central cholinergic activity, as assessed using the SLAI protocol. Thus, stimulation of the dopaminergic system in the central brain areas with a dopamine modulator may have a potential therapeutic role in attenuating AD symptoms. These observations in AD patients using dopamine receptor agonists are further supported by those in similar studies performed in parkinsonian patients [94]. Therefore, both cholinergic and dopaminergic systems may interact in the central nervous system and are involved in the generation of symptoms of neurological diseases.

VTA releases dopamine into neurotransmitters projecting brain areas such as HIP, and dopamine can stimulate signaling pathways in these brain regions and induces various behavioral and biological effects [27, 31, 81, 95, 96]. VTA-HIP dopaminergic circuit is involved in the development of memory and cognition behaviors [25, 31]. Further, levodopa or selegiline treatment restores hippocampal spine density and plasticity at the CA1 hippocampal area, and these effects are associated with improved reward and cognition function, determined by measurement of food-seeking behaviors and memory ability, respectively [25]. A dopamine reuptake blocker was able to restore cortical dopamine levels and improved memory behavioral function [82]. It is difficult to specifically target VTA-HIP dopaminergic pathways; however, modulating dopaminergic system in the mesocorticolimbic system may lead to attenuation of some of the AD symptoms.

5.5 Conclusion

Neurodegeneration is a common manifestation of AD. AD patients have a higher risk for neurodegeneration of the dopaminergic neurons. VTA is the primary source of dopaminergic projections into the NAc, HIP, mPFC, and amygdala. Dopamine stimulates the dopaminergic and other systems such as glutamatergic system in these brain areas and induces behavioral changes such as drug-seeking behaviors. VTA and HIP dopaminergic pathways are critical in AD-associated behaviors such as memory and learning. Therefore, dysfunction of dopaminergic activity in the VTA may lead to dysregulated memory function, which further progresses AD and subsequently induces negative consequences. Activating VTA dopaminergic system is a pharmacotherapeutic strategy which may attenuate behavioral dysfunction in AD models. This approach may not only rescue the dopaminergic system in the VTA but also improve the dopamine function in other brain areas such as the NAc and HIP.

Future research should identify the association between the degree of the VTA dopaminergic neurodegeneration and the stages of AD progression, and the related findings would provide a clear evidence about the efficiency of activating dopaminergic system in the brain in the attenuation of AD symptoms. The amyloid system and tau neurofibrillary tangles play crucial roles in the development of neurodegeneration in AD clinical and preclinical models. Therefore, more pharmacological studies are needed to investigate the effects of Aβ and tau tangles on the VTA dopamine neurons and other brain regions of the mesocorticolimbic system.

Acknowledgment The book chapter was written during the period of fund supported by the International Scientific Partnership Program (ISPP-146) from the Deanship of Scientific Research, King Saud University.

Conflict of Interest The authors declare no conflict of interest.

References

1. Hebert LE, Weuve J, Scherr PA, Evans DA (2013) Alzheimer disease in the United States (2010–2050) estimated using the 2010 census. Neurology 80:1778–1783
2. Duckett L (2001) Alzheimer's dementia: morbidity and mortality. J Insurance Med (New York, NY) 33:227–234
3. Todd S, Barr S, Passmore AP (2013) Cause of death in Alzheimer's disease: a cohort study. QJM Int J Med 106:747–753
4. Hebert LE, Scherr PA, Bienias JL, Bennett DA, Evans DA (2003) Alzheimer disease in the US population: prevalence estimates using the 2000 census. Arch Neurol 60:1119–1122
5. Weuve J, Hebert LE, Scherr PA, Evans DA (2014) Deaths in the United States among persons with Alzheimer's disease (2010–2050). Alzheimer's Dement 10:e40–e46

6. Arriagada PV, Growdon JH, Hedley-Whyte ET, Hyman BT (1992) Neurofibrillary tangles but not senile plaques parallel duration and severity of Alzheimer's disease. Neurology 42:631–631
7. Bancher C, Brunner C, Lassmann H, Budka H, Jellinger K, Wiche G, Seitelberger F, Grundke-Iqbal I, Iqbal K, Wisniewski HM (1989) Accumulation of abnormally phosphorylated τ precedes the formation of neurofibrillary tangles in Alzheimer's disease. Brain Res 477:90–99
8. Brion J-P (1998) Neurofibrillary tangles and Alzheimer's disease. Eur Neurol 40:130–140
9. Metaxas A, Kempf SJ (2016) Neurofibrillary tangles in Alzheimer's disease: elucidation of the molecular mechanism by immunohistochemistry and tau protein phospho-proteomics. Neural Regener Res 11:1579
10. Beharry C, Cohen LS, Di J, Ibrahim K, Briffa-Mirabella S, Alonso Adel C (2014) Tau-induced neurodegeneration: mechanisms and targets. Neurosci Bull 30:346–358
11. Iqbal K, Liu F, Gong CX, Alonso Adel C, Grundke-Iqbal I (2009) Mechanisms of tau-induced neurodegeneration. Acta Neuropathol 118:53–69
12. Cacabelos R, Barquero M, Garcia P, Alvarez XA, de Seijas Varela E (1991) Cerebrospinal fluid interleukin-1 beta (IL-1 beta) in Alzheimer's disease and neurological disorders. Methods Find Exp Clin Pharmacol 13:455–458
13. Griffin WS, Stanley LC, Ling CHEN, White L, MacLeod V, Perrot LJ, White CL 3rd, Araoz C (1989) Brain interleukin 1 and S-100 immunoreactivity are elevated in Down syndrome and Alzheimer disease. Proc Natl Acad Sci 86:7611–7615
14. Ng A, Tam WW, Zhang MW, Ho CS, Husain SF, McIntyre RS, Ho RC (2018) IL-1β, IL-6, TNF-α and CRP in elderly patients with depression or Alzheimer's disease: systematic review and meta-analysis. Sci Rep 8:12050
15. Belkhelfa M, Rafa H, Medjeber O, Arroul-Lammali A, Behairi N, Abada-Bendib M, Makrelouf M, Belarbi S, Masmoudi AN, Tazir M (2014) IFN-γ and TNF-α are involved during Alzheimer disease progression and correlate with nitric oxide production: a study in Algerian patients. J Interferon Cytokine Res 34:839–847
16. Fillit H, Ding W, Buee L, Kalman J, Altstiel L, Lawlor B, Wolf-Klein G (1991) Elevated circulating tumor necrosis factor levels in Alzheimer's disease. Neurosci Lett 129:318–320
17. Komurcu HF, Kilic N, Demirbilek ME, Akin KO (2016) Plasma levels of vitamin B12, epidermal growth factor and tumor necrosis factor alpha in patients with Alzheimer dementia. Int J Res Med Sci 4:734–738
18. Swardfager W, Lanctôt K, Rothenburg L, Wong A, Cappell J, Herrmann N (2010) A meta-analysis of cytokines in Alzheimer's disease. Biol Psychiatry 68:930–941
19. Alasmari F, Alshammari MA, Alasmari AF, Alanazi WA, Alhazzani K (2018a) Neuroinflammatory cytokines induce amyloid beta neurotoxicity through modulating amyloid precursor protein levels/metabolism. Biomed Res Int 2018:3087475
20. Alasmari F, Ashby CR Jr, Hall FS, Sari Y, Tiwari AK (2018b) Modulation of the ATP-binding Cassette B1 transporter by neuro-inflammatory cytokines: role in the pathogenesis of Alzheimer's disease. Front Pharmacol 9:658
21. Shoghi-Jadid K, Small GW, Agdeppa ED, Kepe V, Ercoli LM, Siddarth P, Read S, Satyamurthy N, Petric A, Huang SC, Barrio JR (2002) Localization of neurofibrillary tangles and beta-amyloid plaques in the brains of living patients with Alzheimer disease. Am J Geriatr Psychiatry 10:24–35
22. Forloni G, Demicheli F, Giorgi S, Bendotti C, Angeretti N (1992) Expression of amyloid precursor protein mRNAs in endothelial, neuronal and glial cells: modulation by interleukin-1. Mol Brain Res 16:128–134
23. Walther W, Kobelt D, Bauer L, Aumann J, Stein U (2015) Chemosensitization by diverging modulation by short-term and long-term TNF-alpha action on ABCB1 expression and NF-kappaB signaling in colon cancer. Int J Oncol 47:2276–2285
24. Alberdi E, Sánchez-Gómez MV, Cavaliere F, Pérez-Samartín A, Zugaza JL, Trullas R, Domercq M, Matute C (2010) Amyloid β oligomers induce Ca2+ dysregulation and neuronal death through activation of ionotropic glutamate receptors. Cell Calcium 47:264–272

25. Nobili A, Latagliata EC, Viscomi MT, Cavallucci V, Cutuli D, Giacovazzo G, Krashia P, Rizzo FR, Marino R, Federici M (2017) Dopamine neuronal loss contributes to memory and reward dysfunction in a model of Alzheimer's disease. Nat Commun 8:14727
26. Holtzman-Assif O, Laurent V, Westbrook RF (2010) Blockade of dopamine activity in the nucleus accumbens impairs learning extinction of conditioned fear. Learn Mem 17:71–75
27. Russo SJ, Nestler EJ (2013) The brain reward circuitry in mood disorders. Nat Rev Neurosci 14:609–625
28. Tizabi Y, Bai L, Copeland RL Jr, Taylor RE (2007) Combined effects of systemic alcohol and nicotine on dopamine release in the nucleus accumbens shell. Alcohol Alcohol 42:413–416
29. Tizabi Y, Copeland RL Jr, Louis VA, Taylor RE (2002) Effects of combined systemic alcohol and central nicotine administration into ventral tegmental area on dopamine release in the nucleus accumbens. Alcohol Clin Exp Res 26:394–399
30. Di Chiara G (1998) A motivational learning hypothesis of the role of mesolimbic dopamine in compulsive drug use. J Psychopharmacol 12:54–67
31. Lisman JE, Grace AA (2005) The hippocampal-VTA loop: controlling the entry of information into long-term memory. Neuron 46:703–713
32. Heneka MT, Carson MJ, El Khoury J, Landreth GE, Brosseron F, Feinstein DL, Jacobs AH, Wyss-Coray T, Vitorica J, Ransohoff RM (2015) Neuroinflammation in Alzheimer's disease. Lancet Neurol 14:388–405
33. Yan SD, Chen X, Fu J, Chen M, Zhu H, Roher A, Slattery T, Zhao L, Nagashima M, Morser J, Migheli A, Nawroth P, Stern D, Schmidt AM (1996) RAGE and amyloid-beta peptide neurotoxicity in Alzheimer's disease. Nature 382:685–691
34. Chen WW, Zhang X, Huang WJ (2016) Role of neuroinflammation in neurodegenerative diseases (Review). Mol Med Rep 13:3391–3396
35. Crews L, Masliah E (2010) Molecular mechanisms of neurodegeneration in Alzheimer's disease. Hum Mol Genet 19:R12–R20
36. Toledo EM, Inestrosa NC (2010) Activation of Wnt signaling by lithium and rosiglitazone reduced spatial memory impairment and neurodegeneration in brains of an APPswe/PSEN1DeltaE9 mouse model of Alzheimer's disease. Mol Psychiatry 15(272–285):28
37. Tu S, Okamoto S, Lipton SA, Xu H (2014) Oligomeric Abeta-induced synaptic dysfunction in Alzheimer's disease. Mol Neurodegener 9:48
38. Patel NS, Paris D, Mathura V, Quadros AN, Crawford FC, Mullan MJ (2005) Inflammatory cytokine levels correlate with amyloid load in transgenic mouse models of Alzheimer's disease. J Neuroinflammation 2:9
39. Sivanesan S, Tan A, Rajadas J (2013) Pathogenesis of Abeta oligomers in synaptic failure. Curr Alzheimer Res 10:316–323
40. Dani M, Wood M, Mizoguchi R, Fan Z, Walker Z, Morgan R, Hinz R, Biju M, Kuruvilla T, Brooks DJ (2018) Microglial activation correlates in vivo with both tau and amyloid in Alzheimer's disease. Brain 141:2740–2754
41. Giovannini MG, Scali C, Prosperi C, Bellucci A, Vannucchi MG, Rosi S, Pepeu G, Casamenti F (2002) β-Amyloid-induced inflammation and cholinergic hypofunction in the rat brain in vivo: involvement of the p38MAPK pathway. Neurobiol Dis 11:257–274
42. Yates SL, Burgess LH, Kocsis-Angle J, Antal JM, Dority MD, Embury PB, Piotrkowski AM, Brunden KR (2000) Amyloid β and amylin fibrils induce increases in proinflammatory cytokine and chemokine production by THP-1 cells and murine microglia. J Neurochem 74:1017–1025
43. Perluigi M, Barone E, Di Domenico F, Butterfield DA (2016) Aberrant protein phosphorylation in Alzheimer disease brain disturbs pro-survival and cell death pathways. Biochimica et Biophysica Acta (BBA) Mol Basis Dis 1862:1871–1882
44. Robertson LA, Moya KL, Breen KC (2004) The potential role of tau protein O-glycosylation in Alzheimer's disease. J Alzheimer's Dis 6:489–495
45. Takahashi M, Tsujioka Y, Yamada T, Tsuboi Y, Okada H, Yamamoto T, Liposits Z (1999) Glycosylation of microtubule-associated protein tau in Alzheimer's disease brain. Acta Neuropathol 97:635–641

46. Grundke-Iqbal I, Iqbal K, Tung Y-C, Quinlan M, Wisniewski HM, Binder LI (1986) Abnormal phosphorylation of the microtubule-associated protein tau (tau) in Alzheimer cytoskeletal pathology. Proc Natl Acad Sci 83:4913–4917
47. Inoue H, Hiradate Y, Shirakata Y, Kanai K, Kosaka K, Gotoh A, Fukuda Y, Nakai Y, Uchida T, Sato E (2014) Site-specific phosphorylation of Tau protein is associated with deacetylation of microtubules in mouse spermatogenic cells during meiosis. FEBS Lett 588:2003–2008
48. Bharadwaj PR, Dubey AK, Masters CL, Martins RN, Macreadie IG (2009) Aβ aggregation and possible implications in Alzheimer's disease pathogenesis. J Cell Mol Med 13:412–421
49. Blasko I, Apochal A, Boeck G, Hartmann T, Grubeck-Loebenstein B, Ransmayr G (2001) Ibuprofen decreases cytokine-induced amyloid beta production in neuronal cells. Neurobiol Dis 8:1094–1101
50. Yamamoto M, Kiyota T, Horiba M, Buescher JL, Walsh SM, Gendelman HE, Ikezu T (2007) Interferon-γ and tumor necrosis factor-α regulate amyloid-β plaque deposition and β-secretase expression in Swedish mutant APP transgenic mice. Am J Pathol 170:680–692
51. Caruso G, Fresta CG, Musso N, Giambirtone M, Grasso M, Spampinato SF, Merlo S, Drago F, Lazzarino G, Sortino MA (2019) Carnosine prevents Aβ-induced oxidative stress and inflammation in microglial cells: a key role of TGF-β1. Cells 8:64
52. Morales I, Guzmán-Martínez L, Cerda-Troncoso C, Farías GA, Maccioni RB (2014) Neuroinflammation in the pathogenesis of Alzheimer's disease. A rational framework for the search of novel therapeutic approaches. Fron Cell Neurosci 8:112
53. Lee JW, Lee YK, Yuk DY, Choi DY, Ban SB, Ki Wan O, Hong JT (2008) Neuro-inflammation induced by lipopolysaccharide causes cognitive impairment through enhancement of beta-amyloid generation. J Neuroinflammation 5:37
54. Sutinen EM, Pirttilä T, Anderson G, Salminen A, Ojala JO (2012) Pro-inflammatory interleukin-18 increases Alzheimer's disease-associated amyloid-β production in human neuron-like cells. J Neuroinflamm 9:199
55. Wang Z, Jackson RJ, Hong W, Taylor WM, Corbett GT, Moreno A, Liu W, Li S, Frosch MP, Slutsky I, Young-Pearse TL, Spires-Jones TL, Walsh DM (2017) Human brain-derived Abeta oligomers bind to synapses and disrupt synaptic activity in a manner that requires APP. J Neurosci 37:11947–11966
56. Koffie RM, Meyer-Luehmann M, Hashimoto T, Adams KW, Mielke ML, Garcia-Alloza M, Micheva KD, Smith SJ, Kim ML, Lee VM, Hyman BT, Spires-Jones TL (2009) Oligomeric amyloid beta associates with postsynaptic densities and correlates with excitatory synapse loss near senile plaques. Proc Natl Acad Sci U S A 106:4012–4017
57. Jurgensen S, Antonio LL, Mussi GE, Brito-Moreira J, Bomfim TR, De Felice FG, Garrido-Sanabria ER, Cavalheiro EA, Ferreira ST (2011) Activation of D1/D5 dopamine receptors protects neurons from synapse dysfunction induced by amyloid-beta oligomers. J Biol Chem 286:3270–3276
58. Rapoport M, Dawson HN, Binder LI, Vitek MP, Ferreira A (2002) Tau is essential to beta -amyloid-induced neurotoxicity. Proc Natl Acad Sci U S A 99:6364–6369
59. Iqbal K, Fei L, Gong C-X, Grundke-Iqbal I (2010) Tau in Alzheimer disease and related tauopathies. Curr Alzheimer Res 7:656–664
60. Morales I, Jiménez JM, Mancilla M, Maccioni RB (2013) Tau oligomers and fibrils induce activation of microglial cells. J Alzheimer's Dis 37:849–856
61. Rohn TT, Head E, Joseph HS, Anderson AJ, Bahr BA, Cotman CW, Cribbs DH (2001) Correlation between caspase activation and neurofibrillary tangle formation in Alzheimer's disease. Am J Pathol 158:189–198
62. Kobayashi K, Nakano H, Hayashi M, Shimazaki M, Fukutani Y, Sasaki K, Sugimori K, Koshino Y (2003) Association of phosphorylation site of tau protein with neuronal apoptosis in Alzheimer's disease. J Neurol Sci 208:17–24
63. Wang J-Z, Grundke-Iqbal I, Iqbal K (2007) Kinases and phosphatases and tau sites involved in Alzheimer neurofibrillary degeneration. Eur J Neurosci 25:59–68

64. Croft CL, Kurbatskaya K, Hanger DP, Noble W (2017) Inhibition of glycogen synthase kinase-3 by BTA-EG 4 reduces tau abnormalities in an organotypic brain slice culture model of Alzheimer's disease. Sci Rep 7:7434
65. Oka M, Fujisaki N, Maruko-Otake A, Ohtake Y, Shimizu S, Saito T, Hisanaga S-I, Iijima KM, Ando K (2017) Ca2+/calmodulin-dependent protein kinase II promotes neurodegeneration caused by tau phosphorylated at Ser262/356 in a transgenic Drosophila model of tauopathy. J Biochem 162:335–342
66. Futch HS, Croft CL, Truong VQ, Krause EG, Golde TE (2017) Targeting psychologic stress signaling pathways in Alzheimer's disease. Mol Neurodegener 12:49
67. Vyas S, Rodrigues AJ, Silva JM, Tronche F, Almeida OFX, Sousa N, Sotiropoulos I (2016) Chronic stress and glucocorticoids: from neuronal plasticity to neurodegeneration. Neural Plastic 2016
68. Bachis A, Cruz MI, Nosheny RL, Mocchetti I (2008) Chronic unpredictable stress promotes neuronal apoptosis in the cerebral cortex. Neurosci Lett 442:104–108
69. Carroll JC, Iba M, Bangasser DA, Valentino RJ, James MJ, Brunden KR, Lee VM-Y, Trojanowski JQ (2011) Chronic stress exacerbates tau pathology, neurodegeneration, and cognitive performance through a corticotropin-releasing factor receptor-dependent mechanism in a transgenic mouse model of tauopathy. J Neurosci 31:14436–14449
70. Ennis GE, Yang A, Resnick SM, Ferrucci L, O'brien RJ, Moffat SD (2017) Long-term cortisol measures predict Alzheimer disease risk. Neurology 88:371–378
71. Echouffo-Tcheugui JB, Conner SC, Himali JJ, Maillard P, DeCarli CS, Beiser AS, Vasan RS, Seshadri S (2018) Circulating cortisol and cognitive and structural brain measures: the Framingham Heart Study. Neurology 91:e1961–e1e70
72. Yagami T, Kohma H, Yamamoto Y (2012) L-type voltage-dependent calcium channels as therapeutic targets for neurodegenerative diseases. Curr Med Chem 19:4816–4827
73. Limbrick DD Jr, Churn SB, Sombati S, DeLorenzo RJ (1995) Inability to restore resting intracellular calcium levels as an early indicator of delayed neuronal cell death. Brain Res 690:145–156
74. Hynd MR, Scott HL, Dodd PR (2004) Glutamate-mediated excitotoxicity and neurodegeneration in Alzheimer's disease. Neurochem Int 45:583–595
75. Wildburger NC, Lin-Ye A, Baird MA, Lei D, Bao J (2009) Neuroprotective effects of blockers for T-type calcium channels. Mol Neurodegener 4:44
76. Marder K (2004) Memantine approved to treat moderate to severe Alzheimer's disease. Curr Neurol Neurosci Rep 4:349–350
77. Ferreira IL, Ferreiro E, Schmidt J, Cardoso JM, Pereira CMF, Carvalho AL, Oliveira CR, Rego AC (2015) Aβ and NMDAR activation cause mitochondrial dysfunction involving ER calcium release. Neurobiol Age 36:680–692
78. He Y, Cui J, Lee JCM, Ding S, Chalimoniuk M, Simonyi A, Sun AY, Gu Z, Weisman GA, Wood WG (2011) Prolonged exposure of cortical neurons to oligomeric amyloid-β impairs NMDA receptor function via NADPH oxidase-mediated ROS production: protective effect of green tea (-)-epigallocatechin-3-gallate. ASN Neuro 3:AN20100025
79. Butzlaff M, Ponimaskin E (2016) The role of serotonin receptors in Alzheimer's disease. In: Opera Medica et Physiologica
80. Li Y, Sun H, Chen Z, Xu H, Guojun B, Zheng H (2016) Implications of GABAergic neurotransmission in Alzheimer's disease. Fron Age Neurosci 8:31
81. Stuber GD, Britt JP, Bonci A (2012) Optogenetic modulation of neural circuits that underlie reward seeking. Biol Psychiatry 71:1061–1067
82. Guzmán-Ramos K, Moreno-Castilla P, Castro-Cruz M, McGaugh JL, Martínez-Coria H, LaFerla FM, Bermúdez-Rattoni F (2012) Restoration of dopamine release deficits during object recognition memory acquisition attenuates cognitive impairment in a triple transgenic mice model of Alzheimer's disease. Learn Mem 19:453–460

83. De Marco M, Venneri A (2018) Volume and connectivity of the ventral tegmental area are linked to neurocognitive signatures of Alzheimer's disease in humans. J Alzheimers Dis 63:167–180
84. Hall H, Reyes S, Landeck N, Bye C, Leanza G, Double K, Thompson L, Halliday G, Kirik D (2014) Hippocampal Lewy pathology and cholinergic dysfunction are associated with dementia in Parkinson's disease. Brain 137:2493–2508
85. Lammel S, Lim BK, Ran C, Huang KW, Betley MJ, Tye KM, Deisseroth K, Malenka RC (2012) Input-specific control of reward and aversion in the ventral tegmental area. Nature 491:212–217
86. Leemburg S, Canonica T, Luft A (2018) Motor skill learning and reward consumption differentially affect VTA activation. Sci Rep 8:687
87. Rincón-Cortés M, Grace AA (2017) Sex-dependent effects of stress on immobility behavior and VTA dopamine neuron activity: modulation by ketamine. Int J Neuropsychopharmacol 20:823–832
88. Ambrée O, Richter H, Sachser N, Lewejohann L, Dere E, de Souza Silva MA, Herring A, Keyvani K, Paulus W, Schäbitz W-R (2009) Levodopa ameliorates learning and memory deficits in a murine model of Alzheimer's disease. Neurobiol Aging 30:1192–1204
89. Koch G, Di Lorenzo F, Bonnì S, Giacobbe V, Bozzali M, Caltagirone C, Martorana A (2014) Dopaminergic modulation of cortical plasticity in Alzheimer's disease patients. Neuropsychopharmacology 39:2654
90. Martorana A, Di Lorenzo F, Esposito Z, Giudice TL, Bernardi G, Caltagirone C, Koch G (2013) Dopamine D2-agonist Rotigotine effects on cortical excitability and central cholinergic transmission in Alzheimer's disease patients. Neuropharmacology 64:108–113
91. Martorana A, Mori F, Esposito Z, Kusayanagi H, Monteleone F, Codeca C, Sancesario G, Bernardi G, Koch G (2009) Dopamine modulates cholinergic cortical excitability in Alzheimer's disease patients. Neuropsychopharmacology 34:2323
92. Di Lazzaro V, Oliviero A, Pilato F, Saturno E, Dileone M, Marra C, Daniele A, Ghirlanda S, Gainotti G, Tonali PA (2004) Motor cortex hyperexcitability to transcranial magnetic stimulation in Alzheimer's disease. J Neurol Neurosurg Psychiatry 75:555–559
93. Nardone R, Bergmann J, Kronbichler M, Kunz A, Klein S, Caleri F, Tezzon F, Ladurner G, Golaszewski S (2008) Abnormal short latency afferent inhibition in early Alzheimer's disease: a transcranial magnetic demonstration. J Neural Transm 115:1557–1562
94. Appiah-Kubi LS, Chaudhuri KR (2002) Sustained dopamine agonism with cabergoline in Parkinson's disease. In: Mapping the progress of Alzheimer's and Parkinson's disease. Springer, Boston
95. Juarez B, Han M-H (2016) Diversity of dopaminergic neural circuits in response to drug exposure. Neuropsychopharmacology 41:2424
96. Margolis EB, Lock H, Chefer VI, Shippenberg TS, Hjelmstad GO, Fields HL (2006) κ opioids selectively control dopaminergic neurons projecting to the prefrontal cortex. Proc Natl Acad Sci 103:2938–2942

Part IV

Brain Images and Its Classifications

Substance Dependence: Overview of the Environmental, Genetic, Epigenetic, and Imaging Studies

6

Ranjan Gupta and Arundhati Sharma

Abstract

A complex condition like substance dependence (SD) causes disruption in brain circuits leading to various health issues and conditions like homelessness, crime, and violence. Environmental factors play an important role in substance dependence and genetic causes add to the susceptibility of individuals to these disruptions, thereby leading to dependence. Several studies based on twin pairs, families, and adoption point to the involvement of genetics in substance dependence. Besides genetics, there is evidence highlighting the importance of epigenetic modifications of genes leading to neuroplastic changes in the brain of individuals with substance dependence. Brain imaging techniques like positron emission tomography/single-photon emission computed tomography (PET/SPECT) can be helpful in the identification of neuroadaptive and neurodegenerative changes which in the future may help in the development of pharmacological treatment targets for SD. The progress made in the field of genetics, epigenetics, and imaging in substance dependence and their applications, along with genes whose polymorphisms and methylation status may evolve as potential biomarkers in substance dependence, is highlighted here.

Keywords

Environment interaction · Genetics · Epigenetics · DNA methylation · Imaging

R. Gupta (✉) · A. Sharma
Laboratory of Cyto-Molecular Genetics, Department of Anatomy, AIIMS, New Delhi, India

S. Paul (ed.), *Application of Biomedical Engineering in Neuroscience*,
https://doi.org/10.1007/978-981-13-7142-4_6

6.1 Introduction

Substance use disorder (SUD) is an important issue related to public health, and the term "substance" in this context refers to alcohol or any other psychoactive drug. The World Health Organization (WHO) categorizes substance use disorders into harmful use and dependence. Dependence syndrome is defined as a cluster of behavioral, cognitive, and physiological phenomena that develop after repeated substance use and that typically include a strong desire to take the drug, difficulties in controlling its use, persisting in its use despite harmful consequences, a higher priority given to drug use than to other activities and obligations, increased tolerance, and sometimes a physical withdrawal state (https://www.who.int/topics/substance_abuse/en/). Substance dependence is a relapsing disease having negative effects and shows compulsive behavior leading to long-lasting changes in the brain. Besides dependence, drug use is associated with various adverse conditions like cirrhosis, mental illness, cancer, etc. It also results in violence, homelessness, and crime affecting both the individual and the society. Research indicates that substance dependence alters the brain cell morphology, chemistry, and molecular biology, thereby affecting behavior (Fig. 6.1).

6.2 Intensity of the Problem

Alcohol was rated as the most harmful substance followed by heroin and cocaine by Britain's Independent Scientific Committee on Drugs (ISCD) and the European Monitoring Centre for Drugs and Drug Addiction (EMCDDA). Globally, 140 million people suffer from alcohol dependence (AD), linked to a plethora of health conditions resulting in 3.3 million (5.9%) deaths every year [153]. The WHO also estimates alcohol to be the third most dangerous risk factor impacting health of a population following tobacco and hypertension.

Fig. 6.1 Ramifications of substance dependence

Globally, 0.7% of the adults consume opioids, equivalent to 32.4 million resulting in about 51,000 deaths [34]. Nearly 75% of these deaths occurred due to an overdose of drugs, which were mostly opioids. The United States in 2013 reported the highest death rate due to heroin consumption which increased from 5925 in 2012 to 8257 in 2015 [152].

In India, consumption of alcohol and other drugs varies greatly between different states. Alcohol is presently banned in some states of India such as Manipur, Bihar, and Gujarat, but is legally consumed in a majority of states. According to a survey on prevalence of substance use in India, alcohol was found to be the main substance of use followed by tobacco (21.4%), cannabis (3.0%), and opioids (0.7%) [116]. Another report from the National Family Health Survey (NFHS 2007) recorded an increased use of alcohol in the year 2005–2006 compared to 1998–1999. A WHO study reports 25% of males to be heavy episodic drinkers (Global Status Report on Alcohol and Health 2011; www.who.int). It has also been estimated that at least 66% of alcohol consumption in India is not yet recorded. Another study in Punjab reported an overall substance dependence to be 4.65%, with 1.53% accounting for opioid and 0.91% for injectable opioid dependence among others [7].

Studies show that environmental, genetic, and epigenetic factors (methylation and histone acetylation) have a role in conferring risk to substance dependence (Fig. 6.2), while others reported that environment factors may alter epigenetic mechanisms [4]. Presently, known evidences highlight the importance of methylation changes in substance use/dependence which suggest alteration of gene expression and regulation after substance use. Methylation changes have been the most commonly reported

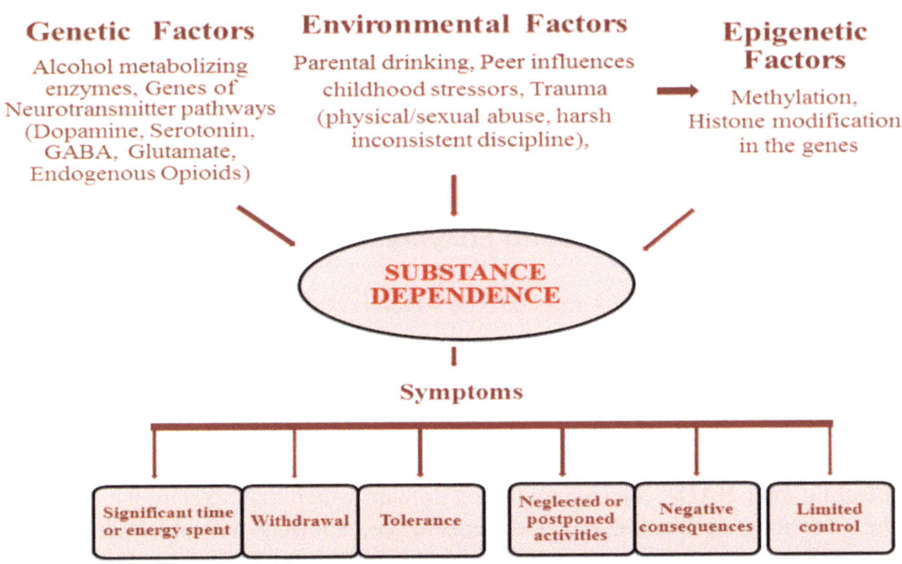

Fig. 6.2 Factors affecting substance dependence

epigenetic changes which influence gene transcription of various brain circuits. This chapter reviews recent findings on substance dependence, genetics, epigenetics, and other molecular mechanisms involved in drug use.

6.3 Genetics

Substance dependence is a complex trait which clusters in families and is influenced by both genetic and environmental factors. Genetic contribution to alcohol dependence (AD) with heritability estimates of about 60% in both the sexes is documented by several studies [93]. Variations at individual level in vulnerability to alcohol are heritable, based on reports of adoption [31] and twin pairs [58, 70, 72]. Use of illicit drugs and dependence on them have also been shown to be influenced by genes to an extent of about 45–79% as established by different studies [3, 71]. A twin pair meta-analysis by Agrawal et al. [2] revealed that inherited factors account for about 66%, 79%, and 59% of alcohol, cocaine, and cannabis dependence, respectively.

Genetic factors in AD can be categorized broadly as genes (1) having a role in personality traits like impulsiveness and sensation seeking making drinkers more prone to excessive consumption, thereby increasing their risk of becoming dependent, (2) genes affecting drug and alcohol metabolism, and (3) genes of the various neurobiological pathways. Single-nucleotide polymorphisms (SNPs) in these genes are recognized which show association with alcohol/drug dependence and speculated to have an interactive role in dependence. Molecular studies prove that variations in multiple genes may determine predisposition of an individual to drug dependence, and these variations can be effective genetic markers in association studies.

Genetic studies identify association of markers with substance dependence phenotypes which may help in delineating the role of a genetic variant acting in a particular biological pathway.

6.4 Alcohol-Metabolizing Enzymes

Alcohol oxidation generates acetaldehyde which is converted to acetate by aldehyde dehydrogenase (ALDH). Functional SNPs in the genes of alcohol dehydrogenase (ADH) and aldehyde dehydrogenase influence the rate of ethanol elimination and the rate of formation and elimination of acetaldehyde in alcohol dependence.

ADH has several isoforms – ADH1A, ADH1B, ADH1C, and ADH4–ADH7. ADH1B and ADH1C exhibit SNPs leading to change in a single amino acid affecting their NAD+ coenzyme binding domain which may explain the effects of the variations on alcohol use and dependence. Subjects with ADH1B*2 allele (Arg48His) variant metabolize alcohol comparatively at a higher rate compared to ADH1B*1. This SNP is commonly found in East Asian populations and confers a protective effect against alcohol use disorders [88, 128, 136]. Studies of European-Americans and Central-West Brazil population also show ADH1B*2 variant with a similar protective role [14, 140]. Another variant ADH1B*3 was found to confer lower risk of alcohol use disorder among the aboriginals and American individuals of African descent [41].

Subsequently, alcohol use and dependence was seen to be associated with SNPs of ALDH isoforms. SNP ALDH2*2 variant rs671 (Glu504Lys), commonly present in East Asians, was seen to metabolize acetaldehyde slowly resulting in alcohol flushing syndrome. A decreased risk of alcohol use disorders was also found in individuals with the variant ALDH2*2 [88, 138].

6.5 Opioid Receptor Mu1 (*OPRM1*)

A G protein-coupled receptor, μ-opioid, is expressed in the brain which binds endogenous and exogenous opioids. *OPRM1* is mostly associated with psychoactive substance disorders. Several reports indicate the possible involvement of OPRM1 SNP rs1799971 in substance dependence [33, 81, 103]. The non-synonymous *OPRM1* exon 1 SNP A118G (asparagine to aspartic acid substitution) was reported to confer risk in Chinese and Indian subjects of heroin dependence [134, 137]. The functional effects of this SNP in altering the expression levels and receptor properties have been demonstrated by several studies [91, 159].

6.6 Polymorphisms of Neurotransmitter Pathway Genes

Studies show the effect of substance use on different neurotransmitter systems in the brain and how the genes encoding these neurotransmitters may have a role in susceptibility to substance dependence. The different neurotransmitters have different functions; dopamine has a key role in movement, reward, punishment, pleasure, etc., whereas glutamate is an excitatory and the most important neurotransmitter for normal functions of the brain like learning, memory, and cognition.

6.7 Dopaminergic Pathway

Substance dependence increases dopamine levels and enhances brain neurotransmission. Dopamine pathway genes which code for dopamine receptors (D1–D5) are mainly implicated in substance dependence. They mediate rewards, increase dopamine levels, and activate neurotransmission [79, 138]. Various reports identify possible polymorphisms in candidate genes of the dopaminergic pathway which increase susceptibility to substance dependence. Association studies on dopamine receptors (D1–D5) have suggested the DRD2 polymorphisms (-141C Ins/Del, TaqI B, and TaqI A) to be involved in susceptibility to alcoholism [47]. Most groups have observed no association, either allelic or genotypic, of TaqI B polymorphism with AD [75, 111], whereas DRD2 ANKK1 gene polymorphism (TaqI A1 allele) was found to confer a higher risk of developing AD.

6.8 Catechol-O-Methyltransferase (COMT)

COMT encodes catechol-o-methyltransferase enzyme which helps dopamine to convert into homovanillic acid. Many *COMT* polymorphisms affecting its activity are reported.

COMT Val158Met SNP (rs4680) is an interesting polymorphism which shows differential activity of the enzyme [29]. Homozygosity of Val158 shows three- to fourfold higher activity of this enzyme, associated with behavior disinhibition leading to impulsive behavior [42]. Better cognitive performance and less stress are reported with the Met158 variant though it also shows an increase in anxiety. Hence, rs4680 SNP is extensively studied in addictive behavior and shows strong association with substance (alcohol/cocaine) dependence ([1, 65, 142, 67, 85]. In certain populations, Val158 is linked to dependence, but some subpopulations show risk with the Met158 allele [37]. In a Finnish report, association of rs4680 was identified with alcohol dependence affecting the patients in such a way that they consume more and more alcohol [73]. Similar studies on the dopaminergic pathway in substance dependence are shown in Table 6.1.

6.9 Serotonin Pathway

Normal development and behavior is a result of the action of serotonin, an important neurotransmitter. Fast synaptic neurotransmission results from utilization of serotonin by the neurons of the central and the enteric nervous systems. Serotonin is responsible for learning, aggression, anxiety, behavior, substance use, depression, and control of obsession. Serotonin transporter (SLC6A4) variants are associated with substance use, major depression, and obsessive compulsive and bipolar disorders.

The gene *SLC6A4* encodes serotonin transporter (SERT), a sodium-dependent protein, involved in reprocessing of serotonin from the synapse into the neuronal cell. A 5-HTTLPR polymorphism of 44bp insertion (L allele) and deletion (S allele) located in the *SLC6A4* promoter is widely reported and seems to influence its transcription. A 50% reduction in the *SLC6A4* expression is seen due to the presence of S allele [90], while the L allele is documented to have a positive role in alcohol dependence [16, 18, 107]. Overall, the association of HTTLPR with alcohol dependence is still questionable due to reports which show positive [146], while others [104] show no association with alcohol intake. Another polymorphism STin2 VNTR acts as a transcriptional regulator of the gene [48] and influences neuron development. The transcriptional activity of the 10-repeat allele is found to be lower compared to the 12-repeat one [90], and AD cases prone to depression are frequently

Table 6.1 Genetic studies of the dopaminergic pathway in substance dependence

S. No.	Gene/polymorphism	Outcome	References
1.	9Ser/Gly DRD3 BalI exon 1 SNP	Positive and negative association	Sander et al. [123], and Heidbreder et al. [59]
2.	DRD4 -521 C/T change	Positive association	Benjamin et al. [11] and Ebstein et al. [39]
3.	DRD2 Taq1A and Taq1B	Positive and negative association	Ponce et al. [112] and Konishi et al. [77]
4.	DRD4 VNTR (long allele)	Strong cue-elicited heroin craving	Shao et al. [126]
5.	DRD1 +1403T/C Bsp 1286 I polymorphism rs686, rs686-rs4532 haplotype	Positive association	Batel et al. [8]
6.	-141C Ins/Del	Positive association with AD	Prasad et al. [113] and Chen et al. [28]
7.	DRD4 VNTR (48bp repeat) exon 3	Positive and negative association	Naka et al. [98] and Du et al.[36]
8.	*COMT* (rs737866)	Significant association with age at onset of heroin use	Li et al. [84]
9.	*DAT1* (40bp VNTR), DRD1-DRD5	Positive association with AD	Hack et al. [57]
10.	DAT rs40184, DRD2-141C Ins/Del	Positive association with AD	Prasad et al. [113]
11.	SNCA-Rep 1, DRD2/ANKK1 TaqIA (rs1800497), and SLC6A3 40 bp-VNTR	Positive and negative association with AD	Janeczek et al. [69]
12.	*DRD3*	Positive association with AD	Agarwal et al. [3]
13.	*COMT* (Val108/158Met), *OPRM1* (A118G)	Positive association with AD and OD	Tiihonen et al. [142] and Tan et al. [137]
14.	SLC6A3 40 bp-VNTR, DRD2/ ANKK1, DRD Ins-141Del, and DAT1 VNTR	Positive and negative association with AD	Vasconcelos et al. [143] and Preuss et al. [114]
15.	*DRD1* (rs5326), *DRD2* (rs1076560, rs2283265, rs2587548, rs1079596), *DRD3* (3773678, rs167771), COMT (rs2239393, rs4818)	Positive association with OD	Levran et al. [83]
16.	*DAT1* (40bp VNTR, rs27072) and dopamine beta hydroxylase gene (rs1611115), *DAT1*	Positive association with AD	Kibitov et al. [74] and Ma et al. [89]
17.	*DRD2* (rs1800497, rs877138, rs10891556, rs4936271, rs17601612), *DRD3, COMT*	Positive and negative association with AD	Celorrio et al. [26]
18.	*DRD2* (rs1800497)	Positive association with AD	Prasad et al. [111]

found to carry the 10-repeat allele [94, 124]. Table 6.2 summarizes the other SNPs of the serotonin pathway.

6.10 GABA Pathway

GABA is an important inhibitory neurotransmitter in the human central nervous system and is known to regulate the behavioral effects of drugs, motor coordination, anxiolysis/sedation, withdrawal signs, and ethanol preference (Buck et al. 1996; Grobin et al. 1998).

The GABAA receptor α1 subunit (GABRA1) and α6 subunit genes (GABRA6) are the most widely studied markers of the GABAergic pathway (summarized in Table 6.3).

6.11 Glutamate Pathway

Glutamate, an important excitatory neurotransmitter, has an important role in the pathogenesis of substance dependence [106, 127]. Glutamate receptors mediate postsynaptic excitation of neural cells and have a key role in learning, neural communication, memory formation, and regulation [5, 64]. Chronic opioid consumption leads to downregulation of glutamate uptake [127] in the development of dependence and withdrawal [97]. GRIN1 is an obligatory subunit in the composition of NMDA receptors found in combinations of different modulatory subunits of GRIN2 (A-D) and GRIN3 (A-B). Studies revealed that GRIN2A subunits of NMDAR have a vital role in memory, learning, plasticity changes involving structure and behavior in drug dependence, cognition, etc. [22, 76, 132]. *GRIN2A* knockout mice showed impairment in reward-related learning response when they failed the conditioned place preference test [22]. *GRIN2A* SNPs have revealed possible association with dependence to heroin and alcohol [82, 160, 161, 162], summarized in Table 6.4.

6.12 Genome-Wide Association Studies (GWAS)

Previously, linkage analyses were used for identifying genes associated with substance dependence. Long et al. [86] performed studies on southwestern Native American population of AD and identified linkage with *GABRB1* gene. Multicenter studies on genetics of alcoholism undertook linkage analysis on multigenerational pedigrees in alcohol dependence subjects and found linkage with loci on chromosome 4 [49, 117]. In African, American, and European-American populations, GWAS on opioid dependence identified significant results in genes KCNC1 and KCNG2, involved in potassium and calcium signaling pathway [52]. GWAS on substance use conducted by Clarke et al. identified significant association with SNPs in 14 loci including genes *ADH1B*, *ADH1C*, *ADH5*, *DRD2*, *PDE4B*, *KLB*, *FAM69C*,

Table 6.2 Genetic studies of the serotonergic pathway in substance dependence

S. No.	Gene/polymorphism	Outcome	References
1.	5-HTTLPR	Positive association of short allele with heroin dependence	Gerra et al. [53]
2.	Serotonin receptor 1B HTR1B 1180A>G	Significant association of allele G with HD	Proudnikov et al. [115]
3.	5HTT-VNTR2	Positive association with AD	Mokrović et al. [94]
4.	5-HTTLPR	Positive association	Pombo et al. [110] and Gokturk et al. [55]
5.	TPH1 rs1799913, TPH2 rs7963720	Positive association with HD	Nielsen et al. [100]
6.	*5-HT2A* (A-1438G)	Positive association with HD	Saiz et al. [121]
7.	*HTR3A* (rs1176724, rs897687)	Positive association with HD	Levran et al. [82]
8.	5HT1B A-161T rs130058	Positive association with AD	Cao et al. [25] and Lee et al. [80]
9.	5-HTTLPR	Positive and negative association with AD	Shin et al. [129] and Wang et al. [148]
10.	*5-HTT*	Negative association	Grochans et al. [56]
11.	5-HTTLPR and 5-HT1B (G861C), Stin-2	Positive and negative association with AD	Wang et al. [148] and Bordukalo-Niksic et al. [21]
12.	HTR1B (rs11568817, rs130058, rs6296, rs13212041)	Positive association with AD	Contini et al. [32]
13.	Serotonin receptor HTR3B rs1176744 gain of function Ser129 allele	Both HTR3B and Ser129 alleles associated with AD	Enoch et al. [43]
14.	5-HTTLPR and 5-HTT, HTR1B (rs130058, rs11568817)	Positive association with AD	Enoch et al. [44]
15.	HTR1B (rs130058, rs11568817)	Positive association	Cao et al. [24]
16.	HTR3B SNP rs11606194	Positive association with HD	Levran et al. [83]
17.	5-HTT, 5-HTTLPR	Positive and negative association with AD	Pombo et al. [111] and Villalba et al. [146]
18.	TPH1 A218C (rs1800532), 5-HTTLPR (rs25531)	Negative association with AD	Wang et al. [148] and Malhotra et al. [92]
19.	HTR3B (rs2000292)	Positive association	Wu et al. [154]
20.	5-HTTLPR, *HTR1A* (rs6295), *HTR1B* (rs13212041)	Positive and negative association with AD	Plemenitas et al. [109]
21.	5-HTTLPR and Stin-2	Positive association with AD	Sahni et al. [120]
22.	*HTR3B* (rs1176746 and rs1185027)	Positive association with HD	Yin et al. [157]

Table 6.3 Genetic studies of the GABAergic pathway in substance dependence

S. No.	Polymorphism/gene	Outcome	References
1.	GABAA 1 (A/G) and GABAA 6 (C/T) receptor genes	Positive association with development of AD	Park et al. [105]
2.	GABAA receptor α2 subunit gene (*GABRA2*)	Positive association with AD	Agrawal et al. [3], Edenberg et al. [40] and Soyka et al. [131]
3.	Pro385Ser variant of GABRA6	Positive association with traits related to alcohol and depression	Iwata et al. [68] and Lucht et al. [87]
4.	*GABRA3* (rs75041)	Positive association with HD	Nielsen et al. [100]
5.	*GABRB3* (rs7165224)	Positive association with HD	Levran et al. [82]
6.	*GABRA2* (rs279871)	Positive and negative association with AD	Kareken et al. [71] and Sakai et al. [122] and Edenberg et al. [40]
7.	*GABBR2*, rs282129, a coding SNP (Met430Thr)	Positive and strong association with early onset of AD	Xuei et al. [156]
8.	*GABRA2* (rs279869, rs279858, and rs279837)	Positive association	Villafuerte et al. [145]
9.	*GABRAG1* and *GABRAG2*, *GRIN2A* (GTn repeat)	Positive association	Ittiwut et al. [66] and Domart et al. [35]
10.	*GABARB* (rs29253), *GABRA2* (rs279858), and *GABRG1* (rs7654165 and rs6447493)	Positive and negative association with AD	Terranova et al. [141] and Arias et al. [6]
11.	*GABRA2* (rs567926, rs279858, and rs9291283)	Positive association with AD	Strac et al. [133]
12.	*GABBR1* (rs29220), *GABRA2*	Positive association with alcohol dependence	Enoch et al. [45, 92]

GCKR, and *CADM2* [30]. Another GWAS on the GABAA receptor polymorphisms associated with alcohol use confirmed the involvement of *GABRA2* gene in its pathogenesis [78].

GWAS of opioid dependence have been published involving large sample sizes. Gelernter et al. [52] studied three populations of US opioid dependent individuals, recruited mostly from treatment facilities, and reported significant associations with SNPs of small effect in the potassium- and calcium-channel subunit genes. Nelson et al. [99] studied Australian opioid dependence samples, drawn from treatment facilities, and confirmed findings in two other populations, including populations analyzed by Gelernter et al. [52]. Another study on African-American methadone-treated OD subjects revealed significant genome-wide association of methadone dosage with an SNP closest to *OPRM1* gene [130].

Table 6.4 Genetic studies of the glutamate pathway in substance dependence

S. No.	Polymorphism/gene	Outcome	References
1.	*GRM8*	Vulnerability to alcoholism	Chen et al. [27]
2.	*NR1*, *GRIN2A*, *NR2B*; *MGLUR5*; *NNOS*; *PRKG2*; *CAMK4*; etc.	Positive association with GRIN2A in AD	Schumann et al. [125]
3.	G2108A polymorphisms of *NMDAR1* gene	Positive association	Wernicke et al. [150]
4.	C2664T/*NMDAR2B* gene	No association observed	Wernicke et al. [150]
5.	Ser310Ala/*GRIK3* gene	Significant role in AD	Preuss et al. [114]
6.	rs1806201/*NR2B* gene (*GRIN2B*)	Positive association with early onset	Tadic et al. [135]
7.	NR1 2108A	Positive association with alcohol withdrawal	Rujescu et al. [119]
8.	*GRIN2A* (rs1650420, rs4587976, 6497730, 1070487)	Positive association with heroin dependence	Levran et al. [82]
9.	*GRIN2A*, *GRIA3*, *GRIN4*, *CAMK2A*, *CAMK4*, etc.	Positive association with *GRIN2A* and *GRIA3* in alcohol dependence	Karpyak et al. [72]
10.	*GRIN2A*, (GT)n repeat	Positive association with *GRIN2A* in alcoholism	Domart et al. [36]
11.	*SLC29A1*, *GRIN2A*, *GABRB*, *CHRM2*, etc.	Positive association with *GRIN2A* in alcohol or drug dependence	Levran et al. [79]

6.13 Epigenetics

Epigenetic changes are DNA modifications which do not cause any structural change in the DNA. Epigenetic information is often transmitted by methylation of cytosine residue in CpG dinucleotides, which are comparatively overrepresented in the promoter regions of about 60% of genes of the mammalian genome [17]. Several studies observed the importance of epigenetic alterations on substance dependence, indicating drug-specific changes in gene expression. Candidate gene methylation studies in humans reported the role of epigenetic modifications influencing substance dependence seen in various genes like *SNCA*, *DAT*, *SERT*, *POMPC*, μ-opioid receptor *(OPRM1)*, etc. (Table 6.5).

6.14 Genome-Wide Methylation Studies in Substance Dependence

Studies in genome-wide methylation have shown association with substance dependence. A genome-wide study investigated methylation quantitative trait loci (mQTLs) and their effect on susceptibility to alcohol dependence which revealed a

Table 6.5 Methylation status of genes and their effect on substance dependence

Name of the gene	Role of the gene	Outcome in substance dependence	References
Proopiomelanocortin *(POMC)*	POMC is modified into ACTH hormone and plays an important role in the regulation of the hypothalamic-pituitary-adrenal (HPA) axis	Alcohol consumption affects function of the HPA axis and DNA methylation status of *POMC* gene at a single CpG site, also found to be associated with alcohol craving	Muschler et al. [96]
N-Methyl-D-aspartate 2b (*NR2B*)	Important role in alcohol consumption and withdrawal	More duration of drinking and elevated alcohol intake daily were negatively correlated with methylation	Biermann et al. [13]
Homocysteine-induced endoplasmic reticulum protein *(HERP)*	HERP is an endoplasmic reticulum protein, which regulates Ca2+ homeostasis that protects endothelial and neuronal cell integrity against oxidative stress	DNA methylation level increases at *HERP* gene promoter in alcohol dependence with high homocysteine levels	Bleich et al. [19], Bönsch et al. [20]
Alpha-Synuclein (*SNCA*)	Function of SNCA protein still elusive; few studies show its role in dopamine synthesis, transmission, storage, release, and reuptake	Increased promoter methylation leads to a downregulation of *SNCA* expression resulting in dysregulation of dopaminergic neurotransmission	Bönsch et al. [20] Fountaine and Wade-Martins, [50]
OPRM1	Involved in the reward pathway of alcohol and drugs	Hypermethylation of CpG sites at positions -18 and + 204 in heroin addicts and reduced *OPRM1* gene expression	Nielsen et al. [101]
		Three CpGs' (80, 71, and 10 bp upstream of the *OPRM1* translation start site) positions were found more methylated in AD	Zhang et al. [158]
		DNA hypermethylation of the *OPRM1* promoter in opium use disorder	Ebrahimi et al. [38]
		Six CpG sites were significantly associated with opioid use	Gao et al. [51]

Table 6.5 (continued)

Name of the gene	Role of the gene	Outcome in substance dependence	References
Dopamine transporter (DAT1/*SLC6A3*)	Dopamine transporter (DAT) encoded by *SLC6A3*. DAT proteins are distributed in the membranes of presynaptic nerve terminals, which mediate synaptic cleft dopamine reuptake	Hypermethylation of the *DAT* promoter positively correlated with AD	Hillemacher et. al. [62]
		No difference between patients and controls. Methylation level of *DAT* increased with age	Nieratschker et al. [102]
		No differences in methylation of *DAT* gene between patients and controls, but one CpG site was significantly more methylated in controls than in alcohol-dependent individuals	Jasiewicz et al. [70]
		DAT methylation associated with alcohol craving in AD	Wiers et al. [151]
		Epigenetic changes due to chronic alcohol exposure modulate striatal neurochemistry affecting reward processing	Muench et al. [95]
SLC6A4	The serotonin transporter (*5HTT* or *SLC6A4*) is a key regulator of serotonergic neurotransmission	Increased CpG methylation and lower mRNA production in females as compared to male subjects with a lifetime history of AD	Philibert et al. [108]
Prodynorphin (*PDYN*)	An opioid polypeptide hormone also known as proenkephalin B. It is involved in chemical signal transduction and cell communication	PDYN mSNPs under influences of environmental factors acting through epigenetic mechanisms may affect PDYN transcription and vulnerability to develop AD	Taqi et al. [139]
Atrial vasopressin peptide (*AVP*)	Vasopressin is known to enhance memory function and has anxiogenic effects	Promoter hypermethylation of vasopressin	Hillemacher et al. [63]

(continued)

Table 6.5 (continued)

Name of the gene	Role of the gene	Outcome in substance dependence	References
Atrial natriuretic peptide (*ANP*)	ANP is known to have strong anxiolytic effects on different levels of the hypothalamic-pituitary-adrenal (HPA) axis	Promoter hypomethylation of ANP precursor genes in patients with AD	Hillemacher et al. [63]
MAOA	Encodes mitochondrial enzymes which catalyze the oxidative de-amination of dopamine, norepinephrine, and serotonin	Methylation linked with AD in women	Philibert et al. [107]
Orexin A	Also known as hypocretins which are neuropeptides derived from the lateral hypothalamus	No significant association with AD symptoms	Bayerlein et al. [9]
Parvalbumin *(PVALB)*	Parvalbumin (PV) is a subgroup of GABAergic neurons	Elevated *PVALB* methylation in methamphetamine dependence/psychosis	Veerasakul et al. [144]
Opioid receptor-like 1 (*OPRL1*)	OPRL1 is a nociceptin opioid peptide receptor	Low methylation levels of *OPRL1* associated with higher frequency of binge drinking	Ruggeri et al. [118]
DRD2	D2 receptor has binding potential to rewarding stimuli	Individuals with promoter methylation responded to alcohol cues	Bidwell et al. [12]

significant association of 282 promoter mQTLs in 82 genes which might confer risk to AD.

In 2013, Zhang et al. generated promoter region methylation data of CpG islands of several genes [158]. This study on African-Americans revealed two CpGs in two genes GABA receptor subunit beta-3 *(GABRB3)* and proopiomelanocortin (*POMC*) which showed significant increase in methylation in AD compared to controls. Another genome-wide DNA methylation study found 865 hypo- and 716 hypermethylated sites in *GABRP*, *GAD1*, *DBH*, *SSTR4*, and *ALDH1L2* genes which suggest their influence on the biological process of alcohol dependence [161].

6.15 Brain Imaging

Various techniques of imaging like magnetic resonance imaging (MRI), functional MRI, single-photon emission computed tomography (SPECT), and positron emission tomography (PET) are used to capture drug effects on functioning of the brain.

These imaging techniques can be used for studying the functional approaches like metabolism and physiology of the brain, while PET analyses have identified neuroadaptive and neurodegenerative changes in levels of receptors and transporters in substance dependence. PET and SPECT imaging studies are important in substance dependence and relatively easy to carry out due to the availability of a variety of radiotracers and scope to design new ones based on need.

6.16 Screening of μ-Opioid Receptors Using PET and SPECT Imaging

Drugs such as naltrexone block the μ-opioid receptors (MORs) and reduce relapse risk in alcohol dependence subjects. PET studies use μ-opioid receptor ligand [11C] carfentanil to confirm the μ-opioid receptor availability in different regions of the brain. A similar study revealed lower MOR binding potential and strong correlation between μ-opioid receptor and craving in AD subjects [10]. Another study which investigated postmortem brain striatal tissue and analyzed MOR ([3H]DAMGO) observed a 23–51% reduction in MOR binding sites and OPRM1 mRNA in the striatum of alcohol users [60].

6.17 Screening of Dopamine Transporter Using PET and SPECT Imaging

Initially, PET was used to measure DAT in blood flow and later was confirmed with SPECT using radiotracers like 99mTc-HMPAO [61]. Studies also evaluated brain dopamine metabolism using [18F]fluoroDOPA and reported reduced dopamine in the brain of cocaine dependents.

Volkow et al. [147] used PET to show that the prefrontal cortex regulates increase in the nucleus accumbens dopamine levels, and this regulation is disrupted in alcohol use and dependence. This disruption may cause decreased sensitivity to rewards in such subjects. Karaken et al. [71] studied alcohol-induced dopamine responses within the striatum to determine the relationship between dopamine responses and alcohol-related behavior. Another studies have shown that sensitization and craving are associated with neuroadaptive phenomenon in the brain reward system [75]. In several psychiatric diseases, association with abnormal function of several genes such as alpha-synuclein was observed, but currently there are no imaging tools to probe the density and function of those genes in vivo. Evaluation of different brain regions with PET radiotracer development could be used to elucidate the association of these genes with diseases and in facilitating drug discovery in future.

6.18 Conclusion

Substance use known to have a multidisciplinary contribution is affected by several factors. Findings till date implicate the involvement of genetic and epigenetic mechanisms which act as powerful tools to delineate the contribution of various factors. Epigenetic marks have potential as biomarkers for research and therapeutics in substance dependence. Effect of substance use and dependence on methylation pattern of genes through various stages of the addiction process and their subsequent expression using PET/SPECT imaging can help in delineating the epigenetic mechanisms and their role in neuroplasticity, thus opening up new paradigms in the treatment and management of substance dependence.

References

1. Abdolmaleky HM, Cheng KH, Faraone SV, Wilcox M, Glatt SJ et al (2006) Hypomethylation of MB-COMT promoter is a major risk factor for schizophrenia and bipolar disorder. Hum Mol Genet 15(21):3132–3145
2. Agrawal A, Budney AJ, Lynskey MT (2012) The co-occurring use and misuse of cannabis and tobacco: a review. Addiction 107(7):1221–1233. https://doi.org/10.1111/j.1360-0443.2012.03837.x
3. Agrawal A, Lynskey MT (2006) The genetic epidemiology of cannabis use, abuse and dependence. Addiction 101(6):801–812. First published: 12 May 2006
4. Alegría-Torres JA, Baccarelli A, Bollati V (2011) Epigenetics and lifestyle. Epigenomics 3 (3):267–277. https://doi.org/10.2217/epi.11.22
5. Ango F, Pin JP, Tu JC, Xiao B, Worley PF, Bockaert J et al (2000) Dendritic and axonal targeting of type 5 metabotropic glutamate receptor is regulated by homer1 proteins and neuronal excitation. J Neurosci 20(23):8710–8716
6. Arias AJ, Covault J, Feinn R, Pond T, Yang BZ, Ge W et al (2014) A GABRA2 variant is associated with increased stimulation and 'high' following alcohol administration. Alcohol Alcohol 49(1):1–9
7. Avasthi A, Basu D, Subodh BN, Gupta PK, Malhotra N, Rani P et al (2017) Substance use and dependence in the Union Territory of Chandigarh: results of a household survey using a multistage stratified random sample. Indian J Psychiatry 59(3):275–283
8. Batel P, Houchi H, Daoust M, Ramoz N, Naassila M, Gorwood P (2008) A haplotype of the DRD1 gene is associated with alcohol dependence. Alcohol Clin Exp Res 32(4):567–572
9. Bayerlein K, Kraus T, Leinonen I, Pilniok D, Rotter A, Hofner B, Schwitulla J, Sperling W, Kornhuber J, Biermann T (2011) Orexin A expression and promoter methylation in patients with alcohol dependence comparing acute and protracted withdrawal. Alcohol 45(6):541–547
10. Bencherif B, Stumpf MJ, Links JM, Frost JJ (2004) Application of MRI-based partial-volume correction to the analysis of PET images of mu-opioid receptors using statistical parametric mapping. J Nucl Med 45(3):402–408
11. Benjamin J, Li L, Patterson C, Greenberg BD, Murphy DL, Hamer DH (1996) Population and familial association between the D4 dopamine receptor gene and measures of Novelty Seeking. Nat Genet 12(1):81–84
12. Bidwell LC, Karoly HC, Thayer RE, Claus ED, Bryan AD, Weiland BJ, YorkWilliams S, Hutchison KE (2018) DRD2 promoter methylation and measures of alcohol reward: functional activation of reward circuits and clinical severity. Addict Biol. https://doi.org/10.1111/adb.12614

13. Biermann T, Reulbach U, Lenz B, Frieling H, Muschler H, Hillemacher T (2009) N-methyl-D-aspartate 2b receptor subtype (NR2B) promoter methylation in patients during alcohol withdrawal. J Neural Transm 116(5):615–622
14. Bierut LJ, Goate AM, Breslau N et al (2012) ADH1B is associated with alcohol dependence and alcohol consumption in populations of European and African ancestry. Mol Psychiatry 17 (4):445–450
15. Bierut LJ (2011) Genetic vulnerability and susceptibility to substance dependence. Neuron 69 (4):618–627
16. Binelli C, Muñiz A, Sanches S, Ortiz A, Navines R, Egmond E, Udina M, Batalla A, López-Sola C, Crippa JA, Subirà S, Martín-Santos R (2015) New evidence of heterogeneity in social anxiety disorder: defining two qualitatively different personality profiles taking into account clinical, environmental and genetic factors. Eur Psychiatry 30(1):160–165
17. Bird A (2002) DNA methylation patterns and epigenetic memory. Genes Dev 16(1):6–21
18. Bleich S, Bonsch D, Rauh J, Bayerlein K, Fiszer R, Frieling H, Hillemacher T (2007) Association of the long allele of the 5-HTTLPR polymorphism with compulsive craving in alcohol dependence. Alcohol Alcohol 42(6):509–512
19. Bleich S, Lenz B, Ziegenbein M et al (2006) Epigenetic DNA hypermethylation of the HERP gene promoter induces down-regulation of its mRNA expression in patients with alcohol dependence. Alcohol Clin Exp Res 30:587–591
20. Bönsch D, Lenz B, Fiszer R, Frieling H, Kornhuber J, Bleich S (2006) Lowered DNA methyltransferase (DNMT-3b) mRNA expression is associated with genomic DNA hypermethylation in patients with chronic alcoholism. J Neural Transm (Vienna) 113 (9):1299–1304
21. Bordukalo-Niksic T, Stefulj J, Matosic A, Mokrovic G, Cicin-Sain L (2012) Combination of polymorphic variants in serotonin transporter and monoamine oxidase-A genes may influence the risk for early-onset alcoholism. Psychiatry Res 200(2-3):1041–1043
22. Boyce-Rustay JM, Holmes A (2006) Genetic inactivation of the NMDA receptor NR2A subunit has anxiolytic- and antidepressant-like effects in mice. Neuropsychopharmacology 31(11):2405–2414
23. Brückmann C, Di Santo A, Karle KN, Batra A, Nieratschkera V (2016) Validation of differential *GDAP1* DNA methylation in alcohol dependence and its potential function as a biomarker for disease severity and therapy outcome. Epigenetics 11(6):456–463
24. Cao J, LaRocque E, Li D (2013) Associations of the 5-hydroxytryptamine (serotonin) receptor 1B gene (HTR1B) with alcohol, cocaine, and heroin abuse. Am J Med Genet B Neuropsychiatr Genet 162B(2):169–176
25. Cao JX, Hu J, Ye XM, Xia Y, Haile CA, Kosten TR et al (2011) Association between the 5-HTR1B gene polymorphisms and alcohol dependence in a Han Chinese population. Brain Res 28(1376):1–9
26. Celorrio D, Muñoz X, Amiano P, Dorronsoro M, Bujanda L, Sánchez MJ et al (2016) Influence of dopaminergic system genetic variation and lifestyle factors on excessive alcohol consumption. Alcohol Alcohol 51(3):258–267
27. Chen AC, Tang Y, Rangaswamy M, Wang JC, Almasy L, Foroud T et al (2009) Association of single nucleotide polymorphisms in a glutamate receptor gene (GRM8) with theta power of event-related oscillations and alcohol dependence. Am J Med Genet B Neuropsychiatr Genet 150B(3):359–368
28. Chen WJ, Chen CH, Huang J, Hsu YP, Seow SV, Chen CC et al (2001) Genetic polymorphisms of the promoter region of dopamine D2 receptor and dopamine transporter genes and alcoholism among four aboriginal groups and Han Chinese in Taiwan. Psychiatr Genet 11(4):187–195
29. Chen J, Lipska BK, Halim N, Ma QD, Matsumoto M, Melhem S et al (2004) Functional analysis of genetic variation in catechol-O-methyltransferase (COMT): effects on mRNA, protein, and enzyme activity in postmortem human brain. Am J Hum Genet 75(5):807–821

30. Clarke M, Fursse J, Brown-Connolly NE, Sharma U, Jones R (2018) Evaluation of the National Health Service (NHS) direct pilot telehealth program: cost-effectiveness analysis. Telemed J E Health 24(1):67–76
31. Cloninger CR, Bohman M, Sigvardsson S (1981) Inheritance of alcohol abuse: cross-fostering analysis of adopted men. Arch Gen Psychiatry 38:861–868
32. Contini V, Bertuzzi GP, Polina ER, Hunemeier T, Hendler EM, Hutz MH et al (2012) A haplotype analysis is consistent with the role of functional HTR1B variants in alcohol dependence. Drug Alcohol Depend 122(1–2):100–104
33. Deb I, Chakraborty J, Gangopadhyay PK, Choudhury SR, Das S (2010) Single-nucleotide polymorphism (A118G) in exon 1 of OPRM1 gene causes alteration in downstream signaling by mu-opioid receptor and may contribute to the genetic risk for addiction. J Neurochem 112(2):486–496
34. Degenhardt L, Whiteford HA, Ferrari AJ et al (2013) Global burden of disease attributable to illicit drug use and dependence: findings from the Global Burden of Disease Study 2010. Lancet 382(9904):1564–1574
35. Domart MC, Benyamina A, Lemoine A, Bourgain C, Blecha L et al (2012) Association between a polymorphism in the promoter of a glutamate receptor subunit gene (GRIN2A) and alcoholism. Addict Biol 17:783–785
36. Du Y, Yang M, Yeh HW, Wan YJ (2010) The association of exon 3 VNTR polymorphism of the dopamine receptor D4 (DRD4) gene with alcoholism in Mexican Americans. Psychiatry Res 177(3):358–360
37. Ducci F, Enoch MA, Funt S, Virkkunen M et al (2007) Increased anxiety and other similarities in temperament of alcoholics with and without antisocial personality disorder across three diverse populations. Alcohol 41:3–12
38. Ebrahimi G, Asadikaram G, Akbari H, Nematollahi MH, Abolhassani M, Shahabinejad G, Khodadadnejad L, Hashemi M (2018) Elevated levels of DNA methylation at the OPRM1 promoter region in men with opioid use disorder. Am J Drug Alcohol Abuse 44(2):193–199
39. Ebstein RP, Novick O, Umansky R, Priel B, Osher Y, Blaine D et al (1996) Dopamine D4 receptor (D4DR) exon III polymorphism associated with the human personality trait of Novelty Seeking. Nat Genet 12(1):78–80
40. Edenberg HJ, Dick DM, Xuei X, Tian H, Almasy L, Bauer LO et al (2004) Variations in GABRA2, encoding the alpha 2 subunit of the GABA(A) receptor, are associated with alcohol dependence and with brain oscillations. Am J Hum Genet 74(4):705–714
41. Edenberg HJ (2007) Role of alcohol dehydrogenase and aldehyde dehydrogenase variants. Lit Rev 30(1):5–13
42. Egan MF, Goldberg TE, Kolachana BS, Callicott JH, Mazzanti CM, Straub RE et al (2001) Effect of COMT Val108/158 met genotype on frontal lobe function and risk for schizophrenia. Proc Natl Acad Sci U S A 98(12):6917–6922
43. Enoch MA, Gorodetsky E, Hodgkinson C, Roy A, Goldman D (2011) Functional genetic variants that increase synaptic serotonin and 5-HT3 receptor sensitivity predict alcohol and drug dependence. Mol Psychiatry 16(11):1139–1146
44. Enoch MA, Hodgkinson CA, Gorodetsky E, Goldman D, Roy A (2013) Independent effects of 5′ and 3′ functional variants in the serotonin transporter gene on suicidal behavior in the context of childhood trauma. J Psychiatr Res 47(7):900–907
45. Enoch MA, Hodgkinson CA, Shen PH, Gorodetsky E, Marietta CA, Roy A et al (2016) GABBR1 and SLC6A1, two genes involved in modulation of GABA synaptic transmission, influence risk for alcoholism: results from three ethnically diverse populations. Alcohol Clin Exp Res 40(1):93–101
46. Feinn R, Nellissery M, Kranzler HR (2005) Meta-analysis of the association of a functional serotonin transporter promoter polymorphism with alcohol dependence. Am J Med Genet B Neuropsychiatr Genet 133B:79–84. https://doi.org/10.1002/ajmg.b.30132
47. Ferguson RA, Goldberg DM (1997) Genetic markers of alcohol abuse. Cliniea Chimica Acta 17(257):199–250

48. Fiskerstrand CE, Lovejoy EA, Quinn JP (1999) An intronic polymorphic domain often associated with susceptibility to affective disorders has allele dependent differential enhancer activity in embryonic stem cells. FEBS Lett 458(2):171–174
49. Foroud T, Edenberg HJ, Goate A et al (2000) Alcoholism susceptibility loci: Confirmation studies in a replicate sample and further mapping. Alcohol Clin Exp Res 24:933–945
50. Fountaine TM, Wade-Martins R (2007) RNA interference-mediated knockdown of alpha-synuclein protects human dopaminergic neuroblastoma cells from MPP(+) toxicity and reduces dopamine transport. J Neurosci Res 85(2):351–363
51. Gao X, Jia M, Zhang Y, Breitling LP, Brenner H (2015) DNA methylation changes of whole blood cells in response to active smoking exposure in adults: a systematic review of DNA methylation studies. Clin Epigenetics 7:113
52. Gelernter J, Kranzler HR, Sherva R, Almasy L, Koesterer R, Smith AH, Anton R, Preuss UW, Ridinger M, Rujescu D, Wodarz N, Zill P, Zhao H, Farrer LA (2014) Genome-wide association study of alcohol dependence: significant findings in African- and European-Americans including novel risk loci. Mol Psychiatry 19:41–49
53. Gerra G, Garofano L, Santoro G, Bosari S, Pellegrini C, Zaimovic A et al (2004) Association between low-activity serotonin transporter genotype and heroin dependence: behavioral and personality correlates. Am J Med Genet B Neuropsychiatr Genet 126B(1):37–42
54. Gianoulakis C, de Waele JP (1994) Genetics ofalcoholism: role ofthe endogenous opioid system. Metab Brain Dis 9:105–131
55. Gokturk C, Schultze S, Nilsson KW, von Knorring L, Oreland L, Hallman J (2008) Serotonin transporter (5-HTTLPR) and monoamine oxidase (MAOA) promoter polymorphisms in women with severe alcoholism. Arch Womens Ment Health 11(5-6):347–355
56. Grochans E, Grzywacz A, Małecka I, Samochowiec A, Karakiewicz B, Samochowiec J (2011) Research on associations between selected polymorphisms of genes DRD2, 5HTT, GRIK3, ADH4 and alcohol dependence syndrome. Psychiatr Pol 45(3):325–335. Polish
57. Hack LM, Kalsi G, Aliev F, Kuo PH, Prescott CA, Patterson DG et al (2011) Limited associations of dopamine system genes with alcohol dependence and related traits in the Irish Affected Sib Pair Study of Alcohol Dependence (IASPSAD). Alcohol Clin Exp Res 35 (2):376–385
58. Heath AC, Bucholz KK, Madden PA, Dinwiddie SH, Slutske WS, Bierut LJ et al (1997) Genetic and environmental contributions to alcohol dependence risk in a national twin sample: consistency of findings in women and men. Psychol Med 27:1381–1396
59. Heidbreder CA, Andreoli M, Marcon C, Thanos PK, Ashby CR Jr, Gardner EL (2004) Role of dopamine D3 receptors in the addictive properties of ethanol. Drugs Today (Barc) 40(4):355–365
60. Hermann D, Hirth N, Reimold M, Batra A, Smolka MN, Hoffmann S et al (2017) Low μ-opioid receptor status in alcohol dependence identified by combined positron emission tomography and post-mortem brain analysis. Neuropsychopharmacology 42(3):606–614
61. Holman BL, Zimmerman RE, Johnson KA, Carvalho PA, Schwartz RB, Loeffler JS et al (1991) Computer-assisted superimposition of magnetic resonance and high-resolution technetium-99m-HMPAO and thallium-201 SPECT images of the brain. J Nucl Med 32(8):1478–1484
62. Hillemacher T, Frieling H, Hartl T, Wilhelm J, Kornhuber J, Bleich S (2009) Promoter specific methylation of the dopamine transporter gene is altered in alcohol dependence and associated with craving. J Psychiatr Res 43(4):388–392
63. Hillemacher T, Frieling H, Luber K, Yazici A, Muschler MA, Lenz B et al (2009) Epigenetic regulation and gene expression of vasopressin and atrial natriuretic peptide in alcoholwithdrawal. Psychoneuroendocrinology 34(4):555–560
64. Huang Z, Davis HH IV, Yue Q, Wiebking C, Duncan NW, Zhang J et al (2015) Increase in glutamate/glutamine concentration in the medial prefrontal cortex during mental imagery: a combined functional mrs and fMRI study. Hum Brain Mapp 36(8):3204–3212

65. Ishiguro H, Haruo Shibuya T, Toru M, Saito T, Arinami T (1999) Association study between high and low activity polymorphism of catechol-O-methyltransferase gene and alcoholism. Psychiatr Genet 9(3):135–138
66. Ittiwut C, Yang BZ, Kranzler HR, Anton RF, Hirunsatit R, Weiss RD, Covault J, Farrer LA, Gelernter J (2012) GABRG1 and GABRA2 variation associated with alcohol dependence in African Americans. Alcohol Clin Exp Res 36(4):588–593
67. Ittiwut R, Listman JB, Ittiwut C, Cubells JF, Weiss RD, Brady K et al (2011) Association between polymorphisms in catechol-O-methyltransferase (COMT) and cocaine-induced paranoia in European-American and African-American populations. Am J Med Genet B Neuropsychiatr Genet 156B(6):651–660
68. Iwata N, Cowley DS, Radel M, Roy-Byrne PP, Goldman D (1999) Relationship between a GABAA alpha 6 Pro385Ser substitution and benzodiazepine sensitivity. Am J Psychiatry 156 (9):1447–1449
69. Janeczek P, MacKay RK, Lea RA, Dodd PR, Lewohl JM (2014) Reduced expression of α-synuclein in alcoholic brain: influence of SNCA-Rep1 genotype. Addict Biol 19(3):509–515
70. Jasiewicz A, Rubiś B, Samochowiec J, Małecka I, Suchanecka A, Jabłoński M et al (2015) DAT1 methylation changes in alcohol-dependent individuals vs. controls. J Psychiatr Res 2015.03.007.
71. Kareken DA, Liang T, Wetherill L, Dzemidzic M, Bragulat V, Cox C et al (2010) Polymorphism in GABRA2 is associated with the medial frontal response to alcohol cues in an fMRI study. Alcohol Clin Exp Res 34(12):2169–2178
72. Karpyak VM, Geske JR, Colby CL, Mrazek DA, Biernacka JM (2011) Genetic variability in the NMDA-dependent AMPA trafficking cascade is associated with alcohol dependence. Addict Biol
73. Kauhanen J, Hallikainen T, Tuomainen TP, Koulu M, Karvonen MK, Salonen JT et al (2000) Association between the functional polymorphism of catechol-O-methyltransferase gene and alcohol consumption among social drinkers. Alcohol Clin Exp Res 24(2):135–139
74. Kibitov AO, Ivashchenko DV, Brodyansky VM, Chuprova NA, Shuvalov SA (2016) Combination of DAT and DBH gene polymorphisms witha family history of alcohol use disorders increases the risk of withdrawal seizures and delirium tremens during alcoholwithdrawal in alchol-dependent men. Zh Nevrol Psikhiatr Im S S Korsakova 116(12):68–80
75. Kienast T, Wrase J, Heinz A (2008) Neurobiology of substance-related addiction: findings of neuroimaging. Fortschr Neurol Psychiatr 76(Suppl 1):S68–S76. https://doi.org/10.1055/s-2008-1038141
76. Kiyama Y, Manabe T, Sakimura K, Kawakami F, Mori H, Mishina M (1998) Increased thresholds for long-term potentiation and contextual learning in mice lacking the NMDA-type glutamate receptor epsilon1 subunit. J Neurosci 18(17):6704–6712
77. Konishi T, Luo HR, Calvillo M, Mayo MS, Lin KM, Wan YJ (2004) ADH1B*1, ADH1C*2, DRD2 (-141C Ins), and 5-HTTLPR are associated with alcoholism in Mexican American men living in Los Angeles. Alcohol Clin Exp Res 28(8):1145–1152
78. Koulentaki M, Kouroumalis E (2018) GABAA receptor polymorphisms in alcohol use disorder in the GWAS era. Psychopharmacology 235(6):1845–1865
79. Koob GF, Le Moal M (1997) Drug abuse: hedonic homeostatic dysregulation. Science 278 (5335):52–58
80. Lee SY, Lin WW, Huang SY, Kuo PH, Wang CL, Wu PL et al (2009) The relationship between serotonin receptor 1B polymorphisms A-161T and alcohol dependence. Alcohol Clin Exp Res 33(9):1589–1595
81. Levran O, Awolesi O, Linzy S, Adelson M, Kreek MJ (2011) Haplotype block structure of the genomic region of the mu opioid receptor gene. J Hum Genet 56(2):147–155
82. Levran O, Londono D, Hara KO, Randesi M, Rotrosen J et al (2009) Heroin addiction in African Americans: a hypothesis-driven association study. Genes Brain Behav 8(5):531–540

83. Levran O, Randesi M, da Rosa JC, Ott J, Rotrosen J, Adelson M, Kreek MJ (2015) Overlapping dopaminergic pathway genetic susceptibility to heroin and cocaine addictions in African Americans. Ann Hum Genet 79(3):188–198
84. Li T, Yu S, Du J, Chen H, Jiang H, Xu K et al (2011) Role of novelty seeking personality traits as mediator of the association between COMT and onset age of drug use in Chinese heroin dependent patients. PLoS One 6(8):e22923
85. Lohoff FW, Weller AE, Bloch PJ, Nall AH, Ferraro TN, Kampman KM et al (2008) Association between the catechol-O-methyltransferase Val158Met polymorphism and cocaine dependence. Neuropsychopharmacology 33(13):3078–3084
86. Long JC, Knowler WC, Hanson RL et al (1998) Evidence for genetic linkage to alcohol dependence on chromosomes 4 and 11 from an autosome-wide scan in an American Indian population. Am J Med Genet 81:216–221. PMID: 9603607
87. Lucht M, Barnow S, Schroeder W, Grabe HJ, Finckh U, John U et al (2006) Negative perceived paternal parenting is associated with dopamine D2 receptor exon 8 and GABA (A) alpha 6 receptor variants: an explorative study. Am J Med Genet B Neuropsychiatr Genet 141B(2):167–172
88. Luczak SE, Glatt SJ, Wall TL (2006) Meta-analyses of ALDH2 and ADH1B with alcohol dependence in Asians. Psychol Bull 132:607–612
89. Ma Y, Fan R, Li MD (2016) Meta-analysis reveals significant association of the 3′-UTR VNTR in SLC6A3 with alcohol dependence. Alcohol Clin Exp Res 40(7):1443–1453
90. MacKenzie A, Quinn J (1999) A serotonin transporter gene intron 2 polymorphic region, correlated with affective disorders, has allele-dependent differential enhancer-like properties in the mouse embryo. Proc Natl Acad Sci U S A 96(26):15251–15255
91. Mague SD, Blendy JA (2010) OPRM1 SNP (A118G): involvement in disease development, treatment response, and animal models. Drug Alcohol Depend 108(3):172–182
92. Malhotra S, Basu D, Khullar M, Ghosh A, Chugh N (2016) Candidate genes for alcohol dependence: a genetic association study from India. Indian J Med Res 144(5):689–696
93. McGue M (1999) The behavioral genetics of alcoholism. Curr Dir Psychol Sci 8:109–115
94. Mokrović G, Matosić A, Hranilović D, Stefulj J, Novokmet M, Oresković D, Balija M, Marusić S, Cicin-Sain L (2008) Alcohol dependence and polymorphisms of serotonin-related genes: association studies. Coll Antropol 32(Suppl 1):127–131
95. Muench C, Wiers CE, Cortes CR, Momenan R, Lohoff FW (2018) Dopamine transporter gene methylation is associated with nucleus accumbens activation during reward processing in healthy but not alcohol-dependent individuals. Alcohol Clin Exp Res 42(1):21–31
96. Muschler MA, Hillemacher T, Kraus C et al (2010) DNA methylation of the POMC gene promoter is associated with craving in alcohol dependence. J Neural Transm (Vienna) 117:513–519
97. Nagy J (2008) Alcohol related changes in regulation of NMDA receptor. Funct Curr Neuropharmacol 6:39–54
98. Naka I, Nishida N, Ohashi J (2011) No evidence for strong recent positive selection favoring the 7 repeat allele of VNTR in the DRD4 gene. PLoS One 6(8):e24410
99. Nelson EC, Agrawal A, Heath AC, Bogdan R, Sherva R (2016) Evidence of CNIH3 involvement in opioid dependence. Mol Psychiatry 21(5):608–614
100. Nielsen DA, Barral S, Proudnikov D, Kellogg S, Ho A, Ott J et al (2008) TPH2 and TPH1: association of variants and interactions with heroin addiction. Behav Genet 38(2):133–150
101. Nielsen DA, Yuferov V, Hamon S, Jackson C, Ho A, Ott J et al (2009) Increased OPRM1 DNA methylation in lymphocytes of methadone-maintained former heroin addicts. Neuropsychopharmacology 34(4):867–873. https://doi.org/10.1038/npp.2008.108
102. Nieratschker V, Grosshans M, Frank J, Strohmaier J, von der Goltz C, El-Maarri O et al (2014) Epigenetic alteration of the dopamine transporter gene in alcohol-dependent patients is associated with age. Addict Biol 19(2):305–311

103. Nishizawa D, Han W, Hasegawa J, Ishida T, Numata Y, Sato T et al (2006) Association of mu-opioid receptor gene polymorphism A118G with alcohol dependence in a Japanese population. Neuropsychobiology 53(3):137–141
104. Oo KZ, Aung YK, Jenkins MA, Win AK (2016) Associations of 5HTTLPR polymorphism with major depressive disorder and alcohol dependence: asystematic review and meta-analysis. Aust N Z J Psychiatry 50(9):842–857
105. Park CS, Park SY, Lee CS, Sohn JW, Hahn GH, Kim BJ (2006) Association between alcoholism and the genetic polymorphisms of the GABAA receptor genes on chromosome 5q33-34 in Korean population. J Korean Med Sci 21(3):533–538
106. Pettorruso M, De Risio L, Martinotti G, Di Nicola M, Ruggeri F, Conte G et al (2014) Targeting the glutamatergic system to treat pathological gambling: current evidence and future perspectives. Biomed Res Int 2014:109786
107. Philibert RA, Gunter TD, Beach SR, Brody GH, Madan A (2008) MAOA methylation is associated with nicotine and alcohol dependence in women. Am J Med Genet B Neuropsychiatr Genet 147B(5):565–570. https://doi.org/10.1002/ajmg.b.30778
108. Philibert RA, Sandhu H, Hollenbeck N, Gunter T, Adams W, Madan A (2008) The relationship of 5HTT (SLC6A4) methylation and genotype on mRNA expression and liability to major depression and alcohol dependence in subjects from the Iowa Adoption Studies. Am J Med Genet B Neuropsychiatr Genet 147B:543–549
109. Plemenitas A, Kastelic M, O Porcelli S, Serretti A, Dolžan V, Kores Plesnicar B (2015) Alcohol dependence and genetic variability in the serotonin pathway among currently and formerly alcohol-dependent males. Neuropsychobiology 72(1):57–64
110. Pombo S, de Quinhones LP, Bicho M, Barbosa A, Ismail F, Cardoso N (2008) Association of the functional serotonin transporter promoter polymorphism (5-HTTLPR) with externalizing and internalizing aggressivity and alcohol abuse. Acta Med Port 21(6):539–546
111. Pombo S, Ferreira J, Cardoso JM, Ismail F, Levy P, Bicho M (2014) The role of 5-HTTLPR polymorphism in alcohol craving experience. Psychiatry Res 218(1-2):174–179
112. Ponce G, Jimenez-Arriero MA, Rubio G, Hoenicka J, Ampuero I, Ramos JA et al (2003) The A1 allele of the DRD2 gene (TaqI A polymorphisms) is associated with antisocial personality in a sample of alcohol-dependent patients. Eur Psychiatry 18(7):356–360
113. Prasad P, Ambekar A, Vaswani M (2010) Dopamine D2 receptor polymorphisms and susceptibility to alcohol dependence in Indian males: a preliminary study. BMC Med Genet 11:24
114. Preuss UW, Koller G, Samochowiec A, Zill P, Samochowiec J, Kucharska-Mazur J et al (2015) Serotonin and dopamine candidate gene variants and alcohol- and non-alcohol-related aggression. Alcohol Alcohol 50(6):690–699
115. Proudnikov D, LaForge KS, Hofflich H, Levenstien M, Gordon D, Barral S, Ott J, Kreek MJ (2006) Association analysis of polymorphisms in serotonin 1B receptor (HTR1B) gene with heroin addiction: a comparison of molecular and statistically estimated haplotypes. Pharmacogenet Genomics 16(1):25–36
116. Ray R (2004) The extent, pattern and trends of drug abuse in India, National Survey, Ministry Of Social Justice and Empowerment, Government of India and United Nations Office on Drugs and Crime, Regional Office for South Asia
117. Reich T, Edenberg HJ, Goate A et al (1998) A genome–wide search for genes affecting the risk for alcohol dependence. Am J Med Genet (Neuropsychiatr Genet) 81:207–215
118. Ruggeri B, Macare C, Stopponi S, Jia T, Carvalho FM, Robert G et al (2018) Methylation of OPRL1 mediates the effect of psychosocial stress on binge drinking in adolescents. J Child Psychol Psychiatry 59(6):650–658
119. Rujescu D, Soyka M, Dahmen N, Preuss U, Hartmann AM, Giegling I et al (2005) GRIN1 locus may modify the susceptibility to seizures during alcohol withdrawal. Am J Med Genet B Neuropsychiatr Genet 133B(1):85–87
120. Sahni S, Tickoo M, Gupta R, Vaswani M, Ambekar A, Grover T et al (2018) Association of serotonin and GABA pathway gene polymorphisms with alcohol dependence: a preliminary study. Asian J Psychiatr pii:S1876-2018(17)30466-5

121. Saiz PA, Garcia-Portilla MP, Florez G, Arango C, Corcoran P, Morales B et al (2009) Differential role of serotonergic polymorphisms in alcohol and heroin dependence. Prog Neuropsychopharmacol Biol Psychiatry 33(4):695–700
122. Sakai JT, Stallings MC, Crowley TJ, Gelhorn HL, McQueen MB, Ehringer MA (2010) Test of association between GABRA2 (SNP rs279871) and adolescent conduct/alcohol use disorders utilizing a sample of clinic referred youth with serious substance and conduct problems, controls and available first degree relatives. Drug Alcohol Depend 106(2–3):199–203
123. Sander T, Harms H, Podschus J, Finckh U, Nickel B, Rolfs A, Rommelspacher H (1995) Schmidt LG Dopamine D1, D2 and D3 receptor genes in alcohol dependence. Psychiatr Genet 5(4):171–176
124. Sarosi A, Gonda X, Balogh G, Domotor E, Szekely A, Hejjas K et al (2008) Association of the STin2 polymorphism of the serotonin transporter gene with a neurocognitive endophenotype in major depressive disorder. Prog Neuro-Psychopharmacol Biol Psychiatry 32(7):1667–1672
125. Schumann G, Johann M, Frank J, Preuss U, Dahmen N et al (2008) Systematic analysis of glutamatergic neurotransmission genes in alcohol dependence and adolescent risky drinking behavior. Arch Gen Psychiatry 65:826–838
126. Shao C, Li Y, Jiang K, Zhang D, Xu Y, Lin L et al (2006) Dopamine D4 receptor polymorphism modulates cue-elicited heroin craving in Chinese. Psychopharmacology (Berl) 186 (2):185–190
127. Shen HW, Scofield MD, Boger H, Hensley M, Kalivas PW (2014) Synaptic glutamate spillover due to impaired glutamate uptake mediates heroin relapse. J Neurosci 34 (16):5649–5657
128. Shen Y-C, Fan JH, Edenberg HJ, Li TK, Cui YH, Wang YF, Tian CH, Zhou CF, Zhou RL, Wang J, Zhao ZL, Xia GY (1997) Polymorphism of ADH and ALDH genes among four ethnic groups in China and effects upon the risk for alcoholism. Alcohol Clin Exp Res 21 (7):1272–1277. 30 May 2006
129. Shin S, Stewart R, Ferri CP, Kim JM, Shin IS, Kim SW et al (2010) An investigation of associations between alcohol use disorder and polymorphisms on ALDH2, BDNF, 5-HTTLPR, and MTHFR genes in older Korean men. Int J Geriatr Psychiatry 25(5):441–448
130. Smith AH, Jensen KP, Li J, Nunez Y, Farrer LA, Hakonarson H et al (2017) Genome-wide association study of therapeutic opioid dosing identifies a novel locus upstream of OPRM1. Mol Psychiatry 22(3):346–352
131. Soyka M, Preuss UW, Hesselbrock V, Zill P, Koller G (2008) Bondy B.GABA-A2 receptor subunit gene (GABRA2) polymorphisms and risk for alcohol dependence. J Psychiatr Res 42 (3):184–191
132. Squire LR (1992) Declarative and nondeclarative memory: multiple brain systems supporting learning and memory. J Cogn Neurosci 4(3):232–243
133. Strac DS, Erjavec GN, Perkovic MN, Sviglin KN, Borovecki F, Pivac N (2015) Association of GABAA receptor α2 subunit gene (GABRA2) with alcohol dependence-related aggressive behavior. Prog Neuropsychopharmacol Biol Psychiatry 63:119–125
134. Szeto CY, Tang NL, Lee DT, Stadlin A (2001) Association between mu opioid receptor gene polymorphisms and Chinese heroin addicts. Neuroreport 12(6):1103–1106
135. Tadic A, Dahmen N, Szegedi A, Rujescu D, Giegling I, Koller G et al (2005) Polymorphisms in the NMDA subunit 2B are not associated with alcohol dependence and alcohol withdrawal-induced seizures and delirium tremens. Eur Arch Psychiatry Clin Neurosci 255(2):129–135
136. Tanaka R, Shiratori Y, Yokosuka O et al (1997) Polymorphism of alcohol-metabolizing genes affects drinking behavior and alcoholic liver disease in Japanese men. Alcohol Clin Exp Res 21:596–601
137. Tan EC, Tan CH, Karupathivan U, Yap EP (2003) Mu opioid receptor gene polymorphisms and heroin dependence in Asian populations. Neuroreport 14(4):569–572
138. Tanda G, Pontieri FE, Di Chiara G (1997) Cannabinoid and heroin activation of mesolimbic dopamine transmission by a common mu1 opioid receptor mechanism. Science 276 (5321):2048–2050

139. Taqi MM, Bazov I, Watanabe H et al (2011) Prodynorphin CpG-SNPs associated with alcohol dependence: elevated methylation in the brain of human alcoholics. Addict Biol 16:499–509
140. Teixeira TM, da Silva HD, Goveia RM, Ribolla PEM, Alonso DP, Alves AA et al (2017) First description and evaluation of SNPs in the ADH and ALDH genes in a population of alcoholics in Central-West Brazil. Alcohol 65:37–43. https://doi.org/10.1016/j.alcohol.2017.04.006
141. Terranova C, Tucci M, Di Pietra L, Ferrara SD (2014) GABA receptors genes polymorphisms and alcohol dependence: no evidence of an association in an Italian male population. Clin Psychopharmacol Neurosci 12(2):142–148
142. Tiihonen J, Hallikainen T, Lachman H, Saito T, Volavka J (1999) Association between the functional variant of the catechol-O-methyltransferase (COMT) gene and type 1 alcoholism. Mol Psychiatry 4:286–289
143. Vasconcelos AC, Neto Ede S, Pinto GR, Yoshioka FK, Motta FJ, Vasconcelos DF et al (2015) Association study of the SLC6A3 VNTR (DAT) and DRD2/ANKK1 Taq1A polymorphisms with alcohol dependence in a population from northeastern Brazil. Alcohol Clin Exp Res 39 (2):205–211
144. Veerasakul S, Watiktinkorn P, Thanoi S, Dalton CF, Fachim HA, Reynolds GP et al (2017) Increased DNA methylation in the parvalbumin gene promoter is associated with methamphetamine dependence. Pharmacogenomics 18(14):1317–1322
145. Villafuerte S, Heitzeg MM, Foley S, Wendy Yau WY, Majczenko K et al (2012) Impulsiveness and insula activation during reward anticipation are associated with genetic variants in GABRA2 in a family sample enriched for alcoholism. Mol Psychiatry 17(5):511–519
146. Villalba K, Attonito J, Mendy A, Devieux JG, Gasana J, Dorak TM (2015) A meta-analysis of the associations between the SLC6A4 promoter polymorphism (5HTTLPR) and the risk for alcohol dependence. Psychiatr Genet 25(2):47–58. https://doi.org/10.1097/YPG.0000000000000078. Review
147. Volkow ND, Wang GJ, Telang F, Fowler JS, Logan J, Childress AR et al (2008) Dopamine increases in striatum do not elicit craving in cocaine abusers unless they are coupled with cocaine cues. Neuroimage 39(3):1266–1273
148. Wang TY, Lee SY, Chen SL, Chang YH, Chen SH, Chu CH et al (2012) Interaction between serotonin transporter and serotonin receptor 1 B genes polymorphisms may be associated with antisocial alcoholism. Behav Brain Funct 8:18
149. Weiss F, Lorang MT, Bloom FE, Koob GF (1993) Oral alcohol self-administration stimulates dopamine release in the rat nucleus accumbens: genetic and motivational determinants. J Pharmacol Exp Ther 267(1):250–258
150. Wernicke C, Samochowiec J, Schmidt LG, Winterer G, Smolka M, Kucharska-Mazur J et al (2003) Polymorphisms in the N-methyl-D-aspartate receptor 1 and 2B subunits are associated with alcoholism-related traits. Biol Psychiatry 54(9):922–928
151. Wiers CE, Shumay E, Volkow ND, Frieling H, Kotsiari A, Lindenmeyer J et al (2015) Effects of depressive symptoms and peripheral DAT methylation on neural reactivity to alcohol cues in alcoholism. Transl Psychiatry 5:e648
152. World drug report, The United Nation Office of Drug and Crime (UNODC) (2015). http://www.unodc.org/unodc/en/data-and-analysis/WDR-2015.html
153. World Health Organisation (1992) The ICD-10 classification of mental and behavioural disorders: clinical descriptions and diagnostic guidelines. World Health Organisation, Geneva
154. Wu LS, Lee CS, Weng TY, Wang KH, Cheng AT (2016) Association study of gene polymorphisms in GABA, serotonin, dopamine, and alcohol metabolism pathways with alcohol dependence in Taiwanese Han Men. Alcohol Clin Exp Res 40(2):284–290
155. Xu J, Wang T, Su Z, Zhou X, Xiang Y, He L, Li C, Yang Z, Zhao X (2018) Opioid exposure is associated with aberrant DNA methylation of OPRM1 promoter region in a Chinese Han population. Biochem Genet 56(5):451–458
156. Xuei X, Flury-Wetherill L, Dick D, Goate A, Tischfield J, Nurnberger J Jr et al (2010) GABRR1 and GABRR2, encoding the GABA-A receptor subunits rho1 and rho2, are associated with alcohol dependence. Am J Med Genet B Neuropsychiatr Genet 153B (2):418–427

157. Yin F, Ji Y, Zhang J, Guo H, Huang X, Lai J, Wei S (2016) Polymorphisms in the 5-hydroxytryptamine receptor 3B gene are associated with heroin dependence in the Chinese Han population. Neurosci Lett 635:123–129
158. Zhang H, Herman AI, Kranzler HR, Anton RF, Simen AA, Gelernter J (2012) Hypermethylation of OPRM1 promoter region in European Americans with alcohol dependence. J Hum Genet 57(10):670–675
159. Zhang H, Luo X, Kranzler HR et al (2006) Association between two mu-opioid receptor gene (OPRM1) haplotype blocks and drug or alcohol dependence. Hum Mol Genet 15:807–819
160. Zhao B, Zhu YS, De Witte W, Cui HM, Wang YP (2013) Analysis of variations in the glutamate receptor, N-methyl D-aspartate 2A (GRIN2A) gene reveals their relative importance as genetic susceptibility factors for heroin addiction. PLoS One 8(8):e70817
161. Zhao R, Zhang R, Li W et al (2013) Genome-wide DNA methylation patterns in discordant sib pairs with alcohol dependence. Asia Pac Psychiatry 5:39–50
162. Zhong HJ, Huo ZH, Dang J, Chen J, Zhu YS (2014) Liu JH. Functional polymorphisms of the glutamate receptor N-methyl D-aspartate 2A gene are associated with heroin addiction. Genet Mol Res 13:8714–8721

Fundamentals of Electroretinogram and Analysis of Retinal Fundus Image

7

Sristi Jha

Abstract

Our sensory organs make us more confident and dependable. Glaucoma which causes damage to the optic nerves is "the sneak thief of sight" as there are no symptoms, and once the vision is lost, then it is permanent. The data from the Glaucoma Research Foundation states that over 60 million people worldwide are affected by glaucoma. This is one of the leading causes of irreversible blindness. Hence, it is very important to detect glaucoma and treat it well at its early stage itself. Electroretinography is one of the methods which can show the status of the cells present in the retina of the eye. Increase in intraocular pressure is a significant information by which one can suspect whether the person has chances for the occurrence of glaucoma or not. Tonometry is used to detect the intraocular pressure. The increase in the pressure has adverse effect on optic disc and optic nerves. Further, using retinal fundus image and applying image processing techniques on that, one can analyse the episode of upcoming eye disorder. The processing of the image includes the preprocessing techniques and segmentation techniques. The active contour method is one of the reliable methods. These procedures can help the population to deal with glaucoma at its early stage itself such that they can get rid of the unwanted blindness.

Keywords

Glaucoma · Optic nerves · Blindness · Electroretinography · Intraocular pressure · Retinal fundus image · Segmentation

S. Jha (✉)
Department of Biomedical Engineering, North Eastern Hill University, Shillong, India

S. Paul (ed.), *Application of Biomedical Engineering in Neuroscience*,
https://doi.org/10.1007/978-981-13-7142-4_7

7.1 Overview and Introduction

Any living organism is a complex organism. It is made up of numerous systems – for example, we, the human body has the cardiovascular system, the nervous system, the digestive system, and the musculoskeletal system. Each system comprises of several other subsystems. Most physiological system is often studied by their signals that determine their nature and activities. These signals are nothing but action potentials generated by the set of cells [1]. Such bio-signals are generated from different parts of the body like:

- Electroneurogram (ENG) which measures the velocity of the propagation of a stimulus in a nerve
- Electromyogram (EMG) which measures the responses generated by the muscles
- Electrocardiogram (ECG) which measures the contractile activity of the heart
- Electroencephalogram (EEG) which represents the electrical activity of the brain and many more

Similarly, there is one more type of bio-signal which provides important information regarding the functioning of the retina (an important part of the eye): electroretinogram (ERG). ERG is a technique that measures the electrical responses of various cell types in the retina which include the rod cells, the cone cells, inner retinal cells, and the ganglion cells [2]. ERG consists of electrical potentials contributed by different types of cells present in the retina, and the stimulus conditions can evoke stronger response from certain components. Mostly, the ERGs are obtained using the surface electrodes from the corneal surface. The standards for the ERG were introduced by the International Society for Clinical Electrophysiology of Vision (ISCEV) in 1989. An ERG can also be used to monitor the progression of the retinal toxicity with various drugs or from a retained intraocular foreign body.

The Swedish physiologist Alarik Frithiof Holmgren performed the first known ERG on an amphibian retina in 1865. Later, James Dewar of Scotland performed the same in humans in 1877; however, its clinical application got generalized in 1941, when American psychologist Lorin Riggs launched the contact lens electrode. Einthoven and Jolly later in 1908 separated the ERG response into three components, a-wave, b-wave, and c-wave, described later in the chapter. Ragnar Granit won the Nobel Prize for Physiology and Medicine in 1967 from which many of the observations served the basis for understanding the ERG. Granit's studies were primarily conducted on dark-adapted, rod-dominated cat retina with which he was able to demonstrate the physiology of the receptor potential of each component of the ERG [2] (Fig. 7.1).

This test is the primary source for detecting the abnormality within the eye. This can be used to explain various diseases related to the retina such as retinitis pigmentosa, choroideremia, night blindness, cone dystrophy, and many more. Apart from just retinal diseases, ERG also provides useful information on other ocular diseases such as diabetic retinopathy, glaucoma, retinal detachment, etc.

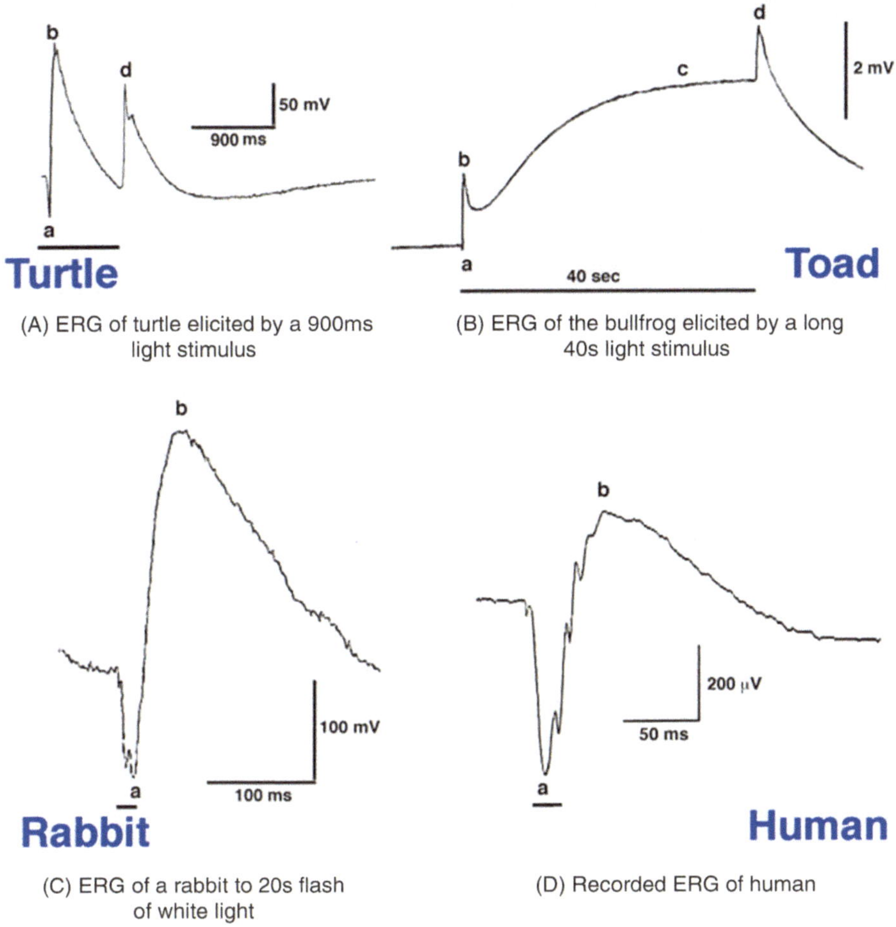

Fig. 7.1 (**a**) ERG of turtle elicited by a 900ms light stimulus. (**b**) ERG of the bullfrog elicited by a long 40s light stimulus. (**c**) ERG of a rabbit to 20s flash of white light. (**d**) Recorded ERG of human Source: https://webvision.med.utah.edu/book/electrophysiology/the-electroretinogram-erg/

This chapter provides fundamental knowledge of the structure of the eye, ERG, and intraocular pressure (IOP). Further, this chapter also contains the overview of glaucoma which is caused by increase in IOP and how we can analyse the retinal fundus images to evaluate the abnormality in the optic disc and optic nerve head (ONH).

7.2 Anatomy and Physiology of an Eye

In humans out of five sensory organs, the eye is one of them. It absorbs light rays from our surrounding and transforms them in the brain such that the information can be processed further. Together the eye and the brain form a unit which develops a visual system.

7.2.1 What Makes Up an Eye?

The human eye is spherical in shape and is also called as the eyeball. It lies in the socket and is “attached” to multiple muscles. These muscles help in the movement of the eyeball in different directions (Fig. 7.2).

Cornea: The eyeball has a circular, transparent part called as cornea which refracts the light onto the lens entering the eye. It focuses the object on the retina. Cornea contains no blood vessels. The curvature of the cornea refracts the light by about 45 dioptres. The cornea also contains tear fluid, which is formed in tear gland which protects the eye.

Pupil: It is the circular opening in the centre of the iris. Light passes into the lens of the eye from here. It controls the amount of the light entering into the eye with the help of the iris.

Iris: It is located in the centre of the cornea. It is coloured and a visible part of the eye in front of the lens. It contains many pathways made up of fine muscles which can contract and expand. The iris changes the size of the pupil.

Lens: It is a transparent liquid sphere-type structure located behind the pupil. It helps to focus the light on the retina. It is responsible for the 15 dioptres of the refractive power which is adjustable.

Choroid: This is the second and middle layer of the eye which is located between the retina (innermost layer) and the sclera (outer layer). As the name suggests, it consists of vessels and capillaries through which the blood supplies the retina with the needed nutrients and oxygen. It also contains a pigment which prevents the blurring of the vision by absorbing the excess of light.

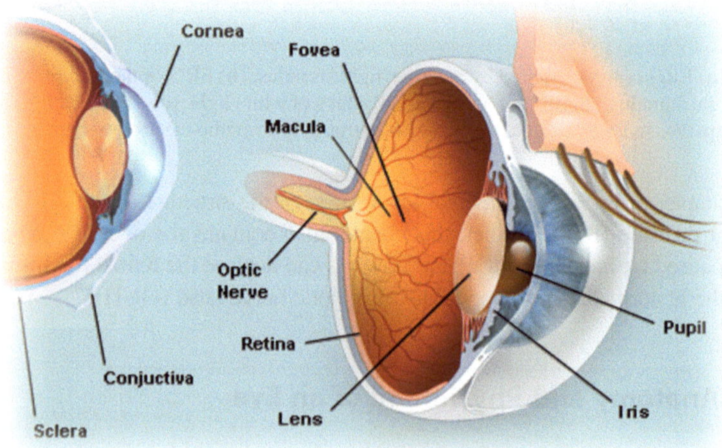

Fig. 7.2 Human eye anatomy
Source: https://www.webmd.com/eye-health/picture-of-the-eyes#1

Ciliary body: This is located behind the cornea which connects the choroid to the iris. In relaxed condition the lens is flat and drawn out such that one can see well in the distance. Whereas, the contraction of ciliary muscles changes its refractive power which allows one to see well in the vicinity. This process is called as accommodation.

Retina: This is the innermost layer of the eye. It consists of light-sensitive cells known as rods and cones. Rods help in seeing in the dim light. On the other hand, cones function best in bright light. The retina works similarly to the film in a camera.

Macula: It is a yellow spot in the retina where the light information is concentrated. This is the area of the sharpest vision as the photoreceptors are densely packed.

Fovea: This forms a small dip in the centre of the macula and is the area where the cones are heavily concentrated. When the eye focuses on any object, the brain accurately registers the part of the image that is focused on the fovea.

Optic disc: It is found in the retina and is the visible portion of the optic that identifies the start of the optic nerve where the messages from rod and cone cells leave the eye via nerve fibres to the brain. This is also called as blind spot.

Optic nerve: It helps in transferring the visual information to the brain.

Sclera: It is the outermost layer of the eyeball. It is whitish in colour in the front open part of the eye in which one can see well.

Rod cells: These cells are light-sensitive cells which are important to the eye to see in the dim light. There are 125 million rod cells.

Cone cells: These are sensitive to light found in the retina of the eye. There are 6–7 million cone cells. They function well in bright light and are essential for receiving sharp and clear vision.

Vitreous chamber: It is the inner space of the eyeball. It consists of a gel-like liquid which helps in the stability of the eyeball. This liquid generates a pressure which is also called as intraocular pressure (IOP). An abnormal IOP is the cause for many eye diseases [3].

7.2.2 Visual Pathway

The visual pathway is made up of series of cells and synapses which carry surrounding information to the brain for further processing. This flow of the information is carried out via the retina, optic nerve, optic chiasm, lateral geniculate nucleus (LGN), optic radiations, and striate cortex (Fig. 7.3). It is initiated by a special sensory cell, the photoreceptor which converts light energy into a neural signal that is later passed to the bipolar cell and the amacrine cell and then to the ganglion cell; all these cells and synapses lie within the retina. The nerve ending of the axon of the cells which leaves the retina is also termed as optic nerve. There is a crossing over that takes place in the optic chiasm from each eye and terminates in the opposite side of the brain. The fibres leave the LGN as the optic radiations that end in the visual cortex of the occipital lobe. In this pathway, information about the visual environment is transferred to related neurologic centres and to visual association areas.

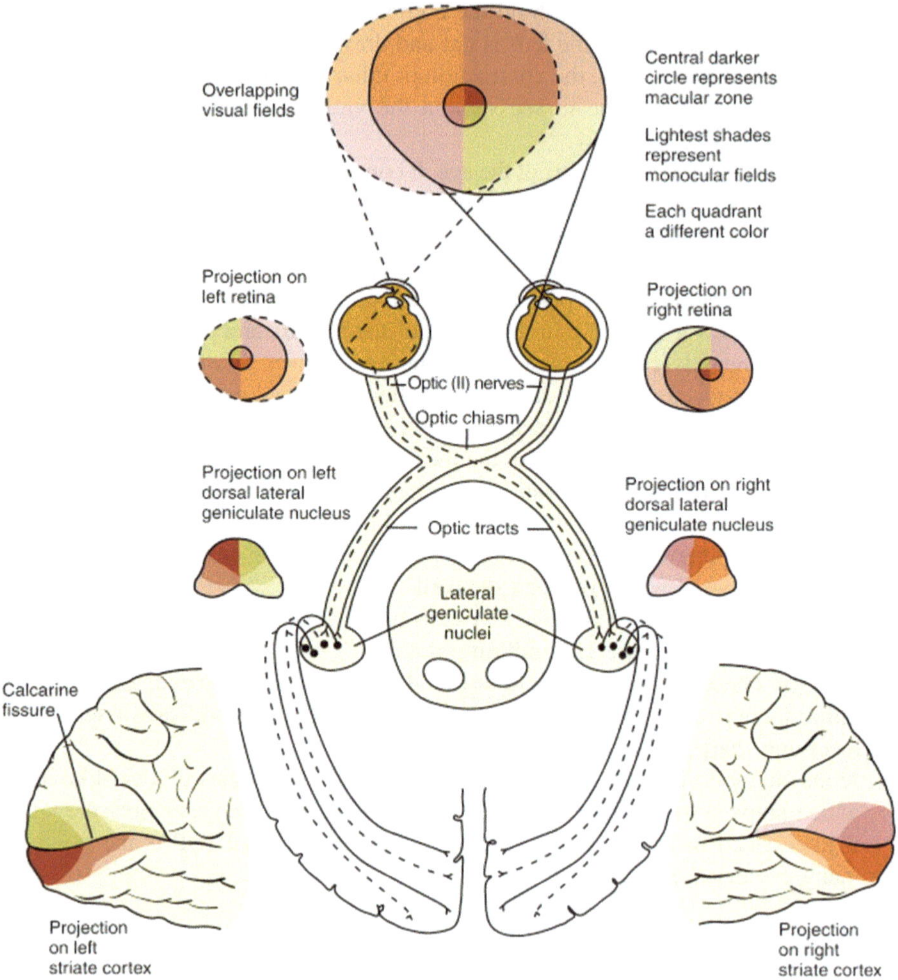

Fig. 7.3 Visual pathway
Source: https://www.sciencedirect.com/topics/neuroscience/visual-pathway

7.3 Electroretinogram Components

As mentioned above, the ERG response has three main components:

1. (a) Wave: It is produced by the external layer of photoreceptor cells due to the corneal retinal deflection. The closure of the sodium ion channels in outer membrane depicts the hyperpolarization of the photoreceptor cells in this wave. Transducin, a G protein, is initiated due to the absorption of light. This prompts the actuation of cyclic guanosine monophosphate phosphodiesterase (cGMP

PDE), causing a decrease in the cGMP inside the photoreceptor. This leads to the closure of the ion channel, resulting in the decrease of sodium channels directed inwardly, or a hyperpolarization of the cell. The a-wave amplitude is measured from baseline to the trough of the a-wave [2].

2. (b) Wave: The depolarization of the post-synaptic bipolar cells is the result of the hyperpolarization of the photoreceptor cells. This wave shows the positive-corneal deflection. The increase in the extracellular potassium is caused by the depolarization of bipolar cells which leads to the formation of transient current. This transient current depolarizes the Muller cells and produces the positive-corneal deflection. The b-wave amplitude is estimated from the trough of the a-wave to the peak of the b-wave. This wave is the most widely recognized segment of the ERG utilized in clinical and exploratory investigation of human retinal capacity [2].
3. (c) Wave: It is drawn from the retinal pigment epithelium (RPE) and photoreceptor cells. The subsequent change in the transepithelial potential because of the hyperpolarization at the apical film of the RPE cells and of the distal segment of the Muller cells gives rise to c-wave. The c-wave peaks within 2 to 10 seconds following a light stimulus, depending upon the intensity. Because of the c-wave reaction developing over several seconds, it is susceptible to influences from electrode drift, eye movements, and blinks [2].

7.3.1 Placement of Electrodes

The recording electrodes are kept over the cornea, the bulbar conjunctiva, or the skin of the lower eyelid. It protects corneal surface with non-irritating ionic conductive solution.

7.3.2 Diagnosis of Glaucoma Using ERG

The pattern of the ERG (PERG) indicates the functioning of the ganglion cells, and this information can be helpful to know about the early damage due to glaucoma. A brief methodology is covered here; more detail can be found in the ISCEV standard [4]. A stimulus is used in a checkerboard pattern which reverses its local luminance while keeping average luminance constant. Therefore, the ERG signals cancel out, and nonlinearities remain which originate in the ganglion cells [5–7]. The potentials in the retina are detected with corneal electrodes. Various types of electrode may be used, such as gold foil or DTL [8]. As reduction in retinal contrast leads to marked reduction of the PERG, it is important to notice that the electrode does not degrade the optical imaging on the retina. A high stability and reproducibility can be obtained with an appropriate technique [9]. The frequency of the checkerboard reversal determines whether the transient response (< 4 rev/s) or the steady-state response (≥ 8 rev/s) is evoked (Fig. 7.4).

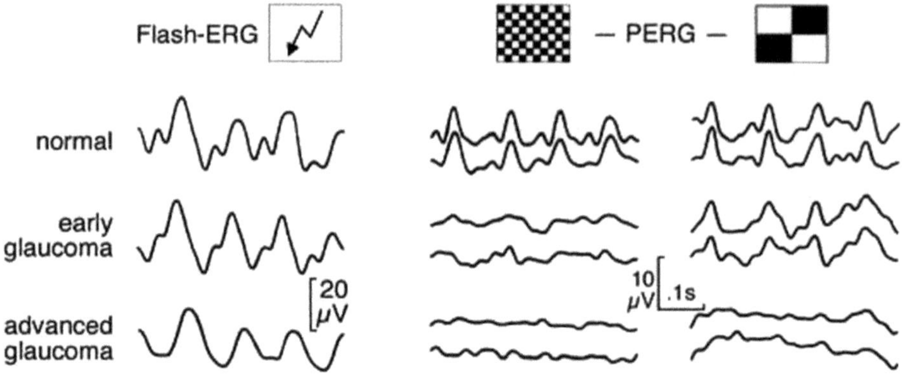

Fig. 7.4 PERG comparison in glaucoma
Source: https://www.researchgate.net/figure/ERG-and-PERG-in-glaucoma-The-flash-ERG-left-column-is-relatively-little-affected-even_fig 17_226008777

7.4 Ocular Hypertension

The reason for the development of ocular hypertension is the increased pressure in the eye than the normal value. This pressure inside the eye is called as intraocular pressure. The unit in which pressure of the eye is measured is in millimetres of mercury (mmHg). The normal range of eye pressure is between 10 and 21 mmHg. Ocular hypertension starts emerging if the IOP becomes greater than 21 mmHg. The condition for the ocular hypertension has the following criteria:

- Raised IOP in one or both of the eyes, i.e. greater than 21 mmHg. The IOP is measured by an instrument known as tonometer.
- Irrespective of the increased IOP, the optic nerve gives a normal impression.
- Evidences of glaucoma are not visible in the visual tests.
- No impression of any other ocular diseases, as there are chances that the IOP increases due to some other eye diseases.

Increased IOP is one of the concern factors for individuals, and those should be observed more closely for the onset of glaucoma. Over the last 20 years, studies have helped to characterize people having ocular hypertension.

- The data from the Ocular Hypertension Study says that the people with ocular hypertension have approximately 10% risk of developing glaucoma over the 5 years. The risk can be decreased to 5% if the IOP is reduced by medication or surgery. The risk can be reduced if the proper treatment is provided to the patients before the loss of vision occurs.

- Patients with thinner cornea may be considered at higher risk for developing glaucoma; therefore, the ophthalmologist uses a device to measure the corneal thickness known as pachymeter.
- In comparison with open-angle glaucoma, ocular hypertension is 10–15 times more likely to occur. However, if about 10 people have higher IOP, only one of those will have glaucoma.

Many research studies have shown that with 21–25 mmHg of IOP, there are 2.6–3% chances of damage due to glaucoma. However, with 26–30 mmHg of IOP, the glaucomatous damage ranges from 12 to 26%, and approximately 42% damage occurs for the patients having IOP greater than 30 mmHg.

7.4.1 Causes (Fig. 7.5)

The reason behind the development of high pressure inside the eye is due to imbalance in the drainage of the fluid inside the eye, i.e. aqueous humour. Generally, the channel named as trabecular mesh helps in time-to-time drainage of the fluid, but sometimes this mesh does not work properly in draining out the fluid. Hence, due to improper drainage channels, more and more fluid gets accumulated in the eye which results in increased pressure. This elevation in pressure damages the optic nerve in the eye.

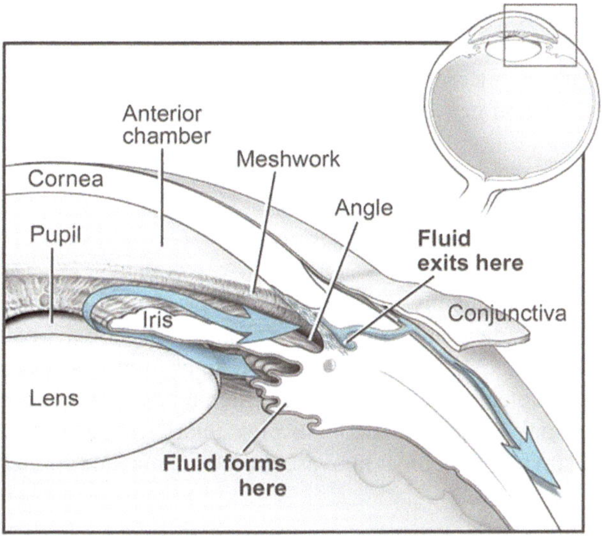

Fig. 7.5 Traverse of fluid starting from aqueous humour to anterior chamber and out through trabecular framework Source: https://www.medicalopedia.org/3183/aqueous-humour-anatomy-physiology/

7.5 Glaucoma

Glaucoma is an eye-related disease which is caused by the degeneration of the optic nerve which over a longer period leads to blindness. The increase in ocular pressure resulting from the improper functioning of the trabecular mesh damages the optic nerves which damages the ganglion cells and the axons [10]. Further progression leads to reduced connection between the photoreceptors and the visual cortex which ultimately reduces the functionality of the retina and hence changes the size of the optic cup. This can be estimated by monitoring the IOP, range of visual fields, and cup-to-disc ratio. Rise in IOP is considered as an important risk factor which leads to the structural changes in the optic nerve head (ONH) that affects the visual field of the patient [11]. Progressive damage of the optic nerve will lead to irreversible reduction in visual acuity, visual field loss, and blindness. When the desirable intraocular pressure cannot be achieved by medication alone, glaucoma surgery may be needed. The level of intraocular pressure is a result of the balance of fluid secretion and outflow inside the eyeball. Glaucoma surgery can lower the level of intraocular pressure either by diminishing the production of fluid inside the eye or by increasing the outflow of fluid away from the eye. Usually, the intraocular pressure can be lowered by the surgery (Fig. 7.6).

Unfortunately even when the intraocular pressure is controlled, the damaged optic nerve and related vision functions cannot be restored. In some circumstances, if the intraocular pressure cannot be controlled after the operation, additional topical or oral anti-glaucoma drugs may be required. If the imbalance of intraocular fluid secretion and outflow is still significant after the operation, you may need another operation to prevent further deterioration. Therefore, it is necessary to detect glaucoma in its early phase so that it can prevent deterioration of retinal nerves and hence vision.

The cup-to-disc ratio (CDR) gives information about the area of the disc that is occupied by the cup, and this detail is accepted for the evaluation of glaucoma [11]. For normal eye 0.3–0.5 is the accepted range [12]. The value of CDR increases with the increase in the neuro-retinal degeneration. As the CDR value reaches 0.8, the vision gets completely lost.

7.5.1 Types of Glaucoma

There are mainly two major forms of glaucoma: open-angle glaucoma and angle-closure glaucoma. Most of the cases of glaucoma are primary open-angle glaucoma (POAG). Majorly others are angle-closure glaucoma. Rare forms of glaucoma include congenital glaucoma, which tends to run in families and is present at birth; normal tension glaucoma; pigmentary glaucoma; pseudoexfoliative glaucoma; traumatic glaucoma; neovascular glaucoma; and iridocorneal endothelial syndrome.

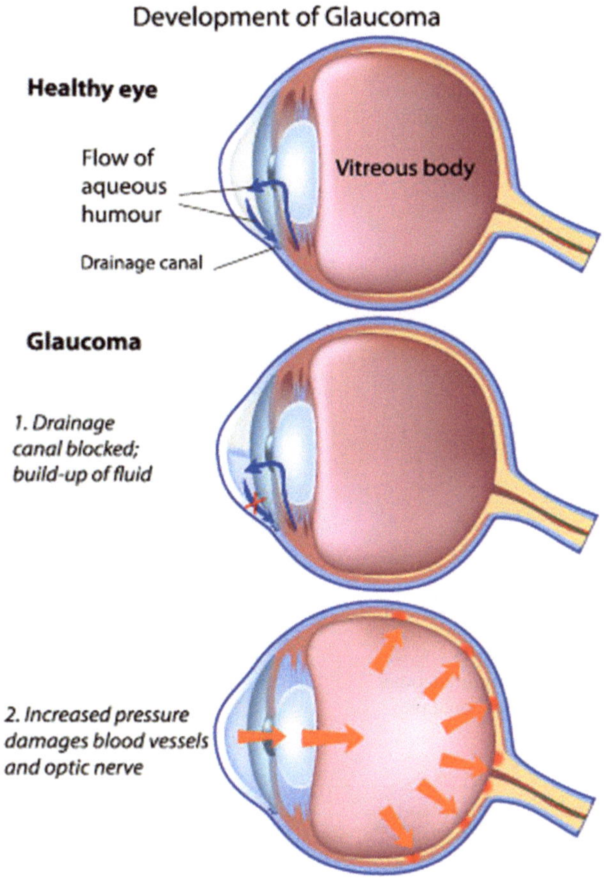

Fig. 7.6 Development of glaucoma
Source: https://www.glaucoma.org/glaucoma/types-of-glaucoma.php

1. Open-Angle Glaucoma

It is the most common form of glaucoma which occurs due to slow clogging of the drainage canals that result into increased eye pressure. It has broader angle between the cornea and the iris. It develops slowly and the damages are not noticeable (Fig. 7.7).

2. Angle-Closure Glaucoma

Compared to open angle, it is a less common form of glaucoma. The main cause for this is the blockage of the drainage canals that results in an unexpected rise in IOP. It has a narrow angle between the cornea and the iris. It grows very quickly, and the damage due to this type is noticeable (Fig. 7.8).

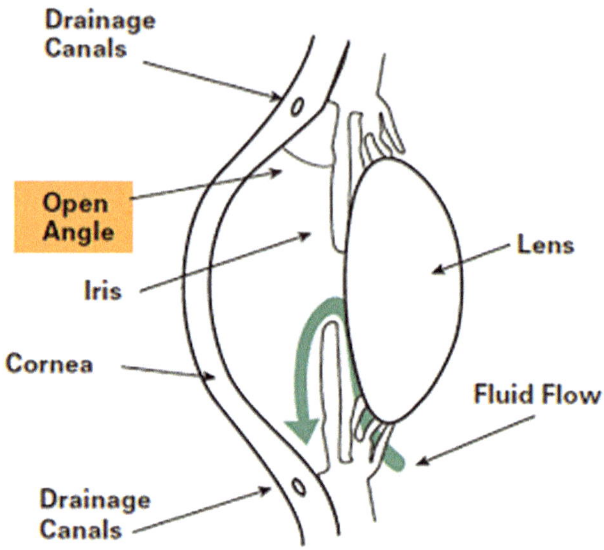

Fig. 7.7 Open-angle glaucoma
Source: https://www.glaucoma.org/glaucoma/types-of-glaucoma.php

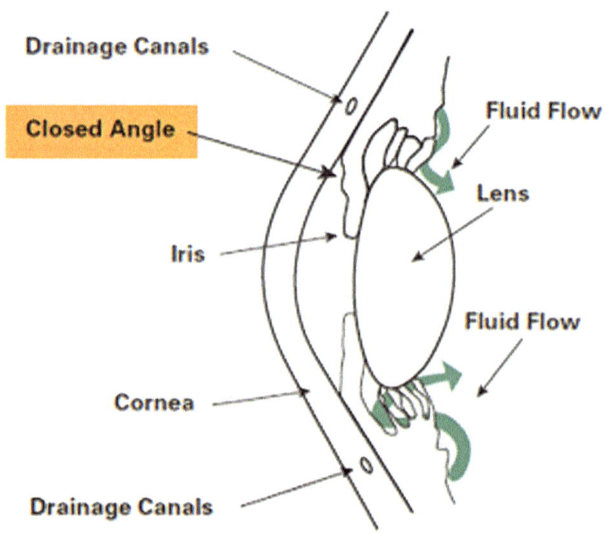

Fig. 7.8 Angle-closure glaucoma
Source: https://www.glaucoma.org/glaucoma/types-of-glaucoma.php

3. Normal Tension Glaucoma (NTG)

In NTG the optic nerve gets damage even if the pressure in the eye is not high. Hence, it is also known as low tension or normal pressure glaucoma. The actual reason behind such kind of damage is still a question. Researchers are still finding the reason behind why some optic nerves get damaged by low eye pressure levels.

4. Congenital Glaucoma

Such type of glaucoma generally occurs in babies due to incomplete development of the drainage canals in the eye during prenatal period. Such cases can be treated with medication and surgery. This is one of the rare conditions which may be inherited.

5. Pigmentary Glaucoma

Although its occurrence is rare, pigment syndrome and pigmentary glaucoma are prone to occur at a younger age than POAG. This happens when pigment granules, which normally cling to the back of the iris, come off in flakes into the clear fluid of aqueous. Sometimes these granules tend to flow towards the drainage canals of the eye which slowly clogs it, resulting in increase in eye pressure which damages the optic nerve. This can be treated with eye drops.

6. Pseudoexfoliative Glaucoma

Patients with exfoliative glaucoma often are seen to have more scenes of high pressure, a greater number of variances, and higher peak pressures than patients with different kinds of glaucoma.

It is caused by the unusual accumulation of protein in the drainage system and other structures of the eye. It is hard to control with medicinal treatment. Patients with this type of glaucoma frequently require an increasingly forceful stepwise treatment and all the more regularly require a laser, or incisional medical procedure.

7. Traumatic Glaucoma

Traumatic glaucoma is any glaucoma caused by damage to the eye. This sort of glaucoma can happen either immediately after the injury to the eye or in later years.

8. Iridocorneal Endothelial Syndrome (ICE)

The occurrence of ICE is rare as compared to other forms of glaucoma. In this condition, cells get spread over the drainage canals on the back surface of the cornea and over the outside of the iris. This causes blockage and increase in eye pressure, which can harm the optic nerve. These cells likewise frame attachments that tie together the iris to the cornea, further obstructing the drainage channels.

ICE occurs more often in light-colour-skinned females. Side effects can incorporate foggy vision after awakening and the presence of halos around lights. ICE is hard to treat, and laser treatment is not a powerful treatment. ICE is normally treated with meds as well as filtering surgery.

7.5.2 Available Diagnostic Methods

Whenever there is some issue with the eye, the ophthalmologist performs tests to measure intraocular pressure such that to rule out the chances of primary open-angle glaucoma or secondary causes of glaucoma [13]. Some of these tests are explained below:

- A slit lamp, a special type of microscope, is used to examine the cornea, iris, anterior chamber, and lens of the eye.
- The pressure inside the eye can be measured by a method known as tonometry. Measurement of both the eyes is taken for at least 2–3 times as IOP varies from hour to hour. This measurement must be taken at different times of the day. If the 3mmHg or more difference of pressure is observed between the two eyes, then the presence of glaucoma can be sensed. In the case of primary open-angle glaucoma, it is observed that the intraocular pressure is constantly increasing.
- For any kind of damage of the eye or any abnormalities, the optic nerves are needed to be examined which requires dilation of pupil. Therefore, fundus images are taken for evaluation of the abnormalities with optic nerves.
- The visual field (side vision) is tested by using a visual field machine. This examination is done to rule out issues in the visual field due to glaucoma. This test should be repeated depending upon the damage due to glaucoma. If the risk is low, then the test is needed to be performed only once a year. If the risk is high, then the test may be required to be performed more frequently.
- To check the drainage angle of the eye, gonioscopy is performed. In this examination a lens is placed in the eye. This test helps in determining if the angles are open, narrowed, or closed and to rule out any other conditions that could cause elevated intraocular pressure.
- Corneal pachymetry is a method which is used to estimate the thickness of the cornea by an ultrasound probe. A thinner cornea can give false reading of low pressure, whereas a thick cornea can give a false reading of high pressure.

Over here, we will be studying about three different image processing techniques using which one can detect the glaucoma and its progressiveness at the early stage. The names of these methods are multi-thresholding technique, active contour model, and region-based segmentation.

7.6 Analysis Techniques for Retinal Fundus Images

7.6.1 Proposed Methods for Detection of Glaucoma

This section explains how to execute retinal fundus image analysis for glaucoma detection. Figure 7.9 shows the proposed flow chart for the analysis of the image consisting of (i) preprocessing, (ii) segmentation of preprocessed image, and (iii) classification based on evaluation of CDR. The segmentation of the preprocessed

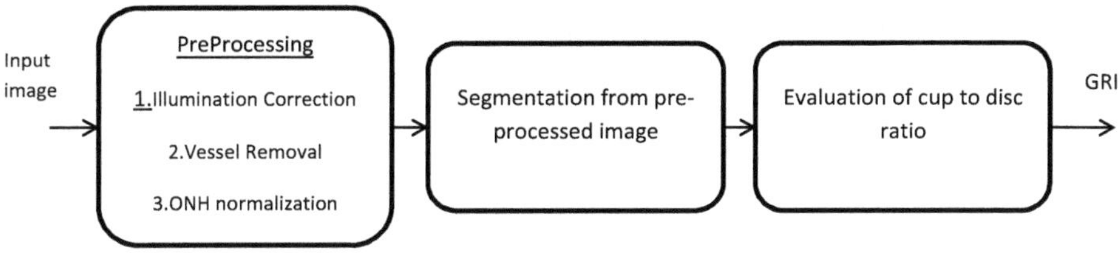

Fig. 7.9 Block diagram showing processing of retinal fundus image

fundus image can be executed by different techniques which are expressed further in the chapter.

7.6.2 Preprocessing

In preprocessing techniques variations in the retinal fundus image are calculated, and the features that do not belong to the glaucoma disease are excluded from the images for emphasizing the desired characteristics. It also removes unwanted noises from the fundus image such that it enhances features of the image. The preprocessing of image can be done in two ways: linear method and nonlinear method [1].

(i) In linear method, algorithm is applied to all pixels without defining the image to be corrupted or uncorrupted image.
(ii) In nonlinear method, algorithm is applied to the pixels by defining which pixel is corrupted or uncorrupted. Later, the corrupted image again goes through another algorithm for filtration, and uncorrupted is retained. Nonlinear filter produces better result as compared to linear filters.

Preprocessing is important to attenuate the variation in the image by normalizing the original retinal image against dataset for processing or analysis [12]. The preprocessing retinal images include the correction for non-uniform illumination, contrast enhancement, and colour normalization.

(a) Adaptive histogram equalization

This method enhances the contrast of the retinal images. Adaptive method calculates the number of histograms, which corresponds to a particular part of the image and redistributes the lightness values of the image. This is used for enhancing the edges in each region of an image. The noisy pixels appear as a background information. Therefore, in this method, contrast between the background pixels and foreground (actual information) leads to enhancement of the noisy. Using contrast limited adaptive histogram equalization (CLAHE), the contrast of the image can be further improved. This algorithm improves the contrast of the blood vessels

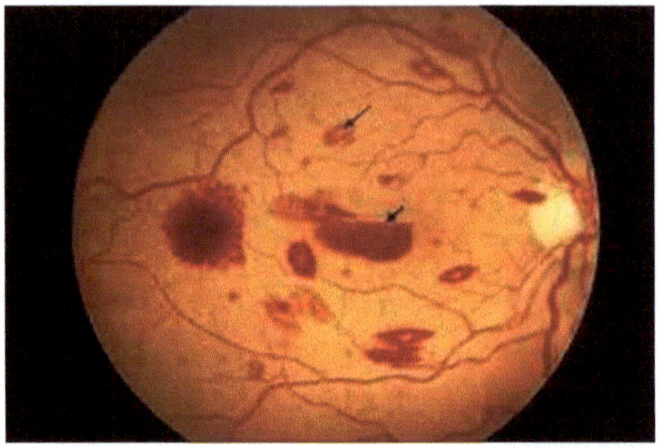

Fig. 7.10 Original image
Source: https://ieeexplore.ieee.org/abstract/document/8073677

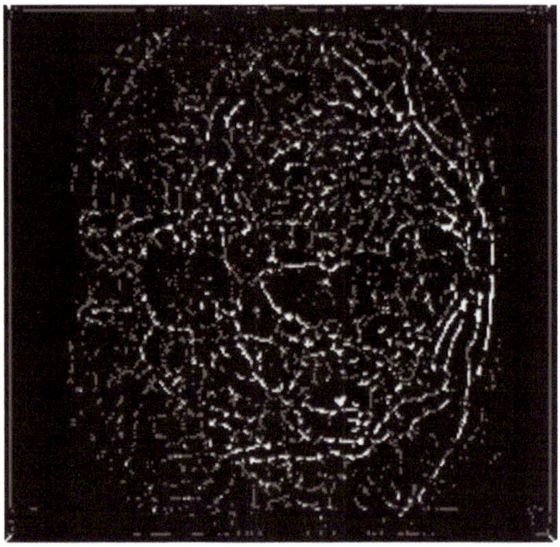

Fig. 7.11 Adaptive histogram equalization
Source: https://ieeexplore.ieee.org/abstract/document/8073677

[14, 15]. Hence, adaptive histogram equalization is used for constant enhancement of the images (Figs. 7.10 and 7.11).

(b) Weiner filter

By using this original image can be recovered from the blurred image by using inverse filtering technique. This method reduces the degradation of image, and it helps in development of restoration algorithm. Weiner filter removes the noise and blurring effect in the fundus image. When a colour image is given as an input, the contrast for each channel (R, G, B) is calculated. Using the low pass filter, the R, G,

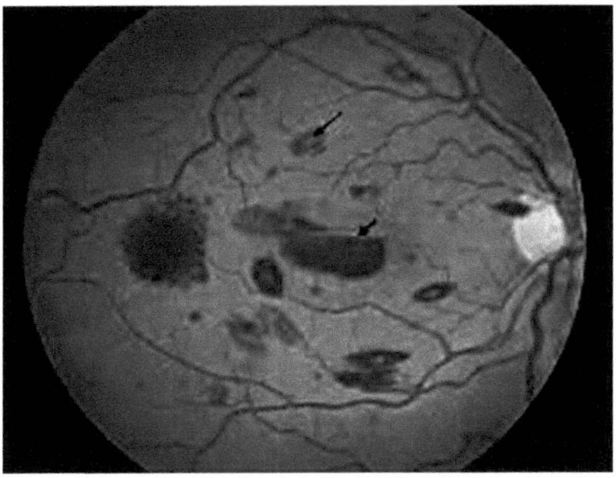

Fig. 7.12 Wiener filter
Source: https://ieeexplore.ieee.org/abstract/document/8073677

B channels can be estimated by Weiner filter. This can be done by measuring the contrast between the original image and the low pass image [16] (Fig. 7.12).

(c) Median filter

A median filter is a nonlinear filter. This is mostly used for removing salt-and-pepper noise. Median is the middle value of the neighbourhood pixel. It is important to keep the sharpness of the image which can be achieved by this while removing the noise. Disadvantage of the median filter is that it removes both noises and details as it cannot understand exact details from noises [25] (Fig. 7.13).

(d) Adaptive median filter

This is a type of linear filter. This method is more advanced as compared to the standard median filter. The main objective of the filter is to remove the impulse noise and smoothen the other noises. Adaptive median filter classifies the pixel in the image by its surrounding neighbour pixels. When in use the adaptive median filter changes the size of the neighbourhood. The main advantage of using this filter is that it preserves the important details of the image [25] (Fig. 7.14).

(e) Gaussian filter

Retinal fundus image contains three channels: red, blue, and green. While observing in green channel, exudates appear brighter as compared to red and blue. Optic disc also appears clear and brighter in green channel. The Gaussian filter smoothens the image by averaging out the neighbouring pixels using Gaussian

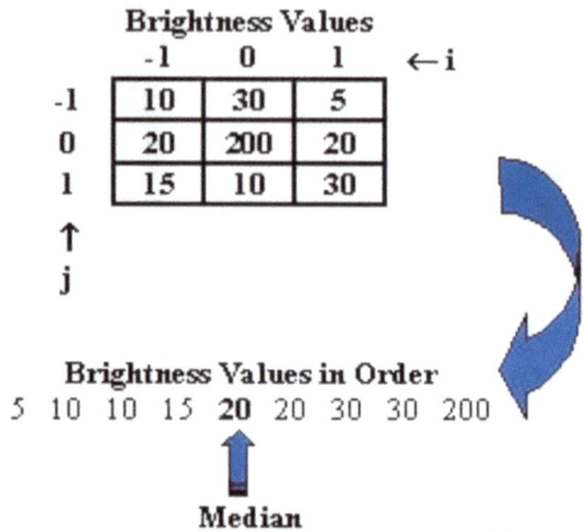

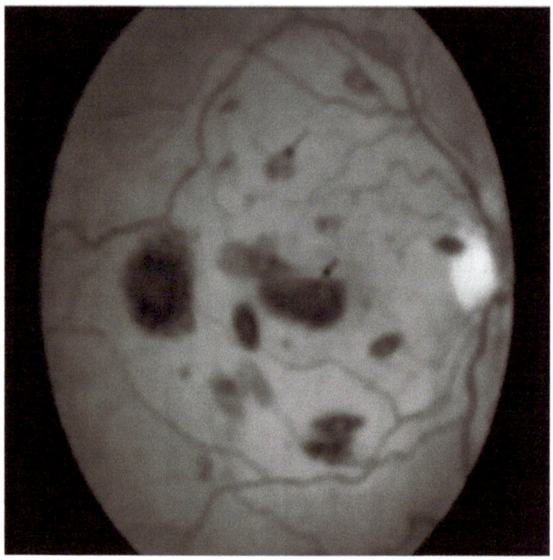

Fig. 7.13 Median filter
Source: https://ieeexplore.ieee.org/abstract/document/8073677

function. This operator removes the effect of noise and other illuminations (Fig. 7.15).

$$Is(x, y) = Ig(x, y) * g(x, y) \tag{7.1}$$

Where $*$ denotes convolution and g(x,y) is a Gaussian function
Ig(x,y) = green channel component
Is(x,y) = Gaussian noise

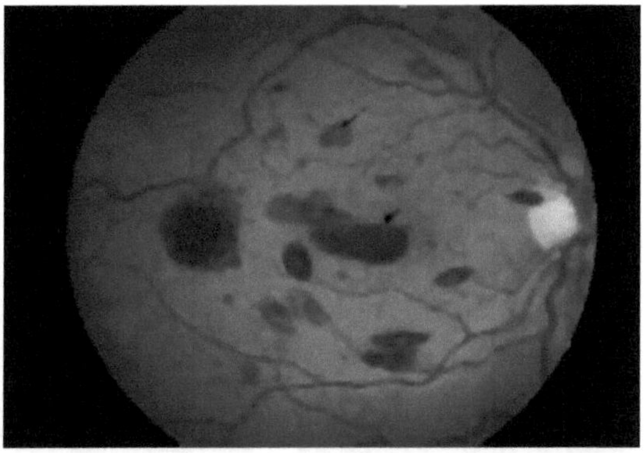

Fig. 7.14 Adaptive median filter
Source: https://ieeexplore.ieee.org/abstract/document/8073677

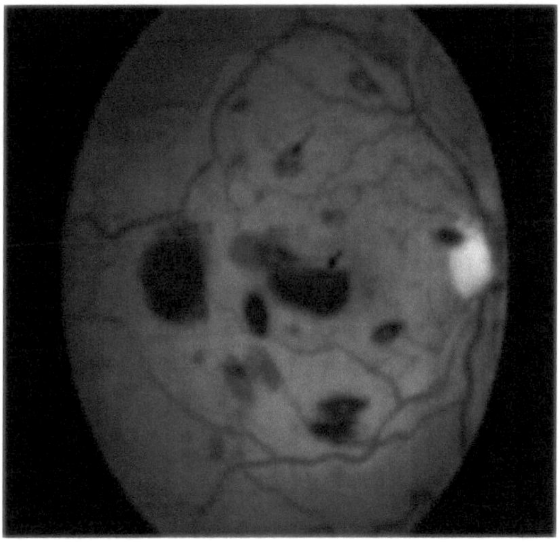

Fig. 7.15 Gaussian filter
Source: https://ieeexplore.ieee.org/abstract/document/8073677

7.6.3 Segmentation from Preprocessed Image

The preprocessed images are used for segmentation of optic disc and cup which helps in classification of glaucoma. Over here we will be discussing three different techniques for glaucoma detection.

7.6.3.1 Multi-thresholding Technique

Talking about one of the easiest method for segmenting cup and disc of the processed fundus image, then multi-thresholding is the best technique. This technique is applied on the preprocessed image containing the optic disc which is converted into the binary image. Using this, the cup can be detected by analysing the brighter region of the optic disc with higher threshold value and the whole disc with lower threshold value. Figure 7.16 shows the detection of the optic disc and cup from the preprocessed fundus image.

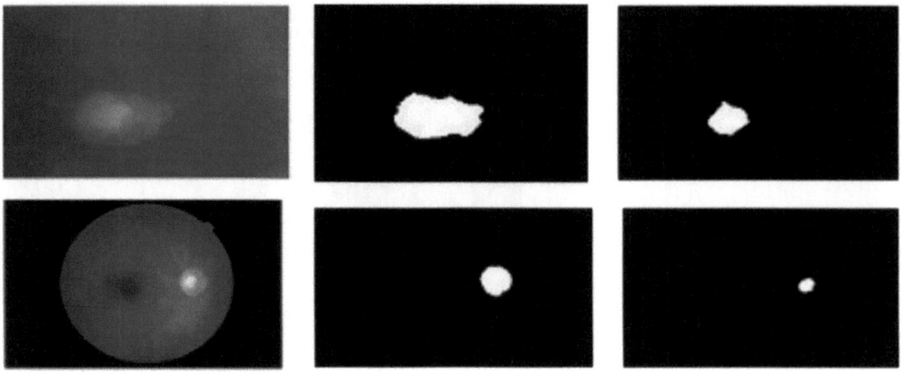

(a) Removal of blood vessels (b) Detection of Optic (c) Cup detection

Fig. 7.16 Results of multi-thresholding technique. (**a**) Removal of blood vessels (**b**) Detection of optic (**c**) Cup detection
Source: https://www.researchgate.net/publication/321614356_Advances_in_Computing_and_Communications_First_International_Conference_ACC_2011_Kochi_India_July_22-24_2011_Proceedings_Part_III

For proper segmentation of the optic disc, it is necessary that original image should be of good quality. The major disadvantage of using this method is that it is based on the values of pixel intensity. Therefore, other techniques like active contour method and region growing segmentation methods are applied on the retinal fundus image for better outcome.

7.6.3.2 Active Contour Method

This model is also called as *snakes***,** a method for describing an outline of an object from a noisy 2D image. A snake is energy minimizing that is influenced by constraint and image forces that pull it towards contours of the object (features like lines and edges) and internal forces that resist deformation. Contours stop at the edges once it detects the desired boundary, where the external energy is minimum [17]. Therefore, active contours are connectivity-preserving relaxation method, applied to the image segmentation problems [18].

Energy Formulation:

Elastic snake is defined by a set of n points v_i where
I = 0. . .n-1,
$E_{internal}$ = internal elastic energy
$E_{external}$ = external edge-based energy

The aim of the internal energy is to control the deformations made to the snake to control the fitting of the contour onto the image. The external energy is usually a combination of the forces due to the image itself E_{image} and the constraint forces

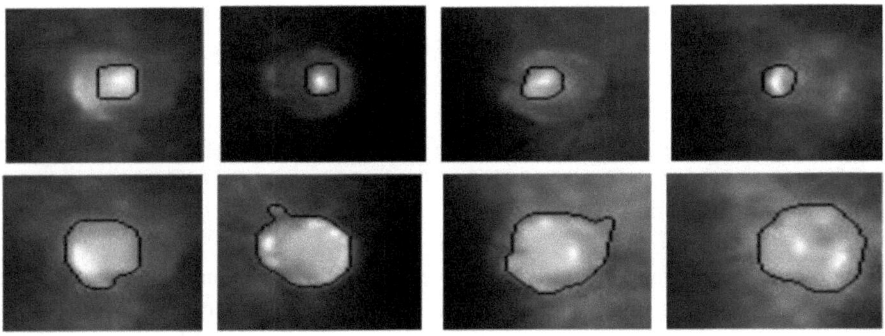

Fig. 7.17 Active contour methods for the detection of cup
Source: https://www.researchgate.net/figure/Results-of-Active-contour-methods-for-cup-and-disc-detection_fig 5_220790076

introduced by the user E_{com}. The internal energy of the snake is composed of the continuity of the contour E_{cont} and the smoothness of the contour E_{curv} [17].

Therefore, the energy function of the snake is the sum of external and internal energy, or

$$\int_0^1 Esnake(v(s))ds = \int_0^1 \times (Einternal(v(s)) + Eimage(v(s)) + Econ(v(s)))ds \quad (7.2)$$

The classical snakes and active contour models uses image intensity (gradient) for the detection of edges to stop evolving curve on the boundary of the desired image, whereas in practical situations, the stopping function is never zero on the edges as the discrete gradients are bounded. In such cases, we use a slightly different active contour model which is not based on gradient of the image but rather is related to segmentation of the image [17] (Fig. 7.17).

7.6.3.3 Region Growing Segmentation

We already know that segmentation is a technique to separate the image into simple regions with homogeneous behaviour. This method computes the neighbouring pixels of the initial seed points and determines whether the pixel neighbours are required to be added to the region or not based on the intensity value and colour similarity [19]. Over here, the maximum value of the image is taken as a seed which is employed in identification of the optic cup. The connection in the neighbourhood pixels is created if they fall below a certain threshold value. Hence, the centroid of the resulting region is calculated, and a disc is estimated that circumscribes the entire region. The area of the cup is hence identified. Further, the disc is grown outwards till a sharp change in intensity is observed. The ratio of the area of cup to disc is evaluated. If it falls below 0.5, the area can be concluded as healthy. If, however, the value exceeds the limit, it is concluded to be defected by some ocular diseases.

7.7 ERG-Based Interfaces for Assisting Disables

7.7.1 Low Vision Aids

The progression in glaucoma affects the eyesight of the individual. There are devices which help people to improve their sight. These aids can be optical lenses, such as telescopes or magnifiers, or no optical devices, such as visors, filters, reading slits, stands, lamps, and large print. The basic principle of these devices is to magnify [20].

There are two types of portable low vision electronic magnifiers. One is the handheld which can be carried with one's own self either in pocket or in handbag and is used to read labels in stores, menu card in restaurants, price tags and more. They are meant for quick reading tasks. The latest models have a camera that takes a still image of the reading material, so it can be viewed either later or at a more convenient angle. Another aid used as a vision magnifier is a camera that is mounted on an adjustable stand. It can focus near or far, so that the person with poor vision can view a reading material on a table, a TV that is across the room, or a distant blackboard on a monitor that is attached to the camera (Fig. 7.18).

7.7.2 A Talking Scanner

It can scan any text document and read it back to the user in seconds. This can help the visually impaired person in gathering any sort of text information [21].

7.8 Conclusions

ERG is the primary level test done for the detection of ocular diseases. Glaucoma is the most common abnormality affecting millions of people worldwide. Once the optic nerves get damaged, it is the vision that cannot be regained. There are various proposed algorithm that can be used for detection of glaucoma. The diagnosis and treatment of eye diseases can be done using digital retinal imaging. It plays a very important role in the evaluation and extraction of clinically useful information. The obtained information may provide a great help to ophthalmologists while monitoring, diagnosing, and treating the ocular diseases. Further, the idea discussed in this paper can also be applied to other images where it is of interest to detect intersections and perform analysis of various vascular objects. For this proposed algorithm, GUIs are designed in MATLAB for retinal blood vessel extraction and detection of statistical features like values of energy, contrast, homogeneity, and entropy. Also

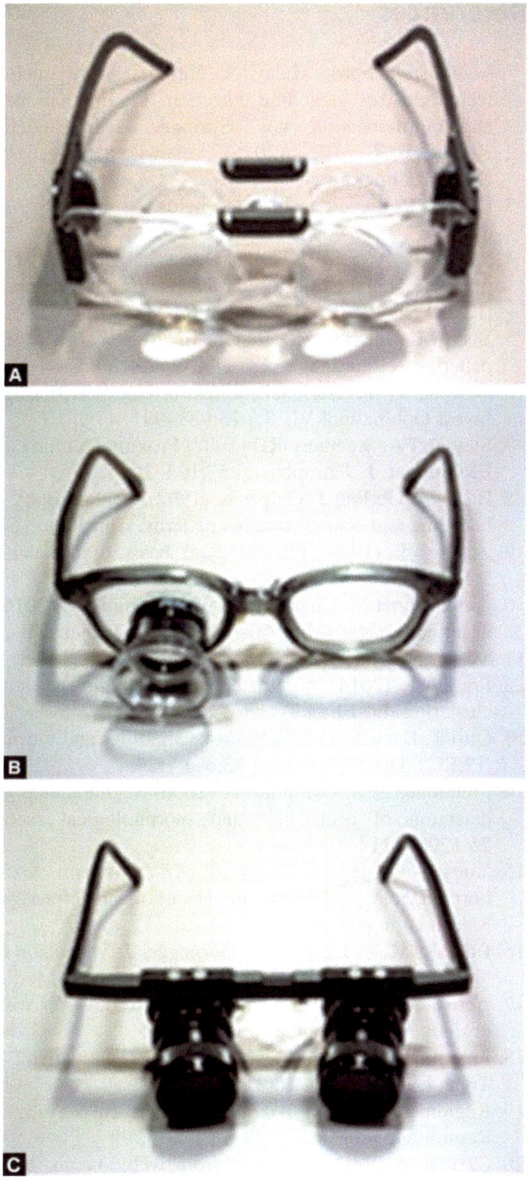

Fig. 7.18 Reading aids: (**a**) magnifying glasses, (**b**) mounted loupes, and (**c**) reading telescopes
Source: https://www.ncbi.nlm.nih.gov/pmc/articles/PMC5159456/

by performing texture analysis, we can find differences in features of normal and abnormal images. Further, in the future, we can perform better feature extraction methods so as to get more than five features. More number of features can give better output and accuracy.

References

1. Madhusudhan M, Malay N, Nirmala SR, Samerendra D (2011, July) Image processing techniques for glaucoma detection. In: International conference on advances in computing and communications, vol 6. Springer, Berlin, Heidelberg, pp 365–373
2. Kumar UR, HL (2015). eyewiki.aao.org. Retrieved from http://eyewiki.aao.org/Electroretinogram
3. (2019). Retrieved from. moorfields.nhs.uk: https://www.moorfields.nhs.uk/content/anatomy-eye
4. Swathi C, Anoop BK, Dhas DAS, Sanker SP (2017, March) Comparison of different image preprocessing methods used for retinal fundus images. In Emerging Devices and Smart Systems (ICEDSS), 2017 conference on (pp. 175–179). IEEE.
5. Bach M, Hawlina M, Holder GE, Marmor M, Meigen T, Vaegan MY (2000) Standard for pattern electroretinography. Doc Ophthalmol 101:11–18
6. Dawson WW, Trick GL, Litzkow CA (1979) Improved electrode for electroretinography. Invest Ophthalmol Vis Sci 18:988–991
7. Sieving PA, Steinberg RH (1987) Proximal retinal contributions to the intraretinal 8-Hz pattern ERG of cat. J Neurophysiol 57:104–20
8. Bach M, Gerling J, Geiger K (1992) Optic atrophy reduces the pattern-electroretinogram for both fine and coarse stimulus patterns. Clin Vision Sci 7:327–333
9. Zrenner E (1989) Physiological basis of the pattern electroretinogram. Progr Retin Res 9:427–64
10. Anto Bennet M, Dharini D, Priyadharshini SM (2016) Detection of blood vessel Segmentation in retinal images using Adaptive filters. J Chem Pharma Res 8(4):290–298
11. Missfeldt, M. (2017). Retrieved from varifocals.net: https://www.varifocals.net/human-eye
12. Prasath R (2014) Automated drusen grading system in fundus image using fuzzy C-means clustering. Int J Eng Technol 2(6):833
13. Otto T, Bach M (1997) Re-test variability and diurnal effects in the pattern electroretinogram (PERG). Doc Ophthalmol 92:311–323
14. Mendonca AM, Campilho A (2006) Segmentation of retinal blood vessels by combining the detection of centerlines and morphological reconstruction. IEEE Trans Med Imaging 25:1200–1213
15. Youssif AA-HA-R, Ghalwash AZ, Ghoneim AASA-R (2008) Optic disc detection from normalized digital fundus images by means of a vessels' direction matched filter. IEEE Trans Med Imaging 27:11–18
16. Bock R, Meier J, Nyúl LG, Hornegger J, Michelson G (2010) Glaucoma risk index: automated glaucoma detection from color fundus images. Med Image Anal 14:471–481
17. Active contour model. In *Wikipedia.* Retrieved November 18, 2018, from https://en.wikipedia.org/wiki/Active_contour_model
18. Fukuda T, Morimoto Y, Morishita S, Tokuyama T (2001) Theory of communication. ACM Trans Database Syst 26(2):179–213
19. Region growing. In *Wikipedia.* Retrieved May 3, 2018, from https://en.wikipedia.org/wiki/Region_growing
20. (2018). WebMD. Retrieved from webmd.com: https://www.webmd.com/eye-health/occular-hypertension
21. Technology and Accessibility. September 14, 2015, Glaucoma Research Foundation

Part V

EEG, EOG and Its Significance

Advanced Approaches for Medical Image Segmentation

8

Sanjay Saxena, Adhesh Garg, and Puspanjali Mohapatra

Abstract

Image segmentation, i.e., dividing an image into its constituent's regions, is a decisive phase in plentiful medical imaging studies to extract meaningful information such as shape, volume, motion, and abnormalities and to quantify changes of the human organs by radiologists and investigators, which can be facilitated by several automated computational procedures. Several efficient approaches for medical image segmentation have been developed till now based on hard and soft computing models such as thresholding, clustering, graph cut approaches, fuzzy-based approaches, neural network approaches, and many more. Tremendous success of deep learning nowadays has achieved state-of-the-art performance for instinctive medical image segmentation. This chapter provides the brief introduction about medical image segmentation and several current researches for the precise dissection. Further, it will provide the information about the deep learning used as an advanced approach presently for accurate segmentation of medical images.

Keywords

Medical image segmentation · Medical imaging · Deep learning · Computational intelligence techniques

8.1 Introduction

Medical imaging is a procedure used to generate images of the human body for clinical purpose. Medical imaging techniques have revolutionized the modern medical era. Medical imaging techniques such as computed tomography (CT), magnetic resonance

S. Saxena (✉) · A. Garg · P. Mohapatra
Department of Computer Science & Engineering, International Institute of Information Technology, Bhubaneswar, India
e-mail: sanjay@iiit-bh.ac.in

S. Paul (ed.), *Application of Biomedical Engineering in Neuroscience*,
https://doi.org/10.1007/978-981-13-7142-4_8

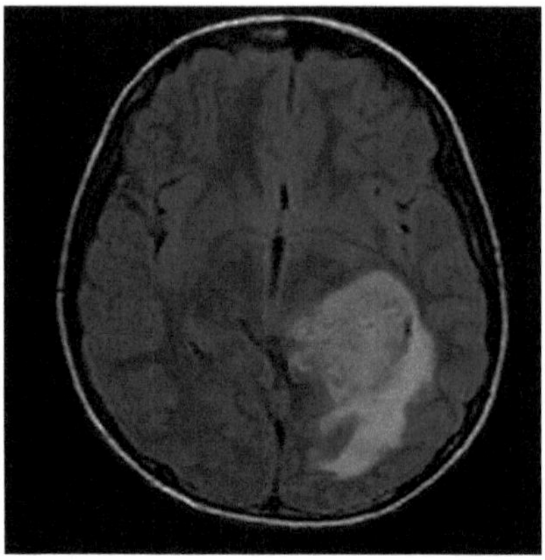

Fig. 8.1 Original brain MR image

imaging (MRI), and ultrasound (US) are widely used for the diagnosis of anomalies like cyst, tumor, etc. CT can generate images at much higher spatial resolution with shorter imaging duration [1]. With expanding utilization of CT and MR imaging for conclusion, treatment arrangement, and clinical investigations, it has turned out to be practically mandatory to utilize computers to help radiological specialists in clinical determination and treatment arrangement. Dependable calculations are required for the outline of anatomical structures and different regions of interest (ROI) [2]. Medical image segmentation is the way toward disengaging a picture it into ample areas as pixels bunches with some comparative highlights such as intensity, texture, color, shape that are easier and more expressive to examines items present in an image [3, 4]. The degree of homogeneousness of the extracted segment can be measured by some property of image (e.g., intensity of the pixel) [5]. Accurate image segmentation is a part of nearly all systems of computer vision, and it is used by way of a step of preprocessing as it is one of the greatest general core technologies and a scientific and perilous issue in medical image analysis [3, 6, 7].

An example of segmented brain tumor MR images taken from [8] has been shown in Figs. 8.1 and 8.2. It presents that by espousing the theory of different segmentation techniques in medical imaging modalities, it becomes very easier to excerpt more information which is useful in the location of tumors, measurement of volumes of tissue, study of different anatomical structures, etc. [9]

Till now, even though wide-ranging researches have been carried out in evolving diverse methods for image segmentation, still no solitary standard process of image segmentation has been developed. Somewhat, there are varieties of several hybrid methods which are used for accurate and precise segmentation. Overall broadly image segmentation methods are classified into two categories, soft and hard computing techniques as discussed by [10–12]. Some of the examples of different hard computing methods are thresholding techniques (based on intensity of pixel), ratio-contour method (RC), normalized cut method (NC), morphological methods (deformable), mean shift method (MS), efficient graph-based method (EG), edge-based

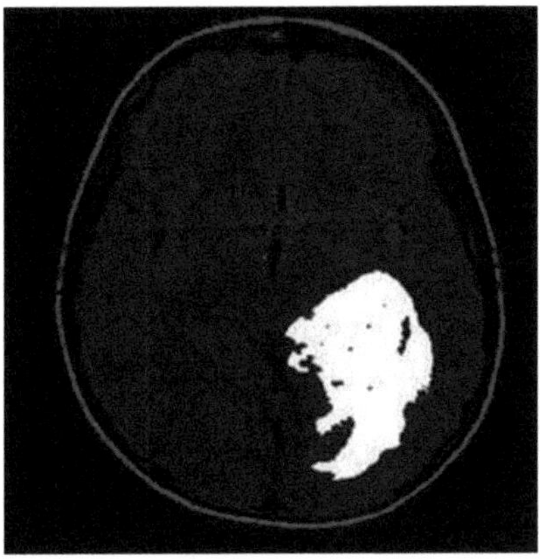

Fig. 8.2 Segmented region

segmentation (boundary localization), level-set method (LS), and many more. Further, soft computing technique includes genetic algorithm (GA), fuzzy logic (FL), artificial neural network (ANN), colony optimization (ACO), ant particle swarm optimization (PSO), and many more [11, 13, 14]. Because of the adaptive nature and accuracy of different soft computing approaches, they are chosen by scientists and researchers currently for image segmentation [3]. They are lenient of inexactitude, vagueness, incomplete veracity, and estimates dissimilar from hard computing. Further, it provides imprecise solutions to the problems [15]. Such kind of approaches have been broadly implemented in the last couple of years in different domains such as medical, scientific research, engineering, management, humanities, etc. [13]. It also plays a very vital role in medical image segmentation.

In this chapter, we will discuss some core approaches based on hard and soft computing used for medical image segmentation. Along with this, we will also discuss the deep learning approaches used for medical image segmentation which is the subset of machine learning and is frequently used currently. If we talk about machine learning, it is characterized as a set of methodologies that robotically perceive patterns in diverse data and then envisage enabling decision to exploit the revealed patterns. Deep learning, which is a part of machine learning, is an unusual kind of neural network that looks like the multilayered human perception system. Currently, deep learning is picking up a great deal of consideration for its use and proficient outcomes in medical image segmentation. Deep learning techniques have showed remarkable performances in imitating humans in numerous fields, especially in medical image analysis, which is one of the distinctive tasks in radiology practice to detect abnormalities and to classify them into different disease categories [16].

After introduction, organization of the chapter follows as Sect. 8.2 familiarizes different hard and soft computing approaches used for medical image segmentation, Sect. 8.3 will discuss about the deep learning as an advanced approach for medical image segmentation, and Sect. 8.4 concludes this chapter followed by references.

8.2 Different Approaches for Medical Image Segmentation

As early as the disease can be detected more are the chances of better treatment and the better probabilities of survival for the patients. As we have already discussed before, image segmentation is a procedure which plays a pivotal job in extraction of information from the medical images, i.e., X-rays, MRI, CT scans, and PET scans, that supports in understanding and better interpretation of the human internal anatomy. The main role of image segmentation is to divide the image in small regions and homogeneous segments with respect to one of the features [17]. Each of these regions can be used to extract information. Efficient image segmentation can be accomplished by identifying all the pixels that belong to the same structure or based on some similar attribute associated with it. It can also helpful in image measurement and compression. It has found many applications as for localization of tumor and microcalcification, delineation of blood cells, tissue classification, and tumor volume estimation. Broadly we can classify approaches for medical image segmentation into two categories, hard and soft computing approaches. As shown in the following (Fig. 8.3), both the approaches are having different characteristics.

Now the brief introduction of the several approaches of hard and soft computing are given below:

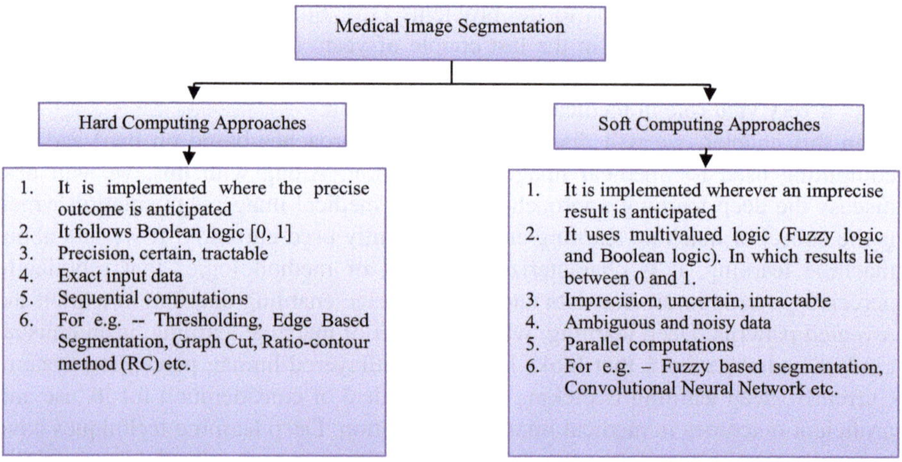

Fig. 8.3 Approaches for medical image segmentation

8.2.1 Hard Computing Approaches

8.2.1.1 Thresholding Method

Thresholding is the one of the meekest segmentation methods. The pixels present in an image are segregated reliant on their values of intensity. In [18], authors have applied segmentation of the brain lesion.

Global thresholding is a type of thresholding which practices an appropriate threshold *T*. General equation of the threshold is given below, in which m(*x*,*y*) is the original image while *n*(*x*,*y*) is the obtained image by threshold *T*.

$$n(x, y) = \begin{cases} 1, & \text{if } m(x, y) > T \\ 0, & \text{if } m(x, y) \leq T \end{cases} \tag{8.1}$$

Variable thresholding is a type of thresholding in which *T* can change over the whole image.

If *T* depends on the neighborhood of the pixel(*x*,*y*), then it is called local thresholding. If *T* is the function of (*x*,*y*), then it is called adaptive thresholding.

$$\text{Multiple thresholding}: n(x, y) = \begin{cases} p, & \text{if } m(x, y) > T2 \\ q, & \text{if } T1 < m(x, y) \leq T2 \\ r, & \text{if } m(x, y) \leq T1 \end{cases} \tag{8.2}$$

Choosing the appropriate threshold is one of the important tasks. The following figures represent histogram of the images with different peaks and valleys that can be helpful in choosing the appropriate threshold(s).

Like in the above figure (Fig. 8.4), *T* can be the appropriate threshold, while in another figure, *T*1 and *T*2 will be the good one. Few important factors that affect the appropriateness of the histogram for guiding the option of the appropriate threshold are:

(a) The separation among peaks
(b) The content of noise present
(c) The comparative size of background and objects
(d) The consistency of the illumination
(e) The evenness of the reflectance

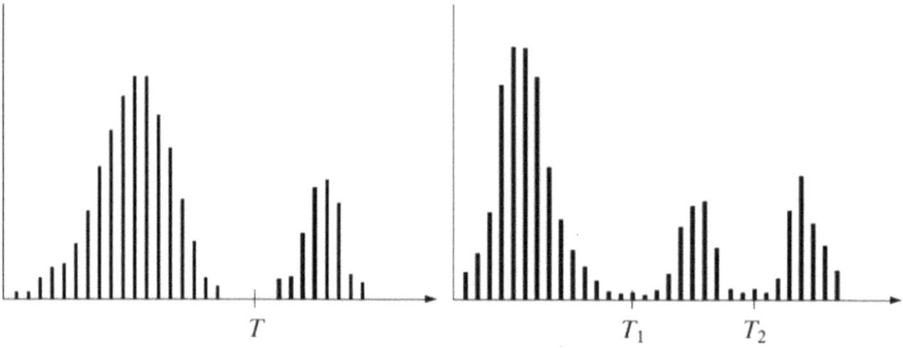

Fig. 8.4 Histograms of different objects with different thresholds

8.2.1.2 Edge-Based Method

Edges are confined changes in the intensity of the image. Edges normally transpire on the boundaries between two distinguished regions. The main features of the object can be computed from the edges of an image. Edge detection has a very significant role in image analysis.

There are numerous edge detection techniques available in the existing literatures. Those techniques are Roberts edge detection, Prewitt edge detection, Sobel edge detection, Marr-Hildreth edge detection, Kirsch edge detection, Robinson edge detection, LoG edge detection, and Canny edge detection. Edge-based segmentation for brain tumor segmentation has been developed by authors discussed in [19, 20].

The general procedure for edge-based segmentation consists the following steps.

1. In order to detect edges of the image, use the different derivative operator.
2. By measuring the amplitude of the gradient, measure the strength of edges.
3. Retain all edge having magnitude greater than threshold T (removal of weak edge).
4. Find the position of crack edges; the crack edge based on the confidence it receives from its predecessor and successor edges is either retained or rejected.
5. Steps 3 and 4 are repeated with different values of threshold to find out the closed boundaries; further, segmentation of an image is achieved.

8.2.1.3 Region-Based Segmentation

The principle of homogeneity gives the region-based approaches – pixels with comparable properties are bunched together to form a homogenous region. The time gray level of pixels gives the criteria for homogeneity, and this criteria can be specified by the following conditions:

$$R1 \cup R2 \cup R3 \cup \ldots \cup Ri = I$$

where

R1, R2, R3, … Ri are the region in the image *I*, and further, $R1 \cap R2 \cap R3 \cap \ldots \cap Ri = 0$.

This is as per the set theory of homogeneity.

On the basis of region, region-based algorithms are following:

1. Region Growing
2. Region Splitting and Merging

Region-growing method exploits the imperative fact that pixels that are close together are having analogous gray values. Start with a pixel, i.e., called seed pixel, and add new pixels. The following is the algorithm of region growing:

1. Pick the seed pixel.
2. Ensure the neighboring pixels and insert them to the region if they are alike to the seed.
3. Repeat step 2 for every recently added pixel; discontinue if further no more pixels can be added.

In [21, 22] authors implemented region-growing approaches for the accurate segmentation using region-based method.

Split and merge method: This is the combination utilizing the advantage of the two methods of splits and merges. This method is based on quad quadrant tree representation of data whereby image segment is split into four quadrants provided the original segment is nonuniform in properties. Depending on the uniformity of the region (segments), after this the four neighboring squares are merged. This split and merge process is proceeded until no additionally split and merge is possible.

The algorithm for split and merge follows the following steps.

1. Define homogeneity criterion and split the image into four square quadrants.
2. Split it further into four quadrants, if any resultant square is not homogeneous.
3. At each level regions satisfying the condition of homogeneity, blend the two or more neighboring.
4. Until no further split and merge of region is possible, continue the split and merge.

8.2.1.4 Watershed Algorithm

Watershed method is a leading mathematical morphological utensil for the partitioning of the image into its constituents' segments. It is more putative in the zones such as computer vision and biomedical image processing. In terms of the geography, it means that the watershed are the ridges that divide regions drained by diverse river systems. Suppose image is observed as geological scenery, the outlines of watershed elect boundaries that split areas of image. The watershed transforms compute ridgelines and catchment basins that are also known as watershed lines, where catchment basins correspond to image regions and ridgelines relate to region boundaries. Segmentation by watershed embodies many of the concepts of the three techniques such as edge-based, threshold-based, and region-based segmentation. In [23] the main improvement of authors' implemented work is that massive quantity of segmented region in edges is shortened by marker-controlled watershed segmentation. However, vanguard objects and the positions of background must be marked previously to get superior segmentation result.

8.2.1.5 Morphological-Based Method

Image morphology is an enormous amount of operations of image processing and computer vision applications that modify the images based on the structures and shapes. It is well thought-out to be one of the data processing procedures. It has many applications like texture analysis, noise elimination, boundary extraction, and many more. Binary images may enclose innumerable defects. In a few circumstances, binary regions constructed by simple thresholding are distorted by some noise and textures. The goal of morphological image processing is to remove all the defects discussed while maintaining the structure of an image. Morphological techniques probe an image with a template or small shape called as a structuring element (SE). The SE is located at all possible locations in the image and compared with the analogous neighborhood of pixels. Some processes check whether the element fits within the neighborhood, while others check whether it hits the neighborhood. Dilation and erosion are the basic two operations in image processing. Other algorithms based on dilation and erosion. The erosion of a binary image f by a SE s (denoted by $f \ominus s$) produces a new binary image $g = f \ominus s$ with ones in all locations (x,y) of a SE's source at which that SE s fits the input image f, i.e., $g(x, y) = 1$ is s fits f and 0 otherwise, repeating for all pixel coordinates (x,y). The dilation of an image f by a SE s (denoted by $f \bigoplus s$) produces a new binary image $g = f \bigoplus s$ with ones in all locations (x,y) of a structuring element's origin at which that structuring element s hits the input image f, i.e., $g(x,y) = 1$ if s hits f and 0 otherwise, repeating for all pixel coordinates (x,y). [24] has implemented a method to segment tumor in low-intensity images. The method consists of several steps for extracting tumor from the MR slices, image enhancement, resampling of image, histogram application, color plane extraction, and advanced morphological operation to segment tumor region. In this method, morphological operations are principally used as the filter to eradicate low-frequency pixels and pixels present in the boundary. Different parameters of the tumor such as area and length are identified effectively for treatment planning and diagnosis of the tumor. The main drawback of this method is that this method uses so many repetitive steps for image segmentation.

8.2.1.6 K-Means Clustering

K-means clustering is a kind of unsupervised learning used to find groups in the unstructured data, with the numeral of groups presented by the variable *K*. This technique works iteratively to allocate each data point to one of the *K* groups based on the features that are given. Data points are clustered based on similarity of feature. Flowchart of the K-means clustering is given below (Fig. 8.5).

The outcome of the *K*-means clustering algorithm are given below:

1. The centroids of the *K* clusters, which can be used to label new data
2. Labels for the training data

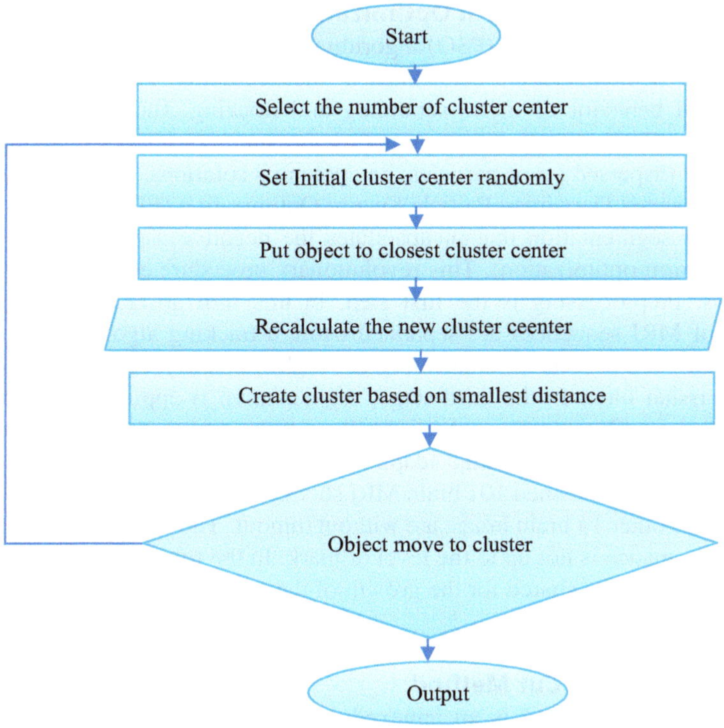

Fig. 8.5 Flowchart of K-means clustering

Rather than defining the groups before looking at the data, clustering allows us to analyze and find the groups that are formed organically [25].

8.2.1.7 Markov Random Field (MRF)

In this, segmentation is restrained in a universal framework, which is called as image labeling, where the predicament is abridged to assigning labels to pixels. If we consider probabilistic approach, dependencies of the label are modeled by Markov random fields (MRF). Further, an optimal labeling is dogged by Bayesian estimation, in meticulous highest a posteriori (MAP) estimation. The major benefit of MRF models is that earlier information can be obligatory close by through clique potentials. MRF model incorporates spatial information into the process of clustering. This deducts segmentation overlap and noise's effect in the segmentation. This special feature motivates different researchers to exploit MRF in segmentation. In [26], authors established an iterative MRF framework to contain voxel-based MRF, adopted MRF, and regional MRF. The applied framework was also used to categorize all the subclass of the medical images. The main restriction of MRF is its computational complexity and effective selection of parameters. However, it is frequently used to model properties of texture and intensity in homogeneity.

8.2.1.8 PSO (Particle Swarm Optimization)-Based Method

Particle swarm optimization (PSO) algorithms are basically population-based and nature-inspired metaheuristic algorithms in the beginning. PSO impersonates the bird's social behavior of fishes schooling and flocking. Basically, PSO tries to enhance the solutions according to a quality measure, i.e., fitness function forms a haphazardly dispersed set of particles, i.e., potential solutions. In [27], authors have applied Enhanced Darwinian Particle Swarm Optimization (EDPSO) for automated brain tumor segmentation that incapacitates the negative aspect of existing PSO (particle swarm optimization). This revolutionary procedure essentially contains of four steps. Preprocessing is the first step. In this film artifacts and superfluous portions of MRI sequences are separated using a tracking algorithm. The second step contains the process of eradicating the high-frequency constituent and noises using Gaussian filter. In the third stage, segmentation is applied using Darwinian Particle Swarm Optimization. Additionally, in the fourth step, classification is applied, which is done by using adaptive neuro-fuzzy inference system. Applied image data set are contained 101 brain MRI slices, which includes 87 tumorous brain images while other 14 brain image are without tumour. The drawback of this method is that performance is not up to the level of mark. In the future, numerous optimization processes can be shared for the growth of the system's performance in rapports of robustness.

8.2.1.9 Normalized Cut Method

A normalized cut method is an annexed of using graph partitioning to achieve efficient and accurate segmentation. Graph partitioning can be implemented by combining a graph hooked on two disjoint sets of vertices. Basically, the grouping is based on the divergence among the two distinct pieces. Further, the cut measure is the sum of the capacities among the two distinct regions under deliberation. The optimal partitioning of the graph minimizes this cut transversely the whole graph, beneath only the least cut measure. In [28], authors have endeavored to grant an application of these performs on brain tumor segmentation from MR Image sequences. Principally, a novel skull stripping process is developed which is based on normalized cut method. Further, histogram classification is executed on skull-free MRI sequences for more accurate segmentation. However, to achieve more accuracy, more data need to be tested.

8.2.2 Soft Computing Approaches

As already discussed the main features of soft computing techniques earlier. Fuzzy, genetic, and artificial neural network are the main pillars of soft computing used for medical image segmentation. Brief introduction of all of them are given below:

8.2.2.1 Genetic algorithm

It is the technique used for optimization and is based on the growth of a population of solutions which beneath the action of a few specific optimized rules. A population of fixed size is manipulated by GA. Chromosomes form this population. Each

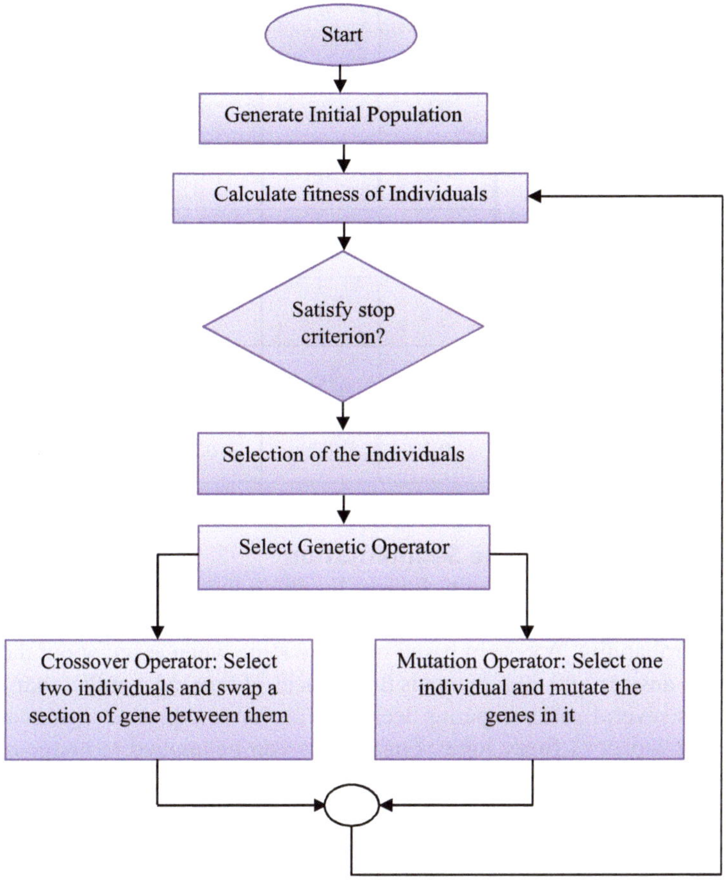

Fig. 8.6 General steps of genetic algorithm

chromosome represents the coding of a potential solution to the problem to be solved; it is formed by a set of genes belonging to an alphabet. At each iteration, by applying the different genetic operators, a new population is formed: selection, crossover, and mutation. The selection of the most pertinent candidates is chosen by GA. Crossover consists in building two new chromosomes from two old ones referred to as the parents. Mutation realizes the inversion of one or several genes in a chromosome. The following figure (Fig. 8.6) shows steps of genetic algorithm GA.

Genetic algorithm is used for optimizing the localized results of brain tumor from MRI sequences, through evaluation criteria that have been implemented in [29]. In the proposed method, clusters of K-means algorithm are used as starting population. Centers that are clustered are evaluated by a fitness function. The main advantage of this method is that it is good in selecting optimal number of region for segmentation. However, the selection of fitness function is difficult.

Fig. 8.7 Image matrix in fuzzy domain [30]

0.1	0.1	0.1	0.1	0.1
0.1	0.2	1	1	0.1
0.2	1	1	0.2	0.1
0.2	1	1	1	0.2
0.2	1	1	1	0.1
0.2	0.2	1	1	0.1
0.2	0.2	0.2	0.1	0.1

8.2.2.2 Fuzzy-Based Image Segmentation

Fuzzy logic is a soft computing technique based on the principal "degree of truth." The idea behind is going more than just 0 or 1 to the degree of the logic. This is very much like probability. A system based on fuzzy logic when asked about the weather can give the answer of 0.53 rain that is the chances of rain today rather than 0 or 1. In recent years several soft computing technologies have improved significantly with the different aspect of fuzzy logic. Fuzzy logic can be applied to image segmentation. Fuzzy logic-based models yield better results as compared to conventional image fusion models. In the recent years many fuzzy-based techniques for image segmentation have been developed. They are fuzzy thresholding, fuzzy integral-based decision-making, fuzzy c-mean clustering, and fuzzy rule-based inferencing scheme. Fuzzy logic is applied as decision operator or feature transformer in image fusion. Before we apply fuzzy processing to the image, image is converted to the fuzzy domain. The fuzzy mapping function is defined such that the features and characteristics of the image can be better represented in the new model.

When image segmentation is done, some pixels are to be made darker or brighter; this problem is fuzzy in nature. Better features can be extracted from the image if the darker and brighter is done with variation as per fuzzy logic. Fuzzy c-means (FCM) algorithm is used to find weighted similarity measure between the pixels in the image and each c cluster center. In contrast with hard clustering, the fuzzy member values range from 0 to 1 as shown in the above figure (Fig. 8.7). Here pixels can belong to more than two clusters with different membership constants. Point on the edge of the cluster may be in the cluster to a lesser degree than the point in the center of the cluster. It is a very important tool for image segmentation. This is the method where one piece of data can belong to two or more clusters. This method (developed by Dunn in 1973 and improved by Bezdek in 1981) is frequently used in pattern recognition. Here, data are bound to each cluster by means of a membership function. Membership function which represents the fuzzy behavior of this algorithm. To do that, we simply must build a suitable matrix named U whose factors are numbers between 0 and 1 and represent the degree of membership between data and centers of clusters. Several iterations are taken to find the best location for the clusters. The iterations are stopped when a max number of iterations are reached or two consecutive iterations improvement is less than specified. The main disadvantage with fuzzy c-mean clustering is computational complexity and increased noise. Several enhanced neuro-fuzzy networks are designed to overcome the flaws and for better results. In [31], authors were modeled the FCM and K-means on T1 contrast axial plane MRI sequences for the extraction of brain tumor with the histogram-guided initialization of the cluster. K-means is capable to cluster the regions relatively superior than FCM. FCM identifies barely three tissue classes; however K-means identifies all the six classes. The main drawback of this method is that few WM (white matter) is characterized as edema and vice versa in using K-means.

8.2.2.3 ANN-Based Image Segmentation

In a machine, implementing the knowledge used by humans during a traditional image segmentation would cost extensive computation time and would require a huge domain knowledge database, which is currently not available. In addition to traditional image segmentation methods, there are some trainable segmentation methods which can model some of this knowledge. Neural network segmentation relies on processing small areas of an image using an artificial neural network or a set of neural networks. Artificial neural network has been developed with large number of applications such as functional approximation, image segmentation, and feature extraction and classification. There real-time application and easy implementation led to the increasing ANN-based segmentation. Some of the most widely used networks in ANN are Hopfield neural network (HNN), cellular neural network (CNN), radial basis function network (RBF), self-organized map (SOM), and pulse-couple neural networks (PCNN).

- Hopfield ANN: In HNN each neuron is liked to each other, and weights are symmetrical. All the neurons have both the functions input and output. The output is a binary number. It is used to map between image pixels and their labels.
- Cellular neural network CNN: In a CNN, the neuron is connected to the nearest neighbor, and the output from the neuron is connected to the input of the neuron in the same neighborhood. CNN is good for noise removal and feature extraction in an image. General structure is shown in the following figure (Fig. 8.8).
- Pulse-coupled neural network (PCNN): PCNN comprises of an accumulation of neurons where each neuron represents a pixel in the image. When the average of stimuli surpasses a dynamic threshold, the neuron becomes active. The pulsing nature of PCNN comes from dynamic threshold. When the segments are of uniform intensity, PCNN functions admirably.
- Self-organized map (SOM): It is a forward feed neural network in which the neurons are orchestrated in a 2D array. To find the optimal weight for each node, an iterative learning process is used. It was designed for dimension reduction, yet

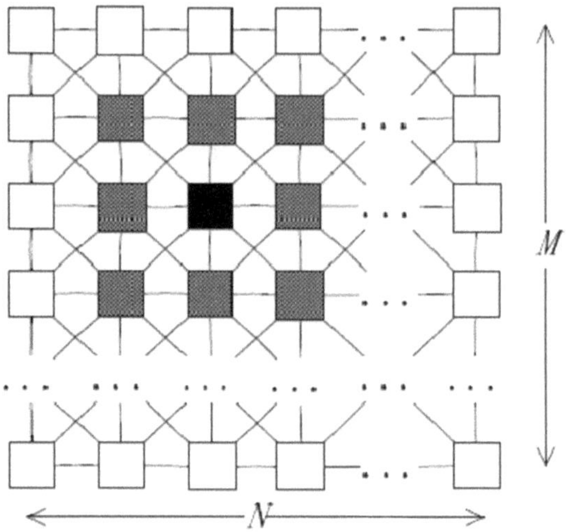

Fig. 8.8 General structure [9]

it is likewise utilized in medical image segmentation. It is utilized for color reduction of colored medical images.

- Radial basis function network (RBF): It is also a forward feed network with activation function as radial. Input layer, hidden layer with radial function, and output layer are the three layers in the network. It is mainly used for the approximation and classification for medical image segmentation.

8.3 Traditional Machine Learning Methods to Deep Learning: An Advanced Approach for Medical Image Segmentation

Deep learning is an artificial intelligence function that imitates the operations of the human cerebrum in data processing and decision-making. It is a subset of machine learning in artificial intelligence (AI). It has networks fit for gaining from data that is unstructured or unlabeled. It uses unsupervised learning. With the increasing popularity of deep learning, semantic segmentation problems are being handled using deep neural network architectures. The most frequently used model is convolutional neural nets, which outperform other methodologies by a substantial edge regarding accuracy and efficiency.

In a conventional neural network, the first layer is the input layer, and the last layer being the output layer, nodes are placed in layers. The input nodes are special such that their outputs are the value of the corresponding features in the input vector.

For instance, in a classification task having a three-dimensional input (x, y, z) and a binary output, one conceivable network design is having three input nodes and one output node. In network design, the input and output layers are usually considered fixed.

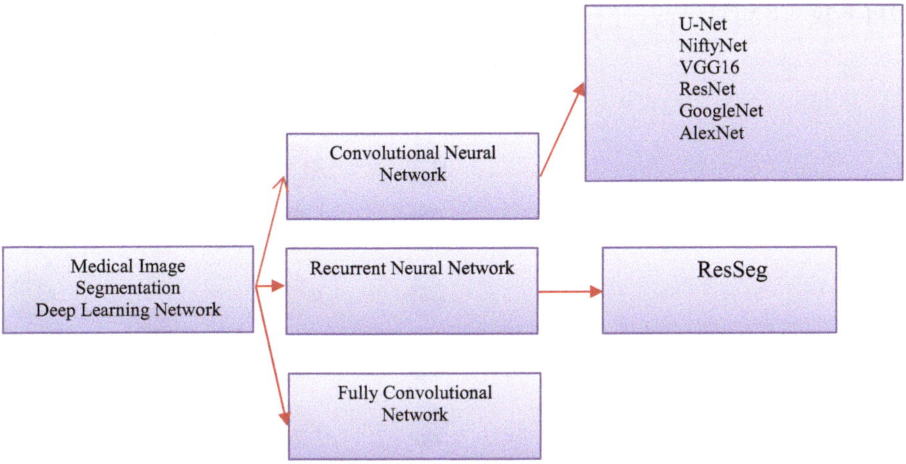

Fig. 8.9 Different deep network

While as the advancements in the field of medical images increased, research on automated analysis has increased. We have come from a sequential application of low-level analysis of pixels to compound rule-based applications. Then supervised learning techniques where data is used to training and develop a system have become well known. Algorithms are designed to determine the optimal decision boundary by extracting discriminant features from the images.

A computer learning from data is the main idea behind deep learning models. Deep learning algorithms composed of multiple layers that transform the input image data by extracting more and more feature in each layer. Image segmentation is the process of partitioning the image into more meaningful data and easy to analyze. The most successful model of deep learning to tackle this and analyze the image data is convolutional neural networks (CNN) as shown in the figure. It classifies each pixel in the image individually by presenting it with the patches extracted from the neighborhood of the pixel. Other deep learning techniques include recurrent neural network (RNN), convolution neural network with multiple pathways, and 3D convolutional networks. These methods can achieve more accurate segmentation results than designed input features explicitly. Several kinds of deep network architectures have been shown in Fig. 8.9. In convolutional neural network, segmentation of substructures such as organs allows quantitative analysis of parameters such as shape or volume which are important in the case of cardiac and brain analysis. Segmentation is defined as the task of identifying the set voxels which make up the region of interest (ROI). There are several CNN architectures used in image segmentation. U-Net is one of the most popular architectures of CNN as shown in Fig. 8.10. U-Net comprises of an equal amount of upsampling and equal amount of downsampling layers. U-Net can process the entire image in one forward pass resulting in segmentation map directly. A variant of U-Net called V-Net is also

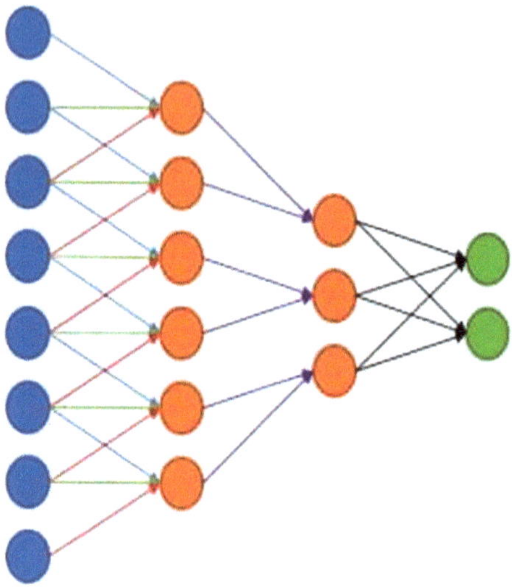

Fig. 8.10 CNN [32]

proposed for the segmentation of 3D images using 3D convolutional layer. Some other popular architectures of CNN that have proved so effective recently include the following (Fig. 8.11):

AlexNet: It won the 2012 ImageNet competition with test accuracy of 84.6%. It uses three convolutional layers.

VGG-16: Oxford's model used a stack of convolutional layers. It won the 2013 ImageNet competition.

GoogleNet: This model uses 22 convolutional layers. It won the 2014 ImageNet competition.

ResNet: This is Microsoft's model. It won the 2016 ImageNet competition.

NiftyNet: It is an open-source neural network. It is a TensorFlow-based framework.

VGG-16: This Oxford's model won the 2013 ImageNet competition with 92.7% accuracy. It uses a stack of convolution layers with small receptive fields in the first layers instead of few layers with big receptive fields. Recurrent neural networks (RNN) as shown in the following figure have become more popular recently for segmentation. Several memories that unite designed especially for them to overcome time problem are long short-term memory (LSTM). A spatial clockwork RNN is used to segment perimysium in H&E-histopathology images. This model considers the prior knowledge of the current batch. The RNN in different orientations is applied four times, and the end results are concatenated. Several models are also proposed which combines LSTM-RNN with U-Net like architecture for segmentation. One of the popular RNN architectures for segmentation is ReSeg [33] based on ReNet image classifier [32–36] (Fig. 8.12).

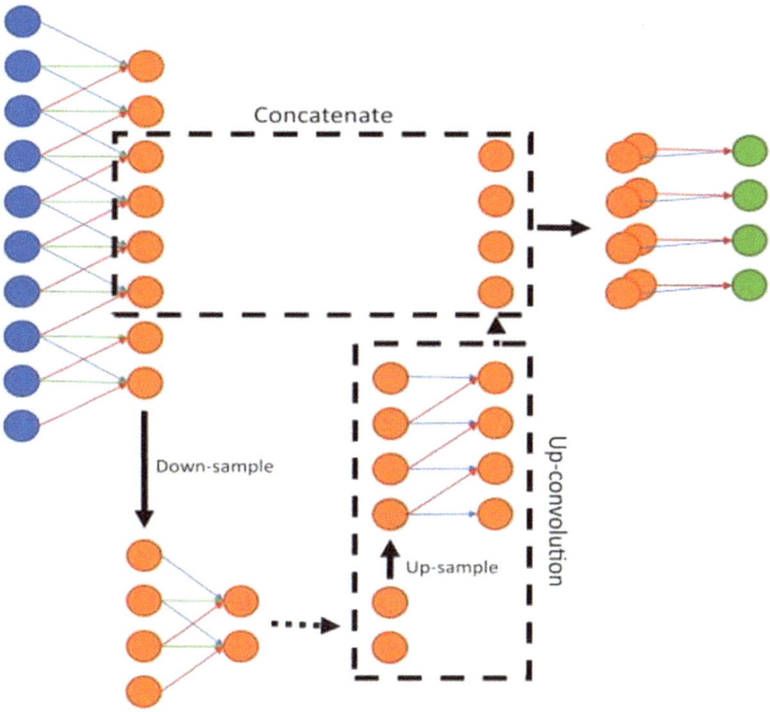

Fig. 8.11 U-Net [32]

Although these two networks have provided excellent results. In the case of CNN, "sliding window" approach is used which produces overlaps of pixels. It does not affect the result but increases the computation time. So recent papers now use a full convolutional neural network (fCNN) as it can input the entire image or volume which reduces redundant computation. fCNN has been also applied on 3D images. Most of the architectures are based on CNN or fCNN only.

8.4 Conclusion

Computer-aided medical image segmentation is a crucial step in finding application in clinical studies and treatment planning by the medical practitioners. In current years an extensive diversity of methods have been developed to segment medical images having their particular advantages and restrictions. This chapter presents several soft and hard computing methods and their exploitations in the field of

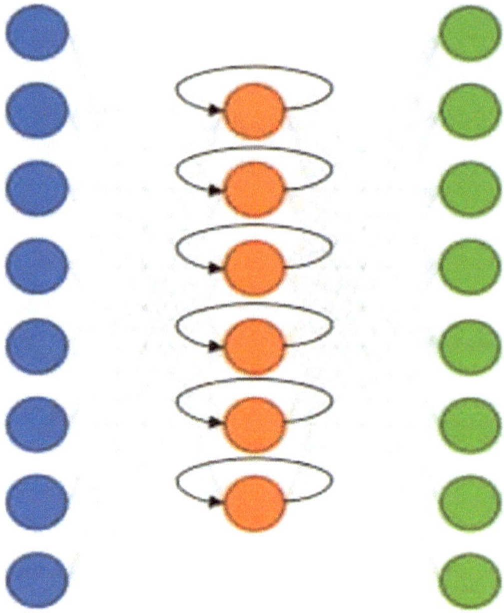

Fig. 8.12 RNN [32]

medical image segmentation. The interest of researchers to use soft computing methodologies especially artificial neural network and deep neural network for medical image segmentation has been increased significantly nowadays as of their efficient results. However, most of these neural network-based approaches and methods necessitate widespread direction and training and their performance be contingent upon the training process and data used for training. Soft computing with its competence for conducts real-life abstruse conditions with flexibility serves the need for construction a system intelligent which could think like human. The methods for image segmentation deliberated in this chapter can be graded on the basis of suitability, performance, and applicability. Hard computing-based segmentation methods based on gray-level techniques, for example, thresholding, and region-based methods are the simplest techniques. However, they are having limited applications. Morphological, edge-based, Markov random field methods give their best in detecting the boundaries of the organs present in the medical images. Genetic, fuzzy, and neural networks as the fundamental approaches of the soft computing techniques are used for medical image segmentation and implemented in a number of applications of medical image segmentation and analysis. Finally, it is concluded that deep learning approaches due to their reliability, accuracy, and robustness perform very well in medical image segmentation and are the main choice of the scientists and researchers presently.

References

1. Reyes Aldasoro C, Bhalerao A (2007) Volumetric texture segmentation by discriminant feature selection and multiresolution classification. IEEE Trans Med Imaging 26(1):1–14. https://doi.org/10.1109/tmi.2006.884637
2. Sharma N, Ray A, Shukla K, Sharma S, Pradhan S, Srivastva A, Aggarwal L (2010) Automated medical image segmentation techniques. J Med Phys 35(1):3. https://doi.org/10.4103/0971-6203.58777
3. Mesejo P, Ibáñez Ó, Cordón Ó, Cagnoni S (2016) A survey on image segmentation using metaheuristic-based deformable models: state of the art and critical analysis. Appl Soft Comput 44:1–29. https://doi.org/10.1016/j.asoc.2016.03.004
4. Choy S, Lam S, Yu K, Lee W, Leung K (2017) Fuzzy model-based clustering and its application in image segmentation. Pattern Recogn 68:141–157. https://doi.org/10.1016/j.patcog.2017.03.009
5. Li Y, Shen Y (2009) An automatic fuzzy c-means algorithm for image segmentation. Soft Comput 14(2):123–128. https://doi.org/10.1007/s00500-009-0442-0
6. Jiao L, Gong M, Wang S, Hou B, Zheng Z, Wu Q (2010) Natural and remote sensing image segmentation using memetic computing. IEEE Comput Intell Mag 5(2):78–91. https://doi.org/10.1109/mci.2010.936307
7. Angel Arul Jothi J, Mary Anita Rajam V (2016) A survey on automated cancer diagnosis from histopathology images. Artif Intell Rev 48(1):31–81. https://doi.org/10.1007/s10462-016-9494-6
8. Saritha S, Amutha Prabha N (2016) A comprehensive review: segmentation of MRI images-brain tumor. Int J Imaging Syst Technol 26(4):295–304. https://doi.org/10.1002/ima.22201
9. Zhao Q, Li X, Li Y, Zhao X (2017) A fuzzy clustering image segmentation algorithm based on Hidden Markov Random Field models and Voronoi Tessellation. Pattern Recogn Lett 85:49–55. https://doi.org/10.1016/j.patrec.2016.11.019
10. Aghajari E, Chandrashekhar G (2017) Self-organizing map based extended Fuzzy C-means (SEEFC) algorithm for image segmentation. Appl Soft Comput 54:347–363. https://doi.org/10.1016/j.asoc.2017.01.003
11. Borges V, Guliato D, Barcelos C, Batista M (2014) An iterative fuzzy region competition algorithm for multiphase image segmentation. Soft Comput *19*(2):339–351. https://doi.org/10.1007/s00500-014-1256-2
12. Sonka M, Hlavac V, Boyle R (1999) Image processing, analysis and machine vision. Thomson Learning, Singapore
13. Bhaumik H, Bhattacharyya S, Nath M, Chakraborty S (2016) Hybrid soft computing approaches to content based video retrieval: a brief review. Appl Soft Comput 46:1008–1029. https://doi.org/10.1016/j.asoc.2016.03.022
14. Jiang X, Wang Q, He B, Chen S, Li B (2016) Robust level set image segmentation algorithm using local correntropy-based fuzzy c-means clustering with spatial constraints. Neurocomputing 207:22–35. https://doi.org/10.1016/j.neucom.2016.03.046
15. Ibrahim D (2016) An overview of soft computing. In: 12th international conference on application of fuzzy systems and soft computing, ICAFS 2016, Vienna, Austria. Proc Comput Sci 102:34–38. , 29–30. https://doi.org/10.1016/j.procs.2016.09.366
16. Lee J, Jun S, Cho Y, Lee H, Kim G, Seo J, Kim N (2017) Deep learning in medical imaging: general overview. Korean J Radiol 18(4):570. https://doi.org/10.3348/kjr.2017.18.4.570
17. Wong K (n.d.) Medical image segmentation: methods and applications in functional imaging. Topics in biomedical engineering international book series handbook of biomedical image analysis, pp 111–182. https://doi.org/10.1007/0-306-48606-7_3

18. Saad NM, Abu-Bakar SA, Muda S, Mokji M (2011) Segmentation of brain lesions in diffusion-weighted MRI using thresholding technique. In: 2011 IEEE International Conference on Signal and Image Processing Applications (ICSIPA). https://doi.org/10.1109/icsipa.2011.6144092
19. Aslam A, Khan E, Beg MS (2015) Improved edge detection algorithm for brain tumor segmentation. Proc Comput Sci 58:430–437. https://doi.org/10.1016/j.procs.2015.08.057
20. Mathur N, Mathur S, Mathur D (2016) A novel approach to improve sobel edge detector. Proc Comput Sci 93:431–438. https://doi.org/10.1016/j.procs.2016.07.230
21. Lin G, Wang W, Kang C, Wang C (2012) Multispectral MR images segmentation based on fuzzy knowledge and modified seeded region growing. Magn Reson Imaging 30(2):230–246. https://doi.org/10.1016/j.mri.2011.09.008
22. Viji KS, Jayakumari J (2013) Modified texture based region growing segmentation of MR brain images. In: 2013 IEEE conference on information and communication technologies. https://doi.org/10.1109/cict.2013.6558183
23. Pandav S (2014) Brain tumor extraction using marker controlled watershed segmentation. Int J Eng Res Technol. ISSN:2278-0181
24. Sudharani K, Sarma T, Prasad KS (2016) Advanced morphological technique for automatic brain tumor detection and evaluation of statistical parameters. Proc Technol 24:1374–1387. https://doi.org/10.1016/j.protcy.2016.05.153
25. Trevino A (n.d.) Introduction to K-means Clustering. Retrieved from: https://www.datascience.com/blog/k-means-clustering
26. Subbanna N, Precup D, Arbel T (2014) Iterative multilevel MRF leveraging context and voxel information for brain tumour segmentation in MRI. In: 2014 IEEE conference on computer vision and pattern recognition. https://doi.org/10.1109/cvpr.2014.58
27. Vijay V, Kavitha A, Rebecca SR (2016) Automated brain tumor segmentation and detection in MRI using Enhanced Darwinian Particle Swarm Optimization(EDPSO). Proc Comput Sci 92:475–480. https://doi.org/10.1016/j.procs.2016.07.370
28. Pezoulas VC, Zervakis M, Pologiorgi I, Seferlis S, Tsalikis GM, Zarifis G, Giakos GC (2017) A tissue classification approach for brain tumor segmentation using MRI. In: 2017 IEEE international conference on Imaging Systems and Techniques (IST). https://doi.org/10.1109/ist.2017.8261542
29. Chandra GR, Rao KR (2016) Tumor detection in brain using genetic algorithm. Proc Comput Sci 79:449–457. https://doi.org/10.1016/j.procs.2016.03.058
30. Chacón M MI (n.d.) Fuzzy logic for image processing: definition and applications of a fuzzy image processing scheme. In: Advances in industrial control advanced fuzzy logic technologies in industrial applications, pp 101–113. https://doi.org/10.1007/978-1-84628-469-4_7
31. Nimeesha KM, Gowda RM (2013) Brain tumour segmentation using Kmeans and fuzzy c-means clustering algorithm. Int J Comput Sci Inf Technol Res Excell 3:60–65
32. Litjens G, Kooi T, Bejnordi BE, Setio AA, Ciompi F, Ghafoorian M, Sánchez CI (2017) A survey on deep learning in medical image analysis. Med Image Anal 42:60–88. https://doi.org/10.1016/j.media.2017.07.005
33. Visin F, Romero A, Cho K, Matteucci M, Ciccone M, Kastner K, Courville A (2016) ReSeg: a recurrent neural network-based model for semantic segmentation. In: 2016 IEEE conference on Computer Vision and Pattern Recognition Workshops (CVPRW). https://doi.org/10.1109/cvprw.2016.60
34. Chen H, Dou Q, Yu L, Qin J, Heng P (2018) VoxResNet: deep voxelwise residual networks for brain segmentation from 3D MR images. NeuroImage 170:446–455. https://doi.org/10.1016/j.neuroimage.2017.04.041
35. Kooi T, Litjens G, Ginneken BV, Gubern-Mérida A, Sánchez CI, Mann R, Karssemeijer N (2017) Large scale deep learning for computer aided detection of mammographic lesions. Med Image Anal 35:303–312. https://doi.org/10.1016/j.media.2016.07.007
36. Milletari F, Ahmadi S, Kroll C, Plate A, Rozanski V, Maiostre J, Navab N (2017) Hough-CNN: deep learning for segmentation of deep brain regions in MRI and ultrasound. Comput Vis Image Underst 164:92–102. https://doi.org/10.1016/j.cviu.2017.04.002

EEG Signal Processing and Its Classification for Rehabilitation Device Control

9

Angana Saikia and Sudip Paul

Abstract

At this present technology-based world, electroencephalography (EEG) instruments have become a tool for various research and diagnoses of different human health disorders. The brain-computer interface (BCI) is an area which is a very much emerging technology that uses the human brain signals to control external devices. For excellent and accurate results, BCI has recognized the need for systems that makes it more user-friendly, real time, manageable, and suited for people like clinical and disabled patients. Thus, this chapter will refer to the processing of the EEG signal and different classification techniques which will further be used to control the rehabilitation devices through BCI system.

A medical diagnostic technique that reads the electrical activity of the scalp which is generated by a human brain is known as electroencephalography, and the recording is called electroencephalogram (EEG). The electrical activity from the scalp of the brain is mainly picked up using metal electrodes having a conductive media. An EEG recording system is a combination of a couple of instruments. They are electrodes consisting of a conductive media, amplifiers and filters, analog-to-digital converters, and recording device/printer. Feature extraction and classification of electroencephalograph signals for human subjects is a challenge for both the engineers and scientists. Mainly fast Fourier transform (FFT), Lyapunov exponent, correlation dimension, and wavelet transformation are the tools for EEG signal processing. There are also various signal processing techniques for classification of nonlinear and nonstationary signals like EEG. Some of the signal processing techniques are support vector machine (SVM) and multilayer perceptron (MLP)-based classifier, back-propagation neural network, self-organizing feature maps followed by an autoregressive modeling and artificial neural network, etc. The classification rate calculated using the various

A. Saikia · S. Paul (✉)
Department of Biomedical Engineering, School of Technology, North-Eastern Hill University, Shillong, Meghalaya, India

S. Paul (ed.), *Application of Biomedical Engineering in Neuroscience*,
https://doi.org/10.1007/978-981-13-7142-4_9

classification techniques can further be used to control the rehabilitation devices like artificial limbs (hand and leg). The advances in brain-computer interface (BCI) research and its applications have given a significant impact to biomedical research. This would be a boon for the person with disability so that they can interact and go through their day-to-day work smoothly with the help of the rehabilitation devices.

Keywords
Electroencephalograph (EEG) · Rehabilitation · Signal processing · Signal classification

9.1 Introduction to EEG Signals

9.1.1 Human Brain

The brain is known as the center for processing information in our body. It is the control and command system of the human body. It is the site for processing of various activities of the human system like vision, hearing, speech, memory, intelligence, emotions, and thoughts. It also controls:

(a) Voluntary movements.
(b) Balance of the body.
(c) Functioning of involuntary organs (e.g., lungs, heart, kidneys, etc.)
(d) Thermoregulation.
(e) Hunger and thirst.
(f) Circadian rhythms of our body.
(g) Activities of several endocrine glands and human behavior.

The skull mainly protects the human brain. The different layers of the brain are as follows:

(a) An outer layer called dura mater.
(b) A middle layer called arachnoid.
(c) An inner layer called pia mater.

Figure 9.1 shows the major parts of the human brain.

The brain is the center of all activity in the human body. Various activities and states of the human brain are monitored by recording the electrical signals from the scalp of the brain. The brain signals resulting out of ionic flow variations within the neurons provide partial information about the physiological state of the human body and can be used for clinical and diagnostic purposes by physicians and researchers. The endocrine system is a nervous and hormonal system which controls and coordinated the human body. The sense organ in our body sends the information in the form of electrical impulses to the spinal cord and brain through the sensory

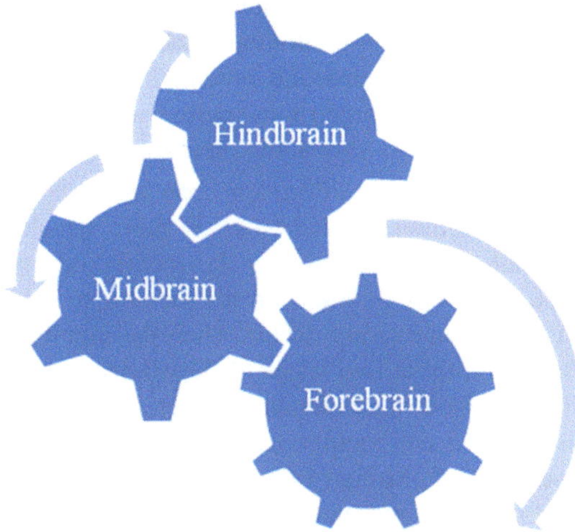

Fig. 9.1 Major parts of the human brain

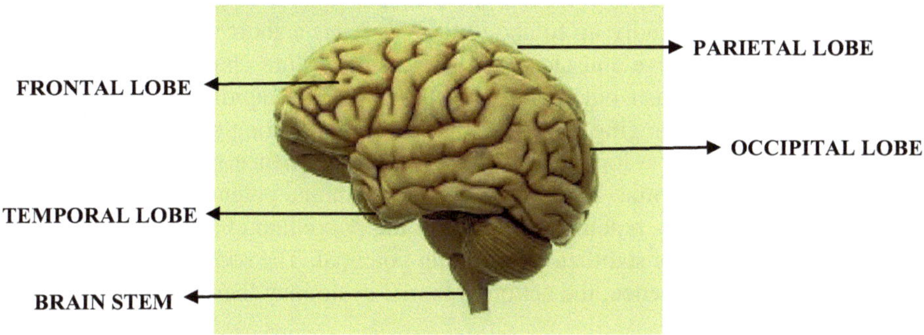

Fig. 9.2 Lobes of human brain

nerves. The other types of nerves called motor nerves then transmit responses from brain and spinal cord effectors [27].

The different lobes present in the human brain are shown in Fig. 9.2.

Some of the activities which take place in the brain are as follows:

(a) *Event-related potential (ERP):* The electrical recording of a sensory and cognitive event of the brain is known as event-related potential which is mainly a stereotyped electrophysiological response to a stimulus. The noninvasive means of evaluating brain function is provided by ERP [16].
(b) *Action potential:* The firing of the neurons leads to an action potential. During the action potential, the movement of positively and negatively charged ions takes place in the cell membrane. The membrane opens, and as result the positively charged ions move inside the cell, whereas the negatively charged ions rush out.

(c) *Resting membrane potential:* Most of the neurons have a negative concentration gradient. This continuous state of a negative concentration of the neurons is known as membrane potential. During this period, the concentration of the positively charged ions increases outside the cell than inside.

Figure 9.3 shows the membrane potential versus time graph.

At a resting condition, a neuron has a resting potential of −60 to −70 mV. As a result, the interior of the cell is more negatively charged than the exterior. The condition of the cell when the membrane potential becomes more negative at a particular point on the neuron's membrane is known as hyperpolarization. The state called depolarization occurs when the membrane potential becomes more positive. Basically, these two states of depolarization and hyperpolarization occur when ion channels in the cell membrane open and close, resulting in the entry and exit of ions through the cell membrane. The binding of neurotransmitters is responsible for the opening and closing of the channels. An action potential occurs when a depolarization increases the membrane voltage so that it crosses the threshold voltage about −55 mV. At this threshold voltage, sodium channels in the cell membrane open up allowing the sodium ions to enter into the cell. This inflow of sodium increases the membrane potential rapidly up to about 40 mV. After a short while, these sodium channels become inactive and close. As a result potassium channels open up. The potassium ions then rush out of the cell. These events rapidly decrease the cell membrane potential, bringing it back toward its normal resting state. The potassium channels stay open for a longer time than needed and hence bring the membrane back to its resting potential. This results in the membrane potential to become more negative than its resting potential. Eventually the potassium channels close and the cell membrane potential stabilizes at a resting potential. The sodium channels return to their normal state. Hence, the action potential cycle may then start again [1].

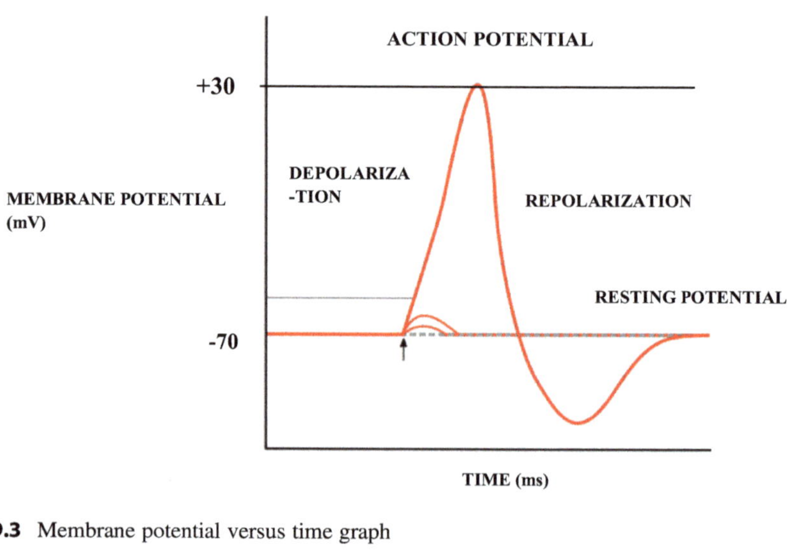

Fig. 9.3 Membrane potential versus time graph

9.1.2 Electroencephalogram (EEG)

Hans Berger in 1929 was the first person to measure the electroencephalogram measured in humans. Electrical impulses generated by nerve firings in the brain can be measured by different types of electrodes placed on the scalp of the human brain. The EEG gives an insight of the neuronal activity and is mostly used for noninvasive studies of the human brain physiology. EEG activity is a small signal, measured in microvolts with the main frequencies of interest up to approximately 30 Hertz (Hz) [32] [3]. Electroencephalography (EEG) is used to diagnose different neurological disorders [2]. A clinician can gain as much information as he can by understanding and investigating this noninvasive technique of EEG signal. EEG techniques used in medical research are not standardized enough for clinical purposes. On the other hand, researches carried out in mental disabilities, like auditory processing disorder (APD), are becoming more dominant and widely known.

Various applications of EEG are as follows:

(a) Control of rehabilitation devices.
(b) Early detection of various neurological disorders.
(c) Therapeutic and recreational use.
(d) EEG-based sensors are nowadays used in medical assessment techniques.
(e) Neuronal behavior of the human brain.

The frequency bands in EEG are shown in Table 9.1 [29].

9.1.3 EEG Artifacts

The cerebral activity of the brain is recorded using EEG. The recorded electrical activity that is not originated from the cerebral is known as EEG artifact. It is mainly divided into physiologic and nonphysiologic artifacts. An artifact is considered as a disturbance in a measured brain signal [7]. They are considered unwanted signals or interference in a signal. Physiologic artifacts are generated from regions other than the brain (i.e., body), whereas nonphysiologic artifacts arise from outside the body. Artifacts can affect many of the EEG features like mean, median, distribution, standard deviation, and signal-to-noise ratio. Some of the ways to minimize the artifacts are proper designing of the EEG test, proper instruction before doing the test, response EMG or force grips, eye tracker, electrode localization equipment, comfortable environment, and good response device.

There are various methods for the removal of EEG artifacts. These are mostly used for clinical diagnosis, research, and brain-computer interface (BCI) applications. Some of them are independent component analysis (ICA) and discrete wavelet transforms (DWT). Independent component analysis such as second-order blind identification (SOBI) is the accurate method to correct the measured EEG.

Table 9.1 EEG frequency band

Type	Frequency (Hz)	Location	Brain state
Delta	Up to 4	Frontal part in adults, posterior half of the children's brain, high amplitude waves	Slow wave sleep in adults and during any kind of continuous attention task in babies
Theta	4–8	Locations other than the human hand task	Drowsiness or arousal conditions in older children and adults
Alpha	8–13	Both sides of the posterior part of the human head	During relaxed states, such as closing of the eyes. Mainly associated with inhibitory controls
Beta	>13–30	Symmetrical distribution on both sides, mostly frontal region and are low amplitude waves	During alert or working states, when the person is active or busy conditions
Gamma	30–100+	Somatosensory cortex	During any sensory processing, matching of recognized objects, sounds, or tactile sensation can also be seen during short-term memory
Mu	8–13	Sensorimotor cortex	Shows resting state of the motor neuron

Table 9.2 EEG artifact

Physiological	Nonphysiological
EMG	60 Hz interference
Eye movement	EEG electrodes
ECG	Environmental sources
Respiration	

Other methods such as extended information maximization (InfoMax) and an adaptive mixture of independent component analyzers (AMICA) can also be used.

Different types of artifacts are shown in Table 9.2.

9.2 Recording Methods

Encephalography is measured using a recording system. Fig. 9.4 shows a schematic diagram of an EEG recording system. In an EEG recording system, the electrodes read the signal from the surface of the head. The proper function of these electrodes is very critical for the acquisition of accurate and high-quality signal [6].

Various types of electrodes used are:

(a) Disposable electrodes.
(b) Reusable disc electrodes.
(c) Headbands and electrode caps.
(d) Saline-based electrodes.
(e) Needle electrodes.

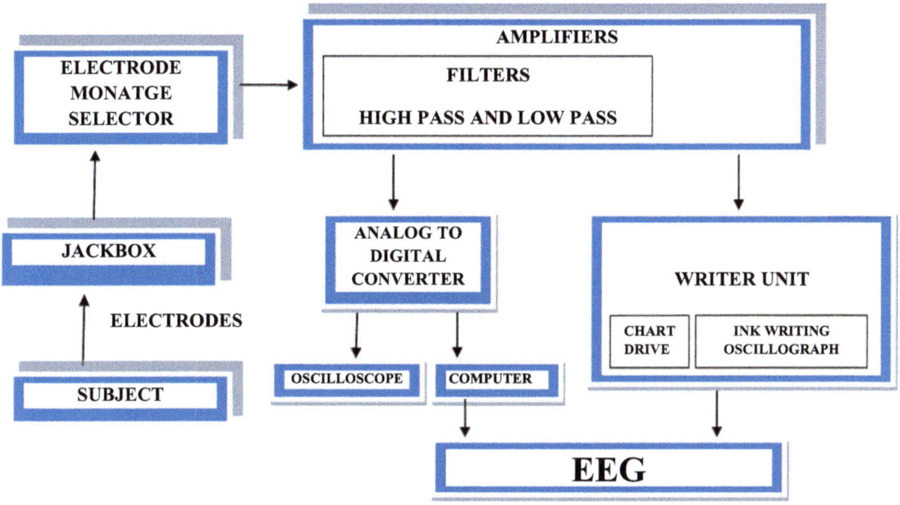

Fig. 9.4 Schematic diagram of an EEG machine

The amplifiers amplify the recorded signal, i.e., it converts the microvolt signals into higher amplitude. The converter changes the analog signal to digital form, and finally the computer stores and displays the required data.

The EEG measures the potential differences between signal (active) electrode and reference electrode over time. The ground electrode is the extra electrode which is used for getting differential voltage. A single-channel EEG system consists of one active electrode, one reference, and one ground electrode, whereas multichannel configurations can comprise up to 128 or 256 active electrodes.

9.2.1 10–20 Electrode Placement System

The 10–20 system or International 10–20 system is an EEG recording method which is internationally accepted. It mainly describes the location of scalp electrodes on the surface of the brain in the context of an EEG experiment. The "10" and "20" refer to the fact that the actual distances between adjacent electrodes are either 10% or 20% of the total front-back or right-left distance of the skull. Each site has a letter to identify the lobe and a number to identify the hemisphere location. The letters F, T, C, P, and O stand for frontal, temporal, central, parietal, and occipital lobes, respectively. Even numbers (2, 4, 6, 8) refer to electrode positions on the right hemisphere, whereas odd numbers (1, 3, 5, 7) refer to those on the left hemisphere. A "z" (zero) refers to an electrode placed on the midline. In addition to these combinations, the letter codes A, Pg, and Fp identify the earlobes, nasopharynx, and frontal polar sites, respectively. Figure 9.5 shows the placement of the electrodes in the skull.

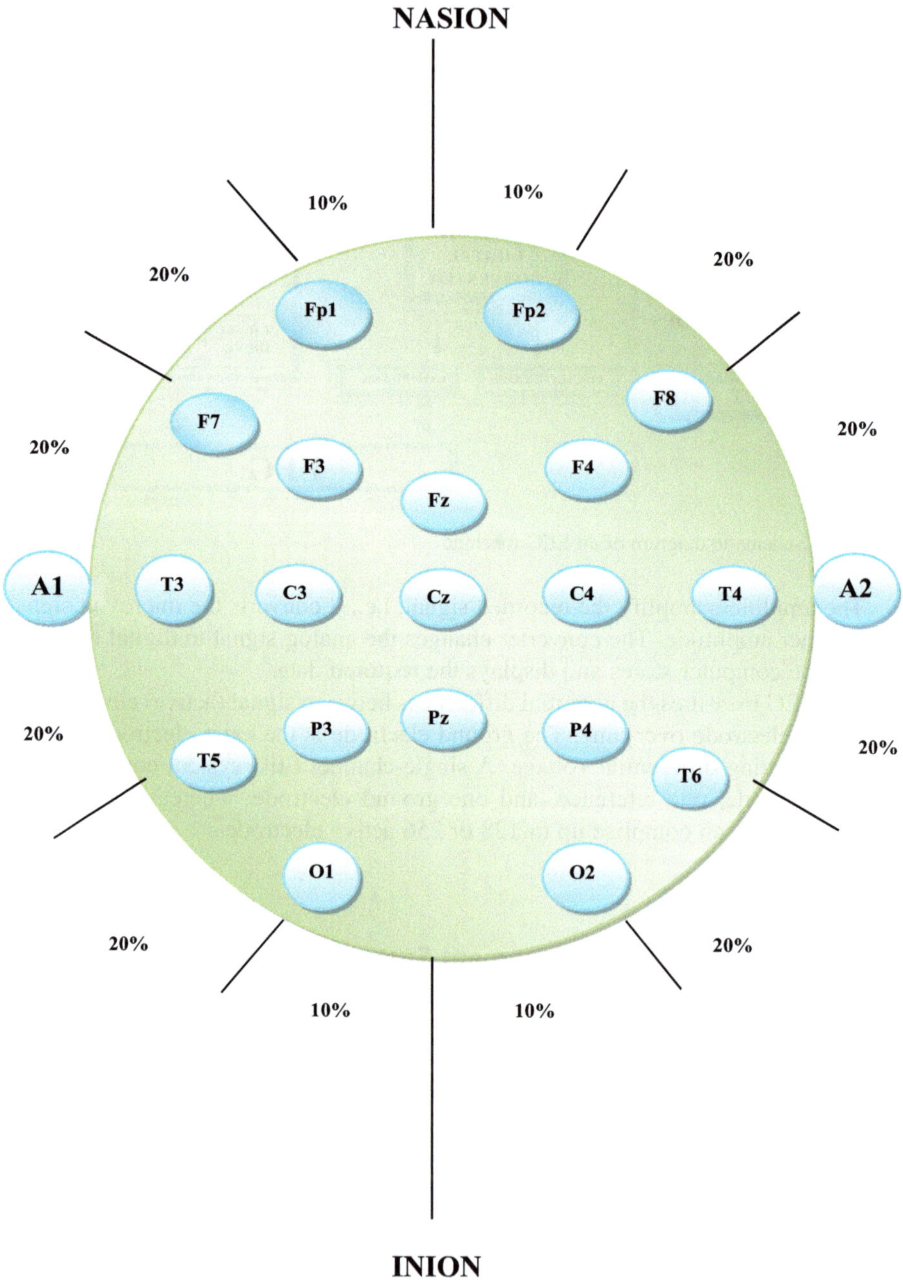

Fig. 9.5 10–20 electrode placement system

9.3 EEG Signal Processing

Preprocessing techniques remove the artifacts from the EEG signal and improve the performance of the system by separating the noise from the raw signal [34] [13]. Followed by a feature extraction block, it extracts the most relevant features from the signal. These features are used in the decision-making mechanism through giving the desired output. A simple block diagram on digitized EEG is shown in Fig. 9.6.

There are various preprocessing techniques. Some of them are as follows:

(a) *Wiener filters or adaptive filters:* It is used for removal of EMG artifacts. Adaptive filtering can be used for removal of EOG artifacts and any background noise in general [23].The most fundamental approach for noise cancelation in the EEG signal is Wiener filter. The Wiener filter is applied to a noise and noisy signals. The noisy signal is mostly a combination of the desired data and noise. A Wiener filter gives the output as the estimation of the noise signal which is further subtracted from the noisy data to yield the error signal.

(b) *Independent component analysis (ICA)*: It is used for the removal of power line noise as well as EOG, EMG, and ECG artifacts [23]. The electrical potentials of the brain are the mixtures of some underlying components of brain activity which can be analyzed and observed using ICA. Interesting information on brain activity are extracted through ICAs.

(c) *Multichannel Wiener filter:* It has been proposed for enhancement of EEG signal. It is a more advanced method than the blind source separation method [20]. The features for brain-computer interfaces are the ERPs of EEG. However, ERPs are hard to observe as EEG signals are easily affected by various artifacts.

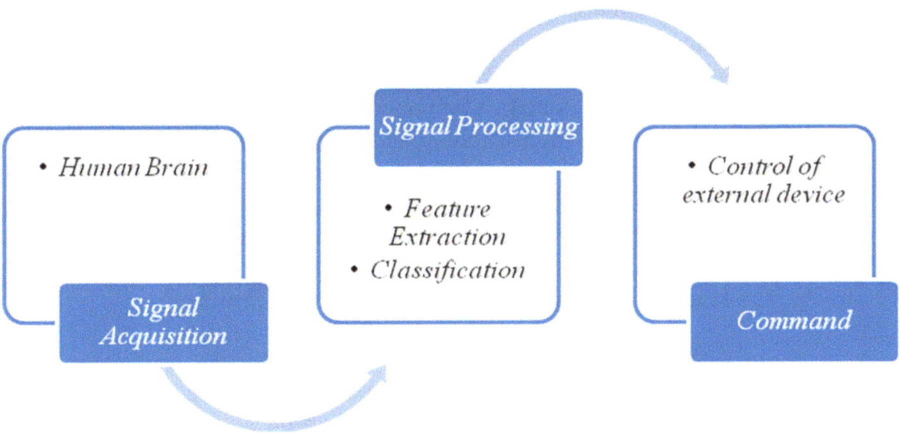

Fig. 9.6 Block diagram of a digitized EEG signal

Multichannel Wiener filter mostly enhances EEG signals and makes ERPs more prominently and accurately diagnosed.

(d) *Joint approximate diagonalization of Eigen-matrices (JADE) method:* It is used to calculate the independent components and hence remove unwanted artifacts from the EEG signal [9]. The basic of JADE is on diagonalization of cumulate matrices.

The matrix W diagonalizes F(M), i.e., WF(M) WT is diagonal. Matrix F is a linear combination of terms W_i W_i^T, where W is a column of W.WF (Mi) W^T which is made as diagonal as possible for different combination of Mi and $i = 1 \ldots k$.

The diagonality of matrix.

$$Q = \mathrm{WF}\,(\mathrm{Mi})\,\mathrm{WT} \tag{9.1}$$

(e) Another method uses an artifact removal technique. It has been proposed that it incorporates *lifting wavelet transform*, rather than wavelet transform, with ICA technique for effective removal of artifacts from EEG signal. This method provides a better and efficient way to eliminate artifacts than traditional ICA method [30].

(f) In some methods the EEG signals were first band-pass filtered to get the desired band of frequency, followed by ICA technique to remove any unwanted artifacts from the signal [35].

(g) *Least mean square algorithm (LMS):* Least mean square algorithm is an adaptive process, where the adaptation is directed by the error signal between the primary signal and the filter output [33]. This algorithm is based on minimum mean square error and hence mostly works with nonstationary signal processing. When there is interference in the output of the analyzed system, this algorithm obtains the optimal solutions of the signal processing problems.

(h) *Discrete Fourier transform (DFT):* This algorithm converts the time domain signal to the frequency domain component which establishes a correlation between the time domain representation and the frequency domain representation. Hence, the frequency analysis of a time domain signal like EEG is done with the DFT algorithm [33]. Discrete Fourier transform can be a continuous Fourier transform for signals known only at N instants with a sample times T. If *f (t)* is the continuous signal, let *N* samples be denoted *f [0], f [1], f [2]. . . f [k]. . . f [N-1]*. The Fourier transform of the original signal *f(t)* will be.

$$F(jw) = \int_{-\infty}^{\infty} f(t) e^{-jwt} dt \tag{9.2}$$

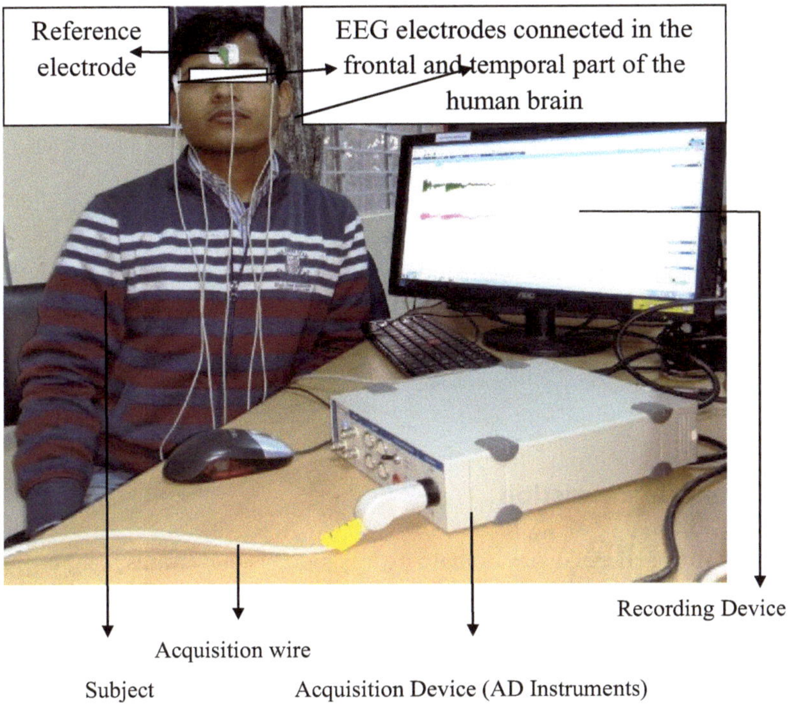

Fig. 9.7 EEG signal acquisition

(i) *Blind source separation (BSS):* Neural signals from EEG data are extracted using this algorithm [18]. BSS mostly extracts a series of codebook vectors. These codebook vectors represent the spectral content underlying the recorded signal. As a result, it becomes possible to extract the particular rhythmic activity underlying the signal recordings.

(j) *Haar wavelet:* This technique is used for the detection of ocular artifacts. The application of Haar wavelet to raw EEG signal helps in the development of a reliable system for the early detection in clinical EEG recording. The de-noising of ocular artifacts from the EEG signals based on the available filtering techniques results in loss of data. Hence, Haar wavelet is used to overcome this problem of loss of data [15].

Figure 9.7 shows the EEG acquisition done using AD instruments in the signal processing laboratory.

Table 9.3 Placement of EEG electrode

Electrode	Part in the brain	Placement	
Channel 1	Frontal	Positive electrode in the right hemisphere of the brain	Negative electrode in the left hemisphere of the brain
Channel 2	Temporal		
Reference	Forehead		

Figure 9.7 shows the EEG acquisition using dual-channel acquisition device from AD instrument. Surface electrodes are used to place the acquisition wire in the different parts of the human brain. The placements of the electrode are shown in Table 9.3.

9.3.1 Feature Extraction

Analysis of time and frequency domain mainly extracts the various features from the signal. These features are defined as parameters that provide information about the characteristics of the particular signal [23].

Some of the time domain features which can be extracted from an EEG signal are as follows:

(a) Root mean square.
(b) Variance.
(c) Waveform length.
(d) Mean.
(e) Standard deviation.

There are also some frequency domain features which can be extracted. They are:

(a) Modified mean frequency.
(b) Modified median frequency.
(c) Median frequency.
(d) Mean frequency.
(e) Power spectrum density.

Time domain features are very quick and easy to implement. They do not require any transformation, and hence the computational complexity is lower as compared to frequency domain features. Frequency domain features are statistical properties of power spectral density of the signals [1].

The time domain features for a signal of length N and let X_n be the nth sample value of the signal are elaborated as follows:

(a) Root mean square (*RMS*): RMS is characterized by the force and non-fatiguing contraction of the signal and defined as.

$$RMS = \left[\frac{1}{N}\sum\nolimits_{n=1}^{N}(X_n)\right]^{1/2} \tag{9.3}$$

(b) Variance (*Var*): It signifies the power of the signal and is the mean value of the square of the deviation of the signal. It is quantified as.

$$Var = \left[\frac{1}{N-1}\sum\nolimits_{n=1}^{N}(X_n)\right]^{1/2} \tag{9.4}$$

(c) Waveform length (*WL*): The cumulative length of the waveform over the time period is given by WL. It is mostly related to the waveform amplitude, frequency, and time. It is given by.

$$WL = \sum\nolimits_{n=1}^{N-1}(X_{n-1} - X_n) \tag{9.5}$$

(d) Standard deviation (*SD*): It is used to find the threshold levels of various human activities through the EEG signals. The SD is expressed as.

$$SD = \left[\frac{1}{N-1}\sum\nolimits_{n=1}^{N}(X_n - \mu)^2\right]^{1/2} \tag{9.6}$$

(e) Mean (*M*): Mean identifies the central location of the data, also known as the average. The mean is calculated as follows:

$$M = \frac{\sum\nolimits_{n=1}^{N}(X_n)}{N} \tag{9.7}$$

The frequency domain features are as follows:

(a) Modified median frequency (*MMDF*): Modified median frequency divides the spectrum into two regions having equal amplitude.

It can be expressed as.

$$MMDF = \left[\frac{1}{2}\sum\nolimits_{j=1}^{M}(A_j)\right] \tag{9.8}$$

where A_j is the EMG amplitude spectrum at frequency bin j.

(b) Modified mean frequency (*MMNF*): MMNF is the sum of the product of the amplitude spectrum and the frequency which is divided by the total sum of spectrum intensity:

$$MMNF = \frac{\sum\nolimits_{j=1}^{M} f_j A_j}{\sum\nolimits_{j=1}^{M} A_j} \tag{9.9}$$

where A_j and f_j are the EMG amplitude spectrum and frequency spectrum at frequency
bin j, respectively.

(c) Mean frequency (*MNF*) and median frequency (*MDF*): Mean frequency (MNF) and median frequency (MDF) are calculated based on the power spectrum of the signal. It is expressed as.

$$MNF = \frac{\sum\nolimits_{j=1}^{M} f_j P_j}{\sum\nolimits_{j=1}^{M} P_j} \tag{9.10}$$

$$MDF = \frac{1}{2}\sum\nolimits_{j=1}^{M}(P_j) \tag{9.11}$$

where P_j and f_j are the EMG power spectrum and frequency spectrum at frequency bin j, respectively.

(d) Power spectrum density (*PSD*): PSD is the Fourier transform of the signal's autocorrelation. This gives the power estimation that each frequency contributes to the signal and is expressed as.

$$PSD = \int_{-\infty}^{\infty} R_x(\tau)e^{-j2\pi ft}\,dt \tag{9.12}$$

where $R_x(\tau)$ is the autocorrelation function for the signal $x\,(t)$ with a time lag τ and is defined as

$$Rx(\tau) = \int_{-\infty}^{\infty} x(t)x * (t - \tau)dt \tag{9.13}$$

Feature extraction is an important step in most classification tasks. Some of the general guidelines to describe about feature extraction are as follows:

(a) Discrimination.
(b) Reliability.
(c) Independence.
(d) Optimality.

EEG feature extraction is used in any kind of BCI-based applications. It helps to extract the most relevant features from the EEG signal, thus giving a more precise description and hence making it suitable for further processing [4]. It is an important aspect in signal processing. Among all the approaches, the principal component analysis is the most widely used. The result obtained from feature extraction is used for signal classification. So, to achieve an optimal classification of signal, feature extraction is very much necessary [4].

There are several feature extraction techniques [34].

(a) *Hilbert-Huang transform:* Hilbert Huang transform decomposes a signal into intrinsic mode functions. It mainly obtains instant frequency data from nonstationary and nonlinear signal. Hilbert-Huang transform analyzes the instantaneous frequency of EEG signals and gives a physical meaning of the signal [25].
(b) *Principal component analysis (PCA):* PCA is the method of analysis of biosignals. It mainly detects the linear combination of a set of variables which has maximum variance and finally removes its effect. It is a dominant tool for analyzing the EEG data and finding patterns in it. Different task can be obtained from EEG signal of a subject by using PCA with neural network [19].
(c) *Local discriminant bases (LDB):* Local discriminant bases (LDB) are based on fully decimated wavelet packet decomposition (WPD). It is only suitable for stationary data. The discriminant features are not consistent in the application of shifted version of the same signal which is due the decimated WPD. They mostly estimate each coefficient individually. The LDB is mostly dependent on the best-basis paradigm. It utilizes a tree adjustment which selects a basis from the dictionary that highlighted the discrepancy among classes [12].
(d) *Autoregressive model (AR):* AR model is the representation of a random process which works on certain time-varying processes in nature. Here, the output variable depends linearly on its own previous values. AR modeling is an efficient algorithm for parameter estimation. Autoregressive models are used for spectral parameter analysis (SPA) of the EEG [24].
(e) *Discrete wavelet transform (DWT):* DWT is an implementation of the wavelet transform using a discrete set of the wavelet scales. It mostly obeys some pre-defined rules during the translation. Consequently, the set of wavelets

forms an orthonormal basis which can be used to decompose the signal. DWT coefficient contains temporal information of the analyzed signal [24].

(f) *Wavelet packet transform:* In wavelet packet decomposition, the discrete-time signal is passed through various filters. Originally it is known as optimal sub-band tree structure. This transform gives an optimal representation of the EEG signal, whose structure provides the flexibility for features. To investigate the time-varying characteristics of the multichannel EEG signals, wavelet packet transform is a useful tool. It allows a sub-band analysis without the constraint of a dyadic decomposition [14].

(g) *Sample entropy:* Sample entropy is a modified feature of approximate entropy which is used for detecting the complexity of physiological time-series signals and hence diagnosing diseased states. It is data length independence and a relatively trouble-free implementation. Sample entropy is more adaptive to the real-time detection and has better ability in accessing the level of consciousness of patients during surgery. The only drawback of sample entropy is the calculated entropy value which is ranged from 0 to 3 variously [5].

9.4 Signal Classification Techniques Applied to EEG Signals

Within the recent years, various EEG classification studies have taken place. Most of these used different classification techniques and compared their performance in terms of classification accuracy. Among these classifiers, k-nearest neighbor (k-NN), linear discriminant analysis (LDA), support vector machine (SVM), and artificial neural network (ANN) are the most popular and extensively used for research purposes [22].

(a) *Support vector machine (SVM):* The method for object classification technique, which relies on the statistical learning theory, is known as the support vector machine. SVM classifiers classify data by finding the best hyperplane that separates all data points of one class from those of the other which applies the concept of decision planes. Decision plane separates a set of features having various class memberships that define decision boundaries. SVM classifier finds the hyperplane which maximizes the separating margin between classes [8].

(b) *K-nearest neighbor (k-NN):* k-NN is a widely used classifier, based on learning. This method is implemented by considering similarity of testing and training data. In this classification process, a sample is classified through a majority favor on it from of its neighbors. When an unknown sample is given, it searches the feature space for the k training samples which are closest to the unknown sample.

(c) *Linear discriminant analysis (LDA):* LDA is widely used in the field of statistics and machine learning. It mainly provides more class separability. It draws a decision region between the given classes and searches for project axes. In the project axes, the data points of the different classes are far from each other, whereas the data points to the same class are to be close to each other [21]. LDA creates a linear combination of data that detects the largest mean differences between the desired classes [26].

(d) *Multilayer perceptron neural network (MLPNN):* Multilayer perceptron includes an input layer, minimum of one hidden layer and an output layer. Each layer has at least one neuron. It performs nonlinear mapping between input space and output space and sends the weights assigned to subsequent layers. Finally, it determines the output and compares it with the desired output. It identifies the error signal which adjusts the connection weights correspondingly [37].

(e) *Decision trees (DT):* Decision tree is a nonparametric supervised learning method which is widely used as a classifier. It provides a model-like tree structure which can be used to detect the class label of a target variable by learning simple decision rules inferred from class-labeled training data. The classification process of testing samples is performed by arranging them based on feature values. Each node in a decision tree denotes a feature, and each branch is generated by comparing testing value to a threshold value of the node. Classification is performed by tracing a path from the root node to a leaf node which has no outgoing edge [17].

(f) *Learning vector quantizer (LVQ):* LVQ is very much related to self-organizing maps and a case of an artificial neural network. It creates different prototypes which make it easy for experts to interpret in their respective domain. In multi-class classification problems, LVQ is mainly applied in a natural way. The network has a competitive layer which is called as subclasses that learns to classify input vectors. A linear layer transforms the competitive layer's classes into target classifications. These are mostly defined by the user and are known as the target classes *[28].*

(g) *Self-organizing feature maps (SOFM):* Self-organizing feature maps (SOFM) classify the input vectors. It depends on their grouping in the input space, as SOFM learns to recognize the neighboring sections of the input space as well as distribution and topology of the input vectors. The neurons in the layer of an SOFM are arranged initially in physical positions according to a topology function [10].

(h) *Back-propagation neural network:* Back-propagation minimizes the total squared error of the output which is computed by the rest. The neural network responds accurately to the input patterns which are used for training. This is known as memorization. The network must also be reasonably to input that is similar to the samples used for training [11]. This is called generalization. The three stages of the training are as follows:
   (i) Feed forward of the input pattern.
   (ii) Calculation and back-propagation of the associated error.
   (iii) Adjustments of the weighs.

(i) *Fuzzy classification:* Fuzziness is a dominantly used technique in object recognition and classification. Fuzzy classifiers are more suitable to real data where the boundaries between subgroups are not well defined. Mostly this classifier is integrated with other classification systems like neuro-fuzzy classifiers, fuzzy decision trees, and fuzzy k-nearest neighbor algorithm for better performance. The analysis of sleep EEGs can be done using fuzzy classifiers [10].

9.5 Design of Brain-Computer Interface Using EEG Signals for Rehabilitation Applications

EEG-based brain-controlled rehabilitation devices are noninvasive technique which can be served as a useful tool for disabled people. Many myoelectric prosthetics are used these days. But it depends on the nerves which are not damaged and they are expensive. This can be overcome by EEG-based brain-controlled prosthetics [36].

Brain-computer interfaces (BCI) are a means of communication between humans and computers. They measure and process brain activity and, after successfully recognizing the input, perform the required actions. BCI systems can be divided into the following subtypes:

(a) *Invasive BCI:* Surgically implanted sensors into the brain.
(b) *Noninvasive BCI:* Sensors are placed on the scalp.
(c) *Dependent BCI:* Requires the use of additional motor movements.
(d) *Independent BCI:* Independent of any muscle activity.
(e) *Synchronous BCI:* Interacts with the system only in specific time frames.
(f) *Asynchronous BCI or "self-paced":* It can be used any time irrespective of other functions.

BCI system consists of EEG sensors to acquire the brain signal, which is further processed using different modules like MATLAB. The signals extracted using the sensors act as command signals which are subsequently transmitted to the microcontrollers. The most suitable way of controlling hand prosthesis would be to detect signals from the sensor motor areas of the brain. The classification rate obtained using different techniques (ANN, SVM, etc.) can be further used to control the different functions of the rehabilitation devices. BCI systems are mostly used by people with disabilities as it can help them to perform certain activities without muscle movements. Initially BCI was used for medical reasons, but nowadays with advanced technologies, BCI systems are also being developed for general population. In a brain-computer interface (BCI), commands that an individual generates via the brain are transferred to a computer and finally used to control an external device. For example, by using EEG signals generated by the brain, a person can control the direction of movement of the wheelchair, the position of the cursor in the computer screen, etc. BCI operation depends on the interaction of a user and a system where the user must maintain proper correlation between his or her activity and the signal features used by the BCI. It extracts only those features that are controlled by the user. Further it translates those features into device commands accurately. EEG-based BCI has a number of applications in today's world. Several assistive technologies have been developed based on EEG BCI. BCIs are mainly used for managing and regulating movements, for example, the motorized wheelchair or a

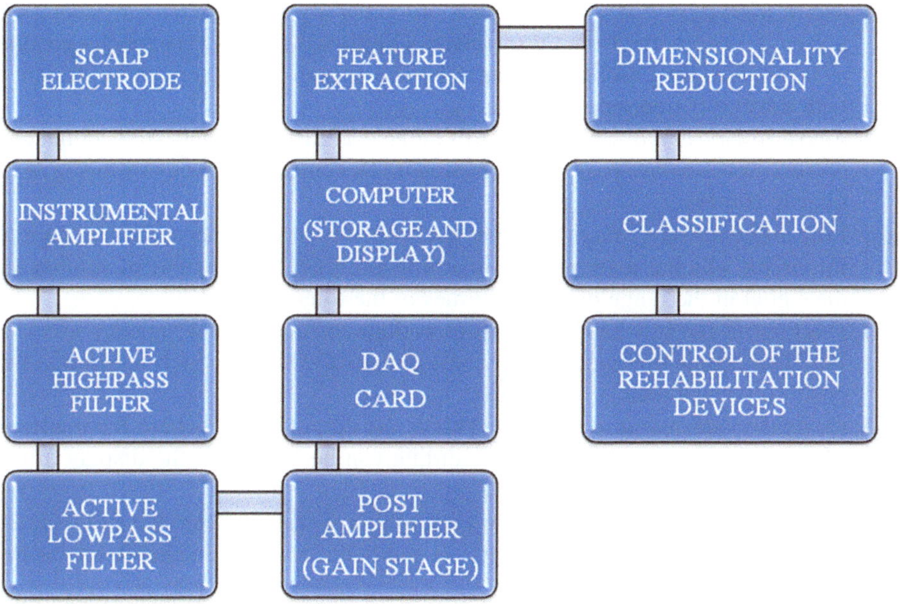

Fig. 9.8 BCI system

prosthetic limb; restoring the mobility of a paralyzed limbs by electrically stimulating the damaged muscles; controlling various home appliances, lights, television, and room temperature; operating the door just by thinking; controlling a robotic car; playing computer games; decoding brain activity to reproduce movements in robotic arms; controlling elements in virtual reality; walking in a virtual street by thought; typing a message on computer screen by concentrating on the display; controlling a computer cursor, for spelling words; etc. [31].

BCI system can be explained using the following block diagram as shown in Fig. 9.8.

(a) *Scalp electrode:* It is a small device that is attached to the scalp of the human brain to detect the electrical activities of the brain. These electrodes are the means through which the activities are communicated to the input circuits of the amplifier. Though there are different types of electrode, metal interface is the common component for all. The electrode is composed of a metal, and the electrolyte is a conducting solution in the form of gel or paste. This metal-electrolyte interface makes the current flow inside the brain.

(b) *Instrumental amplifier:* An instrumental amplifier is a type of differential amplifier that has been fitted with input buffer amplifiers. It eliminates the need for input impedance matching. This makes the amplifier suitable for use in measurement and test equipment.

(c) *Active high-pass filter:* EEG signals are generally mixed with low-frequency noises. To overcome from these noises, a high-pass filter is very much essential as it removes the baseline drifting. The active high-pass filter can be designed with the cutoff frequency of 1 Hz.
(d) *Active low-pass filter:* When an EEG signal is disturbed with high-frequency noises, a proper low-pass filter removes it. The cutoff frequency of this type of filter is 100 Hz.
(e) *Post-amplifier (gain stage):* A post-amplifier circuit provides additional gain to the circuit which is a simple non-inverting amplifier. It is capable of amplifying the signal in the range of 1–1000 times of the filtered signal.
(f) *DAQ card:* The raw EEG information is digitized using the designed hardware. The data acquisition card (DAQ card) obliges the signal which is to be in the positive voltage area.
(g) *Computer (storage and display):* This is the output device where the digitized EEG signal is stored. The EEG recording can be viewed in the display of the computer.
(h) *Feature extraction:* It extracts the most effective features from the EEG signal, thus giving a more precise description and hence making it suitable for further processing.
(i) *Classification:* It is a process in a system which divides things into groups or types. In EEG different mental tasks can be classified, and furthermore these classified tasks can be used to control different rehabilitation devices like artificial hand.
(j) *Dimensionality reduction:* Converting a set of data having vast dimensions into lesser dimensions is known as dimensionality reduction. It fastens the time required for performing the process and also removes redundancy.

9.6 Future Research Directions

EEG is a novel technique for the early diagnosis of neurodegenerative disorders. Techniques like CT/MRI take more time and are expensive. Hence, EEG is a cost-effective tool for the diagnosis of the neurological disorders. Mathematical modeling designed using EEG can be a useful tool for the researchers and clinicians for the detection of the disease. In the near future, this mathematical model can be made as application based on mobile phones and tablets. This would be an advanced diagnostic technique. Different electrophysiological processing techniques are available which in the future can be made more advanced and further be used in controlling advanced functions of the rehabilitation devices. Recent studies are confined with only the control of few functions through EEG. But once it becomes more advance like if the mathematical model becomes application based, it would be more user-friendly. Hence, people would carry with them the mobile phone containing the application and can use it wherever they want. Artificial neural network and much other classification techniques will be very much useful for detecting the level of different neurological disorders which in turn will help in the early detection of disease. This early detection will act as a boon for the patients to get diagnosed in the early stage of the disease. As a result they will be medicated by the clinicians at the earlier stage so that the disease side effects can be stopped. Various types of interactive games can be designed through EEG control which will further help the people with cerebral palsy and other disabilities to control and improve their oral motor and sensory disorders.

The recent advancements in BCI research have given a significant boost to BCI research which surrounds different neuroscientists, engineers, mathematicians, and clinical rehabilitation specialists. BCI systems can not only be used for people with severe disabilities but also for common masses. Hence, in the coming years, BCI systems will become a new means of man-machine interface.

9.7 Conclusion

Electroencephalography (EEG) is extensively used recording technique of the human brain. All kinds of brain-computer interface systems generally depend on the EEG signals. Hence, further it can control rehabilitation devices like artificial limbs. Artifacts are the most common in EEG recordings. Significant research has focused on identifying ways for minimizing such artifacts at the preprocessing stage of the EEG signal processing. Therefore, EEG preprocessing is an important parameter in all the brain-computer interface systems. This chapter presents a review of the EEG signal processing and classification techniques for all EEG-based applications and analysis.

Some of the conclusions drawn from the chapter are as follows:

(a) Biosignal analysis requires better time and frequency resolution. The different signal processing techniques used in EEG are Wiener filters or adaptive filters, least mean square algorithm, joint approximate diagonalization of Eigen-matrices method, independent component analysis discrete Fourier transform, wavelet packet transform, Haar wavelet and lifting wavelet transform, etc.
(b) Feature extraction is very much necessary to study the EEG signals and its characteristics. Features can be both time domain and frequency domain. These features are very much important to study the different frequency bands of EEG signals. Some of the features are mean, median, standard deviation, variance, median frequency, mean frequency, power, modified mean and median frequency, etc. Some of the feature extraction techniques are Hilbert-Huang transform, principal component analysis, local discriminant bases, autoregressive model, discrete wavelet transform, sample entropy, etc.
(c) To design a brain-computer interface model classification of EEG signal plays an important role. There are different types of classifiers used. Basically, k-nearest neighbor (k-NN), linear discriminant analysis (LDA), support vector machine (SVM), and artificial neural network (ANN) are more popular.

Therefore, this chapter is expected to be helpful for the researchers who would like to work on different EEG preprocessing and classification techniques and also willing to work on the control of different rehabilitation devices through EEG signals.

Acknowledgments This study has been ethically approved by the Institutional Ethical Committee, NEHU, Shillong, vide no: IECHSP/2017/42 and also from the collaborating institute: "North Eastern Indira Gandhi Regional Institute of Health & Medical Science," Shillong, vide no: NEIGR/IEC/M6/F13/18.

References

1. Al-Fahoum AS, Al-Fraihat AA (2014) Methods of EEG signal features extraction using linear analysis in frequency and time-frequency domains. ISRN Neurosci 2014:7. https://doi.org/10.1155/2014/730218
2. Alix JJP, Ponnusamy A, Pilling E, Hart AR (2017) An introduction to neonatal EEG. Paediatr Child Health *27*(3):135–142. https://doi.org/10.1016/j.paed.2016.11.003
3. Atwood HL, MacKay WA (1989) Essentials of neurophysiology. Decker, Toronto
4. Azlan WAW, Low YF (2014) Feature extraction of electroencephalogram (EEG) signal – a review. Paper presented at the 2014 IEEE conference on Biomedical Engineering and Sciences (IECBES), December 8–10
5. Bruce EN, Bruce MC, Vennelaganti S (2009) Sample entropy tracks changes in EEG power Spectrum with sleep state and aging. J Clin Neurophysiol 26(4):257–266. https://doi.org/10.1097/WNP.0b013e3181b2f1e3
6. Chi YM, Jung TP, Cauwenberghs G (2010) Dry-contact and noncontact biopotential electrodes: methodological review. IEEE Rev Biomed Eng 3:106–119. https://doi.org/10.1109/RBME.2010.2084078
7. Croft RJ, Barry RJ (2000) Removal of ocular artifact from the EEG: a review. Neurophysiologie Clinique/Clin Neurophysiol 30(1):5–19. https://doi.org/10.1016/S0987-7053(00)00055-1
8. El-Naqa I, Yongyi Y, Wernick MN, Galatsanos NP, Nishikawa RM (2002) A support vector machine approach for detection of microcalcifications. IEEE Trans Med Imaging 21 (12):1552–1563. https://doi.org/10.1109/TMI.2002.806569
9. Guerrero-Mosquera C, Vazquez AN (2009) New approach in features extraction for EEG signal detection. In: Conference proceedings : Annual international conference of the IEEE Engineering in Medicine and Biology Society. IEEE Engineering in Medicine and Biology Society. Annual conference, 2009, 13–16. https://doi.org/10.1109/iembs.2009.5332434
10. Guler I, Derya Ubeyli E (2005) Adaptive neuro-fuzzy inference system for classification of EEG signals using wavelet coefficients, vol 148
11. Hsin HC, Li CC, Sun M, Sclabassi RJ (1992) An adaptive training algorithm for back-propagation neural networks. Paper presented at the [Proceedings] 1992 IEEE International Conference on Systems, Man, and Cybernetics, October 18–21
12. Ince NF, Tewfik A, Arica S (2005). Classification of movement EEG with local discriminant bases. Paper presented at the proceedings. (ICASSP '05). IEEE International Conference on Acoustics, Speech, and Signal Processing, 2005, March 18–23
13. Islam MK, Rastegarnia A, Yang Z (2016) Methods for artifact detection and removal from scalp EEG: a review. Neurophysiologie Clinique/Clin Neurophysiol *46*(4):287–305. https://doi.org/10.1016/j.neucli.2016.07.002
14. Jian-Zhong X, Hui Z, Chong-Xun Z, Xiang-Guo Y (2003) Wavelet packet transform for feature extraction of EEG during mental tasks. Paper presented at the proceedings of the 2003 International Conference on Machine Learning and Cybernetics (IEEE Cat. No.03EX693), November 2–5

15. Kalpakam NV, Venkataramanan S (2004) Haar wavelet decomposition of EEG signal for ocular artifact de-noising: a mathematical analysis
16. Kok A (1997) Event-related-potential (ERP) reflections of mental resources: a review and synthesis. Biol Psychol 45(1):19–56. https://doi.org/10.1016/S0301-0511(96)05221-0
17. Kotsiantis SB (2007) Supervised machine learning: a review of classification techniques. Paper presented at the proceedings of the 2007 conference on Emerging Artificial Intelligence Applications in Computer Engineering: Real Word AI Systems with Applications in eHealth, HCI, Information Retrieval and Pervasive Technologies
18. Li K, Sun G, Zhang B, Wu S, Wu G (2009) Correlation between forehead EEG and sensorimotor area EEG in motor imagery task. Paper presented at the 2009 Eighth IEEE International Conference on Dependable, Autonomic and Secure Computing, December 12–14
19. Makeig S, Jung T-P (1996) Changes in alertness is principal component of variance in the EEG spectrum, vol 7
20. Maki H, Toda T, Sakti S, Neubig G, Nakamura S (2015) EEG signal enhancement using multichannel wiener filter with a spatial correlation prior. Paper presented at the 2015 IEEE international conference on Acoustics, Speech and Signal Processing (ICASSP), 19–24 April 2015
21. Martinez AM, Kak AC (2001) PCA versus LDA. IEEE Trans Pattern Anal Mach Intell 23 (2):228–233. https://doi.org/10.1109/34.908974
22. McFarland DJ, Wolpaw JR (2017) EEG-based brain–computer interfaces. Curr Opin Biomed Eng 4:194–200. https://doi.org/10.1016/j.cobme.2017.11.004
23. Motamedi-Fakhr S, Moshrefi-Torbati M, Hill M, Hill CM, White PR (2014) Signal processing techniques applied to human sleep EEG signals—a review. Biomed Signal Process Control 10:21–33. https://doi.org/10.1016/j.bspc.2013.12.003
24. Ocak H (2009) Automatic detection of epileptic seizures in EEG using discrete wavelet transform and approximate entropy. Expert Syst Appl *36*(2, part 1):2027–2036. https://doi.org/10.1016/j.eswa.2007.12.065
25. Oweis RJ, Abdulhay EW (2011) Seizure classification in EEG signals utilizing Hilbert-Huang transform. Biomed Eng Online 10:38–38. https://doi.org/10.1186/1475-925X-10-38
26. Panahi N, Shayesteh MG, Mihandoost S, Varghahan BZ (2011) Recognition of different datasets using PCA, LDA, and various classifiers. Paper presented at the 2011 5th international conference on Application of Information and Communication Technologies (AICT), 12–14 Oct 2011
27. Petersen SE, Posner MI (2012) The attention system of the human brain: 20 years after. Annu Rev Neurosci 35:73–89. https://doi.org/10.1146/annurev-neuro-062111-150525
28. Pregenzer M, Pfurtscheller G, Flotzinger D (1996) Automated feature selection with a distinction sensitive learning vector quantizer. Neurocomputing 11(1):19–29. https://doi.org/10.1016/0925-2312(94)00071-9
29. Roman-Gonzalez A (2010) Communication technologies based on brain activity. Paper presented at the 2010 World Congress in Computer Science, Computer Engineering and Applied Computing – WORLDCOMP 2010, July 26, Las Vegas
30. Sai CY, Mokhtar N, Arof H, Cumming P, Iwahashi M (2018) Automated classification and removal of EEG artifacts with SVM and wavelet-ICA. IEEE J Biomed Health Inf PP(99):1–1. https://doi.org/10.1109/JBHI.2017.2723420
31. Sarma P, Tripathi P, Sarma MP, Sarma KK (2016) Pre-processing and feature extraction techniques for EEG-BCI applications-a review of recent research. ADBU J Eng Technol 5(1)
32. Schomer DL, Lopes da Silva FH (2018) Niedermeyer's electroencephalography: basic principles, clinical applications, and related fields
33. Tan L (2008) Digital signal processing fundamentals and applications. From http://www.books24x7.com/marc.asp?bookid=28057
34. Tatum WO, Rubboli G, Kaplan PW, Mirsatari SM, Radhakrishnan K, Gloss D, Beniczky S (2018) Clinical utility of EEG in diagnosing and monitoring epilepsy in adults. Clin Neurophysiol 129(5):1056–1082. https://doi.org/10.1016/j.clinph.2018.01.019

35. Wang Y, Makeig S (2009) Predicting intended movement direction using EEG from human posterior parietal cortex. Paper presented at the Foundations of Augmented Cognition. Neuroergonomics and Operational Neuroscience, Berlin/Heidelberg
36. Wolpaw JR, Birbaumer N, McFarland DJ, Pfurtscheller G, Vaughan TM (2002) Brain-computer interfaces for communication and control. Clin Neurophysiol 113(6):767–791. https://doi.org/10.1016/S1388-2457(02)00057-3
37. Zurada JM (1992) Introduction to artificial neural systems. West Publishing Company, St. Paul

Computer-Aided Diagnosis of Epilepsy Using Bispectrum of EEG Signals

10

Rahul Sharma, Pradip Sircar, and Ram Bilas Pachori

Abstract

This chapter aims to analyze the dynamics of brain activity from the electroencephalogram (EEG) signals and classify the seizures that are responsible for epilepsy. Hence, seizure classification is the primary task of this article. In this study, a nonlinear higher-order spectral method is proposed that significantly explore the underlying dynamics of nonstationary EEG signals. Various statistical parameters are measured from the principal region of the higher-order spectra that are subjected to the data reduction technique of the locality sensitive discriminant analysis (LSDA). The LSDA maps the measured features at higher dimensional space and ranks them according to the probability of discrimination. The ranked features are then used as inputs to the support vector machine (SVM) classifier with radial basis function kernel. The proposed algorithm is simulated on the web-available Bonn university database that achieved excellent seizures classification accuracy.

Keywords

EEG signals · Higher-order spectra · Locality sensitive discriminant analysis · Support vector machine · Radial basis kernel

R. Sharma (✉) · P. Sircar
Department of Electrical Engineering, Indian Institute of Technology Kanpur, Kanpur, UP, India
e-mail: rahuls@iitk.ac.in

R. B. Pachori
Discipline of Electrical Engineering, Indian Institute of Technology Indore, Indore, MP, India

S. Paul (ed.), *Application of Biomedical Engineering in Neuroscience*,
https://doi.org/10.1007/978-981-13-7142-4_10

10.1 Introduction

The World Health Organization (WHO) report states that more than 2% of the world population are suffering from epilepsy each year worldwide; out of them, nearly 90% of people belong to the developing countries [90]. The reasons may be lack of awareness, cure, diet, neutrinos, vitamins supplements, etc. It can be observed that the proper diet plan and meditation cure most types of seizures.

The human brain commands the human nervous system that controls most of the human activities. It is a cluster of billions of nerve cells or neurons that are connected via a very complex network. Usually, the information exchange among the neurons takes place through electrical and chemical impulse signals that have a swift response time. It collects information from the sensory organs and transfers to the muscles.When the neurons clusters are hyper-synchronously fired at a time, this sudden surge of electrical activity causes involuntary and temporary disturbance of regular neural activity which can lead to seizures, or in other sense, anything that disturbs the regular pattern of neuron activity can lead to seizures. The recurrent unprovoked seizures are known as epilepsy [10].

The abnormality in brain activity is not always a sign of a seizure. It may happen that those persons are diagnosed with some specific syndrome. Otherwise, in some conditions, a seizure-affected person may continue to show standard electrical activity patterns. Hence, the seizures can be classified depending on the starting area of the disturbance inside the brain and how far it spreads. A single seizure may occur due to high fever or head injury, and this does not indicate that a person has epilepsy. Only when a person suffers from two or more seizures, they may be considered as symptoms of epilepsy. Although the seizures are the predominant symptom of epilepsy, but having a seizure does not necessarily mean that a person is affected by epilepsy [57, 90].

Epilepsy can happen from a variety of causes such as brain tumors, traumatic brain injury, strokes, heart attacks, genetically, abnormal blood vessel formation, etc., which do not lead to epileptic seizures. In most cases, the seizures are controlled with diet, exercise, and medication as many types of seizures do not cause brain destruction. When the seizures are occurring very frequently, that may cause brain damage that leads to epilepsy. Those seizures cannot be controlled with medication [10, 57]. The area of the brain where seizures occur first is known as the focus area or epileptic zone area. By performing surgery, the doctors remove the defined focus area of the brain where seizures are originated. Before performing surgery, the epileptic person is monitored intensively to pinpoint the exact location in the brain as the focus area [33].

There are numerous neuroimaging techniques that are used to explore the dynamics of brain activities at the pre-surgical stage. The computed tomography (CT), single-photon emission computed tomography (SPECT), positron emission tomography (PET), functional magnetic resonance imaging (fMRI), diffusion optical tomography (DOT), magnetic resonance imaging (MRI), magnetoencephalogram (MEG), and more are employed to reveal the structural abnormalities of the brain. These techniques map the damaged areas or abnormalities that may be the origin of seizures. Despite of rapid advancement of the neuroimaging techniques, the analysis of electroencephalogram (EEG) signals is the most frequently used technique for diagnosis of neurological diseases [33, 90].

The EEG signals are recorded in two ways, namely, extracranial and intracranial. The extracranial EEG signals are recorded by noninvasive electrodes placed on the scalp with the international 10/20 system, whereas the small and deep penetrating electrodes are used to record the intracranial signals. The intracranial EEG signals having highly nonstationary and nonlinear behavior including impulses and sharp spikes as compared to the extracranial EEG signals. The prediction, detection, and occurrences of epilepsy by visualization of lengthy EEG signals are a soporific and time-consuming process for Neuro-experts. Hence, the prime challenge is to develop a signal processing algorithm which can extract substantial information about the nonlinear dynamics of the brain's electrical activities. In this chapter, a computer-aided nonlinear higher-order statistics (HOS)-based method has been proposed that leads to an excellent seizure classification accuracy on the Bonn university EEG database [7].

The layout of the present chapter is as follows: Sect. 10.2 contains the literature study about the existing seizures classification algorithms. The studied database is briefly illustrated in Sect 10.3. Section 10.4 contains the proposed methodology which includes brief descriptions about higher-order spectra, feature extraction, dimension reduction algorithm, and the support vector machine (SVM) classifier. The discussion on the achieved results is given in Sect. 10.5, whereas the conclusion is included in Sect. 10.6.

10.2 Related Work

The literature reveals that many research papers have been published on automated classification of epileptic seizure based on the EEG signals.

Earlier the linear prediction (LP) technique is used for seizure classification. The LP algorithm computes the future values which depend on the present and past values behaviors. Pradhan et al. [63] used the LP algorithm for data compression, storage, and transmission of EEG signals. It is relevant to extract the significant features that reveal the underlying nonlinearity and complexity of the EEG signals. Altunay et al. [6] computed the LP error for the epileptic EEG signal detection. In the series of LP-based EEG signal classification algorithm, Joshi et al. [34] proposed the fractional LP method. The energy of the original signal and LP fractal model error are used as measurements for the classification of ictal and seizure-free EEG signals. However, the proposed features on the LP algorithm are not suitable to capture the required information about the nonstationary EEG signals, which yields a low classification accuracy. Some model-based algorithms are also proposed for the EEG signals classification [36, 37, 83]. The autoregressive (AR) model deals with the prediction of the future values based on the present and previous values, which also helps to compute the transfer function. The AR coefficients of the EEG signals can be used as inputs to different classifiers. Ubeyli et al. [83] used the least square-SVM (LS-SVM) classifier to classify the EEG signal based on AR coefficients.

Usually, the local binary pattern (LBP) is applied for image classification or in pattern recognition. It extracts some suitable patterns that explore the pattern characteristics [40]. Kaya et al. [35] extracted the uniform and non-uniform 1D-LBP features from the studied signals and employed those attributes as inputs to different classifiers to evaluate the performance efficiency. The authors used five different classifiers, namely, the BayesNet, artificial neural network (ANN), functional tree (FT), SVM, and logistic regression (LR). They got maximum classification accuracy with the BayesNet classifier. Tiwari et al. [80] employed the 1D-LBP for automated epilepsy detection. They detect the key-points features of the EEG signals by applying the difference of Gaussian (DoG) filter at multiple scales. The 1D-LBP algorithm is applied at these detected key-points, and the histogram of the LBPs are employed as features. The SVM classifier with radial basis (RBF) kernels is used to distinguish the seizure classes. The proposed algorithm achieved excellent classification accuracy in both binary and three seizures classes classification cases.

Many researchers proposed the spectral analysis techniques to explore the nonlinear dynamics of EEG signals. Polat and Gunes [60] used the non-parametric spectral analysis technique of EEG signals for the purpose of epilepsy detection. The spectral estimation had been performed with various methods, and the spectral coefficients are discriminated with the decision tree classifier. They extended their work with same parameters and classified them with the artificial immune recognition systems [61]. Faust et al. [19] proposed the frequency domain parametric method for automatic identification of epileptic seizure based on EEG signals.

The frequency domain techniques alone are not often suitable for analysis of non-stationary signals. As it does not reveal the time and frequency information simulteneously. To overcome this problem, many researchers proposed the time-frequency (TF) analysis methods that allow an analysis of both parameters (time and frequency) in a single frame. The TF techniques are used by numerous authors [32, 65, 69, 75, 81, 82] for analysis of the EEG signals vis-a-vis the brain activities. The TF methods decompose the signal at time and frequency axes simultaneously. Samiee et al. [65] proposed the rational discrete short-time Fourier transform (DSTFT) for epileptic seizure classification. They measured five absolute values of the statistical parameters (standard deviation, mean, median, minimum value, maximum value) from the decomposed DSTFT coefficients. They achieved 98.1% classification accuracy with the multilayer perceptron (MLP) classifier to classify sezures and non-sezures EEG signals. In the DSTFT, a fixed-length window does not provide an accurate analysis of time in frequency frame and vice versa. Sharma and Pachori [69] proposed the improved eigenvalue decomposition of the Hankel matrix and Hilbert transform (IEVDHM-HT) technique for the TF representation. The studied signal is decomposed into sub-signals by using the IEVDHM with a specific criterion such as eigenvalue selection, number of iterations, etc. The Hilbert transform (HT) algorithm is applied on each sub-band to extract the instantaneous amplitude and frequency parameters to represent the studied signal on the TF plane. The Rényi entropy is measured from the TF matrix, and it is used as features for seizure classification. They achieved 99.41% classification accuracy for seizure-free and seizure classes when the length of the studied signals is 4080 points. Fu et al. [21] used the Hilbert marginal spectrum (HMS) analysis technique for seizure classification. The HMS is derived from the empirical mode decomposition (EMD) method which decompose the non-stationary multicomponent signals into a group of sub-band signals known as the intrinsic mode functions (IMFs). The spectral entropies and energy are measured from these sub-bands and used as inputs to the SVM classifier for EEG signal classification.

Iscan et al. [32] combine time domain and frequency domain attributes for seizures classification. The cross-correlation (CC) and the power spectral density (PSD) are computed from the studied signals. The four statistical features, namely, the peak value, equivalent width, centroid and mean square abscissa, are measured from the CC of the analysis signals, while the total power is computed for each EEG sub-band signal (δ, θ α, β, γ). Hence, the combined features are used for the classification of normal and epileptic EEG signals. Various classifiers, namely, the Naive Bayes (NB), k-nearest neighbor (KNN), SVM, least square support vector machine (LS-SVM), Parzen, Fisher discriminant analysis (FDA), quadratic classifiers, binary decision tree (BDT), and nearest mean, are used to classify the labeled combined attributes. They achieved 94.94%, 97.97%, 88.88%, 100%, 91.91%, 89.89%, 100%, 98.98%, and 91.91% classification accuracy, respectively. Tzallas et al. [81, 82] used the time-frequency (TF) analysis techniques for automatic EEG signal classification for different problems such as seizure classification with the ANN [81] and epileptic seizure detection [82].

With regard to the TF resolution property, the wavelet transform (WT) is a powerful tool for investigation of non-stationary signals. It provides a time-varying decomposition of EEG signals which makes possible to capture the transient features of the studied signals. Hence, various wavelet-based techniques and their variants are proposed for exploration of nonlinear dynamics of the EEG signals. Subasi used the discrete wavelet transform (DWT) for the EEG signals classification. The mixture of the expert model is used as a classifier to classify the seizures-labeled wavelet coefficients [77]. They extended their work and applied various data dimensionality reduction techniques such as the principal component analysis (PCA), independent component analysis (ICA), and linear discriminative analysis (LDA) for reducing the dimensionality of the wavelet coefficients. The resultant wavelet coefficient matrix is fed to the SVM classifier to discriminate seizures classes [78]. Acharya et al. [1] computed different nonlinear features from the DWT coefficients that are evaluated by the SVM classifier. The multifractal analysis technique and WT had been proposed for classification of EEG signals [84], while the phase-space reconstruction (PSR) with Euclidean distance (on wavelet coefficients) methods is used to distinguish healthy and epileptic EEG signals [43]. Kumar et al. [39], Guo et al. [26], and Ocak et al. [50] computed the approximation entropy (ApEn) from the wavelet coefficients. The ANN and the optimal threshold value classifiers are used to classify the computed features. The results obtained in [50] reveal that the detail coefficients of the first level provided the best classification accuracy of 96.65% using the ApEn as a feature.

Guo et al. [24, 25] proposed various measurements for the seizures-labeled EEG signal classification. The relative wavelet energy [24] and line length [25] attributes are measured from the decomposed signals that are classified by the ANN classifier.

The wavelet packet entropy in [86] and the mixed band wavelet chaos neural network in [22] are used for epileptic seizure detection. Orhan et al. [51] proposed the DWT for EEG signal decomposition into sub-bands. The K-mean clustering techniques are employed to separate the features according to the labeled classes. These labeled clustered data are classified using the MLP neural network (MPNN) model. Zhou et al. [95] decomposed the EEG signals by using the DWT method up to three, four, and five levels of decomposition. The different features, namely, lacunarity and fluctuation indexes, are measured from these decomposed scaled signals. The lacunarity is a scale-dependent feature that measures heterogeneity, while fluctuation index reflects the intensity fluctuation of EEG signals. The Bayesian linear discriminant analysis classifier is used for labeled feature classification. The proposed method achieved an excellent classification accuracy of the order of 96.67% for intracranial EEG signal classification.

Many authors proposed multiple variants of the WT, namely, the multiwavelet transform [26], orthogonal wavelet filter banks [11, 12], tunable Q-wavelet transform [14, 29, 58, 68], wavelet packet decomposition (WPD) [4], dual-tree complex wavelet transformation (DTCWT) [17, 59], empirical wavelet transform [13], flexible analytic wavelet transform [70], and new frequency slice wavelet transform [79], etc., for seizure classification based on the EEG signals. Many variants of the WT extract relevant information from the nonlinear and nonstationary EEG signals. However, the obtained results may vary with a change of the basis wavelet; also the computational complexity increases with an increase of decomposition levels.

Similar to the WT, the EMD is another nonlinear algorithm that identifies the subtle information regarding seizure classification and it is easy to use [5, 8, 20, 44, 52, 53, 67]. The EMD is a data decomposition method that decomposes an arbitrary signal into the IMFs. It is a data-dependent TF analysis method that is characterized by the Hilbert-Huang transform (HHT) algorithm [31]. Fu et al. [20] used the HHT-based TF representation (TFR) method for seizure and non-seizure EEG signal classification. The TFR is simulated as a time-frequency image (TFI), and the segmentation of these images has been done based on the frequency bands of rhythms presented in the EEG signals. The different statistical features up to fourth order are measured from the pixel intensity of the TFI. These attributes are evaluated by the SVM classifier with RBF kernel function. They reported 99.125% classification accuracy with the θ-rhythm of EEG signals. Alam et al. [5] computed the same features as in [20] directly from the IMFs. The collected features are classified by the ANN classifier that yields 100% classification accuracy. Bajaj and Pachori [8] proposed the EMD-based algorithm for epileptic seizure detection. The instantaneous signal area is computed from the trace of the windowed IMFs. They achieved 90% sensitivities with 24.25% error rate detection. Pachori [53] uses the EMD algorithm for ictal and seizure-free EEG signal classification, where the decomposed IMFs are transformed in terms of the Fourier Bessel (FB) expansion, and the mean frequency of FB expansion is considered as features.

Pachori et al. [54–56] extracted different features from IMFs and inputted them into various classifiers to classify labeled EEG signals. The second-order difference plot (SODP) is measured from the IMFs, and the area of 95% confidence ellipse (measured from the SODP) is considered as a feature [55]. They obtained 97.72% classification accuracy to discriminate epileptic seizure from seizure-free EEG signals. They extended their work in [56], where they measured two areas, first one corresponding to the SODP, while the other for the graph obtained when the IMFs of the analytic signal are represented in the complex plane. Sharma et al. [67] classify the epileptic seizures based on the phase-space representation (PSR) of IMFs. Oweis et al. [52] used the HHT algorithm to decompose the EEG signal into the IMFs, and the instantaneous frequency and amplitude are computed from each decomposed signal. They observed that the starting three or four modes are highly determinant by using the hypothesis testing. The hypothesis testing used the t-test with two different p-values that allow reducing the insignificant feature space. The proposed algorithm is evaluated on 25 healthy and 25 seizure EEG signals. The ANN is used as a classifier that yields 94% classification accuracy, while the 80% accuracy is obtained with the multivariate empirical mode decomposition (MEMD) method. However, the EMD does not correctly extract the low-energy components from the nonlinear EEG signals; also, the EMD suffered from mode mixing problem that can be temporarily reduced by the ensemble empirical mode decomposition (EEMD) [91]. As compared to the EMD and EEMD, the complete ensemble empirical mode decomposition with adaptive noise (CEEMDAN) gives an accurate reconstruction of the EEG signal without mode mixing that allows a better spectral separation of the mode functions. Hassan et al. [28] proposed the CEEMDAN method for decomposition of the EEG signals into the IMFs. The different statistical features, namely, the average, variance, and kurtosis are computed from these IMFs. The ANN classifier is used to separate the labeled classes. They achieved the classification accuracy of order 98.87% for normal (eye open) and non-seizure including normal eye closed and 86.37% for three categories, e.g., seizure, seizure-free, and healthy EEG signals. The variational mode decomposition (VMD) algorithm is proposed in [94] for automated seizure detection based on the EEG signals. The VMD can extract relevant low-energy component information from the raw EEG signals by decomposing it into a fixed number of band-limited IMFs (BLIMFs).

To explore the subtle information about the underlying nonlinear dynamics from the EEG signals, many authors used various nonlinear parameters extraction methods with different algorithms. The recurrence quantification analysis (RQA) visualizes the recurrence behavior of the phase-space trajectory of dynamical systems. Acharya et al. [3] used the RQA for automatic epilepsy EEG signal detection. They extended their work with another nonlinear parameter called the Hurst parameter [2]. The Hurst parameter customarily used to measure the long-term memory of the nonlinear signal. The short-term maximum Lyapunov exponent (STLmax) that measures the rate of divergence, is used to know the dynamic behavior of the EEG signals [23, 45]. To measure the complexity of the studied signals, various entropies, such as the ApEn, Shannon entropy, Fuzzy entropy Renyi entropy, log energy entropy, and more are used. Srinivasan et al. [76] and Zhang et al. [93] used the ApEn as a feature, and the labeled features are classified by the ANN and the SVM classifiers, respectively. Guo et al. [25]

proposed the line length features derived from the DWT coefficients. The line length features are used to capture the changes in the waveform dimensionality. These are very sensitivity for signal amplitude and frequency variations. The EEG signals are decomposed up to five levels of decomposition with the DWT corresponding to various frequency bands. The line length features are measured from each sub-signal and classified by using the ANN classifier. The proposed algorithm achieved 99.60% for normal (eye open) and seizure, 97.75% for seizure and seizure-free including eye open, and 97.77% for seizure and non-seizure classification accuracy for different EEG signal sets.

Wang et al. [87] proposed the adaptive learning algorithm, whereas Shin et al. [73] used the sparse representation model for EEG signal classification. Besides the single-scale entropy measurement, the multiscale entropy is also registered in literature for the EEG signal based seizures classification. The single-scale entropy measurement is not able to quantify the underlying dynamics of the nonlinear signals typically when the signals are generated from the biological system. Labate et al. [41] proposed the multiscale entropy as a feature for measuring the complexity in Alzheimer's disease. They extended the same concept for the EEG signal classification with the multiscale permutation entropy. The features are classified by the SVM classifier with various kernel functions such as linear, RBF, and sigmoidal, and they obtained 78%, 95%, and 91% classification accuracy, respectively [42]. Guo et al. [27] proposed the genetic programming (GP) algorithm for the EEG signal classification. The GP automatically extracted the features from the original feature database. It not only improves the performance of discrimination of the classifier but at the same time reduces the dimensionality of feature space. The DWT is applied on the raw EEG signals, and the mean, standard deviation, energy, curve length, and skewness are measured as original features from each sub-band signal that are fed to the GP algorithm. The GP generated the new dimension features having a substantial probability of discrimination capability. These GP generated features are classified with the KNN classifier which results in 99.20% classification accuracy. There are various learning based algorithms, namely, the random forest, extreme machine learning [92], ANN [88], recurrent neural network (RNN) [23], etc., that are also used for the EEG signal classification.

10.3 Database

In the study, the web-available EEG signal database has been considered that is recorded in the University of Bonn, Germany [7]. The database is the collection of five sets of EEG signals, namely, Z, O, N, F, and S. Each set contains 100 single-channel EEG time series of 23.6 sec duration. These signals are the subset of 128-channel EEG signals that are selected by the neuro experts. All these signals are digitalized at 173.61 Hz sampling rate using 12-bit resolution. A band-limited filter (0.53–40 Hz) is used to remove the artifact due to muscle activities. The EEG time series sets Z and O are recorded extracranially on five healthy persons during the relaxed state with eyes open (set Z) and eyes closed (set O) condition, respectively. However, the remaining time series sets N, F, and S are recorded intracranially from five epileptic patients in their pre-surgical evaluation of epilepsy. Sets N and F consist of seizure-free time series. The signals of set N are collected from the hippocampal formation of the opposite hemisphere of the brain, while the set F time series are recorded from the epileptogenic zone, whereas the intracranial time series of set S are collected from the epileptic zone during seizure activity. All signals are collected using the standard 10–20 electrode placement scheme.

In brief, the different EEG signals groups can be summarized as:

O – Normal signal set, persons are relaxed and in the awake state with eyes closed.
Z – Normal signal set, persons are relaxed and in the awake state with eyes open.
N – Seizure-free signal set recorded from the hippocampal formation of the opposite hemisphere of the brain.
F – Seizure-free signal set recorded from the epileptogenic zone.
S – Signal set collected from the epileptic zone during seizure activity.

In this chapter, the three classification problems have been considered as follows:

Case 1. Seizure (S) versus normal eyes-open (Z)
Case 2. Seizure (S) versus all non-seizure (FNZO)
Case 3. Seizure (S) versus seizure-free (FN) versus normal (ZO)

Figure 10.1 shows the EEG time series from each set. The Z and O are the healthy, F and N are the seizure-free, while S represents the seizure EEG signal, respectively. In the broad scene, S is known as the seizure EEG signal, while the remaining (ZOFN) is known as the non-seizure EEG signals.

10.4 Proposed Methodology

In this study, a nonlinear higher-order spectra (HOS) based algorithm is proposed for the seizure-labeled EEG signal classification. Figure 10.2 depicts the block diagram representation of the proposed algorithm. The EEG signals are highly nonlinear in nature. Hence, to extract the relevant information from such signals, the higher-order spectruma, namely, bispectrum, is used. The bispectrum reveals the nonlinearity present in EEG signals and may better illustrate the dynamics of the nervous system. It helps to the diagnosis of brain diseases by understanding of neurological processes. The bispectrum contains the six symmetries along with various directions. Thus, the principal region has been extracted to avoid the irrelevant mathematical calculation. The numerous features are extracted from the principal region that capture the underlying dynamics of the studied signals. As the studied dataset contains five sets of different classes, each consisting of a hundred long EEG time series,hence, the dimension of the extracted features is much high; also the extracted features may contain some irrelevant features that may degrade the algorithm

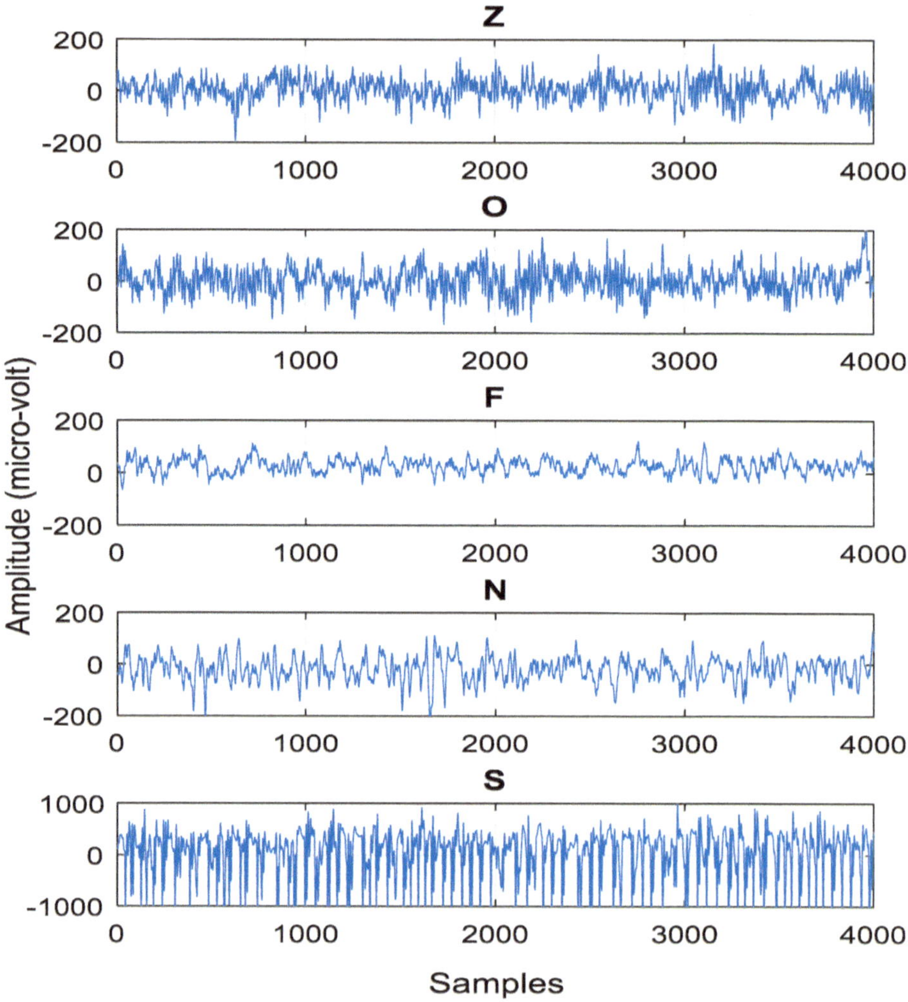

Fig. 10.1 Signals from each set of database

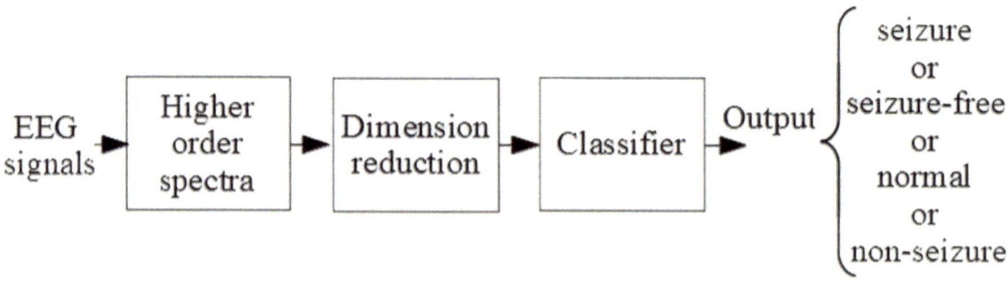

Fig. 10.2 Block representation of proposed algorithm for epileptic seizure detection

performance. A dimensionality reduction technique known as the locality sensitive discriminant analysis (LSDA) is used to moderate feature space dimensionality. It ranked the features according to the probability of separation. These ranked features are then subjected to the supervised learning SVM classifier with RBF kernel.

10.4.1 Higher-Order Spectra

The nonlinearity analysis of a system operating under a random input has been quite extensively done for many years [15]. The HOS are initially generated for the ergodic random processes, and later, it has been expanded to the deterministic signals [15, 64]. In the real-time signal processing, the researchers are mainly dealing with the nonlinear, non-Gaussian, and nonstationary signals. The second- or lower-order statistics do not adequately analyze these types of signals. Also, the considered noise in signal processing is assumed to be Gaussian distributed and additive. The literature reveals that the HOS of order higher than two is commonly used for the analysis of nonstationary and non-Gaussian signal [15, 48, 64]. The HOS can reveal in terms of moments and cumulant functions. The cumulant function is defined as the logarithm of the moment function. For a zero mean signal, both are equal up to order three. The HOS have some unique properties that help for the analysis of non-Gaussian signals; such as, the higher-order (three or more) cumulants of the Gaussian distribution are equal to zero [47, 49]. Figure 10.3 illustrates the bispectrum of each set of the EEG signals and the principal region of bispectrum. The third-order cumulant (ToC) is infinitely differentiable and convex, that allows analysis of the non-minimum-phase and phase couple signals; also, it is scale-invariant. Hence, the ToC of a non-Gaussian signal with an additive uncorrelated Gaussian noise filters out the Gaussian noise part, and represents only the ToC of the signal [47]. Sharma et al. [71] used the residual-based higher-order statistics algorithm for the classification of focal and nonfocal EEG signals. The proposed method measured the disturbance in the labeled EEG signals with various statistical attributes. They extended their work and proposed a center slice algorithm of higher-order statistics for automated glaucoma detection based on fundus images [72].

Let $y(\tau)$ be a zero mean random process where $\boldsymbol{\tau} = 1, 2, \ldots L$. The ToC can be represented as follows [47]:

$$\begin{aligned} C_y^3(a,b) &= E[y(\tau)^* \, y(\tau+a) \, y(\tau+b)] \\ &= E[y(\tau) \, y(\tau+a)^* \, y(\tau+b)] \\ &= E[y(\tau) \, y(\tau+a) \, y(\tau+b)^*] \end{aligned} \tag{10.1}$$

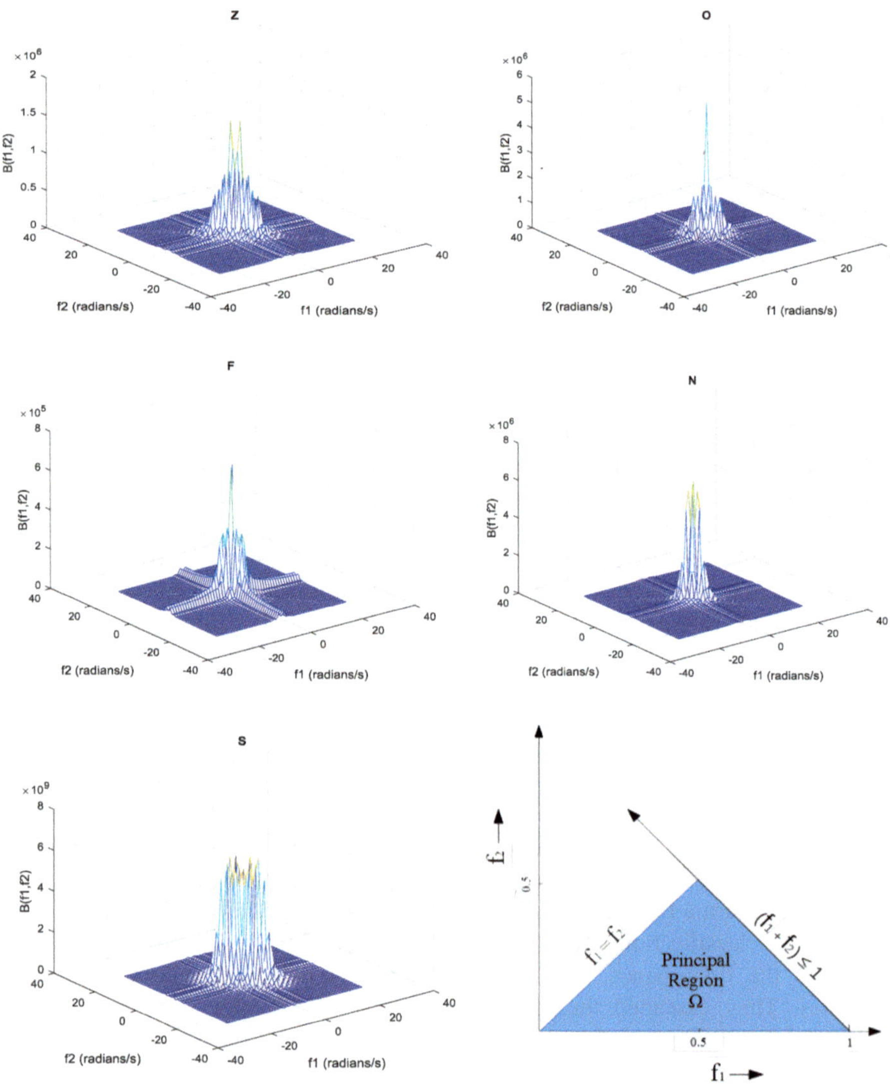

Fig. 10.3 Bispectrum magnitude corresponding to the various set of EEG signals and the principal region of the bispectrum

where $\boldsymbol{E}$ and "$*$" denote the expectation operator and complex conjugate, respectively. The Fourier transform of ToC is defined as bispectrum [49] and is given below:

$$B(\zeta_1, \zeta_2) = FT\left[C_y^3(a, b)\right] \tag{10.2}$$

$$or$$

$$B(\zeta_1,\zeta_2) = E[Y(\zeta_1)^*\, Y(\zeta_2)\, Y(\zeta_1+\zeta_2)]$$

where *FT* denotes the Fourier transform, $Y(\zeta)$ is the discrete-time Fourier transform of $y(\tau)$, and $\zeta_1$and $\zeta_2$are the normalized frequency components that lie between 0 and 1. The bispectrum of a real signal is uniquely defined with the triangle $\mathbf{0 \le \zeta_2 \le \zeta_1 \le (\zeta_1 + \zeta_2) \le 1}$, known as the principal domain [64].

10.4.2 Feature Extraction

The suitable feature extraction is an essential part of any classification algorithm. Otherwise, it may lead to underperformance of the classifier. To explore the characteristics of EEG signals, various parameters, namely, the maximum value, energy, mean, variance, normalized bispectrum entropy (NbE) up to order three, interquartile distance (IqD), and spectral flatness are measured from the principal domain of the bispectrum [38, 66, 74]. These statistical features are very significant and less computationally complex.

Let z_i be the random variable that denotes the elements of bispectrum principal domain, and let $p(z_i)$; $i = 0, 1, \ldots, \eta - 1$ be the probability of occurrence of z_i, where η is the number of distinct random values. Then, the features are defined as follow:

$$Maximum\ value\ of\ z_i = Max\ (\mathrm{z}_i) \tag{10.3}$$

$$Energy = \sum_{i=1}^{\eta} |z_i|^2 \tag{10.4}$$

$$Mean = \sum_{i=1}^{\eta} z_i\, p(z_i) \tag{10.5}$$

$$Variance = \sum_{i=1}^{\eta} (z_i\text{-Mean})^2\, p(z_i) \tag{10.6}$$

$$NbE1 = -\sum_i p_i\ \log(p_i);\quad p_i = \frac{|B(\zeta_1,\zeta_2)|}{\sum_\Omega |B(\zeta_1,\zeta_2)|} \tag{10.7}$$

$$NbE2 = -\sum_i q_i\ \log(\mathrm{q}_i);\quad \mathrm{q}_i = \frac{|B(\zeta_1,\zeta_2)|^2}{\sum_\Omega |B(\zeta_1,\zeta_2)|^2} \tag{10.8}$$

$$NbE3 = -\sum_i r_i\ \log(\mathrm{r}_i);\quad \mathrm{r}_i = \frac{|B(\zeta_1,\zeta_2)|^3}{\sum_\Omega |B(\zeta_1,\zeta_2)|^3} \tag{10.9}$$

The IqD is a measure of variability, based on dividing the dataset into quartiles [74] and can be expressed as follows:

$$IqD = \gamma - \nu \tag{10.10}$$

where $P(\gamma) = \frac{3}{4}$, $P(\nu) = \frac{1}{4}$, and P is the sample cumulative distribution function of the diagonal of the center slice.

$$Spectral\ flatness = \frac{GM(z_i)}{AM(z_i)} \tag{10.11}$$

where GM and AM represent geometric and arithmetic mean, respectively [74].

10.4.3 Locality Sensitive Discriminant Analysis (LSDA)

The LSDA is a supervised dimensional reduction technique. It is based on the concept of maximizing the local margin between different classes while minimizing the local margin within the classes. It splits the data into the within-class and between-class graphs according to its class labels using the nearest neighbor graph (NNG) [16]. The transformation or mapping from the original feature space to a new feature space is allowed to maximize the probability of separation among different classes. Hence, each separated class data is mapped by using a linear transformation matrix such that it preserves the local as well as discriminant neighborhood information. Let l data points $\{x_1, x_2, \ldots, x_l\} \subset R^n$ be sampled from a large dataset. Assume that $N_s(x_i)$ is a group of neighbors having the same label, while $N_d(x_j)$ is a group of neighbors having different labels such that $N_s(x_i) \cap N_d(x_j) = \phi$; *where* ϕ *is the null space* and $N_s(x_i) \cup N_d(x_j) = N(x_l)$. The LSDA ranked the resultant mapped features and arrange them from the highest probability of discrimination to the lowest probability of discrimination.

10.4.4 Support Vector Machine

The SVM is a most popular supervised machine learning technique commonly used for data classification and regression. It maximizes the distance between the support vectors such that the probability of data points error becomes minimum. The slack variable is introduced for compensating the data points error. The support vector of a class represents the boundary that passes through the far most data point that is well separated from another class data [18, 30, 85]. The hyperplane is a virtual perpendicular plane between the support vectors. The SVM is the most popular machine learning algorithm having high generalization performance without any additional a priori knowledge even in high dimensional input space. Different linear or nonlinear kernels are used to map the old feature into a new feature space to achieve the lowest probability of misclassification error [89].

Let X_j be a training vector such that $X_j \in R^n; j = 1, 2, \ldots, l$ in two classes, labeled as$y \in \{+1, -1\}^l$. The SVM separates the training vector such as [18]:

$$\begin{aligned} &\min_{\omega, b, \xi} \quad \frac{1}{2}\omega^T\omega + C\sum_{j=1}^{l}\xi_j \\ &\textit{subject to} \quad y_j\left(\omega^T\phi(X_j) + b\right) \geq 1\text{-}\xi_j \end{aligned} \tag{10.12}$$

where ϕ is defined as mapping space with an error cost,$C \succ 0$, $X_j \geq 0$. The above equation is a constraint optimization problem that can be solved with the Lagrangian method. Hence, the optimization function can be written as follows:

$$\begin{aligned} &\min_{\lambda} \quad f(\lambda) = \frac{1}{2}\lambda^T Q\lambda \text{ - } e^T\lambda \\ &\textit{subject to} \quad 0 \leq \lambda_j \leq C; \quad y^T\lambda = 0 \end{aligned} \tag{10.13}$$

where $Q_{jk} \equiv y_j\ y_k\ \phi(X_j)^T\ \phi(X_k)$ and e is the vector of all ones. Let $K(X_j, X_k) = \phi(X_j)^T\ \phi(X_k)$ be the kernel function; then, the generalized function of the SVM classifier can be defined as [85]:

$$\beta_\kappa = sign\left[\sum_{j\in l}\lambda_j\ y_j\ K\left(X_j, X_k\right) + b\right] \tag{10.14}$$

where λ_j is a nonzero Lagrange multiplier and b denotes the bias.

10.5 Results and Discussion

In this study, a nonlinear third-order spectrum, namely, bispectrum algorithm, is proposed for seizures classification based on the EEG signals. To explore the seizure dynamics, the EEG signals undergo the bispectrum analysis, and the bispectrum of the EEG signal reveals that significant alteration takes place with the EEG signals classes. The magnitude of the bispectrum of seizure-EEG signal (S) is more spread than other classes bispectrum magnitude, also its magnitude is of the order of 10^9 which is high as compared to the bispectrum magnitude graph of seizure-free and normal EEG signals. It can be illustrated that spectra magnitude of standard EEG signal has significant high peaks at a higher frequency as compared to the spectra of seizure-free EEG signals.

From the principal region of bispectrum-magnitude of the EEG signals, several statistical features are computed to extract the relevant information about the brain electrical activities. In this study, two binary-classes classification and one three-classes classification have been performed. The extracted features are subjected to the LSDA algorithm to reduce the feature matrix dimension and also to rank the features from the higher value to lower value subjected to the probability of discrimination. The ranked features are fed to the SVM classifier with RBF and ten-fold cross-validation strategy. The SVM classifier with RBF kernels achieved the maximum accuracy in all cases which are 98.67% for S-Z, 96.30% for S-FN-ZO, and 97.53% for S-FNZO.

In literature, various algorithms are listed for seizure classification. Table 10.1 illustrates various seizure classification methods to investigate brain electrical activities. Peker et al. [59] used the DT-CWT for decomposition of the EEG signals into low-dimensional wavelet coefficients. These features are discriminated with the complex-valued neural network. Their proposed technique is suited for binary as well as three-classes classification. They achieved 100% for S-Z, 99.15% for S-FNOZ classes, and 98.28% and 99.30% classification accuracy for three-classes, namely, S-FN-ZO and S-F-Z, respectively.

Bao et al. [9] employed the power spectral features, Higuchi fractal dimension, Petrosian fractal dimension, and Hjorth parameters as features that were classified with probabilistic neural network (PNN) classifier. They registered 97.00% classification accuracy for differentiating S-from NF-labeled signals. The LP method for seizure classification is used by Joshi et al. [34] and Altunay et al. [6]. The LP energy is used as a feature in [6], while the fractional LP error energy and fractal model signal energy are computed from the EEG signals to be used as features in [34]. They obtained 94% and 95.33% classification accuracy for seizures and seizure-free (S-NF) datasets with the threshold value classifier and the SVM classifier, respectively. The permutation entropy (PE) is used as a measure to distinguish the epileptic EEG signals from the seizure-free EEG signals [46]. They classified subsets S and N with the SVM classifier to the obtained classification accuracy of order 88.83%. Polat and Gunes [60] proposed the non-parametric spectral estimation method for the EEG signal classification. They reported 98.72% classification accuracy with the decision tree (DT) classifier, while the same group achieved a maximum classification accuracy of 100% when the principal component analysis (PCA) algorithm is implemented for dimension reduction of the fast Fourier transform (FFT) coefficient with the artificial immune recognition system classifier [61]. In [62], the authors proposed the AR model for EEG signal representation. The PSD vector is considered as a feature. The distance-based data reduction (DBDR) algorithm is applied to reduce the dimension of the attribute vectors. The obtained features are classified with the C4.5 decision tree classifier. They achieved 99.32% classification accuracy for S-O datasets. Tiwari et al. [80] proposed the 1D-LBP algorithm to extract the relevant information for seizure classification. They obtained 99.45% for S-NF, 99.31% for S-FNZO, and 98.80% for ZO-FN-S classification accuracy when the sparse LBP features are classified by the SVM classifier. Kaya et al. [35] proposed the uniform and non-uniform 1D-LBP feature extraction method from the raw EEG signals. These extracted features are applied to different classifiers to evaluate algorithm performance. They obtained 99.50% classification accuracy with the functional tree (FT) classifier for S-O classes classification.

Table 10.1 A comparison of performances of the various methods for seizure classification

References	Methods and features	Classifier	Classes	Accuracy (%)
Peker et al. [59]	DT-CWT	Complex-valued neural networks	S-Z	100
			S-NFZO	99.15
			S-F-Z	99.30
			S-FN-ZO	98.28
Bao et al. [9]	Power spectral features, Petrosian fractal dimension and Higuchi fractal dimension, Hjorth parameters	PNN	S-NF	97
Altunay et al. [6]	LP error energy	Thresholding	S-NF	94.00
Joshi et al. [34]	FLP error energy and signal energy	SVM	S-NF	95.33
Nicolaou and Georgiou [46].	Permutation entropy	SVM	S-N	88.83
Polat and Gunes [62]	AR	C4.5 decision tree classifier	S-O	99.32
Polat and Gunes [60]	FFT	DT	S-Z	98.72
Polat and Gunes [61]	FFT	The artificial immune recognition system	S-Z	100
Tiwari et al. [80]	1D LBP	SVM	S-NF	99.45
			S-FNZO	99.31
			S-FN-ZO	98.80
Kaya et al. [35]	1D LBP	Functional tree	S-O	99.50
Sharma and Pachori [68]	TQWT and FD	LS-SVM	S-FNZO	99.60
			S-Z	100
			S-O	100
			S-ZO	100
			S-FN	99.67
Bhattacharyya et al. [14]	TQWT-based multiscale KNN entropy	SVM	S-FN-ZO	98.60
			S-Z	100
			S-FNZO	99
Ubeyli [83]	Burg AR	LS-SVM	S-O	99.56
Bhati et at. [11]	Three-band synthesis wavelet filter bank	MLPNN	S-NF	99.66
Subasi [77]	DWT	A mixture of expert model	S-Z	95.00

(continued)

Table 10.1 (continued)

References	Methods and features	Classifier	Classes	Accuracy (%)
Subasi and Gursoy [78]	DWT-PCA	SVM	S-Z	98.75
	DWT-ICA			99.50
	DWT-LDA			100
Acharya et al. [4]	WPD-PCA	GMM	S-NFZO	99.31
Tzallas et al. [81]	TF analysis	ANN	S-Z	100
			S-NFZO	97.73
			S-F-Z	99.28
			S-FN-ZO	97.72
Samiee et al. [65]	DSTFT	MLP	S-Z	99.30
			S-NFZO	98.10
Srinivasan et al. [75]	Time and frequency features	RNN	S-Z	99.60
Orhan et al. [51]	DWT, k-means clustering	MLPNN	S-Z	100
			S-NFZO	99.60
			S-F-Z	96.67
			S-FN-ZO	95.60
Guo et al. [27].	Genetic algorithm	KNN	S-Z	99.20
			S-F-Z	93.50
			S-FN-ZO	93.50
Ocak [50]	DWT	ApEn optimal threshold value	S-ZNF	96.65
Sharma and Pachori [67]	95% confidence area measure of 2D PSR of IMFs, IQR of Euclidian distances of 3D PSR of IMFs	LS-SVM	S-NF	98.67
Fu et al. [20]	Hilbert-Huang transform	SVM	S-Z	99.13
Proposed work	Higher-order spectra	SVM	S-Z	98.67
			S-FN-ZO	96.30
			S-FNZO	97.56

The TQWT is proposed in Sharma and Pachori [68] and Bhattacharyya et al. [14] for the seizure EEG signal classification. Sharma and Pachori [68] computed the signal fractal dimension as a feature on each sub-band. These features are subjected to the LS-SVM classifier for the seizure classification. They achieved 100% classification accuracy for the individual and combine normal classes with seizure class (S-O, S-Z, and S-ZO), and 99.67% classification accuracy for S-NF while 99.60% for the binary class (S-FNZO).

Bhattacharyya et al. [14] measured the multiscale KNN entropy as features for seizure classification on the TQWT decomposed sub-bands. The multiscale entropy measures the complexity of multivariate EEG signals over different scales. These mixed labeled features are classified by the SVM classifier and obtained 98.60% to 100% classification for different seizure classes. The Burg AR coefficients are used for the EEG signal classification [83]. They reported 99.56% classification accuracy. Bhati et al. [11] proposed a TF localized three-band wavelet filter-bank technique for the EEG signal classification. The proposed technique is implemented by using the semidefinite relaxation and nonlinear least square method. They obtained classification accuracy of the order of 99.66% for seizure and seizure-free (S-NF) classes with the MLP neural network (MLPNN) classifier.

Subasi [77] used the DWT for EEG signal decomposition. The author computed four statistical features. Three out of four features, namely, the mean, average power, and the standard deviation, are computed from each sub-band, while the fourth feature is the ratio of the absolute mean values of the adjacent sub-bands. These four statistical features from the DWT coefficients were classified by a modular neural network called the mixture of experts (MEs) and reported classification accuracy of order 95.00%. The author extended the wavelet-based signal classification technique [78]. In this study, the different data dimension reduction techniques such as the principal component analysis (PCA), independent component analysis (ICA), and linear discriminant analysis (LDA) are used to reduce the dimension of the wavelet coefficients, and the resultant coefficients were classified by the SVM classifier. They achieved 98.85% using PCA, 99.5% using ICA, and 100% using LDA, classification accuracy with the SVM classifier. Acharya et al. [4] used the wavelet packet decomposition (WPD) algorithm for the seizure classification. The method decomposed the EEG signals into higher and lower frequency bands simultaneously, and the PCA is simulated on computed feature vectors from each sub-band. For the binary (S-NFZO) seizure classification, they reported 99.31% classification accuracy with the Gaussian mixture model classifier.

Tzallas et al. [81] used the TF analysis technique for automated EEG signal classification. The fractional energy is computed as features from the selected segments obtained with the TF-EEG signal decomposition method. The ANN classifier is used to classify the computed features, and has obtained from 97.72% to 100% classification accuracy for separating different seizure classes. Samiee at al. [65] used the DSTFT for EEG signal decomposition on the TF axis. They computed different statistical features, namely, the average, median, maximum, standard deviation, and minimum, from the rational DSTFT coefficients. They obtained 98.10% for S-NFZO and 99.30% for S-Z, the order of accuracy using the MLP classifier. The TF feature-based classification methodology is also proposed in Srinivasan et al. [75]. They achieved classification accuracy of 99.60% for S-Z classes separation with the recurrent neural network (RNN) classifier. Orhan et al. [51] measured various nonlinear parameters from the DWT coefficients that capture the subtle information from the EEG signals. They reported 96.67% to 100%

accuracy with the MLPNN classifier for separating the different classes of seizures. The authors recorded 99.20% to 99.50% classification accuracy with the features extracted from the genetic programming, while these features are subjected to the KNN classifier to separating the seizures [27]. The ApEn used to compute the complexity of the nonlinear signal. Ocak et al. [50] computed the ApEn from the DWT coefficients for the S-ZNF seizure classification. They achieved 96.65% classification accuracy with the optimal threshold value-based classification algorithm.

The EMD method-based seizure classification algorithms are listed in the literature. Sharma et al. [67] used the EMD algorithm that utilized the Hilbert-Huang transform for signal decomposition into IMFs for complexity computation at high scales. They measured 95% confidence area of 2D and 3D PSR of each IMF. These computed areas are used as features to register 98.67% classification accuracy for binary (S-NF) seizure classification with the LS-SVM classifier. Fu et al. [20] registered 99.13% binary (S-Z) classification accuracy with the EMD algorithm and the SVM classifier.

10.6 Conclusion

As epilepsy occurs, due to episodic nature of the seizures, the detention and prediction of the seizures with visualization of the EEG signals by Neuro-experts are a very tedious and time-consuming process. The EEG signals are practically very lengthy. Hence, the prime challenge is to develop signal processing algorithms to explore the underlying behavior of the nonstationary and nonlinear EEG signals. Many nonlinear classification methods have been proposed that are able to capture the significant underlying information about the dynamics of the brain's electrical activities. In this study, an elegant and convenient nonlinear HOS algorithm is proposed to extract relevant information from the EEG signals. The various statistical features are measured from the bispectrum of the EEG signals. These extracted measurements are input to the LSDA for discrimination of features according to the class label. The SVM with the different kernels (to maximize the probability of separation) are used for signal classification. The proposed method has achieved an excellent classification accuracy to classify seizures in various categories. The achieved results show the effectiveness of the proposed algorithm.

References

1. Acharya UR (2013) Automated diagnosis of epileptic electroencephalogram using independent component analysis and discrete wavelet transform for different electroencephalogram durations. Proc Inst Mech Eng H J Eng Med 227(3):234–244
2. Acharya UR, Chua CK, Lim TC, Dorithy, Suri JS (2009) Automatic identification of epileptic EEG signals using nonlinear parameters. J Mech Med Biol 9(04):539–553

3. Acharya UR, Sree SV, Chattopadhyay S, Yu W, Ang PCA (2011) Application of recurrence quantification analysis for the automated identification of epileptic EEG signals. Int J Neural Syst 21(03):199–211
4. Acharya UR, Sree SV, Alvin APC, Suri JS (2012) Use of principal component analysis for automatic classification of epileptic EEG activities in wavelet framework. Exp Syst Appl 39 (10):9072–9078
5. Alam SS, Bhuiyan MIH (2013) Detection of seizure and epilepsy using higher order statistics in the EMD domain. IEEE J Biomed Health Inform 17(2):312–318
6. Altunay S, Telatar Z, Erogul O (2010) Epileptic EEG detection using the linear prediction error energy. Expert Syst Appl 37(8):5661–5665
7. Andrzejak RG, Lehnertz K, Mormann F, Rieke C, David P, Elger CE (2001) Indications of nonlinear deterministic and finite-dimensional structures in time series of brain electrical activity: dependence on recording region and brain state. Phys Rev E 64(6):061907
8. Bajaj V, Pachori RB (2013) Epileptic seizure detection based on the instantaneous area of analytic intrinsic mode functions of EEG signals. Biomed Eng Lett 3(1):17–21
9. Bao FS, Lie DYC, Zhang Y (2008) A new approach to automated epileptic diagnosis using EEG and probabilistic neural network. In: Tools with artificial intelligence, 2008. ICTAI'08. 20th IEEE international conference on, vol 2. IEEE, Dayton, pp 482–486
10. Bassett DS, Sporns O (2017) Network neuroscience. Nat Neurosci 20(3):353
11. Bhati D, Pachori RB, Gadre VM (2017a) A novel approach for time-frequency localization of scaling functions and design of three-band biorthogonal linear phase wavelet filter banks. Digital Signal Processing 69:309–322
12. Bhati D, Sharma M, Pachori RB, Gadre VM (2017b) Time-frequency localized three-band biorthogonal wavelet filter bank using semidefinite relaxation and nonlinear least squares with epileptic seizure EEG signal classification. Digital Signal Processing 62:259–273
13. Bhattacharyya A, Pachori RB (2017) A multivariate approach for patient-specific EEG seizure detection using empirical wavelet transform. IEEE Trans Biomed Eng 64(9):2003–2015
14. Bhattacharyya A, Pachori RB, Upadhyay A, Acharya UR (2017) Tunable wavelet transform based multiscale entropy measure for automated classification of epileptic EEG signals. Appl Sci 7(4):385
15. Brillinger DR (1965) An introduction to polyspectra. Ann Math Stat 36:1351–1374
16. Cai D, He X, Zhou K, Han J, Bao H (2007) Locality sensitive discriminant analysis. IJCAI 2007:1713–1726
17. Chen G (2014) Automatic EEG seizure detection using dual-tree complex wavelet-Fourier features. Exp Syst Appl 41(5):2391–2394
18. Cortes C, Vapnik V (1995) Support-vector networks. Mach Learn 20(3):273–297
19. Faust O, Acharya UR, Min LC, Sputh BH (2010) Automatic identification of epileptic and background EEG signals using frequency domain parameters. Int J Neural Syst 20(02):159–176
20. Fu K, Qu J, Chai Y, Dong Y (2014) Classification of seizure based on the time-frequency image of EEG signals using HHT and SVM. Biomed Signal Process Control 13:15–22
21. Fu K, Qu J, Chai Y, Zou T (2015) Hilbert marginal spectrum analysis for automatic seizure detection in EEG signals. Biomed Signal Process Control 18:179–185
22. Ghosh-Dastidar S, Adeli H, Dadmehr N (2007) Mixed-band wavelet-chaos neural network methodology for epilepsy and epileptic seizure detection. IEEE Trans Biomed Eng 54 (9):1545–1551
23. Güler NF, Ubeyli ED, Güler I (2005) Recurrent neural networks employing Lyapunov exponents for EEG signals classification. Expert Systems with Applications 29(3):506–514
24. Guo L, Rivero D, Seoane JA, Pazos A (2009) Classification of EEG signals using relative wavelet energy and artificial neural networks. In: Proceedings of the first ACM/SIGEVO Summit on Genetic and Evolutionary Computation. ACM, New York, pp 177–184
25. Guo L, Rivero D, Dorado J, Rabunal JR, Pazos A (2010a) Automatic epileptic seizure detection in EEG based on line length feature and artificial neural networks. J Neurosci Methods 191 (1):101–109

26. Guo L, Rivero D, Pazos A (2010b) Epileptic seizure detection using multiwavelet transform based approximate entropy and artificial neural networks. J Neurosci Methods 193(1):156–163
27. Guo L, Rivero D, Dorado J, Munteanu CR, Pazos A (2011) Automatic feature extraction using genetic programming: An application to epileptic EEG classification. Expert Syst Appl 38 (8):10425–10436
28. Hassan AR, Haque MA (2015) Epilepsy and seizure detection using statistical features in the complete ensemble empirical mode decomposition domain. In: TENCON 2015-2015 IEEE Region 10 Conference. IEEE, pp 1–6. https://doi.org/10.1109/ICMEW.2016.7574707-8
29. Hassan AR, Siuly S, Zhang Y (2016) Epileptic seizure detection in EEG signals using tunable-q factor wavelet transform and bootstrap aggregating. Comput Methods Progr Biomed 137:247–259
30. Hsu CW, Chang CC, Lin CJ et al (2003) A practical guide to support vector classification. University of National Taiwan, Taipei
31. Huang NE, Shen Z, Long SR, Wu MC, Shih HH, Zheng Q, Yen NC, Tung CC, Liu HH (1998) The empirical mode decomposition and the Hilbert spectrum for nonlinear and non-stationary time series analysis. In: Proceedings of the Royal Society of London A: mathematical, physical and engineering sciences, vol 454. The Royal Society, London, pp 903–995
32. Iscan Z, Dokur Z, Demiralp T (2011) Classification of electroencephalogram signals with combined time and frequency features. Expert Syst Appl 38(8):10499–10505
33. Jirsa VK, Stacey WC, Quilichini PP, Ivanov AI, Bernard C (2014) On the nature of seizure dynamics. Brain 137(8):2210–2230
34. Joshi V, Pachori RB, Vijesh A (2014) Classification of ictal and seizure-free EEG signals using fractional linear prediction. Biomed Signal Process Control 9:1–5
35. Kaya Y, Uyar M, Tekin R, Yıldırım S (2014) 1d-local binary pattern based feature extraction for the classification of epileptic EEG signals. Appl Math Comput 243:209–219
36. Khamis H, Mohamed A, Simpson S (2009) Seizure state detection of temporal lobe seizures by autoregressive spectral analysis of scalp EEG. Clin Neurophysiol 120(8):1479–1488
37. Kim SH, Faloutsos C, Yang HJ (2013) Coercively adjusted autoregression model for forecasting in epilepsy EEG. Comput Math Methods Med 2013. https://doi.org/10.1155/2013/545613
38. Kreyszig E (2011) Advanced engineering mathematics. J. Willey & Sons Inc, New York, London
39. Kumar Y, Dewal M, Anand R (2014) Epileptic seizures detection in EEG using DWT-based ApEn and artificial neural network. Signal, Image, Video Process 8(7):1323–1334
40. Kumar TS, Kanhangad V, Pachori RB (2015) Classification of seizure and seizure-free EEG signals using local binary patterns. Biomed Signal Process Control 15:33–40
41. Labate D, La Foresta F, Morabito G, Palamara I, Morabito FC (2013a) Entropic measures of EEG complexity in Alzheimer's disease through a multivariate multiscale approach. IEEE Sensors J 13(9):3284–3292
42. Labate D, Palamara I, Mammone N, Morabito G, La Foresta F, Morabito FC (2013b) SVM classification of epileptic EEG recordings through multiscale permutation entropy. In: Neural Networks (IJCNN), The 2013 International Joint Conference on. IEEE, San Diego, pp 1–5
43. Lee SH, Lim JS, Kim JK, Yang J, Lee Y (2014) Classification of normal and epileptic seizure EEG signals using wavelet transform, phase-space reconstruction, and Euclidean distance. Comput Methods Progr Biomed 116(1):10–25
44. Li S, Zhou W, Yuan Q, Geng S, Cai D (2013) Feature extraction and recognition of ictal EEG using EMD and SVM. Comput Biol Med 43(7):807–816
45. Mammone N, Principe JC, Morabito FC, Shiau DS, Sackellares JC (2010) Visualization and modeling of STLmax topographic brain activity maps. J Neurosci Methods 189(2):281–294
46. Nicolaou N, Georgiou J (2012) Detection of epileptic electroencephalogram based on permutation entropy and support vector machines. Expert Syst Appl 39(1):202–209
47. Nikias CL, Mendel JM (1993) Signal processing with higher-order spectra. IEEE Signal Process Mag 10(3):10–37

48. Nikias CL, Petropulu AP (1993) Higher-order spectra analysis: a nonlinear signal processing framework. PTR Prentice Hall, Englewood Cliffs
49. Nikias CL, Raghuveer MR (1987) Bispectrum estimation: A digital signal processing framework. Proc IEEE 75(7):869–891
50. Ocak H (2009) Automatic detection of epileptic seizures in EEG using discrete wavelet transform and approximate entropy. Expert Syst Appl 36(2):2027–2036
51. Orhan U, Hekim M, Ozer M (2011) Eeg signals classification using the K-means clustering and a multilayer perceptron neural network model. Expert Syst Appl 38(10):13475–13481
52. Oweis RJ, Abdulhay EW (2011) Seizure classification in EEG signals utilizing Hilbert-Huang transform. Biomed Eng Online 10(1):38
53. Pachori RB (2008) Discrimination between ictal and seizure-free EEG signals using empirical mode decomposition. Res Lett Signal Process 2008:14
54. Pachori RB, Bajaj V (2011) Analysis of normal and epileptic seizure EEG signals using empirical mode decomposition. Comput Methods Prog Biomed 104(3):373–381
55. Pachori RB, Patidar S (2014) Epileptic seizure classification in EEG signals using a second-order difference plot of intrinsic mode functions. Comput Methods Prog Biomed 113 (2):494–502
56. Pachori RB, Sharma R, Patidar S (2015) Classification of normal and epileptic seizure EEG signals based on empirical mode decomposition. In: Complex system modeling and control through intelligent soft computations. Springer, Cham, pp 367–388
57. Pati S, Alexopoulos AV (2010) Pharmacoresistant epilepsy: from pathogenesis to current and emerging therapies. Cleve Clin J Med 77(7):457–567
58. Patidar S, Panigrahi T (2017) Detection of epileptic seizure using kraskov entropy applied on tunable-q wavelet transform of EEG signals. Biomed Signal Process Control 34:74–80
59. Peker M, Sen B, Delen D (2016) A novel method for automated diagnosis of epilepsy using complex-valued classifiers. IEEE J Biomed Health Inform 20(1):108–118
60. Polat K, Güneş S (2007) Classification of epileptiform EEG using a hybrid system based on decision tree classifier and fast Fourier transform. Appl Math Comput 187(2):1017–1026
61. Polat K, Güneş S (2008a) Artificial immune recognition system with fuzzy resource allocation mechanism classifier, principal component analysis, and FFT method based new hybrid automated identification system for classification of EEG signals. Expert Systems with Applications 34(3):2039–2048
62. Polat K, Güneş S (2008b) A novel data reduction method: distance-based data reduction and its application to classification of epileptiform EEG signals. Appl Math Comput 200(1):10–27
63. Pradhan N, Dutt DN (1994) Data compression by a linear prediction for storage and transmission of EEG signals. Int J Bio-Med Comput 35(3):207–217
64. Rosenblatt M, Van Ness JW et al (1965) Estimation of the bispectrum. Ann Math Stat 36 (4):1120–1136
65. Samiee K, Kovacs P, Gabbouj M (2015) Epileptic seizure classification of EEG time-series using rational discrete short-time Fourier transform. IEEE Trans Biomed Eng 62(2):541–552
66. Shannon C (1948) A mathematical theory of communication. Bell Syst Tech 27:379–423
67. Sharma R, Pachori RB (2015) Classification of epileptic seizures in EEG signals based on phase space representation of intrinsic mode functions. Expert Syst Appl 42(3):1106–1117
68. Sharma M, Pachori RB (2017a) A novel approach to detect epileptic seizures using a combination of tunable-q wavelet transform and fractal dimension. J Mech Med Biol 17(07):1740003
69. Sharma RR, Pachori RB (2017b) Time-frequency representation using IEVDHM-HT with application to classification of epileptic EEG signals. IET Sci Meas Technol 12(1):72–82
70. Sharma M, Pachori RB, Acharya UR (2017) A new approach to characterize epileptic seizures using analytic time-frequency flexible wavelet transform and fractal dimension. Pattern Recogn Lett 94:172–179
71. Sharma R, Sircar P, Pachori RB (2019a) A new technique for classification of focal and nonfocal EEG signals using higher-order spectra. J Mech Med Biol 19(01):1940010

72. Sharma R, Sircar P, Pachori RB, Bhandary SV, Acharya UR (2019b) Automated glaucoma detection using a center slice of higher order statistics. J Mech Med Biol 19(01):1940011
73. Shin Y, Lee S, Ahn M, Cho H, Jun SC, Lee HN (2015) Noise robustness analysis of sparse representation based classification method for nonstationary EEG signal classification. Biomed Signal Process Control 21:8–18
74. Spiegel RM, Murray R, Stephens LJ (1999) Theory and problems of statistics, 3rd edn. McGraw-Hill, New York
75. Srinivasan V, Eswaran C, Sriraam N (2005) Artificial neural network based epileptic detection using time-domain and frequency-domain features. J Med Syst 29(6):647–660
76. Srinivasan V, Eswaran C, Sriraam N (2007) Approximate entropy-based epileptic EEG detection using artificial neural networks. IEEE Trans Inf Technol Biomed 11(3):288–295
77. Subasi A (2007) Eeg signal classification using wavelet feature extraction and amixture of expert model. Expert Syst Appl 32(4):1084–1093
78. Subasi A, Gursoy MI (2010) Eeg signal classification using PCA, ICA, LDA and support vector machines. Expert Syst Appl 37(12):8659–8666
79. Tao Z, Wan-Zhong C, Ming-Yang L (2016) Automatic seizure detection of electroencephalogram signals based on frequency slice wavelet transform and support vector machine. Acta Phys Sin 65(3):038703
80. Tiwari AK, Pachori RB, Kanhangad V, Panigrahi BK (2017) Automated diagnosis of epilepsy using key-point based local binary pattern of EEG signals. IEEE J Biomed Health Inform 21 (4):888–896
81. Tzallas AT, Tsipouras MG, Fotiadis DI (2007) Automatic seizure detection based on time-frequency analysis and artificial neural networks. Comput Intell Neurosci 2007:80510
82. Tzallas AT, Tsipouras MG, Fotiadis DI (2009) Epileptic seizure detection in EEGs using time-frequency analysis. IEEE Trans Inf Technol Biomed 13(5):703–710
83. Ubeyli ED (2010) Least squares support vector machine employing model-based methods coefficients for the analysis of EEG signals. Expert Syst Appl 37(1):233–239
84. Uthayakumar R, Easwaramoorthy D (2013) Epileptic seizure detection in EEG signals using multifractal analysis and wavelet transform. Fractals 21(02):1350011
85. Vapnik V (2013) The nature of statistical learning theory. Springer Science &Business Media, New York
86. Wang D, Miao D, Xie C (2011) Best basis-based wavelet packet entropy feature extraction and hierarchical EEG classification for epileptic detection. Expert Syst Appl 38(11):14314–14320
87. Wang S, Chaovalitwongse WA, Wong S (2013) Online seizure prediction using an adaptive learning approach. IEEE Trans Knowl Data Eng 25(12):2854–2866
88. Webber W, Lesser RP, Richardson RT, Wilson K (1996) An approach to seizure detection using an artificial neural network (ann). Electroencephal Clin Neurophysiol 98(4):250–272
89. Weston J, Watkins C et al (1999) Support vector machines for multi-class pattern recognition. Esann 99:219–224
90. WHO (2015) Epilepsy report by world health organization. URL http://www.who.int/en/news-room/fact-sheets/detail/epilepsy
91. Wu Z, Huang NE (2009) Ensemble empirical mode decomposition: a noise-assisted data analysis method. Adv Adapt Data Anal 1(01):1–41
92. Yuan Q, Zhou W, Li S, Cai D (2011) Epileptic EEG classification based on extreme learning machine and nonlinear features. Epilepsy Res 96(1-2):29–38
93. Zhang Z, Zhou Y, Chen Z, Tian X, Du S, Huang R (2013) Approximate entropy and support vector machines for electroencephalogram signal classification. Neural Regen Res 8(20):18–44
94. Zhang T, Chen W, Li M (2017) Ar based quadratic feature extraction in the VMD domain for the automated seizure detection of EEG using random forest classifier. Biomed Signal Process Control 31:550–559
95. Zhou W, Liu Y, Yuan Q, Li X (2013) Epileptic seizure detection using lacunarity and Bayesian linear discriminant analysis in intracranial EEG. IEEE Trans Biomed Eng 60(12):3375–3381

Electroencephalogram: Expanded Applications in Clinical and Nonclinical Settings

11

Shivadata Prabhu

Abstract

Electroencephalography (EEG) has been used for the understanding of brain functions and in clinical neurosciences for more than eight decades. Its importance in applied fields related to cognition has assumed more importance in spite of advances in functional neuroimaging in recent years. This article reviews methods in EEG analysis and functional significance of oscillatory synchrony in different bands as related to cognition. It then further mentions the potential role of EEG as biomarker and its use in studying consumer behaviour and effects of meditation by mentioning a few examples.

Keywords

Cognition · Electroencephalogram · Oscillatory synchrony · EEG methods · EEG biomarker · Consumer neurosciences · Meditation

11.1 Introduction

Ever since Berger found out the modulation of alpha waves of electroencephalogram (EEG) by mental tasks, the notion that EEG may be able to represent the cognition as it happens in brain got formed. The idea that studying EEG would provide an opportunity to study cognition has interested neuroscientists. This interest over decades has been waxing and waning. From the 1980s onwards, the great strides in functional neuroimaging like fMRI, PET and SPECT ushered an era where EEG became secondary and mainly confined to its application in clinical domains like epilepsy. However, recent advances in understanding of neuronal functioning and improved use of computational methods, applied to neural systems, have brought EEG again in prominence in applied fields like consumer neurosciences and

S. Prabhu (✉)
School of Nanoscience and Technology, Shivaji University, Kolhapur, India

S. Paul (ed.), *Application of Biomedical Engineering in Neuroscience*,
https://doi.org/10.1007/978-981-13-7142-4_11

meditation. This mini review aims to document these methods and recent use of EEG in these fields.

The recent advancements in EEG application have been made possible by developments in several fields that occurred simultaneously.

- *Digital EEG*
  - The days when EEG trace used to be an ink and paper faithful recording of the voltage across the scalp are gone. Now the hardware and software are completely digitized. This has increased recording ease, removal of artefacts. Analysis of signals by manipulation of the on screen records digitally is now simpler. Putting high and low band filters, getting topographic maps, getting information about voltage and frequency and power spectrum analysis all have become easy in digital age.
- *Multi-electrode EEG hardware*
  - Now the hardware has also improved. Multi-electrode array with 64,128 channels are available to record electroencephalogram. This improves spatial resolution and also helps in EEG analysis like source localization. The electrodes, conducting gel and amplifiers are also modern, making the signals noise- and artefact-free.
- *Increased understanding of brain functions*
  - Our understanding of brain processing as a computational framework and in terms of information flow in the organ has increased substantially in recent years. Though not complete, it is much advanced than before. This has improved the applicability of EEG in understanding brains in diseased state and in its normal functions.
- *Development of AI*
  - Application of artificial intelligence and big data analysis has made it easy to analyse EEG signals in an automated, predictable and faithful way. The AI also helps in chunking large data like EEG waveform and establish relationships hidden in it. The algorithms generated are validated by applying them to known cases. As they are self-learning they become more and more robust as time advances.

11.1.1 Steps in EEG Signal Processing

The signal processing usually is carried out through following general steps.

- *Pre-processing*
  - The raw EEG data is not suitable for immediate analysis. Cleaning is required to get data of interest. Filters such as high-pass filters at cut-off 1 Hz usually remove direct current DC components. Similarly, low-pass filter also is used to remove vary high gamma band frequencies. Gamma frequencies are actually receiving renewed attention as carrier waves. Similarly, artefacts generated by eye blinking, muscle contraction and eye movements need to be removed.

Such clean EEG is then cut off in time (lasting few seconds) called epoch; then, the next stage is applied.

- *Feature extraction*
  - The feature extraction is done to gather overall characteristics of the EEG signal. Usually time domain features like mean frequency, amplitude and entropy are studied. Frequency and amplitude are simple terms well defined in physics. Entropy is a term borrowed from thermodynamics . Itrepresents instability and uncertainty in system. Applied to EEG analysis, it represents the probabilistic functions of uncertainty in signals. Frequency domain features like Fast Fourier Transdformation (FFT) and wavelet analysis are also used. Both these techniques help in analysis of rapidly changing transient signal like EEG at each electrode on the scalp. Similarly, a topographic map of brain indicating several regions of brain being active in synchrony can be generated. A complex network analysis also is possible which informs us about relationships between different scalp electrodes representing certain groups of cortical neurons firing together from different regions of the brain.
- *Feature selection*
  - This step selects a particular feature and separates it among the large sets of features that ae generated in step 2. The statistical methods such as principle component analysis are used in this step.
- *Classification*
  - After the feature selection, machine learning (artificial intelligence) is used to generate and train a classifier which recognises selected feature. It puts them in a certain class or condition. This helps in the final correlation between EEG signals and the conditions it represents in the brain. For example, characteristic EEG signatures of diseases like Alzheimer's and states of normal brain like emotion, memory and attention in consumer neurosciences and meditative states can be documented. Figure 11.1 illustrates these steps.

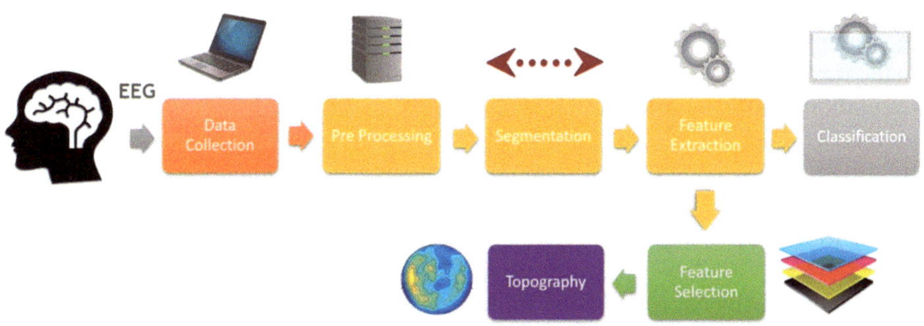

Fig. 11.1 Steps in EEG signal processing

11.1.2 EEG And Cognition: The Concept Development From Waves to Oscillations

EEG has a low spatial resolution but high temporal correlation to neuronal activities. The scalp EEG represents summation of activities of group of neurons in the cortex. In the early years of 1950s and 1960s, the studies were restricted to understanding waves of EEG based on their frequency and change in the waveform as external stimuli were processed by the brain. Hence, the popular terms like 'beta band are waves of attention' and 'theta band represents waves of sleep' were in vogue at that time.

Advances in microelectrode recording led to direct recording of cortical neurons in vivo. And it was discovered that external events indeed lead to definite reproducible responses in single cortical neurons [1]. Soon the theory of 'receptive fields, 'which maintained that action potentials generated in single neurons represent the presence or absence of external stimuli in the particular location gained prominence. Experiments were designed to validate this theory better. In this period, the temporal dynamics of neuronal activities were rather neglected in preference to individual neuronal firing patterns. These firing patterns of individual and group of neurons was thought to be a real representation of cognition as it happens in brain. So the interactions between neuronal ensembles was largely ignored [2].

But cognition is a complex process. It refers to the mental process by which external or internal input is transformed, stored, recovered and used. As such, it involves many things like perception, attention, memory, decision-making, reasoning, problem solving, imaging, planning and executing actions [3]. Mere understanding of activity of group of neurons was insufficient to explain cognition completely. But the studies in neuronal firing led to the understanding that there are patterns in the activities of neurons. These patterns change over time (temporal dynamics) and across regions of brain (spatial dynamics) when processing stimuli (the 'cognition'). Temporal dynamics were then considered the major mechanism of information exchange in brain regions and the fundamental basis of cognition.

The Computational Brain

The popular notion that brain works like a computer is not very far from the truth. The graphic below will illustrate that the brain and digital computers work almost on the same principles. For digital computers, the information is coded and transmitted in the form of binary codes, but in the brain it is the neuronal firing in synchrony and oscillations that carry information. Figure 11.2 explains this concept. Both digital computer and brain receive inputs from the environment, process them, store information and then provide outputs. Of course, the similarity ends here. The brain is a much more sophisticated processor with larger memory capacity and ability to use fuzzy logic more than the digital computer.

The Firing Patterns: Synchrony and Oscillations

The firing patterns of individual neurons or group of neurons can be described in following terms:

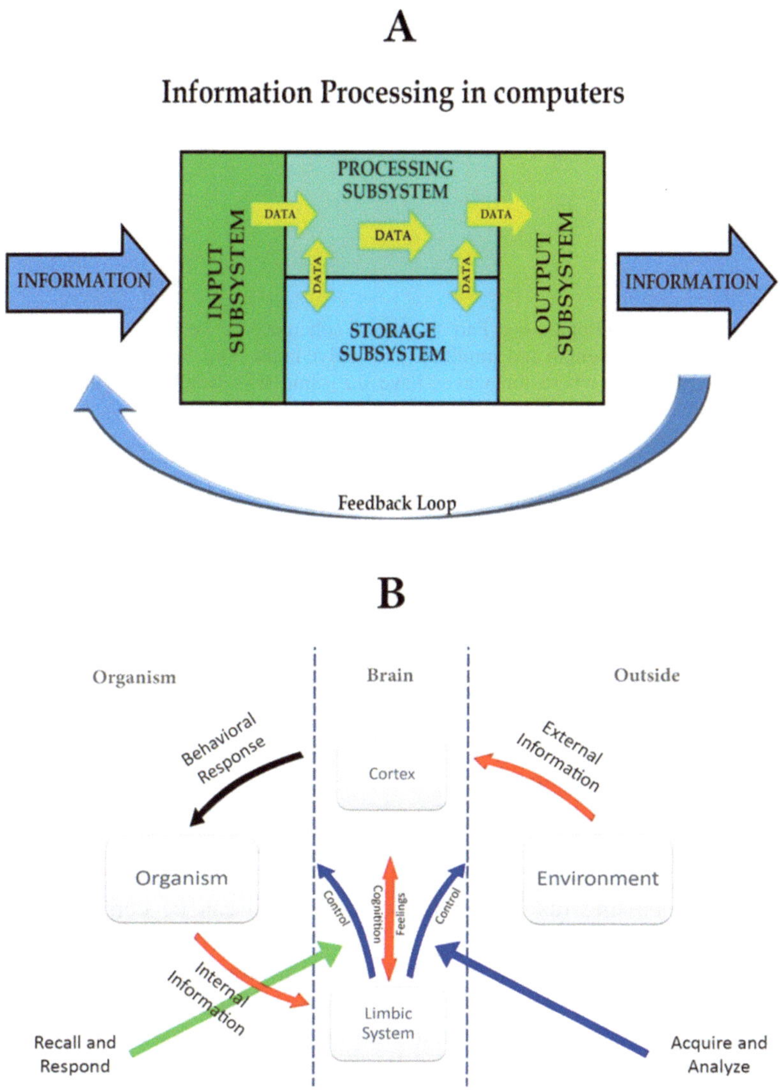

Fig. 11.2 Schematic diagram of the brain and computer working. Both electronic computer and brain work in similar ways where an input is processed, and an output is given. Many times the output acts as a feedback for input for future processing. In computers, the information flow is by binary codes, while in brain the flow is by neural firing representing a code

Oscillations: The neuronal firing pattern and frequency vary with time. The spike timing of neurons at rest and during stimulation vary periodically, and this is termed as oscillation.

Synchrony. This is the tendency of different groups of neurons to fire at the same time.

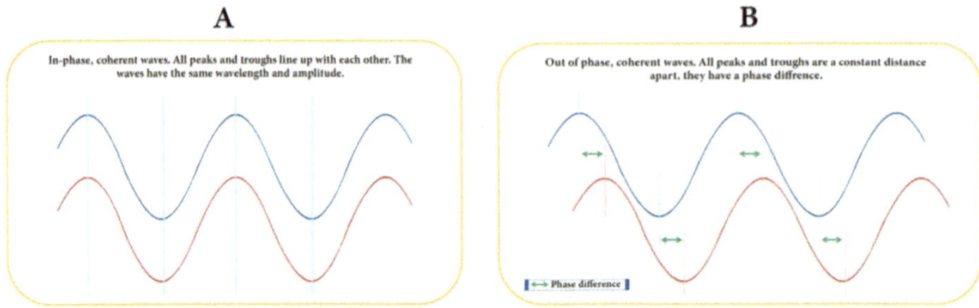

Fig. 11.3 Coherence and phase lag. Part 'A' shows coherence between waves when both waves have the exact same wavelength and amplitude and evolve in the same time domain. Part 'B' shows a phase difference (lag) where the waves have the same wavelength and amplitude, but the red-coloured wave appears more lagged than the wave coloured blue

Oscillatory synchrony: When neuronal ensembles located in different space fire at the same time with same frequency, the phenomenon is called oscillatory synchrony.

While analysing EEG signals, two more terms became relevant: coherence and phase.

EEG coherence. It is a measure of the degree of association or coupling of frequency spectra between two different time series.

Phase: EEG phase delays are a measure of the temporal leading or lagging of spectra. Following figure illustrates these basic concepts. Figure 11.3 illustrates concepts of coherence and phase.

If we consider the brain as a network of neurons, then these spontaneous 'rhythms' of EEG can be modelled as network oscillations. The role of network oscillations in information processing (as an implicit of cognition) can be considered in three possible ways.

First possibility is that network oscillations contribute to the representation of information. Second possibility is that oscillations and synchrony regulate the flow of information in neural circuits. And the third possibility is that oscillations assist in the storage and retrieval of information in neural circuits. All these possibilities can be true simultaneously (Sejnowski and Paulsen 2006).

11.2 Theoretical Considerations in Understanding Synchrony as a Communication Tool

11.2.1 The CTC Model

Oscillatory synchrony between different population of neurons serves as communication and information transfer in the brain. The theoretical framework to understand this phenomenon was laid by many workers but especially by Fries [4, 5] in his theory of 'communication through coherence' (CTC) model.

CTC postulates that a group of neuron, when activated, engages itself into rhythmic synchronization. This synchronization produces sequences of excitation and inhibition. These excitations and inhibitions in turn modify both spike output and sensitivity to synaptic input for very short time windows. The rhythmic changes in postsynaptic excitability produce rhythmic modulations in synaptic input gain. Inputs that consistently arrive at moments of high input gain lead to enhanced effective connectivity. Thus, strong effective connectivity requires rhythmic synchronization within pre- and postsynaptic groups.

The postsynaptic group of neurons would essentially receive many presynaptic inputs. This is related to the fact that synapsis are complex and receive multiple inputs from other neurons converging on the presynaptic area. These postsynaptic neurons respond more to the specific presynaptic group to which it has more coherence. Thus, selective coherence is the basis of selective communication among the neuronal groups. It explains why a smooth flow of information happens in the brain despite the complex system of billions of neurons and trillions of synapses in the brain.

All these three postulates of CTC theory have some experimental proofs.

The postulate of presynaptic modulation by postsynaptic coherence is experimentally validated [6]. The other postulate that strong communication requires strong coherence is also supported by experimental evidence [7]. Finally the postulate that selective communication requires selective coherence is also substantiated by experimental evidence especially in the domain of visual attention and selective movement intention [8].

11.2.2 Information Flow in the Brain

Fngelkurts and Fingelkurts have explained the information flow in the brain elegantly in a review [9]. They explain that brain is a non-linear, nonstable system. That means that the changes in outputs of system need not always be proportional to changes in the inputs. The information flow in the brain as small perturbations in the neural system. These transient instabilities take place at all scales (micro, meso and macro) in a nested hierarchy of multiple coordinative processes. Nested hierarchy means groups of related things share common shared characteristics. So brain functioning can be considered as a harmonic orchestra where beautiful music is created by different instruments playing in related form in specific time interval. The

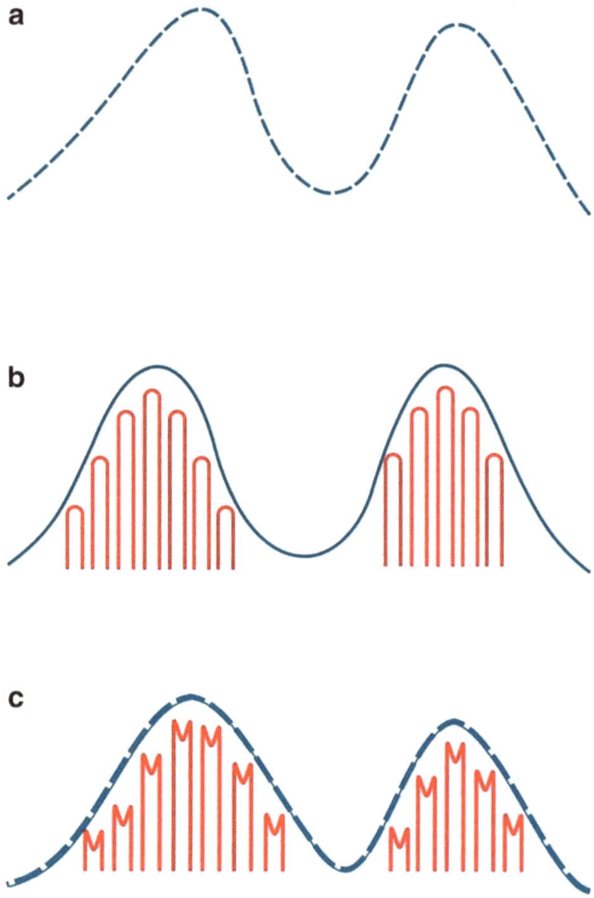

Fig. 11.4 Schematic illustration of information flow in the brain by perturbations. (**a**) The figure represents a metastable brain state where at large-scale network level neuronal assemblies are stable. (**b**) Each metastate is composed of meso- and microstates indicated by red bars with round tops. (**c**) Small perturbations in microstates represented by jagged tops of red bars is the event that represents information flow. The information flow is achieved by both autonomous and synchronous firing of group of neurons

autonomous tendencies coexist together with the interdependent tendencies of the network [10]. It maintains that population of neurons can engage and disengage in self-assembly mode to perceive external word as information, mental imagery, etc. When these perturbations finally bind them all together into a coherent whole, then a new principle of metastability of brain functioning emerges.

Metastability in the brain refers to the competition of complementary tendencies of cooperative integration and autonomous fragmentation among multiple distributed neuronal assemblies [11]. In simple terms, it means the neuronal population can synchronize for cooperative function and also independently. The interplay of these two tendencies (autonomy and integration) constitutes the metastable regime of brain functioning [10, 11]. This concept is illustrated well in Fig. 11.4.

11.2.3 Tools to Understand EEG Synchrony and Coherence

The dynamics of EEG can be studied in two ways:

1. The spontaneous coherence between different brain regions in different frequencies that occur because of a hidden process
2. Oscillatory changes that happen by evoked stimuli either sensory (sensory-evoked potentials) and cognition-related stimuli (event-related potentials)

Both, but especially later, methods are useful when EEG is applied in domains of neurological diseases and consumer neurosciences and for understanding meditation.

In order to study the synchrony between neuronal signals, as cognition happens, it is necessary to analyse EEG signals. Advances in designing algorithms based on non-linear physics have helped in making EEG analysis reliable and speedy. The dynamic network phenomenon are best understood by these mathematical methods.

Non-linear Dynamics and EEG

A dynamic system is any system which has 'states' that change more or less systematically over time. A dynamic system is deterministic when its future values can be predicted by having sufficient information about past states. A non-deterministic or stochastic system is one where it is impossible to predict an exact future value even if one knows its entire past history. EEG essentially is a stochastic process, and concepts of non-linear or chaos theory can easily be applied to the EEG.

Time domain and frequency domain are two ways of looking at the same dynamic system. Roughly speaking, in the time domain, we measure how long something takes, whereas in the frequency domain, we measure how fast or slow it is. These are just two ways of looking at the same data. The time domain is a record of the response of a dynamic system, as indicated by some measured parameter, as a function of time. This is the traditional way of observing the output of a dynamic system.

Fast Fourier Transformation (FFT) Jean Baptiste Fourier showed that any waveform that exists in the real world can be generated by adding up sine waves. By picking the amplitudes, frequencies and phases of these sine waves, one can generate a waveform identical to the desired signal. Every such sine wave generated from input appears as a vertical line. Its height represents its amplitude and its position represents its frequency. So input signal can thus be transformed in the frequency domain. This frequency domain representation of signal is called the spectrum of the signal. Each sine wave line of the spectrum is called a component of the total signal. Since any real-world signal can be represented by the sum of sine waves, we can predict the response of a system, for example the brain oscillations by processing these waves and studying their frequency, amplitude and phase angle mathematically.

11.2.4 Predicting Cognition

Using these mathematical tools, EEG signal can be studied for understanding cognition. There are several methods to analyse EEG signal. The details of each method are outside the aims of this review. We mention here some of these methods which are found robust and predictable.

Technique	Description	Usefulness
Power spectral density	Describes the power present in the signal as a function of frequency, per unit frequency	Marker of strength of activity of group of neurons
Evoked spectra	FFT analysis of the sensory-evoked potential elicited by simple light, tone signal, etc.	Documents early processing of sensory information
Event-related spectra	FFT analysis of an event-related potential (ERP), e.g. target or non-target signal during an oddball paradigm	Documents early response to target words and images which represent cognition
Phase-locking, phase synchrony	Used to investigate task-induced changes in long-range synchronization of neural activity	Indicates changes in system by a specific task
Cross-correlation	*Cross-correlation* measures the similarity between two discrete time series, say X and Y. Here X and shifted (lagged) copies of Y are studied as a function of the lag	Indicates evolution of cognitive processing as time passes in different brain regions
EEG coherence	Quantifies the association between pairs of signals as a function of frequency	Useful for measuring changes in EEG topography related to cognitive tasks
Inter-electrode correlation	Quantifies the association between pairs of signals as a function of frequency between different electrodes located on the scalp	Indicates how different brain regions 'cross-talk' to make a holistic cognition

Combining Both Analysis Methods Frequency domain analysis is ideal for signals that are stationary over long duration of time. But practically, EEG signals, in general, represent a nonstationary process. This means that it will be insufficient to use frequency components from the classical spectrum estimators. In order to overcome this problem, the nonstationary EEG data can be divided into short durations over which the signal is considered stationary and then time series analysis can be used. Several joint time–frequency representation techniques have been proposed; among them, the short-time Fourier transform (STFT) and the wavelet transform (WT) have been found to be useful.

Why So Many Methods? We need different mathematical techniques to understand cognition by EEG signal analysis. This is because of the fact that the responses of neuronal groups on receiving stimulus are selectively distributed. The amplitude in different frequencies and phase locking also varies according to type of stimulus.

Also, the different electrodes placed at different locations show variable latency in response to stimulus. Hence, in order to understand integrative brain functions, different mathematical methods are necessary.

11.3 All Frequencies and All the Evidence for Cognition

We now look at traditional frequencies of EEG and its relation to cognition as documented by experiments. The number of experiments available are large and just few representative samples have been mentioned. Cognition is a complex interplay of different individual constituent components like emotions, memory and attention. Hence, we find many changes in EEG signals in all frequencies of EEG that reflect some parts of cognition.

Delta: This band has a frequency of 3.5 Hz and below. It appears highest in amplitude and the lowest in frequency. It is the dominant rhythm in infants up to 1 year and in stages 3 and 4 of sleep [12].

Theta: This band has a frequency of 3.5–7.5 Hz and is classified as 'slow' activity. It is normal in children up to 13 years and in sleep but abnormal in awake adults.

Alpha: This band has a frequency between 7.5 and 12.5 Hz and is usually best seen in the posterior regions of the head on each side, with the higher amplitude on the dominant side. It appears when closing the eyes and relaxing, and disappears when opening the eyes or when concentrating on any mechanism. It is major rhythm noted in normal adults in relaxed states

Beta: This band has a frequency between 12.5 and 30.0 Hz and is usually seen on both sides in a symmetrical distribution; it is most evident frontally. It is the dominant rhythm in adults who are alert, noxious or have their eyes open.

Gamma: This band has a frequency range of 31 Hz and greater. It is thought that it reflects the mechanism of consciousness. Beta and gamma waves together have been associated with attention, perception and cognition [13, 14].

Evidence from Experimental Studies

Delta

Oscillations in delta band and their implications in cognitive processing have been studied experimentally. Researchers have shown that that delta oscillatory responses were involved in 'perception, attention learning, decision and working memory' [15].

Theta

Experiments indicate that there is increase in the frontal theta power during bimodal sensory stimulation; thus, demonstrating that complex events increase the frontal processing in the theta range [16]. Event-related oscillations in the theta band are prolonged or have a second time window approximately 300 ms after the presentation of the target in oddball experiments. This prolongation can be considered a marker of selective attention. In another experiment, it was demonstrated that increased coherence in theta power activity of an array of left-frontal, left-parietal and bilateral-inferior-temporal cortices may assist retrieval during sentence comprehension. It would also have implication in verbal working memory and long-term memory [17]. In yet another experiment, transient increase in theta coherence in left temporal region was shown to aid in rapid memory stabilization [18].

Alpha Band

It was thought initially that alpha waves represent 'idling of brain', but in the 1980s, Basar demonstrated the functional relevance of alpha activity [19]. He continued with his work that led to more understanding of role of alpha in cognition. In experiments with invasive microelectrode recording, it was shown that auditory and visual stimulations elicit alpha responses in the auditory and visual pathways of the hippocampus and reticular formation. Basar [20] maintained that it represented sensory functions of alpha rhythm. Oscillatory potentials related to movement in alpha band have been demonstrated in some experiments. Further role of alpha activity correlating strongly with working memory and probably with long-term memory engrams was also documented [21, 22].

Many investigators have found hemispheric differences in alpha power in tasks that use predominantly one hemisphere for processing. Diminution of alpha power over the left hemisphere was found during mental arithmetic (Butler and Glass 1987; [23]), word search tasks [24], spatial imagery [25] and music processing [26].

Gamma Fast and Ultra-Fast Bands

Gamma band was previously considered to be around 40 Hz frequency. This was primarily due to limitations of recording instruments. With improved amplification and digitalization, many higher frequencies are now considered part of gamma band. Gamma band, and its synchrony, is attracting attention of researchers because, not only on EEG, but also in invasive microelectrode recording of brains this band features prominently. Gamma waves and fast activity have been studied in all sensory modalities, especially visual systems, providing a mechanism for awareness and processing of visual objects. In invasive micro-recordings, it is found that in sensory cortex gamma power increases with sensory drive [27] and with a broad range of cognitive phenomena, including perceptual grouping [28] and attention [29]. It is postulated that gamma oscillations aid selective and flexible coupling of neighbouring or distant cortical regions and help communication (Varela et al. 2001; [30])

Buzsáki and Schomburg [31] have taken an in-depth review of endogenous gamma oscillations and their role in interregional communication. Scalp EEG in the future might be able to detect these findings with advancement in recording instruments.

Event-Related Potentials

Event-related potentials (ERP) are another method to understand cognition through EEG. ERP is the measured brain response that is the direct result of a specific sensory, cognitive or motor event. Here, the stimulus-specific changes in waveform are extracted from the EEG. Due to their precise temporal resolution, the event-related potential (ERP) technique has proven particularly valuable for testing theories of perception and attention [32]. Since the various components of ERP waveform are precisely time locked to the stimulus, a fair inference of how brain processes these stimuli can be understood with ERP.

Up till now, we have documented the EEG, its oscillatory and synchronous character and methods to study the EEG waveforms. We now mention how these methods are used for applied cognition.

11.4 EEG as a Biomarker in Clinical Settings

A biomarker is defined as a naturally occurring substance (molecule, gene) or set of characteristic features by which a particular pathological or physiological process, disease, etc., can be identified. For example, high level of low-density lipoproteins (LDL) in blood is correlated well with chances of getting ischaemic heart diseases. Here LDL is a biomarker of heart disease and can be used to document the disease status. EEG as a neurophysiological signature can be used to classify and monitor disease process, and its response to the therapeutic interventions. This assumes importance since molecular pathologies of many brain diseases are known but biomarkers are not established well. Obviously, invasive sampling (biopsy) of human brain for documenting disease is too risky. The imaging of brain using techniques like MRI does not always correlate well with the disease status or progress. For example, many neurodegenerative disorders affect structures of brain differently, changing relative ratio of these structures to each other in the brain. Yet these structural findings do not reflect disease status or progress faithfully. In such cases, EEG can become a useful tool as all such diseases ultimately affect global brain functions in spite of different molecular pathologies. EEG can establish the global cerebral dysfunction accurately. It also records changes in EEG as time progresses and that can reflect the disease progress or response to treatment or interventions.

Several candidate disease conditions can be studied with EEG as a biomarker. For present discussion, only a few examples of developmental and degenerative brain diaseases where structural changes are not really predictive of disease status would be mentioned.

Autism Spectrum Disorders

Autism spectrum disorders (ASD) are a group of developmental brain disorders with complex presentations. ASD affects approximately 1 in 88 children and 1 in 54 males [33]. Autism involves the wide spectrum of developmental disorders characterized by cognitive impairments in behavioural domains: These domains range from social interaction, language, communication and imaginative play and range of interests and activities. An autistic child can have deficits in establishing meaningful relations with other humans; his language can be limited and communication skills poor. They also exhibit restricted imagination. They often look withdrawn from the normal world and show stereotype behavioural pattern. A review of EEG abnormalities found that in ASD relatively high power in low and high frequencies with relatively low power in mid frequencies as compared to control group of healthy population was a constant feature of ASD [34].

Alzheimer's Disease

Alzheimer's disease (AD) is a neurodegenerative disease and main contributor to dementia. The disease burden is growing very fast all over the world. By 2050, the prevalence is expected to quadruple, so that 1 in 85 persons will be living with the disease [35]. Even in developing country like India, the increase in life expectancy is reflected as increased incidence of Alzheimer's in aging population. A study conducted for 10 years in India found that the incidence rates per 1000 person-years for AD was 11.67 for those aged ≥55 years and higher (15.54) for those aged ≥65 years [36].

Alzheimer's disease progresses through clinical stages. Stage one is characterized by early dementia evident in forgetfulness (amnesia). Next the mild to moderate stages show a substantial decline in wide range of cognitive functions and independence. For example, the affected persons may show flat affect or less robust emotional interactions. They may show deficits in higher cognitive domain like understanding humour and comprehending literature, poems, etc. They may have developed deficit in understanding subtle, contextual meanings of sentences in text or in conversations. Further advanced stages are characterized by patients complete dependence on caregivers for daily activities. Finally a stage comes where complete destruction of personality and severe dependence on caregivers for basic needs like food and maintained of personal hygiene occurs.

Few studies have found that in EEG power spectrum analysis, the increased power in low-frequency bands like theta and delta and decreased power in high bands like alpha and beta are noted in this disease. The neuronal synchrony was found abnormal in patients of Alzheimer's [37]. Other studies reported other electroencephalographic abnormalities in these patients. In the patients suffering from AD, information transmission between cortical neuronal networks was found abnormal. Some models have found out neuronal connectivity abnormalities [38, 39] and the sleep EEG is found to be altered in these patients [40]. Normal sleep is important in healthy memory, and perhaps sleep abnormalities in EEG may explain the predominant memory deficits noted in patients of AD.

Parkinson's disease is a neurodegenerative disorder marked by loss of dopaminergic neurons in substantia nigra region of the brain. Its features are movement difficulties mainly but also include cognitive decline. A study using signal processing of EEG by power spectrum and wavelet function methods found out that there was relative decrease in the alpha and beta band power with subsequent increase in theta and delta band power. The wavelet analysis indicated increased wavelet entropy in gamma band in patients suffering from Parkinson's disease [41]. As mentioned earlier, entropy in EEGs represents inherent instability and chaos of dynamic state of signals. This may mean that information flow in these patients is not smooth compared to normal brains. Cognition has many components like emotions and memory but ultimately cognition depends on working of different parts of brain in harmony, This harmony is dependent on smooth information flow across different brain regions.

Huntington's disease is another neurodegenerative genetic disorder characterized by movement abnormalities and cognitive dysfunctions like difficulty in impulse control and focus. The EEG has been used to map cognitive dysfunction in this disease. A study using power spectrum analysis of frontal and temporal activity found decreased power of alpha and theta band with subsequent increase in power of beta and delta band [42].

Another pilot study demonstrated the use of EEG as possible biomarker for Huntington's disease. In this pilot, EEG from healthy and diseased participants were collected and analysed using machine learning techniques. An index was generated and then applied to larger dataset. The classification index had a specificity of 83%, a sensitivity of 83% and an accuracy of 83% [43].

11.5 EEG in Consumer Neurosciences

Now, we will discuss the use of EEG in nonclinical settings. Finding markers of consumer choices and their decision-making has always been important for marketers. The traditional methods of surveys and focused group discussions have limitations because they are at the expressed state of thoughts at the time of interviews. The person might express a different opinion in stated response compared to actual true response in his brain. For example, when asked a respondent might say that he liked the product and brand, and he has every intention of buying it. But in his implicit mind, he might like the product or brand less. So as evolved species, we humans have this ability to express something different from true emotions or thoughts that occur in our brain. With explosion in marketing messages, it has become far more important to accurately know the mind of the consumers. Facial coding and peripheral biomarkers like heart rate varience, galvanic skin response and pupillometry have been used to address this need. But these methods essentially are surrogate indicators of what actually happens in the brain. These methods are downstream in time frame to actual neuronal activities. EEG is increasingly being used in the fields of consumer neurosciences. We now see few of the cases to illustrate this statement.

As EEG methodologies and computational algorithms became refined, their use in consumer field increased. The pioneering studies were published in the 1990s. EEG has been applied to assess marketing stimuli such as media involvement [44], the processing of TV commercials [45] and the prediction of memory for components of TV commercials [46].

A study of TV commercial using EEG revealed how one scene may efficiently prime other scenes on a neurophysiological level [47]. Another study found out that consumers indeed use metacognition during decision-making, and math anxiety affects buying decision when pricings are instrumental in decision-making. This study used EEG methods of ERP in perceptual and conceptual domain [48]. Another study by Astolfi [49] using EEG and the topology of the cortical fields found that the cortical activity and connectivity elicited by the viewing of the TV commercials that were remembered by the experimental subjects were markedly different from the brain activity elicited during the observation of the TV commercials that were forgotten. Now such findings open up an applied field where performance of programme and episode can be predicted. When large datasets are available, then using machine learning might become possible to understand which themes are more liked or remembered. Then programmes can be tailored to suit these preferences increasing their chances of getting more commercial success like increase viewership. In modern era, multiple channels offering variety of programmes are available. In such clutter, this EEG tool can be an advantage.

Another study using ERP found out reliable neural signature of emotional conflict in counter conformity book purchase on online shopping [50]. Several studies have been able to determine implicit like and dislike by viewers using EEG. One study using 3D shapes onscreen could predict like and dislike as expressed in explicit answers by feature extractions in frontal electrodes [51]. Cheung and colleagues [52] using EEG found out that during aesthetic judgments, long-range theta coherence increased in both hemispheres, and more positive frontal alpha asymmetry was found when the styles were judged to be beautiful. Aesthetics is an important part of implicit decision-making for consumers. With so many products with same functional attributes, the choice is likely to depend upon perceived aesthetics of design of the product. It is a well-known fact that emotions drive memory and attention and influence decision-making. So an insight into these implicit activities of the brain might correlate well with buying decisions and can be predictive of actual consumer behaviour.

Gender differences to subconscious processing of advertisements are also being explored by EEG. Using EEG, Cartocci and others [53–55] could demonstrate reliable differences in genders for processing same commercials. Such differences may not come out well in surveys and focussed discussions during marketing research in traditional way. The implications of these findings are significant. If a particular gender does not like the specific advertisement implicitly, then they might have less connect with the product depicted in the advertisement. Since usually only one advertisement is created and aired on TV, it is important to know the gender difference early so that market success is ensured.

EEG has been successfully used to determine neural patterns that predict audience preferences of TV programmes. Here, a small population tested using these TV programmes as stimuli could reliably predict audience preference by inter-subject correlation of EEG patterns. This prediction matched the tweets and opinions shared on social media which is considered as the 'expressed' preferences [56].

Online shopping is on the rise. Whether people trust a product or are satisfied with the buying, is often difficult to predict. These two things are important for predicting repeat buying and establishing loyalty to a particular online shopping platform. The trust as indicated by country of origin and quality of online buying experience can be predicted by using EEG measures [57, 58]. This might aid in understanding preferences and decision-making of largely anonymous faceless population of online buyers.

Based on memory and attention and global power as documented by EEG spectrum, an impression index was conceived by Kong and team to classify video commercials into liked and disliked scenes [59]. Such understanding could serve to optimize the video into shorter edits that could be more effective and less expensive to air on media.

The use of EEG is being extended to newer fields like politics and movie viewership. One study by Vechiato demonstrated possibility of using EEG to know political preferences in a simulated election based on judgement of trustworthiness of face of politicians. Use of EEG as a predictor of movie preferences was demonstrated by Boksem and colleagues [60]. The possibility of using EEG to understand subconscious reaction to brand extensions was explored by Jin [61]. EEG has also been used to study impact of illumination of stores during buying of food in retail environment [62].

11.6 EEG in Understanding Meditation

Meditation is as old as human existence. With developing frontal lobe and neocortex during evolution, human species acquired the capacity to imagine, reflect and do abstract thinking. This capacity to reflect on 'self' and the 'outside world' are hallmarks of meditation. There are different methodologies and practises of meditation. Each one appears different and claims are made for superiority of one method over other.

There are several well-documented benefits of meditation. The most obvious is that of stress reduction. Modern life generates stress, and this is reflected in increased incidences of psychosomatic disorders and conditions like anxiety and depression. Meditation might be useful in setting the hypothalamic–pituitary–adrenal axis (HPA axis) in the body. This axis explains brain body connection through endocrine glands. Several diseases are now considered to have abnormalities of the HPA axis as a contributing cause to the development of disease. In such scenarios, meditation might be a useful adjuvant non-pharmacological therapy. Besides, the benefit of meditation is known to increase attentional control, emotional regulation and generate better cognitive control over stressful and conflicting situations [63].

Though there are different varieties of meditative techniques available, all can be classified in two distinct ways: focused attention (FA) method and open monitoring (OM) method. In FA, selective attention is directed towards an object or thought and any distractions; attentional shifts from that focus are removed out to bring back the focused attention back to the object. The OM method is more elaborate. It does not direct or sustain attention on a specific object, but open monitoring of thoughts, sensations and experiences is practised. In normal circumstances, individual 'react' to the changes in these things. In OM, the reaction is avoided and 'reflection' is encouraged [64].

In recent time, effect of meditation on brain structures has generated tremendous interest. Few decades ago, the effect of meditation were thought to be limited to peripheral nervous system essentially on the sympathetic tone. However, the progress in functional imaging of brain made it possible to document effects of meditation on the brain. But functional neuroimaging like fMRI has its own problems like selectivity to be used in research regarding meditation [65].

Here, we present some studies that have used EEG to understand brain functions during meditation. A review of various studies by Cahn and Polich [66] mentioned overall slowing with theta and alpha activation in EEG after meditation. It indicated that cognitive event-related potential evaluation of meditation leads to changes in attentional allocation. The most dominant effect standing out in the majority of studies on meditation is a state-related slowing of the alpha rhythm (8–12 Hz) in combination with an increase in the alpha power [67, 68]. The beginners in meditative techniques usually concentrate on an image, word or thought, utilizing an increasing amount of attentional resources. This is reflected in alpha band changes as alpha oscillations are known to arise from an increase of internal attention [69]. A general increase of theta activity during meditation has been reported in a large number of studies as mentioned in a review [66]. Theta waves are markers of stage 1 of sleep and relaxed state. Meditation in early stages might be reducing anxiety and producing a relaxed state.

High-amplitude gamma band oscillations occur during meditation for well-experienced practitioners of meditation. In a study of Buddhist Lama, a meditation practitioner for 20 years, it was documented that altered state of consciousness could be inferred from altered patterns of Gamma band activity [70]. Long-term practitioners of meditation are reported to be able to self-induce synchronous gamma band oscillation. Synchronised gamma activity is highly relevant for neural plasticity and the implementation of new processing circuits [71]. This neural plasticity might explain the structural and functional changes noted on fMRI studies in meditators.

Even short-term meditators have changes in the sleep EEG. During non-REM sleep, low-frequency oscillatory activities (1–12 Hz, centred around 7–8 Hz) over prefrontal and left parietal electrodes were noted in meditators compared to nonmeditators [72]. Authors maintain that this might represent the neural plasticity occurring in wakeful period of meditation.

A study of EEG and heart rate variability before and after 8 weeks training of mindfulness-based stress reduction (MBSR) showed that meditation could reduce the chaotic activities of both EEG and heart rate as a change of state. This study used wavelet entropy analysis to arrive at this conclusion [73].

Meditation has been implicated in changes in the default mode network (DMN) of the brain. Several functional imaging studies have reported this finding. In an EEG study using novice meditators and training them for 4 months, Fingelkurts and his colleagues found changes in the DMN. They found that strength of operational synchrony of EEG was decreased in posterior module of the DMN. The posterior module of DMN is considered involved in self-referential status of 'I' to be embodied self within the body space. This finding may explain why meditators have less awareness of 'self' as distinct from others and they are considered more compassionate.

A review of 56 papers published between 1966 and Aug 2015 about EEG and mindfulness meditation indicated that consistent findings were that of an increase in alpha and theta power. In other frequencies, the findings were not conclusive. The authors concluded that the state of relaxed alertness indicated by these findings might aid the overall mental health [74].

11.7 Conclusion

Electroencephalogram is a well-established technique to study brain functions. Usefulness of EEG to understand cognition is increasing. This is largely due to better understanding of process underlying how neuronal assemblies work together in cognition. Use of computational methods has allowed facile manipulation of EEG signals to generate quick and reliable insights into cognition. These findings are being extrapolated to newer applied fields of biomarkers of neurological diseases as well as non-clinical fields of consumer neurosciences and meditation. Finding biomarkers for neurological disorders will help documentation of stages of disease and response to therapeutic interventions. These diseases have very limited characteristic structural abnormalities definable on imaging. This field of use of EEG as biomarkers is still nascent. More robust findings of EEG changes in various neurological disorders and possible explanation of the altered cognition in these diseases are required. Studying consumer behaviour and effects of meditation on the brain have assumed importance in modern times. Both these fields uphill now lacked serious insights into neural phenomenon that are underpinnings of behaviour related to these things. However, published research shows that it is still a developing field and virtually all components of EEG are found representative of a part of cognition. Before using these methods in applied domains, a careful look at the results of basic research in terms of methodology, reproducibility and reliability needs to be undertaken. Further progress in the subject may aid in developing a better and holistic picture of how cognition happens based on its neural correlates.

References

1. Hubel DH (1982) Cortical neurobiology: a slanted historical perspective. Annu Rev Neurosci 5 (1):363–370
2. Sturm A, König P (2009) Time–frequency analysis methods to quantify the time-varying microstructure of sleep EEG spindles: possibility for dementia biomarkers? J Neurosci Methods 185(1):133–142
3. Brandimonte MA, Bruno N, Collina S, Pawlik P, d'Ydewalle G (eds) (2006) Psychological concepts: an international historical perspective. Psychology Press, Hove
4. Fries P (2005) A mechanism for cognitive dynamics: neuronal communication through neuronal coherence. Trends Cogn Sci 9:474–480

5. Fries P (2015) Rhythms for cognition: communication through coherence. Neuron 88(1):220–235. https://doi.org/10.1016/j.neuron.2015.09.034
6. Cardin JA, Carlen M, Meletis K, Knoblich U, Zhang F, Deisseroth K, Tsai LH, Moore CI (2009) Driving fast-spiking cells induces gamma rhythm and controls sensory responses. Nature 459:663–667
7. Womelsdorf T, Schoffelen JM, Oostenveld R, Singer W, Desimone R, Engel AK, Fries P (2007) Modulation of neuronal interactions through neuronal synchronization. Science 316:1609–1612
8. Schoffelen J, Poort J, Oostenveld R, Fries P (2011) Selective movement preparation is subserved by selective increases in corticomuscular gamma-band coherence. J Neurosci 31(18):6750–6758. https://doi.org/10.1523/jneurosci.4882-10.2011
9. Fingelkurts AA, Fingelkurts AA (2017) Information flow in the brain: ordered sequences of metastable states. Special issue "Symmetry and information" (ISSN:2078-2489). Information 8(1):22. https://doi.org/10.3390/info8010022
10. Fingelkurts AA, Fingelkurts AA (2004) Making complexity simpler: multivariability and metastability in the brain. Int J Neurosci 114:843–862
11. Kelso JAS (1995) Dynamic patterns: the self-organization of brain and behaviour. The MIT Press, Cambridge, MA
12. Hammond DC (2005) What is neuro feedback? J Neurother 10(4):25–36. https://doi.org/10.1300/J184v10n04_04
13. Rangaswamy M (2002) Beta power in the EEG of alcoholics. J Soc Biol Psychiatry 51:831–842
14. Schomer DL, da Silva FHL (1999) Niedermeyer's electroencephalography: basic principles, clinical applications, and related fields, 4th edn. Lippincott Williams & Wilkins, Philadelphia. Ed. A. Glass, pp 103–120. Plenum, New York
15. Başar E, Stampfer HG (1985) Important associations among EEG-dynamics, event related potentials, short-term memory and learning. Int J Neurosci 26:161–180
16. Basar E (1999) Brain function and oscillations. II. Integrative brain function. Neurophysiology and cognitive processes. Springer, Berlin/Heidelberg
17. Meyer L, Grigutsch M, Schmuck N, Gaston P4, Friederici AD (2015) Frontal-posterior theta oscillations reflect memory retrieval during sentence comprehension. Cortex 71:205–218. https://doi.org/10.1016/j.cortex.2015.06.027
18. Thézé R, Guggisberg AG, Nahum L, Schnider A (2016) Rapid memory stabilization by transient theta coherence in the human medial temporal lobe. Hippocampus 26:445–454. https://doi.org/10.1002/hipo.22534Varela
19. Başar E (1980) EEG–brain dynamics. Relation between EEG and brain evoked potentials. Elsevier, Amsterdam
20. Başar E, Schürmann M (1997) Functional correlates of alphas Panel discussion of the conference 'Alpha edited by processes in the brain'. Int J Psychophysiol 26(1–3):455–474. https://doi.org/10.1016/s0167-8760(97)00782-4
21. Başar E, Yordanova J, Kolev V, Başar-Eroglu C (1997) Is the alpha rhythm a control parameter for brain responses? Biol Cybern 76(6):471–480. https://doi.org/10.1007/s004220050360
22. Klimesch W, Schimke H, Schwaiger J (1994) Episodic and semantic memory: an analysis in the EEG theta and alpha band. Electroencephalogr Clin Neurophysiol 91(6):428–441. https://doi.org/10.1016/0013-4694(94)90164-3
23. Morgan AH, Macdonald H, Hilgard AE (1974) EEG alpha: lateral asymmetry related to task, and hypnotizability. Psychophysiology 11(3):275–282. https://doi.org/10.1111/j.1469-8986.1974.tb00544.x
24. McKee G, Humphrey B, McAdam DW (1973) Scaled lateralization of alpha activity during linguistic and musical tasks. Psychophysiology 10(4):441–443
25. Rebert CS, Low DW (1978) Differential hemispheric activation during complex visuomotor performance. Electroencephalogr Clin Neurophysiol 44:724–734
26. Duffy FH, Bartels PH, Burchfiel JL (1981) Significance probability mapping: an aid in the topographic analysis of brain electrical activity. Electroencephalogr Clin Neurophysiol 51(5):455–462

27. Henrie JA, Shapley R (2005) LFP power spectra in V1 cortex: the graded effect of stimulus contrast. J Neurophysiol 94:479–490
28. Tallon-Baudry C, Bertrand O (1999) Oscillatory gamma activity in humans and its role in object representation. Trends Cogn Sci 3:151–162
29. Fries P (2001) Modulation of oscillatory neuronal synchronization by selective visual attention. Science 291(5508):1560–1563. https://doi.org/10.1126/science.10554651563
30. Engel AK, Fries P, Singer W (2001) Dynamic predictions: oscillations and synchrony in top–down processing. Nat Rev Neurosci 2:704–716. 4
31. Buzsáki G, Schomburg EW (2015) What does gamma coherence tell us about inter-regional neural communication? Nat Neurosci 18(4):484–489. https://doi.org/10.1038/nn.39521
32. Woodman GF (2010) A brief introduction to the use of event-related potentials in studies of perception and attention. Atten Percept Psychophy 72(8):2031–2046
33. CDC(Centers for Disease Control and Prevention) et al (2012) MMWR Surveill Summ 61 (3):1–19
34. Wang J, Barstein J, Ethridge LE, Mosconi MW, Takarae Y, Sweeney JA (2013) Resting state EEG abnormalities in autism spectrum disorders. J Neurodev Disord 5(1):24. https://doi.org/10.1186/1866-1955-5-24
35. Brookmeyer R, Johnson E, Ziegler-Graham K, Arrighi HM (2007) Forecasting the global burden of Alzheimer's disease. Alzheimers Dement 3:186–191
36. Mathuranath PS, George A, Ranjith N et al (2012) Incidence of Alzheimer's disease in India: a 10 years follow-up study. Neurol India 60(6):625–630. https://doi.org/10.4103/0028-3886.105198
37. Dauwels J, Vialatte F, Musha T, Cichocki A et al (2010) NeuroImage 49(1):668–693. https://doi.org/10.1016/j.neuroimage.2009.06.056
38. Jeong J, Gore JC, Peterson BS (2001) Mutual information analysis of the EEG in patients with Alzheimer's disease. Clin Neurophysiol 112(5):827–835
39. Stam CJ, de Haan W, Daffertshofer A et al (2009) Graph theoretical analysis of magnetoencephalographic functional connectivity in Alzheimer's disease. Brain 132 (1):213–224
40. Ktonas PY, Golemati S, Xanthopoulos P et al (2009) Time-frequency analysis methods to quantify the time-varying microstructure of sleep EEG spindles: possibility for dementia biomarkers? J Neurosci Methods 185(1):133–142
41. Han CX, Wang J, Yi GS, Che YQ (2013) Investigation of EEG abnormalities in the early stage of Parkinson's disease. Cogn Neurodyn 7(4):351–359. https://doi.org/10.1007/s11571-013-9247-z
42. Bylsma FW, Peysers CE, Folstein SE, Ross C et al (1994) EEG power spectra in Huntington's disease: clinical and neuropsychological correlates. Neuropsychologia 32(2):137–150
43. Odish OFF, Johnsen K, van Someren P, Roos RAC, van Dijk JG (2018) EEG may serve as a biomarker in Huntington's disease using machine learning automatic classification. Sci Rep 8:16090. https://doi.org/10.4329/wjr.v6.i7.471
44. Swartz BE (1998) Timeline of the history of EEG and associated field. Electroencephalogr Clin Neurophysiol 106:173–176
45. Rothschild M, Thorson E, Reeves B, Hirsch J, Goldstein R (1986) EEG activity and the processing of television commercials. Commun Res 13(2):182–220
46. Rothschild M, Hyun YJ (1990) Predicting memory for components of TV commercials from EEG. J Consum Res 16(4):472–478
47. Ohme R, Matukin M, Szczurko T (2010) Neurophysiology uncovers secrets of TV commercials. Markt 49:133. https://doi.org/10.1007/s12642-010-0034-7
48. Jones WJ, Childers TL, Jiang Y (2012) The shopping brain: math anxiety modulates brain responses to buying decisions. Biol Psychol 89(1):201–213
49. Astolfi L, Fallani FD, Cincotti F, Mattia D, Bianchi L, Marciani M, Babiloni F (2008) Neural basis for brain responses to TV commercials: a high-resolution EEG study. IEEE Trans Neural Syst Rehabil Eng 16(6):522–531. https://doi.org/10.1109/tnsre.2008.2009784

50. Chen M, Ma Q, Li M, Lai H, Wang X, Shu L (2010) Cognitive and emotional conflicts of counter-conformity choice in purchasing books online: an event-related potentials study. Biol Psychol 85(3):437–445. https://doi.org/10.1016/j.biopsycho.2010.09.006
51. Chew LH, Teo J, Mountstephens J (2015) Aesthetic preference recognition of 3D shapes using EEG. Cogn Neurodyn 10(2):165–173. https://doi.org/10.1007/s11571-015-9363-z
52. Cheung M, Law D, Yip J (2014) Evaluating aesthetic experience through personal-appearance styles: a behavioral and electrophysiological study. PLoS One 9(12):e115112. https://doi.org/10.1371/journal.pone.0115112
53. Cartocci G, Cherubino P, Rossi D, Modica E, Maglione AG, Flumeri GD, Babiloni F (2016) Gender and age related effects while watching TV advertisements: an EEG study. Comput Intell Neurosci 2016:1–10. https://doi.org/10.1155/2016/3795325
54. Vecchiato G, Maglione AG, Cherubino P, Wasikowska B, Wawrzyniak A, Latuszynska A, Babiloni F (2014a) Neurophysiological tools to investigate consumer's gender differences during the observation of TV commercials. Comput Math Method Med 2014:1–12. https://doi.org/10.1155/2014/912981
55. Vecchiato G, Toppi J, Maglione AG, Olejarczyk E, Astolfi L, Mattia D, Babiloni F (2014b) Neuroelectrical correlates of trustworthiness and dominance judgments related to the observation of political candidates. Comput Math Method Med 2014:1–19. https://doi.org/10.1155/2014/434296
56. Dmochowski JP, Bezdek MA, Abelson BP, Johnson JS, Schumacher EH, Parra LC (2014) Audience preferences are predicted by temporal reliability of neural processing. Nat Commun 5. https://doi.org/10.1038/ncomms5567
57. Min B, Cho K, Sung J, Cho E (2014) Neurophysiological evidence for the country-of-origin effect. NeuroReport 25(4):274–278. https://doi.org/10.1097/wnr.0000000000000102
58. Wang J, Han W (2014) The impact of perceived quality on online buying decisions. NeuroReport 25(14):1091–1098. https://doi.org/10.1097/wnr.0000000000000233
59. Kong W, Zhao X, Hu S, Vecchiato G, Babiloni F (2013) Electronic evaluation for video commercials by impression index. Cogn Neurodyn 7(6):531–535. https://doi.org/10.1007/s11571-013-9255
60. Boksem MAS, Smidts A (2015) Brain responses to movie trailers predict individual preferences for movies and their population-wide commercial success. J Mark Res 52(4):482–492
61. Jin J, Wang C, Yu L, Ma Q (2015) Extending or creating a new brand. NeuroReport 26 (10):572–577. https://doi.org/10.1097/wnr.0000000000000390
62. Berčík J, Horská E, Wang RW, Chen Y (2016) The impact of parameters of store illumination on food shopper response. Appetite 106:101–109. https://doi.org/10.1016/j.appet.2016.04.010
63. Tang YY, Holzel BK, Posner MI (2015) The neuroscience of mindfulness meditation. Nat Rev Neurosci 16:213–225. https://doi.org/10.1038/nrn3916
64. Lutz A, Slagter HA, Dunne JD, Davidson RJ (2008) Attention regulation and monitoring in meditation. Trends Cogn Sci 12(4):163–169. https://doi.org/10.1016/j.tics.2008.01.005
65. Marchand WR (2014) Neural mechanisms of mindfulness and meditation: evidence from neuroimaging studies. World J Radiol 6(7):471–479. https://doi.org/10.4329/wjr.v6.i7.471
66. Cahn BR, Polich J (2006) Meditation states and traits: EEG, ERP, and neuroimaging studies. Psychol Bull 132(2):180–211. https://doi.org/10.1037/0033-2909.132.2.180
67. Hirai T (1974) Psychophysiology of Zen. Tokyo, IgakuShoin. https://sites.google.com/a/perthgrammar.co.uk/physics/courses/higher/particles-and-waves/35-interference-and-diffraction/351-interference
68. Taneli B, Krahne W (1987) EEG changes of transcendental meditation practitioners. Adv Biol Psychiatry 16:41–71
69. Ray W, Cole H (1985) EEG alpha activity reflects attentional demands, and beta activity reflects emotional and cognitive processes. Science 228(4700):750–752. https://doi.org/10.1126/science.3992243
70. Lehmann D, Faber P, Achermann P, Jeanmonod D, Gianotti LR, Pizzagalli D (2001) Brain sources of EEG gamma frequency during volitionally meditation-induced, altered states of consciousness, and experience of the self. Psychiatry Res Neuroimaging 108(2):111–121. https://doi.org/10.1016/s0925-4927(01)00116-0

71. Lutz A, Greischar LL, Rawlings NB, Ricard M, Davidson RJ (2004) Long-term meditators self-induce high-amplitude gamma synchrony during mental practice. Proc Natl Acad Sci 101(46):16369–16373. https://doi.org/10.1073/pnas.0407401101
72. Dentico D, Ferrarelli F, Riedner BA, Smith R, Zennig C, Lutz A, Davidson RJ (2016) Short meditation trainings enhance non-REM sleep low-frequency oscillations. PLoS One 11(2). https://doi.org/10.1371/journal.pone.0148961
73. Gao J, Fan J, Wu BW, Zhang Z, Chang C, Hung Y, Sik HH (2016) Entrainment of chaotic activities in brain and heart during MBSR mindfulness training. Neurosci Lett 616:218–223. https://doi.org/10.1016/j.neulet.2016.01.001
74. Lomas T, Ivtzan I, Fu C (2015) A systematic review of the neurophysiology of mindfulness on EEG oscillations. Neurosci Biobehav Rev 57:401–410. https://doi.org/10.1016/j.neubiorev.2015.09.018

Computational Mechanisms for Exploiting Temporal Redundancies Supporting Multichannel EEG Compression

12

M. S. Sudhakar and Geevarghese Titus

Abstract

Multichannel electroencephalogram (MCEEG) recordings generally result in humongous volume of data that places constraint on space and time. Online transmission of such data demands schemes rendering significant performance with lesser computations. To compact such data, numerous compression algorithms have been introduced in the literature. Heretofore single channel algorithms when extended to multichannel applications do not accomplish remarkable results. If achieved it generally results in higher computational cost. Much of this chapter deals with the development of computationally simple algorithms that aim to reduce the computational aspects without comprising on the compression and decompression performance. Also, the amicability of this implementation supporting efficient storage with low bandwidth is performed. The objective of this chapter is to introduce some of the basic supporting concepts for exploiting data representation redundancies that aid compression. Accordingly, simple and novel compression schemes with its simulation comparison are presented in this chapter. The potency of the direct domain compression model is assessed in terms of compression ratio and reconstruction error between the original and reconstructed dataset. A significant compression with substantially low Percent Root-mean-square Difference (PRD) is accomplished by the novel compression schemes, thereby upholding diagnostic information of EEG for telemedicine applications with higher reconstruction accuracy and reduced computational load.

M. S. Sudhakar (✉)
SENSE, Vellore Institute of Technology, Vellore, India
e-mail: sudhakar.ms@vit.ac.in

G. Titus
Amal Jyothi College of Engineering, Kanjirapally, Kerala, India
e-mail: geevarghesetitus@amaljyothi.ac.in

S. Paul (ed.), *Application of Biomedical Engineering in Neuroscience*,
https://doi.org/10.1007/978-981-13-7142-4_12

Keywords
Compression ratio · Decoding · Encoding · Multichannel EEG · Percent root-mean-square difference

12.1 Introduction

EEG signal is a complex signal with both stochastic and deterministic properties. The redundancies present in the EEG or any biomedical signal can be removed without loss of signal quality and remain the basic principle supporting compression. In this context there are three forms of compression algorithms, namely lossless, lossy, and near lossless. For most critical clinical analysis of EEG, a lossless compression is the best bet, but the degree of compression is too low such that there is no significant reduction in storage space. Furthermore, most of such compression algorithms have complex architecture. The next best choice is to construct a near lossless compression system having higher compression with significant space saving. One critical factor to consider in such algorithms is the quality of the decompressed signal. Different quality parameters need to be defined like PSNR, PRD, and MSE, that will quantify the level of distortion in the decompressed signal. In other words, the features in the EEG should be retained such that any classification system can work properly irrespective of whether the process is automated or manual.

Another perspective of signal compression is by removing unwanted signals like noises and different forms of artifact from the EEG signal and then storing the clean signal. Different parameters like PSNR can be employed to quantify the quality of the denoised signal. Dimensionality reduction techniques fall under another class of method that can reduce the size of the data that needs to be stored. The subsequent section includes discussions on various time domain strategies and transform domain and compressive sensing methods for compressing the biomedical signal. Most of the methods are supported by coding schemes like Arithmetic, Huffman, Lempel-Ziv, and Lempel-Ziv-Welch that are generally classified as lossless coders, and other coders like SPIHT are EBCOT are lossy coders. These coders exploit the redundancies in the data thereby contributing to data compression.

EEG compression is an active research area for more than a decade. Various compression algorithms have been developed that are broadly classified into different groups based on their operational mode to achieve either lossless, lossy or near lossless compression. These are direct and transform mode based compression schemes. In the direct mode the compression is achieved by exploiting redundancies in the temporal or acquired domain itself. Whereas in transform mode, the compression is achieved by transforming the signal using various transforms and exploiting the redundancies in the transformed domain. These generic compression modes assist in data compression and specifically MCEEG compression. To further enhance the prevailing compression methods, they can be combined together resulting in next-generation compression algorithms.

The neural network-based predictor system of [22] defined the context of the EEG signal, based on which the corresponding coefficients are entropy coded to achieve compression. The MCEEG compression scheme of [21] packed EEG data in 2D form and later processed them using 2D Integer Lifting Transform. The produced coefficients were SPIHT coded along with the residue. Bazn-Prieto et al. [3] employed Cosine Modulated Filter Banks to perform domain transformation. The transformed coefficients were later quantized and entropy coded. A MCEEG compression system described in [21] represents MCEEG signal using tensor representation on which 2D wavelet transform is performed followed by uniform quantization and arithmetic coding of residues. This method tries to maintain an upper bound on the distortion level. Dauwels et al. [7] extended the earlier version using various tensor decomposition methods like SVD and PARAFAC along with residual coding to limit the error introduced in the system.

The compression scheme of [20] initially performed PCA on the EEG data, followed by Wavelet Packet Transform (WPT). The resulting high and low frequency coefficients were coded using wavelet-based and fractal coding techniques, respectively. These coded coefficients were finally entropy coded. Finally, genetic algorithm-based optimization technique was employed for enhancing compression performance. Lin et al. [18] combined dimensionality reduction methods such as the PCA with ICA to compress the MCEEG data represented as a tensor, and the resulting coefficients were then SPIHT coded.

The 1.5D compression algorithm of [24] employed 1D discrete wavelet transform on the 2D MCEEG data followed by No List Set Partitioning in Hierarchical Trees (NLSPIHT) coding. Another MCEEG compression algorithm by [4] employed Fast Discrete Cosine Transform (FDCT) followed by lossy coding. A near lossless MCEEG compressor by [11] used DCT and artificial neural network-based coder. Another compression scheme from [10] performed DPCM on the EEG channels that are then clustered using k-nearest neighbor (kNN). The resulting clusters were entropy coded using arithmetic coders.

To achieve compression, the above methods rely on computationally intensive progressive operations, limiting its use in real-time hardware critical applications. Some algorithms like [4] claim to be computationally fast but with compromise in either compression or distortion levels. So, prospective hardware-friendly compression scheme without significant signal deterioration needs to explored, such that the signal can be used in diagnosis or related applications.

12.1.1 MCEEG Compression Methods

In this chapter the concept of introducing and exploiting the data representation redundancies in the data for achieving compression is discussed. The proposed Logarithmic Spatial Pseudo CODEC (LSPC) methodology negates the necessity for domain transformation, for exploring redundancies, hence, leading to faster computations. The methodology was tested on different database using various performance parameters and discussed in subsequent sections. The proposed near lossless compression model illustrated in Fig. 12.1 is a two-stage process where the MCEEG signal is normalized using Translation Logarithmic (TL) block which is subsequently coded in the Integer Fraction Coder (IFC).

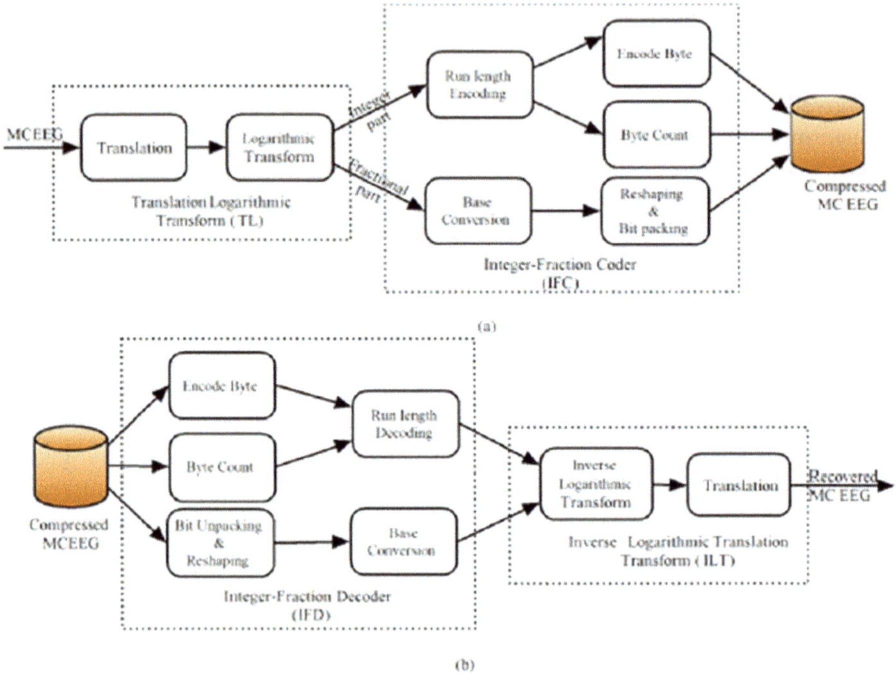

Fig. 12.1 Proposed LPSC system. (**a**) Encoder section. (**b**) Decoder section

12.1.2 Encoding Process

The LSPC encoding process illustrated in Fig. 12.1 a commences with the organization of MCEEG data in 2D form as represented in Eq. 12.1, where M and N correspond to the number of channels and samples per channel, respectively. Each of these sets is processed and stored as frames as illustrated in Fig. 12.2.

$$x = \begin{bmatrix} X_{11} & X_{12} & \dots & X_{1N} \\ X_{21} & X_{22} & \dots & X_{2N} \\ . & . & . & \\ . & . & . & \\ X_{M1} & X_{M2} & \dots & X_{MN} \end{bmatrix} \tag{12.1}$$

To exploit the contribution of all samples and suit them for subsequent processing, the data is pre-processed by the TL block by translation and logarithm transformation. To ascertain the resulting normalized samples to be real and positive,

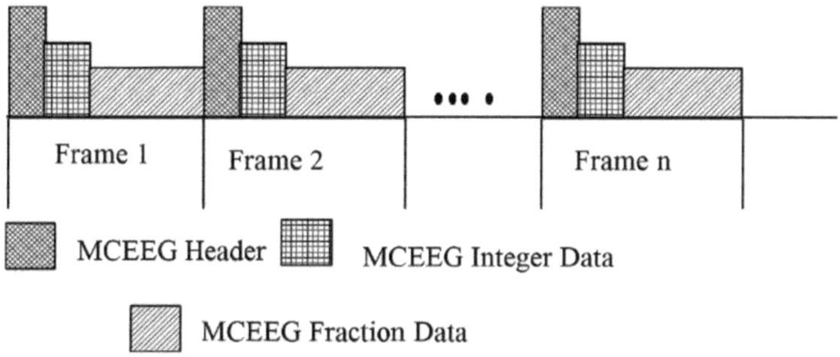

Fig. 12.2 Framing process in MCEEG compression

the MCEEG samples are translated by a factor f, which depends on the gain of the employed MCEEG recorders, and the operation is defined in Eq. 12.2.

$$g(x) = T(x + f) \tag{12.2}$$

Natural logarithm is employed in the LSPC for normalization; while other bases do not contribute to any significant improvement in compression and signal recovery, hence are not discussed here. The normalization process is defined by Eq. 12.3.

$$h(x) = \ln g(x) \tag{12.3}$$

The normalized data $h(x)$ is split into integer $I_r(x)$ and fractional $I_f(x)$ parts using Eqs. 12.4 and 12.5, respectively.

$$I_r(x) = h(x) \tag{12.4}$$

$$I_f(x) = h(x) - I_r(x) \tag{12.5}$$

$I_r(x)$ and $I_f(x)$ are encoded separately using the IFC. Later, $I_r(x)$ can be encoded using any spatial coding schemes that exploit the redundancies introduced by the TL transform. In the current architecture, Run Length Encoding (RLE) of [2] is employed because of its simplicity in implementation. The encoded data is represented as $(D_1, C_1)(D_2, C_2) \ldots (D_i, C_i) \ldots (D_n, C_n)$, where D_i and C_i are the ith distinct integer and their occurrence (count), respectively. As arithmetic and Huffman coders introduce complexity in the system with only a marginal improvement in compression, they are not considered suitable in the proposed architecture.

Subsequently, $I_f(x)$ is encoded to an equivalent representation with error deviation depending on the bit depth d of the base converter. First, the converter converts $I_f(x)$ to its equivalent binary stream b using Eq. 12.6, which is the generalized relation for converting fractions to or from any base.

$$b = a_0 m^{-1} + a_1 m^{-2} + a_2 m^{-3} + \cdots + a_{k+1} m^{-k} \tag{12.6}$$

where m is the base, k is the resolution, and a takes values 0, 1, ..., $m - 1$. For example, for converting the fraction to binary, m is 2, and a takes value of either 0 or 1.

Data representations for faster encoding and decoding emphasized by [17] include methods like Variable Byte, Byte-Oriented Encoding, Binary Packing, Binary Packing with Variable-Length Blocks, and Patched Coding. Such representations are also useful in size reduction and faster access. A similar packing strategy of the binary stream has been employed in this architecture.

For data representation in the proposed algorithm, the binary stream b from base converter is reshaped and packed into groups of eight bits to form an integer equivalent representation of the fractional data called Pseudo Integers (PI). This representation supplemented with spatial coding technique helps the algorithm to achieve compression.

12.1.3 Decoding Process

The decoder of the SPC system illustrated in Fig. 12.1b takes RLE encoded data and PI as input. This is first processed by the Integer Fractional Decoder (IFD) wherein the RLE data is unpacked by interpolating the coded integer D_i by the parameter C_i to obtain $I_f(x')$. Subsequently, the PI are binarized and reshaped according to the bit depth value used for encoding, which is retrieved from MCEEG header. The reshaped data is converted to fractional base using Eq. 12.6 to obtain $I_f(x')$. Both processes are performed in parallel, thus reducing the decoding time. $I_f(x')$ is added with $I_r(x')$ resulting in the reconstructed normalized signal $I(x')$; this operation is represented in Eq. 12.7.

$$I(x') = I_r(x') + I_f(x') \tag{12.7}$$

Inverse transformation of $I(x')$ is processed by the Inverse Logarithmic Translation (ILT) block where an exponential operation defined by Eq. 12.8 is performed.

$$h(x') = \exp I(x') \tag{12.8}$$

Subsequently, the reconstructed MCEEG signal x' is obtained by translating $h(x')$, given by Eq. 12.9, where factor f is the same as that used in the encoder.

$$x' = T(h(x') - f) \tag{12.9}$$

Finally, the overall encoding and decoding performance of the LSPC on MCEEG signals were observed using the publicly available EEG datasets namely Berlin BCI Competitions [23], Swartz Center for Computational Neuroscience (SCCN) [5], PhysioNet [6], UCI MLR UCI machine learning repository [19], DREAMS sleep spindles database [15], the Bern-Barcelona EEG database [16], European Epilepsy Database [8], Child Mind Institute – Multimodal Resource for Studying Information Processing in the Developing Brain (MIPDB) Database [9], DREAMER Database [1], Stanford Research Data, and Australian Electroencephalography Database (AED) [12].

All channels of the EEG data sets were taken for simulation, though increasing or reducing the number of channels or samples per channel did not have any significant impact on the degree of compression. To evaluate the efficacy of the proposed scheme, few data sets have been chosen, labeled as data set 1 to 8, from each of these standard databases illustrated in Table 12.1.

The database is made up of MCEEG recording of different subjects performing various motor or imagery tasks, subjected to various constraints and stimuli. The performance metrics used in this study are CR, LAE, PAE, MAE, PRD, and PSNR [3] and are computed using Eqs. 12.10, 12.11, 12.12, 12.13, 12.14, and 12.15.

$$CR = \frac{Bits_{\mathrm{orig}}}{Bits_{\mathrm{comp}}} \tag{12.10}$$

Where, *Compression Ratio (CR)* represents the ratio of the number of bits of the uncompressed or original signal to the number of bits of the compressed signal. The terms $Bits_{\mathrm{orig}}$ and $Bits_{\mathrm{comp}}$ correspond to the number of bits of the uncompressed and compressed signal, respectively. Apart from CR, the impact of distortion in the decompressed EEG has a profound effect. Different quality indicators are used that are able to quantify this distortion introduced by the compression system. As a single indicator alone does not effectively quantify the distortions introduced, more than one of such measures and their relations needs to be investigated.

Local Absolute Error (LAE) is the absolute difference between the actual and the reconstructed value and is given in Eq. 12.11.

Table 12.1 MCEEG datasets employed for performance analysis

Dataset	No. of channels	Sampling frequency (in Hz)	Dataset source	Resolution (bits)
1	22	250	BCIC IV Set 2a	16
2	31	500	SCCN	16
3	64	256	UCI MLR	16
4	28	1000	BCIC II Set 4a	16
5	28	100	BCIC II Set 4b	16
6	64	240	BCIC III Set 2	16
7	64	1000	BCIC IV Set 1	16
8	64	80	Physionet	12

BCIC: Brain Computer Interface Competitions
SCCN: Swartz Center for Computational Neuroscience
UCI MLR: UCI Machine Learning Repository

$$LAE = abs\left(X_{\text{ori}}(i) - X_{\text{rec}}(i)\right) \tag{12.11}$$

The terms $X_{\text{ori}}(i)$ and $X_{\text{rec}}(i)$ correspond to the actual data and the decompressed data, respectively.

Peak Absolute Error (PAE) provides an indication of the maximum error occurring in the reconstructed signal or it is the maximum of the absolute difference between the actual and reconstructed value, given in Eq. 12.12.

$$PAE = max\left(abs(X_{\text{ori}}(i) - X_{\text{rec}}(i))\right) \tag{12.12}$$

MeanAbsoluteError(MAE) is the mean of the maximum absolute error difference between the actual and reconstructed signal and is given by Eq. 12.13

$$MAE = \frac{max\left(abs(X_{\text{ori}}(i) - X_{\text{rec}}(i))\right)}{N} \tag{12.13}$$

with N representing the data length

Percentage Root mean square Difference (PRD) given in Eq. 12.14 provides information on the amount of error in the decompressed signal.

$$PRD = \sqrt{\frac{\sum\left[X_{\text{ori}}(i) - X_{\text{rec}}(i)\right]^2}{\sum X_{\text{ori}}(i)^2}} \tag{12.14}$$

Root Mean Square Error (RMSE) is the measure of the differences between actual and the decompressed signal or in other words it is the square root of the average of squared errors. It is sometimes referred to as *RootMean Deviation(RMSD)*. As the effect of the error is proportional to the value of squared error, it can be observed as large variations in RMSD and can be computed using Eq. 12.15.

$$RMSE = \sqrt{\frac{\sum\left[X_{\text{ori}}(i) - X_{\text{rec}}(i)\right]^2}{n}} \tag{12.15}$$

n – Total number of samples

Finally, *Peak Signal to Noise Ratio (PSNR)* is mathematically the ratio between the maximum signal power to the noise power of the recovered signal and is represented in Eq. 12.16.

$$PSNR = 20 * \log\frac{\max\left(X_{\text{ori}}\right)}{RMSE} \tag{12.16}$$

Based on the above metrics, further analysis is presented below. At the onset, the impact of logarithmic normalization is witnessed in Fig. 12.3. From Fig. 12.3b it can be clearly observed that the data is within the range of 3 and 4, with only few values lower than four.

An illustration of the original, reconstructed, and error signal at bit depths of 3, 4, and 5 is shown in Fig. 12.4. The bit depth of five can be considered as an optimal

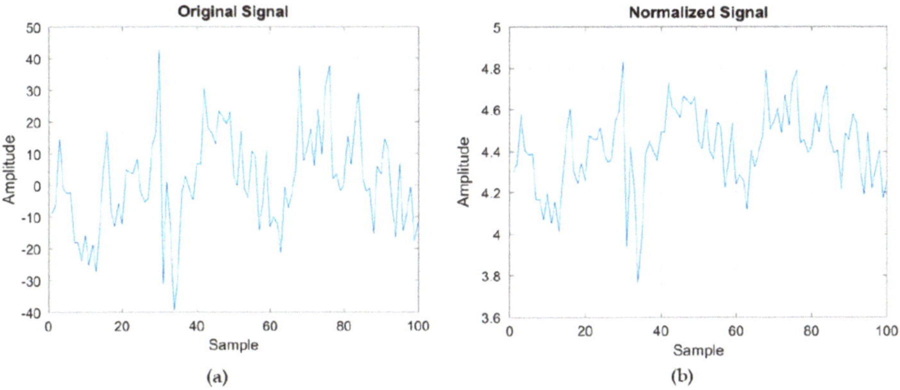

Fig. 12.3 Illustration of logarithmic normalization

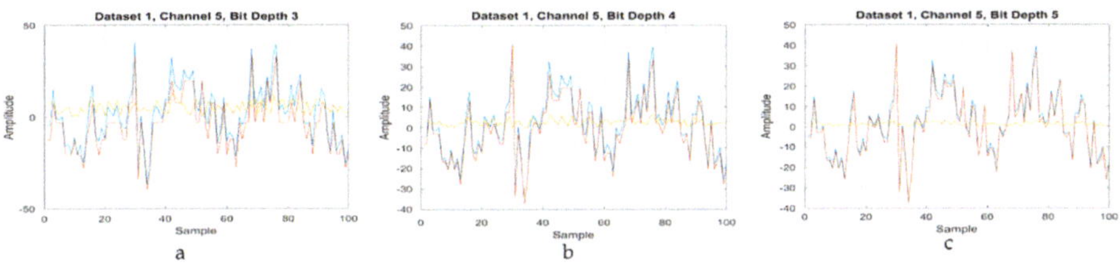

Fig. 12.4 Superimposed original, reconstructed, and residual error signals at a bit depth of 3, 4, and 5

choice as it gave almost similar and relatively good CR and distortion measures for all the data sets, and this finding is validated in subsequent discussions. The coded MCEEG is stored as frames as depicted in Fig. 12.2, having three fields MCEEG Header, Integer Data, and Fraction Data.

Figure 12.5 illustrates the encoding of four samples of 2D MCEEG signals represented by Distinct Integer, Integer Occurrence, and Pseudo Integers. The normalization process is fully reversible. The reconstructed signal quality was visually validated and quantified numerically using the LAE distortion parameter. The maximum error in the reconstructed signal was in the order of 10^{-7} to 10^{-9}.

12.1.4 Inferences from LSPC

The performance of the proposed algorithm for different data sets is illustrated in Figs. 12.6, 12.7, and 12.8. The bit depth d of base converter contributes largely to the compression as well as distortion in the recovered signal, as illustrated in Fig. 12.6a–d.

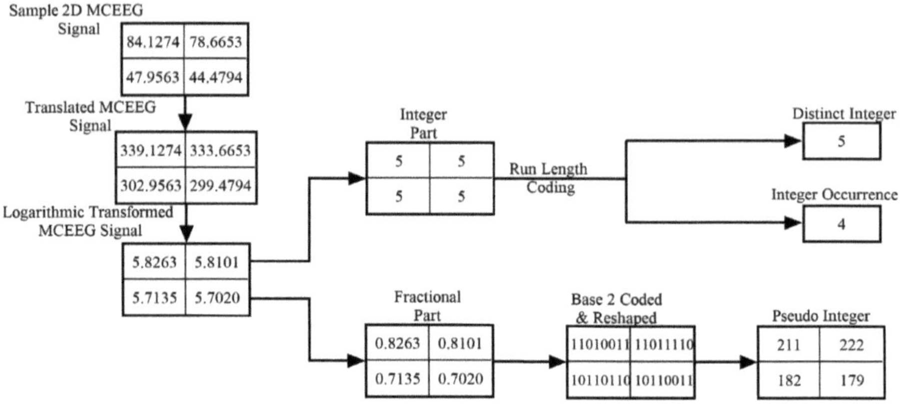

Fig. 12.5 Sample MCEEG encoding process

The lower value of bit depth results in higher compression, with larger distortion in the recovered or decompressed signal and vice versa. Hence, choice of optimal value of bit depth depends on the acceptable amount of distortion in the reconstructed signal. This value was found out with visual scoring along with numerical values quantified by the distortion metrics PRD, RMSE, PAE, MAE, and PSNR as elaborated in the subsequent paragraphs.

It was observed that at higher bit depths distortion is negligible and CR is comparatively less. For example, from Figs. 12.6a, 12.7a, d, CR at a bit depth of eight is nearly two, with PRD value very close to zero, and PSNR above 27 dB. The performance of the algorithm at different sampling frequencies (100 Hz, 1000 Hz) is almost the same as exemplified by the distortion indicators corresponding to data sets 4 and 5 in the performance plots. Another observation is that the resolution of the original data does play a significant role in determining the achievable CR. It can be concluded that higher CR can be achieved at higher resolutions of the ADCs and vice versa. This is a significant observation as currently available ADCs are operating at higher resolutions than those employed in this study.

Distortion performance indicators vary among the data sets at the same bit depth. PAE value indicates the largest difference between the original and reconstructed signal. MAE, on the other hand, indicates the mean error in the reconstructed signal. RMSE value, apart from being an error measure also serves as a performance indicator of the outliers. There was a fourfold increase in the RMSE value and a twofold increase in the PAE and MAE values for a successive reduction in bit depth.

For a bit depth of five, the acceptable average CR of 3.16 occurs at average MAE value of 3.88. Moreover, from Fig. 12.8b, at the average value of MAE, the maximum variation of PAE ranges from 10 to 15.

This provides an upper limit to the sample error at the current CR. Furthermore, average PSNR of 23.07 dB and PRD of 1.39 are observed at the specified CR. Hence, based on the visual scoring and critical inference from Fig. 12.6, 12.7, and 12.8, it can be concluded that the optimal bit depth across different data sets can

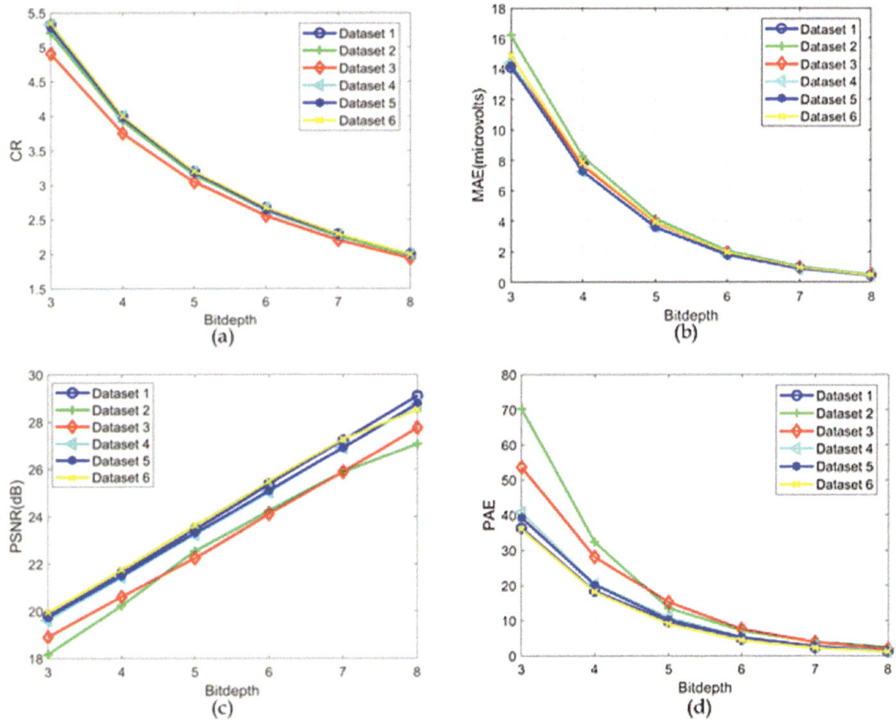

Fig. 12.6 Performance of LSPC on different MCEEG datasets for varying bit depths

be taken as five, though in some data sets lower bit depths can also be considered having similar distortion parameters.

Table 12.2 illustrates the encoding and decoding time of the proposed algorithm for different number of channels and samples per channel. Computations were performed on a computer having an Intel Core2Duo processor operating at 1.8 GHz with 2 GB RAM.

From Table 12.2, the average encoding and decoding times per sample can be computed, which are 0.3 and 0.04 ms respectively. This is a clear indicator that the algorithm is computationally simple and fast when compared to other progressive computation algorithms. Use of public and common data sets is required for relative performance analysis of novel compression algorithms. In this work, data sets 4, 7, and 8 were used for performance comparison as they were the ones used in recent compression algorithms discussed in [21, 4, 10]. The comparison is illustrated in Table 12.3, taking PRD and PSNR as reference quality indicators.

The comparative results signify that the performance of the proposed method in terms of compression and distortion is comparable with recent work but with a much lower time complexity. The main highlight of the proposed algorithm is its simplicity in compressing and decompressing the MCEEG data. Birvinskas et al. [4] claims to be computationally light, but it is not superior to the proposed algorithm, as

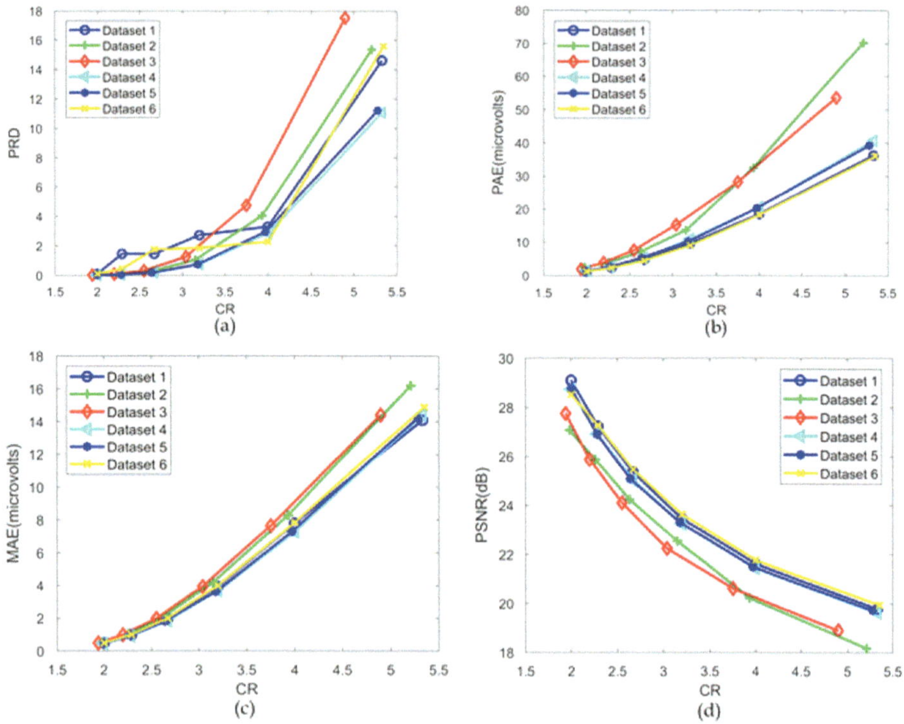

Fig. 12.7 LSPCs performance on different MCEEG datasets for varying CR

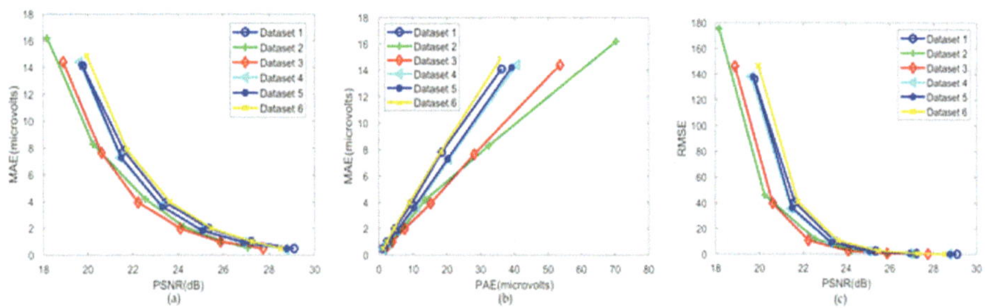

Fig. 12.8 LSPC analyzed using quantitative and qualitative metrics on diverse datasets

observed from the table, in terms of compression and distortion parameters of the reconstructed signal. Furthermore, the proposed algorithm outperforms [10] in terms of CR, with minimal distortion in the decompressed signal.

Extension of the aforesaid LSPC is further done for introducing and exploiting the data representation redundancies in the data for enhancing compression. The LSPC system based on logarithm-based normalization produces random outliers in the decompressed signal, due to the inherent nature of the logarithmic operation.

Table 12.2 Computation time of the proposed scheme, for different sample sizes

No. of channels	Number of samples per channel			
	800		400	
	Encoding time (s)	Decoding time (s)	Encoding time (s)	Decoding time (s)
10	3.38	0.23	2.06	0.1
20	6.03	0.58	3.37	0.23
30	8.58	1.09	4.74	0.39
40	11.26	1.76	6.02	0.58
50	13.76	2.56	7.28	0.82
60	16.48	3.52	8.67	1.1
61	16.87	3.63	8.64	1.1

Table 12.3 Relative performance analysis of LSPC

Data set	Algorithm	CR	PRD	PSNR	Optimal criteria
4	Fast DCT [4]	4	11.09	****	BinDCT
	LSPC	4	2.86	21.44	Bit depth of 4
7	Wavelet image and volumetric coding [21]	2.56	1.72	37.95	Wavelet-s/s/t
	Clustering method [10]	2.67	****	****	DPCM-kNN
	LSPC	3.20	4.43	23.63	Bit depth of 5
8	Wavelet image and volumetric coding [21]	6.63	9.21	28.92	Wavelet-s/s/t
	LSPC	3.89	2.43	20.32	Bit depth of 4

****The metric has not been evaluated in the algorithm

Furthermore, an additional translation process is required in the LSPC to handle negative samples, which requires the estimation of the maximum value, to guarantee that the signals after translation lie above zero. To overcome the aforesaid issues Min-Max normalization is used to replace the logarithm normalization. The scheme was tested on different database using various performance parameters and is discussed in subsequent sections.

12.2 MMSPC: Encoding Process

The modified scheme illustrated in Fig. 12.9 is able to achieve lossy to near lossless compression by a two-stage process. The signal is initially normalized and subsequently coded using the Integer Fraction Coder (IFC). In the encoder of the Min-Max Spatial Pseudo CODEC (MMSPC) system [14] illustrated in Fig. 12.9a, the raw MCEEG data is arranged in a 2D structure as shown in Eq. 12.1, where M and N correspond to the number of channels and samples per channel, respectively.

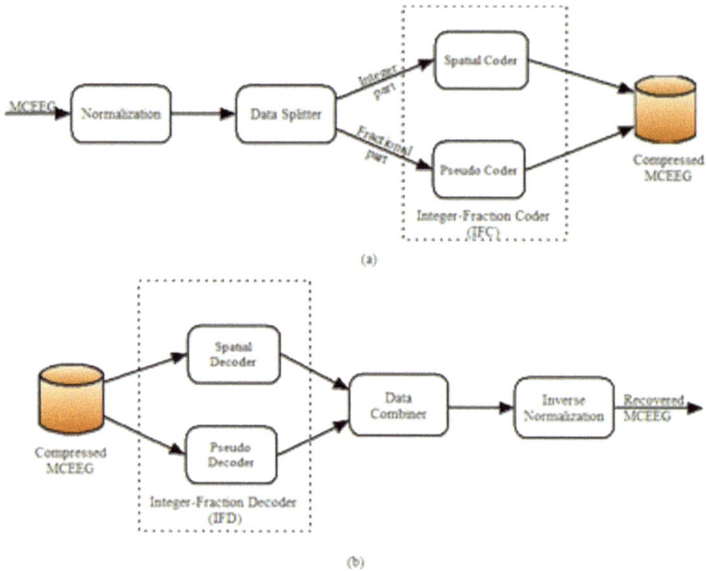

Fig. 12.9 Proposed MMSPC system. (**a**) Encoder section. (**b**) Decoder section

To increase the computational stability and memory requirement and maintain the contextual content, large sequence of MCEEG data are split into sets of maximum 1000 samples per channel, and each of these is processed and stored as frames as illustrated in Fig. 12.2. The next step is to normalize the MCEEG data using Feature Scaling (FS) that restricts the values between two arbitrary points a and b, and the normalized value is represented as $h(x)$ in Eq. 12.17

$$h(x) = a + \frac{(X - X_{\min}) * (b - a)}{X_{\max} - X_{\min}} \tag{12.17}$$

where X is the sample data, and $X_{\min}$ and $X_{\max}$ are the minimum and maximum of each channel. Size of $X_{\min}$ and $X_{\max}$ will be $M \times 1$. a and b take on values 0 and 1 in the proposed method. The normalized data $h(x)$ is split into integer $I_r(x)$ and fractional $I_f(x)$ parts using Eqs. 12.4 and 12.5 which are coded separately by the IFC.

Subsequently, $I_r(x)$ can be encoded using any spatial coding schemes that exploit the redundancies introduced by FS normalization. In the proposed system, Run Length Encoding (RLE) is employed because of its simplicity in implementation. The encoded data S presented in Eq. 12.18 is a sequence of codes representing a byte and the number of times it occurs consecutively.

$$S = (D_1, C_1)(D_2, C_2)\ldots(D_i, C_i)\ldots(D_n, C_n) \tag{12.18}$$

where D_i and C_i are the ith distinct integer and its occurrence, respectively. Alternatively, arithmetic coders could have been employed instead of RLE, but the coding efficiency was only marginally better at the expense of a slightly higher computational complexity.

Next a two-stage process encodes $I_f(x)$ and represents it by a novel PI representation. Initially a lossy process binarizes $I_f(x)$ based on the required resolution k. A base converter converts $I_f(x)$ to its equivalent binary stream b using Eq. 12.6.

Subsequently, in the second stage is a lossless process, based on the concept of data representation [17] for faster encoding and decoding is performed on the binary stream. The binary stream b from base converter is reshaped and packed into groups of eight bits to an equivalent PI representation. Outputs from the spatial coder and pseudo coder constitute the compressed MCEEG, and the resulting data is stored as frames as depicted in Fig. 12.2.

12.2.1 Decoding Process

The strength of the algorithm is its simple decoding architecture. In the decoder depicted in Fig. 12.9b, the coded MCEEG comprising the header, integer, and fractional data undergoes two processes, namely Spatial and Pseudo decoding, and the unit is collectively grouped as the Integer Fractional Decoder (IFD). In the Spatial Coder, the spatial coded data S is unpacked to obtain the integer part $I_r(x')$, based on the attributes in Eq. 12.18, where the coded integer D_i is repeated several times defined by C_i. Simultaneously, in the Pseudo Coder, the pseudo integers are binarized as normal unsigned integers. The binarized data is then reshaped with the same bit depth value used in the encoder, the value of which is retrieved from the corresponding header. This is followed by a fractional base conversion using Eq. 12.6 to obtain $I_f(x')$. The recovered integer and fractional parts are combined to reconstruct the normalized signal $I(x')$ presented in Eq. 12.7

Subsequently, inverse normalization of $I(x')$ defined by Eq. 12.19 results in reconstructing the MCEEG signal x'.

$$x' = X_{\min} + \frac{(I(x') - a) \times (X_{\max} - X_{\min})}{b - a} \tag{12.19}$$

where $X_{\min}$, $X_{\max}$, a, and b are values from the encoder.

Performance metrics on datasets that were discussed earlier for LSPC analysis are extended here for investigating the effectiveness of MMSPC for MCEEG compression. The effect of Min-Max normalization is illustrated in Fig. 12.10. Accordingly, the following observations were made:

It can be clearly observed that the data is now within the range of 0 and 1, depicting data representation redundancies. In Feature Scaling, normalization arbitrary values of a and b can be taken. It was observed that for all the variations of a, b other than $a = 0$ and $b = 1$ (Min-Max Normalization), the signal compression and reconstruction quality reduces. The normalization process is fully reversible, and the reconstructed signal follows the actual signal with visual validation. For numerical validation, Local Absolute Error (LAE) was computed, which is the deviation of

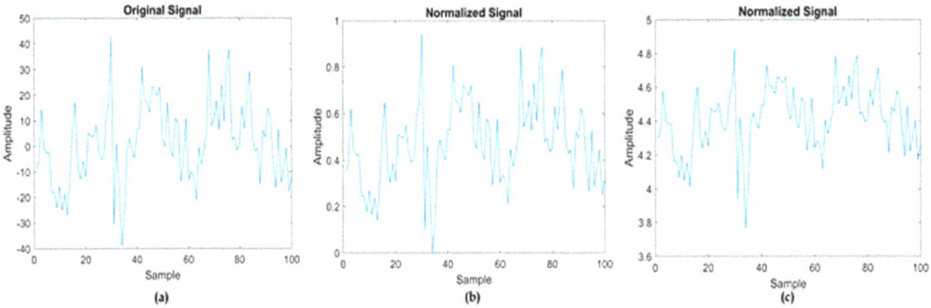

Fig. 12.10 Illustration of different normalization techniques. (**a**) Original, (**b**) Min-Max normalized, (**c**) logarithmic normalized

the decompressed sample from the original and was found to be in the order of 10^{-8} to 10^{-9}.

The performance of the compression algorithm at near optimal bit depths on the different MCEEG datasets has been numerically quantified in Table 12.4.

From the table for better signal quality a bit depth of four is better than its lower ranges, but results in lower CR, whereas lower bit depth of three provides a better CR with marginal increase in signal distortion in some of the datasets. An illustration of the original, decompressed and residual error signals at bit depths of 3, 4, and 5 is illustrated in Fig. 12.11.

It can be observed from the plots that at bit depths greater than 4 the error in the decompressed signal is significantly less. It can be observed that the algorithm is invariant to the data representation format and the performance metrics are correlated to the bit depth only. The choice of optimal bit depth is based on trade-off between acceptable amount of distortion in the decompressed signal and CR. This value can be found out with visual inspection along with numerical values of the distortion metrics PRD, RMSE, PAE, MAE, and PSNR in Fig. 12.12.

Furthermore, Fig. 12.12a–d, justifies that the performance metrics of the decompressed signal is highly correlated to the bit depth *d*. PRD provides an indication on the amount of error in the decompressed signal. For different data sets, at a bit depth of 3, the value varies from a minimum of 0.62 to a maximum of 12.01, while at a higher bit depth of 4, it varies from 0.15 to 2.98, indicating a higher signal quality. PAE which indicates the maximum absolute difference between the actual and decompressed sample varies from 11.73 to 61.02 at a bit depth of 3 and from 6.12 to 21.52 at bit depth of 4. MAE which provides a measure of closeness of the reconstructed sample varies from 3.15 to 9.93 and from 1.58 to 4.91 at bit depths 3 and 4, respectively. PSNR which is the ratio between the maximum signal power and the noise power of the recovered signal varies from 18.55 to 23.18 dB at bit depth of 3 and from 20.39 to 24.66 dB at bit depth of 4.

The distortion introduced in the decompressed signal depends on the bit depth or resolution of the base converter. At lower bit depths, distortions tend to increase but are complimented with larger signal compression. It can be observed from

Table 12.4 Performance evaluation of the proposed scheme at different bit depth

Data set	Min-Max normalization					
	Bit depth = 4			Bit depth = 3		
	CR	PRD	PSNR(dB)	CR	PRD	PSNR(dB)
1	3.38	0.23	24.06	5.03	12.01	22.21
2	6.03	0.58	20.39	5.12	7.06	18.55
3	8.58	1.09	21.32	5.12	4.21	19.48
4	11.26	1.76	24.63	4.66	0.78	22.79
5	13.76	2.56	24.66	2.60	0.62	23.18
6	16.48	3.52	24.22	5.10	7.31	22.33

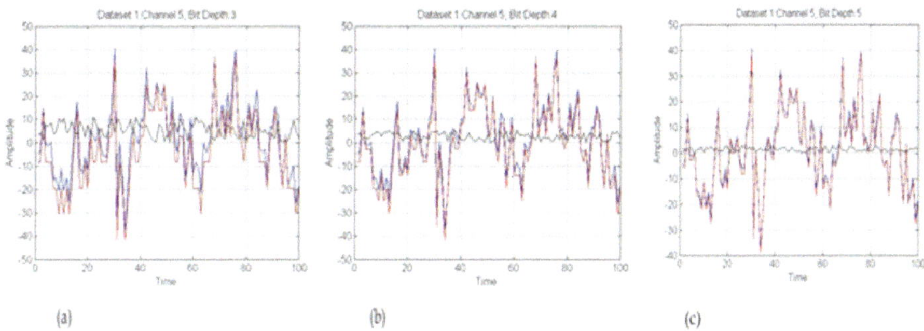

Fig. 12.11 Superimposed original, decompressed, and residual error signals at bit depth of three, four and five. (**a**) Signal compression at bitdepth of 3. (**b**) Signal compression at bitdepth of 4. (**c**) Signal compression at bitdepth of 5

Fig. 12.13a–d that the distortion parameters vary greatly among different data sets for varying CRs. Performance of the distortion parameters other than CR is almost the same for data sets 4 and 5, which are similar MCEEG signals with different sampling rates (1000 Hz and 100 Hz). The performance in terms of CR for data set 5 is comparatively lower than data set 4 as there were only 50 samples per channel, with the header information X_{min} and X_{max} having $M \times 1$ samples occupying 4% of the space. The header information for data set 4 with 500 samples per channel takes only 0.4% of the space, leading to higher CR. For MCEEG signals having longer duration, the overhead due to this header becomes insignificant.

The distortion introduced in the reconstructed signal depends on the bit depth or resolution of the base converter. At lower bit depths, distortions tend to increase, but are complimented with larger signal compression. It can be observed from Fig. 12.13a–d, that the distortion parameters vary greatly among different data sets for varying CRs. Figure 12.12a–d justify our initial theory that the performance metrics of the reconstructed signal are highly correlated to the bit depth *d*.

The relation between PSNR and PAE with MAE is depicted in Fig. 12.14a–b which can be used to decide on the choice of bit depth. From Fig. 12.14b the MAE can vary between 2 and 4 for an acceptable PAE of 10, which corresponds to the

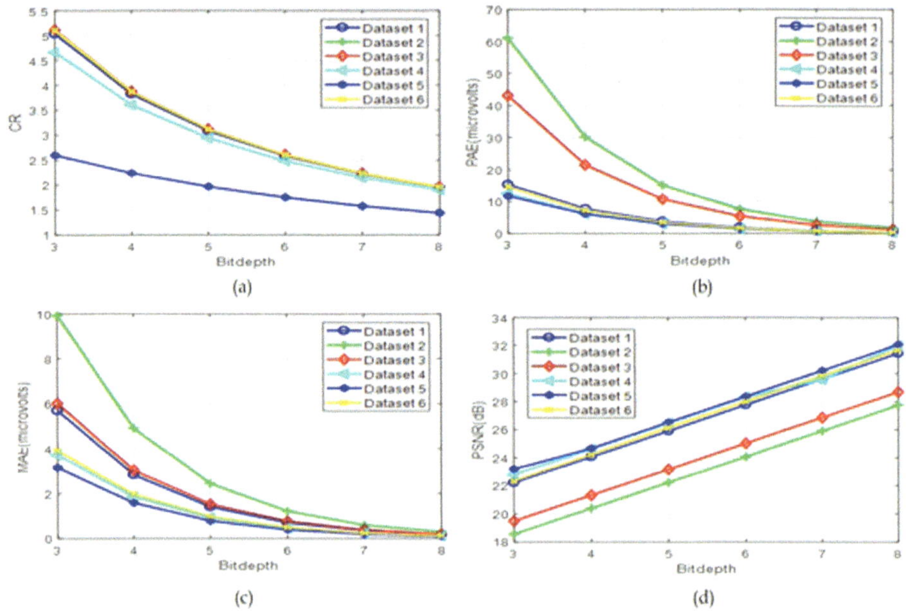

Fig. 12.12 Performance illustration of proposed scheme of MCEEG data sets for varying bit depth. (**a**) Bit depth versus CR. (**b**) Bit depth versus PAE. (**c**) Bit depth versus MAE. (**d**) Bit depth versus PSNR

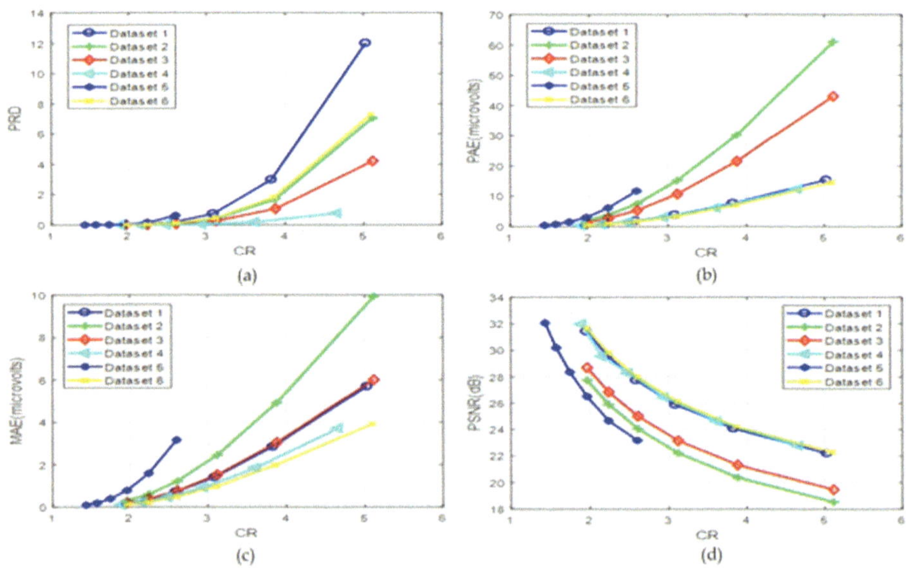

Fig. 12.13 Performance illustration of proposed scheme of MCEEG data sets for varying CR. (**a**) CR versus PRD. (**b**) CR versus PAE. (**c**) CR versus MAE. (**d**) CR versus PSNR

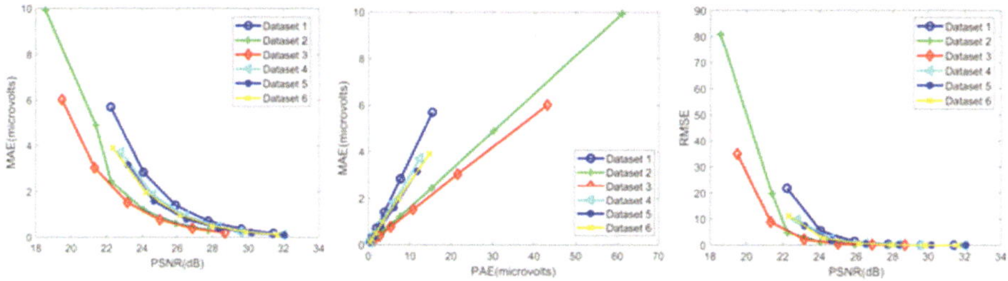

Fig. 12.14 Quantitative and qualitative analysis of MMSPC on different datasets

maximum error amplitude. On correlating this range of MAE with bit depth from Fig. 12.13c, an optimal bit depth can be set at 4, where the PAE is within 10 and MAE is within 4 in most of the data sets.

A relative comparison between LSPC and the proposed MMSPC system is illustrated in Figs. 12.15 and 12.16. It can be observed from the figures that MMSPC is better than LSPC in all grounds with reference to both CR and the degree of distortion introduced, with similar computational complexity of $O(zN)$.

Table 12.5 illustrates the time required for the encoding and decoding operations of MMSPC based on the variations in number of channels and samples per channel. Similar to LSPC, MMSPC computations were also performed on Intel Core2Duo system with two cores operating at 1.8 GHz, with 2 GB RAM. The average encoding and decoding time per sample was found to be 0.3 and 0.04ms, respectively, thereby strengthening the fact that the proposed algorithm is computationally fast.

Relative performance comparison of the proposed system with recent MCEEG compression algorithms [7, 4, 11, 13] is illustrated in Table 12.6 which uses CR, PRD, and computational complexity as measures.

CR of the proposed algorithm as observed from the table is greater when compared with other algorithms. The distortion indicator PRD is the lowest for the proposed algorithm suggesting that the proposed algorithm is less lossy than the other methods. Another highlight of the proposed algorithm is that it is computationally much lighter than all other algorithms, with a complexity of $O(zN)$, except for [4, 13] which claim to be computationally simple with comparable complexity as indicated in Table 12.6. But in both cases signal quality of the decompressed signal is more deteriorated with larger values of PRD as illustrated in the Table 12.6. One critical observation is that MMSPC performance is directly related to the bit resolution of the ADC, i.e., larger the bit resolution, the larger will be the CR with better decompressed signal quality and vice versa. This factor led to a lower performance of the algorithm for data set 8 which was recorded using a 12-bit ADC.

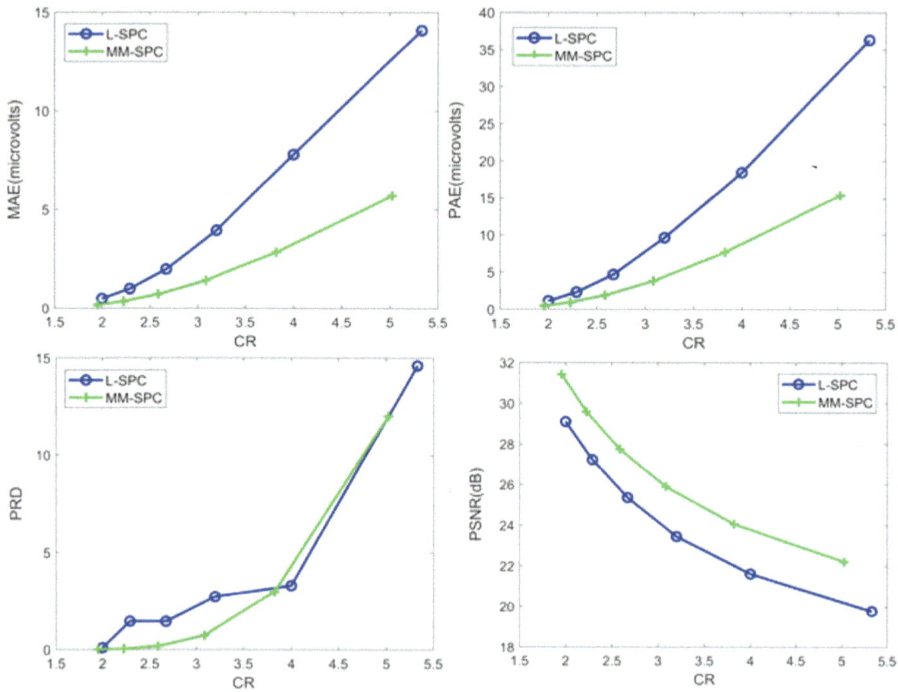

Fig. 12.15 Relative performance comparison of LSPC and MMSPC system under different CR

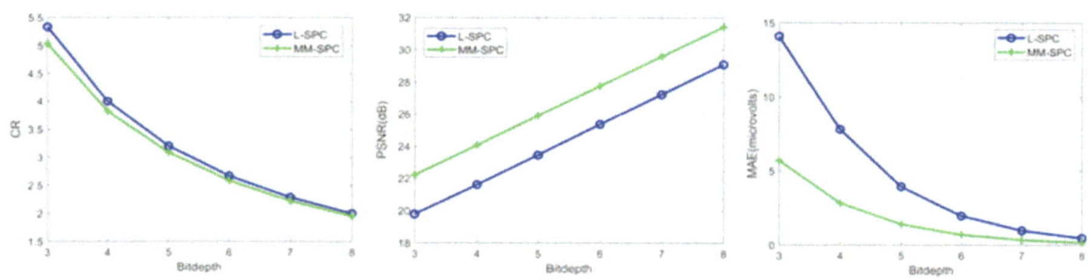

Fig. 12.16 Relative performance comparison of LSPC and MMSPC system for different bit depth

12.2.2 Complexity Analysis

In this section, an approximate estimation of performance of the MMSPC & LSPC is analyzed in terms of memory requirement and time complexity.

12.2.2.1 Memory Requirement Analysis

Lemma 1 *Let* $S = s_1, s_2 \ldots, s_n$ *be sample space of length n, with each sample represented by b bits. Then the total space B will be* $n \times \log \frac{2^b}{n}$

Table 12.5 Computation time of MMSPC for different sample sizes

No. of channels	Number of samples per channel			
	800		400	
	Encoding time (s)	Decoding time (s)	Encoding time (s)	Decoding time (s)
10	3.16	0.21	1.97	0.1
20	5.98	0.54	3.25	0.21
30	8.03	1.02	4.54	0.32
40	11.07	1.46	6.00	0.55
50	13.55	2.41	7.17	0.78
60	16.28	3.33	8.26	1.09
61	16.11	3.25	8.61	1.05

Table 12.6 Relative performance comparison of the compression algorithms

Authors & year	Dataset	CR	PRD	Complexity
Dauwels et al. [7]	7& 8	4.59	5.46	$O(3mN2)$
Birvinskas et al. [4]	4	4	11.09	$O(cN)$
Hejrati et al. [11]	7	2.67	****	$O(TKN)$
Titus and Sudhakar [13]	1 to 6	3.61	8.73	$O(zN)$
MMSPC	1 to 7	6.78	5.33	$O(zN)$

**** The metric has not been evaluated in the algorithm

Proof Consider n distinct b bit resolution samples. By the rules of information theory, the number of bits to represent the data is given by

$$B = \log \binom{2^b}{n} \tag{12.20}$$

This can be approximated using the property in combination to get the size of n elements or samples

$$B = n \times \log \frac{2^b}{n} \tag{12.21}$$

Spatial Coder returns two values namely distinct integer and its occurrence count of I_r defined in Eq. 12.3. The memory requirement of the coder is denoted by B_s

Lemma 2 *For given n samples with b bits per sample, there exists c set of distinct integers with size l bits per sample, such that* $1 \le c \le n$, $l < b$ *then from Lemma* 1 B_s *can be generalized by Eq.*12.20 ,

$$B_s \leq 2 \times c \times \log \frac{2^l}{c} \tag{12.22}$$

Pseudo Coder returns the integer representation of I_f defined in Eq. 12.4. Here the data is binarized using d bits defined by bit depth, and the number of samples n is taken such that product of $d \times n$ is divisible l the size of the integer. The memory requirement of the coder denoted by B_p

Lemma 3 *For a given n samples with b bits per sample, there exists n samples represented by d bits per sample such that* $d < b$ *then from Lemma* 1 B_p *can be generalized by Eq.*12.21 ,

$$B_p = k \times \log \frac{2^l}{k} \tag{12.23}$$

where $k = \frac{d \times n}{l}$

Lemma 4 *Given* B_s *and* B_p, *the space requirements of spatial and pseudo coders, respectively, the total space requirement for the proposed system* B_t *will be the sum of* B_s *and* B_p.

Proof Given B_s and B_p from Lemmas 2 and 3 B_t, the total space complexity of the proposed algorithm is defined by Eq. 12.22.

$$B_t = 2 \times c \times \log \frac{2^l}{c} + k \times \log \frac{2^l}{k} \tag{12.24}$$

Lemma 5 *Let* B, B_t *be the space requirements of the original and compressed signal, then* $B_t << B$.

Proof Given B, B_t the space requirements from Lemmas 1 and 4 subject to the conditions $c \leq n$, $d < b$ and $l < b$, then

$$n \times \log \frac{2^b}{n} >> 2 \times c \times \log \frac{2^l}{c} + \frac{d \times n}{l} \times \log \frac{2^l}{\frac{d \times n}{l}} \tag{12.25}$$

Time Complexity

Majority of EEG compression algorithms available in the literature commonly employ PCA, Fast ICA, and compressed sensing to get an equivalent structure of the actual data. To the best of our knowledge the lowest computational complexity that can be attained by any MCEEG compression algorithm is above $O(N^2)$, where N is the number of samples.

Lemma 6 *For a given sample space S, the complexity of the proposed system is O (zN)*

Proof Given N the number of samples per channel, M the number of channels, and d the bit depth. The complexity of the system is $O(NMd)$. As the number of channels M and precision or bit depth d requirement are fixed, their contribution to the complexity is minimal. Thus, the algorithm has a linear dependence with the number of samples taken over a period with a complexity of $O(zN)$, where $z = d \times M$ and $z << N$.

12.3 Conclusion

The significance of this study is that it provides a computationally simple model for compressing MCEEG signals. In the proposed LSPC system, the signal is normalized using the TL transform and the resulting coefficients are coded using IFC, ensuing in data compression. The proposed scheme achieved an average CR of 3.16 with a computational complexity of only $O(zN)$ and average encoding and decoding times per sample of 0.3 and 0.04 ms, respectively. This is reasonably better than the other methods which rely on computationally intensive operations, to achieve similar CR. Also, an extension of LSPC in the form of MMSPC is proposed. Upon FS normalization, the normalized coefficients are coded in the IFC, resulting in data compression. The proposed scheme was able to achieve an average CR of 3.54 for a decent average PSNR of 23.38 dB with a computational complexity of $O(zN)$ only. The average encoding and decoding time for the scheme per sample is 0.3 and 0.04 ms. The signal compression achieved is at par or reasonably better than the other compression schemes employing computationally intensive operation. The compression range is highly related to the recording resolution; more the resolution, the better will be the compression, while maintaining better signal quality. Moreover, the distortion level indicators such as PRD, MAE, PAE, and PSNR showed promising results to substantiate that the algorithm is suitable for MCEEG compression.

References

1. Andrzejak RG *et al* (2012) Nonrandomness, nonlinear dependence, and nonstationarity of electroencephalographic recordings from epilepsy patients. Phys Rev E 86:1–17
2. Antoniol G, Tonella P (1997) EEG data compression techniques. IEEE Trans Biomed Eng 44 (2):105–114
3. Bazn-Prieto C *et al* (2012) Retained energy-based coding for EEG signals. Med Eng Phys 34 (7):892–899
4. Birvinskas D *et al* (2015) Fast DCT algorithms for EEG data compression in embedded systems. Comput Sci Inform Syst 12:49–62

5. Blankertz B *et al* (2002) Classifying single trial EEG: towards brain computer interfacing. In: Dietterich TG, Becker S, Ghahramani Z (eds) Advances in neural information processing systems, vol 14. MIT Press, Cambridge, pp 157–164
6. Blankertz B *et al* (2006) The BCI competition III: validating alternative approaches to actual BCI problems. IEEE Trans Neural Syst Rehabil Eng 14(2):153–159
7. Dauwels J *et al* (2013) Near lossless multichannel EEG compression based on matrix and tensor decompositions. IEEE J Biomed Health Inform 17:708–714
8. Delorme A, Makeig S (2004) EEGLAB: an open source toolbox for analysis of single-trial EEG dynamics including independent component analysis. J Neurosci Methods 134(1):9–21
9. Dheeru D, Karra Taniskidou E (2017) UCI machine learning repository
10. Hejrati B *et al* (2017a) Efficient lossless multichannel EEG compression based on channel clustering. Biomed Signal Process Control 31:295–300
11. Hejrati B *et al* (2017b) A new near-lossless EEG compression method using ANNbased reconstruction technique. Comput Biol Med 87:87–94
12. Hunter M *et al* (2005) The Australian EEG database. Clinical EEG Neurosci 36(2):76–81
13. Titus G, Sudhakar MS (2017) A simple and efficient algorithm operating with linear time for MCEEG data compression. Australas Phys Eng Sci Med 40(3):759–768
14. Titus G, Sudhakar MS (2018) A simple but efficient EEG data compression algorithm for neuromorphic applications. IETE J Res 1:1–12
15. Goldberger AL *et al* (2000) Physiobank, physiotoolkit, and physionet: components of a new research resource for complex physiologic signals. Circulation 101(23):215– 220
16. Katsigiannis S, Ramzan N (2018) DREAMER: a database for emotion recognition through EEG and ECG signals from wireless low-cost off-the-shelf devices. IEEE J Biomed Health Inform 22(1):98–107
17. Lemire D, Boytsov L (2012) Decoding billions of integers per second through vectorization. ArXiv e-prints 1209.2137, 1–30
18. Lin L *et al* (2015) Multichannel EEG compression based on ICA and SPIHT. Biomed Signal Process Control 20:45–51
19. Naeem M *et al* (2006) Seperability of four-class motor imagery data using independent components analysis. J Neural Eng 3:208–216
20. Saeedi J *et al* (2014) Hybrid fractal-wavelet method for multi-channel EEG signal compression. CSSP 33(8):2583–2604
21. Srinivasan K *et al* (2013) Multichannel EEG compression: Wavelet-based image and volumetric coding approach. IEEE J Biomed Health Inform 17:113–120
22. Sriraam N (2009) Context based near lossless compression of EEG signals using neural network predictors. Int J Electron Commun (AE) 63:311–320
23. Tangermann M *et al* (2012) Review of the BCI competition IV. Front Neurosci 6:1–31
24. Xu G *et al* (2015) A 1.5-D multi-channel EEG compression algorithm based on NLSPIHT. IEEE Signal Process Lett 22(8):1118–1122

An Adaptive Approach of Fused Feature Extraction for Emotion Recognition Using EEG Signals

13

Sujata Bhimrao Wankhade and Dharmapal Dronacharya Doye

Abstract

Emotion recognition is a basic part towards complete correspondence among human and machine. Recently, research efforts in Human Computer Interaction (HCI) are focused to allow personal computers (PCs) to analyze human feelings. Although, some researchers are trying to realize human-machine interfaces with an emotion understanding capability. In this study, an effective emotion recognition framework is proposed dependent on the support vector machine (SVM) classifier. The proposed classifier employs a versatile technique for combined element extraction, which uses the methods of empirical mode decomposition (EMD) and kernel density estimation (KDE). The technique used in the proposed classifier decomposes the signal and accomplishes feature extraction with diminished computational unpredictability. The re-enactment results show that the proposed acknowledgement framework can accomplish 92.991% precision, and the correlation with some ordinary acknowledgement framework is additionally given.

Keywords

Emotion recognition · EEG signal · Independent component analysis (ICA) · Fused feature extraction · Empirical mode decomposition (EMD) · Kernel density estimation (KDE)

S. Bhimrao Wankhade (✉)
Computer Science & Engineering Department, Shri Guru Gobind Singhji Institute of Engineering and Technology, Nanded, Maharashtra, India

D. D. Doye
Department of Electronic and Telecommunication Engineering, Shri Guru Gobind Singhji Institute of Engineering and Technology, Nanded, Maharashtra, India

S. Paul (ed.), *Application of Biomedical Engineering in Neuroscience*,
https://doi.org/10.1007/978-981-13-7142-4_13

13.1 Introduction

The sense of feeling is of prevalent significance in normal and astute reasoning [1]. Feeling is associated with a social event of structures in the focal point of the brain, which consolidates amygdale, thalamus, nerve centre and hippocampus [2]. Overseeing impact in sight and sound, three substitute perspectives of sentiments explicitly incorporate imparted sentiments, felt sentiments and expected feelings [3]. The assessment of emotion is principal to the comprehension of human behavior [4]. In human computer interaction (HCI), researchers are analysing a passionate condition of subject can extraordinarily improves the communication quality [5]. The examination of enthusiastic states reveals more informationon the HCI from the perspective of clinical application [6]. Furthermore, this understanding about feelings can help psychiatrics in the treatment of mental issues, such as extraordinary autism spectrum disorders (ASD), attention deficiency hyperactivity disorder (ADHD) and tension issues [7].

Emotional intelligence is the ability to distinguish, evaluate, and control feelings of one self and of others; to accept a critical part in learning shapes; and to extent of withdrawing the information that is most basic [8]. Discoveries in enthusiastic intelligence have uncovered the significance of its knowledge in accomplishing individual yearnings as well as achievements through sound social cooperation [9]. Electroencephalography (EEG) measures voltage fluctuations resulting from the ionic current flow inside the neurons of the cerebrum [10]. These EEG signals could reflect the "internal" certifiable feelings of the subject that could be essential in medicinal applications or item showcasing applications [11].

In the cerebrum, EEG signals provide valuable characteristics in response to the eager states [12]. In the frontal cerebrum, electrical activity was identified with the standard of positive and negative sentiments [13].To significantly grasp the mind, response under the condition of various feelings can in general advance the models for feeling acknowledgement [14]. The connections between the human emotions and EEG signals are examined with quick progression of wearable devices and dry cathode techniques [15]. EEG signals are usually asymmetry and non-stationary. Recently, new methods for the examination of non-stationary and non-direct flags are proposed, which are mostly dependent on empirical mode decomposition (EMD) [16]. The investigative intrinsic mode functions (IMFs) (Hilbert–Huang Transform (HHT)) for seizure arrangement in EEG signals are used as part. Hilbert Huang Transform allows discriminating instantaneous frequencies and locating patterns not detectable by traditional systems of signal analysis [17]. IMFs weighted frequencies have been used to perceive seizures in EEG signals [18, 19]. The higher request measurements of the IMFs are adequate for the portrayal of EEG signals. Excluding the characteristics of highlight extraction methods identified with immediate frequencies (IF), the extraction of IF is increasingly important when the IMFs separated from the EEG signals are monopart [20].

To recognize human emotions from EEG signal in this research work we present a novel approach of feature extraction, which combines the IMFs of EMD and the attributes of wavelet change to characterise the component vector sets utilized for the classification. This study is organised as follows: Sect. 13.2 reviews some current works identified with emotion; Sect. 13.3 portrays the proposed technique of face recognition system; Sect. 13.4 discusses the simulation consequences of the proposed study; and Sect. 13.5 concludes this study.

13.2 Related Research

Linet al. [21] expected an EEG-supported ER algorithm to determine the associates among passionate states and cerebrum movement. Here, an artificial intelligence (AI) calculation was performed to mark EEG elements presenting to passionate states. An EEG-based ER framework was proposed that optimize by pursuing emotion-specific EEG features and discover the classifiers. SVM was engaged to classify the four emotional states and found the averaged classification accuracy. Regardless of the way this work, it does not evaluate the specific association between EEG components, eager responses and music structures to isolate the cerebrum responses to the music perception, music appreciation and music-induced sentiments.

Zhang and Lee [22] introduced a feeling liberal system and Gwangju Institute of Science and Technology (GIST) to collect feelings imitated by ordinary scenes. Due to distorted relationship between human inclination and cerebrum movement, reverberation imaging techniques, such as functional magnetic resonance imaging (fMRI) and electroencephalogram (EEG),were used to separate and orchestrate excited states. The "GIST" was used to mine visual low-level highlights that are used as input signals to a classifier for achieving the abnormal state enthusiastic significance. .However, this work doesn't research the more elevated amount of features, i.e., the setting data and also doesn't consider on powerful attributes.

Garrett et al. [23] exhibited strategies for highlight extraction from EEG signals utilising EMD. Its utilisation was driven by EMD that provide a viable time–frequency examination. This work employed linear discriminate analysis (LDA) to produce models of likelihood thickness capacities by the information produced from each class. Moreover, artificial neural network (ANN) was introduced to create non-linear arrangement limits. It is then portrayed about SVM that dealt to solve the drawbacks of ANN, i.e. with the need to choose proper number of hidden units. Subsequently, genetic improvement calculation was also introduced for highlight determination.

Shahabi and Moghimi [24] researched the cerebrum systems related to blissful, melancholic and impartial music. Availability model between EEG terminals was inferred by multivariate autoregressive displaying. Melodic determinations were portrayed standing to the subjects' normal self-calculation results. Availability lattices were investigated to perceive deviations in the network lists related to arranged concentrates. At the point when subjects heard blissful pieces, network was expanded in the frontal and frontal-parietal locales.

Rahnuma et al. [25] portrayed an elective strategy in dissecting and understanding pressure, utilising the four fundamental feelings such as glad, quiet, miserable and dread. EEG signals were recorded from the scalp of the brain and estimated in responds to various stimuli from the four basic emotions to stimulating stress based on the IAPS emotion stimuli. Highlights from the EEG signals were separated utilising the kernel density estimation (KDE). This technique is utilized to extract signal features by computing density estimate using kernel-smoothing method and classified using the Multilayer Perceptron (MLP) neural network classifier to obtain accuracy of the subject's emotion leading to stress. Results have showed the capability of utilising the fundamental feeling premise capacity to envision the pressure observation as an elective instrument for designers and clinician.

Jie et al. [26] showed the application of sample entropy (SampEn)-based feeling affirmation approach. The SampEn impacts of EEG channels screened by K-S test were upheld to the support vector machine (SVM)–weight classifier. One is to isolate constructive and contrary inclination and the elective individual emotions. They were associated to custom the information vectors of SVM-weight classifier. Sensible acknowledgement precision appeared thorough approval methodology. Be that as it may, the work did not focus because of EEG information length and recurrence groups. Further, the performance of sample entropy was not improved better.Seyyed Abed Hosseini and Mohammad Ali Khalilzadeh [27] presented a proficient obtaining protocol to secure the EEG and psychophysiological signal under pictures generation environment for members. Subjective and quantitative assessment of psychophysiological sign have been attempted to choose appropriate portions of EEG signal for improving effectiveness and execution of emotional stress identification framework. After pre-processing the signals, both Linear and nonlinear features were utilized to separate the EEG parameters. Wavelet coefficients and disorderly invariants like fractal measurement by Higuchi's calculation and correlation measurement were utilized to separate the qualities of the EEG signal which demonstrated that the accuracy in two emotional states was 82.7% utilizing the Elman classifier.

13.3 Fused Feature Extraction-Based Emotion Recognition

An adaptive element extraction system is proposed in this chapter for the signal order of EEG. This system consists of two phases. The first phase includes the EMD calculation with EEG flag, giving an IMF set. The second phase incorporates highlight extraction done by registering the worldly and unearthly attributes of the IMFs. This is the principal commitment of this work. The temporal and spectral features are acquired from the kernel density of estimated IMFs that expel the DC offset from the spectral substance. The strategy of the proposed ER framework is mentioned in Fig. 13.1.

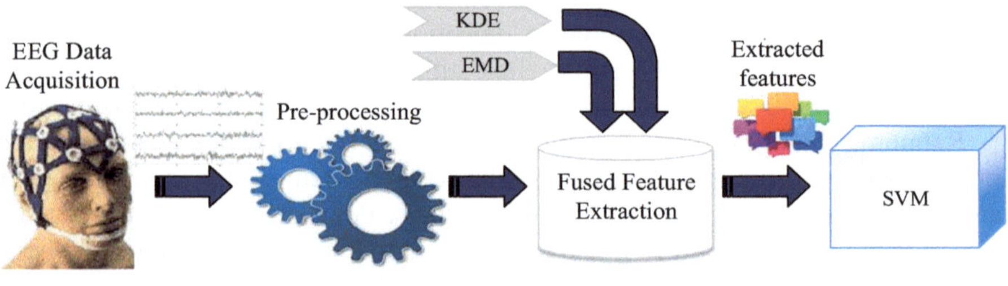

Fig. 13.1 Process of proposed emotion recognition system

The utilisation of SVM classifier is energised for feeling of grouping on account of its great speculation properties. The basic preparing show of SVM is to find the ideal hyper-plane where the normal arrangement blunders of tests are limited. The ideal hyper-plane exaggerates the edges, which grows the speculation capacity.

13.3.1 Initial Data Preparation

EEG signals are recorded from the various patients by setting number of electrodes on the scalp is input dataset, which is stored in matrix form of $X = \{x_1, x_2, x_n\}$ with n observed EEG signals. The EEG flags also involve noise and ancient rarities that are detected on the scalp other than the brain.

13.3.2 Pre-processing

To remove noise and artefacts from the EEG data, FastICA is used to eliminate noise caused by (i) muscles movement, (ii) heartbeat rate, (iii) eyeball movement and (iv) eyelid movement. FastICA [28] works by separating data between noise and original EEG data, and then by extracting valuable data.

13.3.2.1 Signal Separation

The EEG signal X, comprising mixed EEG data and their noise, is stored in array form and can be developed as

$$X = AS, \tag{13.1}$$

where A denotes the $[n \times m]$ mixing matrix, and S is the original source EEG signal represented as $S = [s_1, s_2, \ldots s_n]^T$, which is a lattice of m hidden signs containing independent components (ICs). Independent component analysis (ICA) searches for S by assessing framework W, which is the opposite of matrix A in request to invert the blending impact. In this way, ICA processes the first source EEG information as

$$S = WX. \tag{13.2}$$

1. Here, the estimation of matrix W is evaluated since the value of matrix A is unknown. It is assessed by boosting non-Gaussian esteem. Before this estimation centring and brightening process are finished, the accompanying two procedures must be done before the estimation of W.
2. *Centring*

The average value of each of all the channels in the emotion recognition system is forced to make equal to zero. For focusing, subtract X by its mean vector given as in this manner making X a zero-mean variable. After estimation of the blending grid

A with focused information in (1), the estimation is settled by expansion of mean vector of S back to the jogged appraisals of S. The mean vector of S is given by

$$m = E(X). \tag{13.3}$$

3. *Whitening*

By brightening process, signs are made uncorrelated and have the change esteem equivalent to one. By this procedure, the watched vector X is changed over into another vector $\tilde{X}$, which is white. This should be possible by figuring eigenvalue and eigenvector because eigenvalue decomposition (EVD) is utilised. Brightening is then communicated as

$$\tilde{X} = ED^{-1/2}E^T X, \tag{13.4}$$

where E is the orthogonal matrix of eigenvectors of $E\{XX^T\}$ and D is the eigenvalue in diagonal.

13.3.2.2 Component Extraction

Component extraction is performed in each channel with the quantity of wanted segment C to evaluate W, which is unknown by augmenting non-Gaussian. This can be determined as

$$W'_p = \frac{1}{N}\left[X_g\left(W_p^T\tilde{X}\right)^T - g'\left(W_p^T\tilde{X}\right)W_p\right], \tag{13.5}$$

$$W_p = \frac{W'_p}{\|W'_p\|}, \tag{13.6}$$

where W'_p is assigned by random vector with length N, p is an index iteration from 1 to C, g is *tanh* function and g' is *1-tanh*^{2} function. Then $\tilde{X}$ the noise-free EEG signal is given as input to fused feature extraction by EMD. Thus, the input dataset X in matrix form with n experiential EEG signals is given as $\tilde{X} = [\tilde{x}_1, \tilde{x}_2, \ldots \tilde{x}_n]^T$.

13.3.3 Fused Feature Extraction

The goal of highlight extraction is to secure reliable information for feeling location. In this study, EMD is implemented in highlight extraction to break down EEG signals into IMF arrangement. Then, piece densities of IMFs are assessed for highlight vectors development and are sustained into an SVM classifier to see enthusiastic states. A versatile element extraction calculation called a combination of EMD with KDE is given. By this combination, calculation proficient element extraction is acquired and is also clarified tentatively.

13.3.3.1 Empirical Mode Decomposition (EMD)

EMD is an information-driven flag taking care of examination technique [30]. The idea of EMD is to separate the non-direct and non-stationary EEG movements into different movements called IMFs on a couple of repeat scales. Each IMF fulfils two objectives: (1) the amount of outrageous and the amount of zero-crossing points ought to either be proportionate or shift greatest by one and (2) the mean incentive between the upper envelope and the lower envelope must be zero at each point. Mostly, these IMFs delineate nearby traits of the one of a kind flag. The change among the first flag and the IMFs is shown as lingering. The proposed combination estimation is given as follows:

Algorithm 1: Fusion of EMD and KDE
Input: EEG signal $\tilde{X}(t)$.
Output: Decomposed IMFs.
Begin
Calculate IMF of $\tilde{X}(t)$.
Estimate temporal features by EMD
Estimate PSD by spatial representation by EMD
Estimate density function by KDE
Concatenate all the results
End.

The non-stationary signal $\tilde{X}(t)$ is given as

$$\tilde{X}(t) = \sum_{m=1}^{M} IMF_M(t) + r_M(t), \tag{13.7}$$

where

$$r_1(t) = \tilde{X}(t) - IMF_1(t), \tag{13.8}$$

$$IMF_1(t) = h_1(t), \tag{13.9}$$

$IMF_m(t)$ is the mth extracted IMF, M is the number of IMFs, $r_M(t)$ is the final residue, $h_1(t)$ is the smallest first temporal scale in $\tilde{X}(t)$ and $r_1(t)$ is a residual signal.

1. *IMF Analytic Representation*

After IMF extraction, its diagnostic portrayal is procured. This portrayal expels the direct current (DC) offset from the spectral component of the signals, which is a significant angle to make up for the non-stationarity of the signs. Given an IMF $C_M(t)$, its analytic representation is

$$y(t) = IMF_M(t) + iH\{IMF_M(t)\}, \tag{13.10}$$

where $H\{IMF_M(t)\}$ the Hilbert is transformed to $IMF_M(t)$ and is the *m*th IMF extracted from the signal $\tilde{X}(t)$.

2. *IMF Temporal Representation*

The IMFs procured from sound and epilepsy subjects through interracial and octal subjects after Hilbert change uncovered that they are assorted from each another. These distinctions are fittingly caught utilising the insights of the IMFs. For an IMF, these statistics can be obtained by the accompanying quantities:

$$\mu_t = \frac{1}{N} \sum_{i=1}^{N} y_i, \tag{13.11}$$

$$\sigma_t = \frac{1}{\sqrt{N}} \left[\sum\nolimits_{i=1}^{N} (y_i - \mu_t)^2 \right]^{1/2}, \tag{13.12}$$

$$\beta_t = \frac{1}{N} \sum_{i=1}^{N} \left(\frac{y_i - \mu_t}{\sigma_t} \right)^3, \tag{13.13}$$

where μ_t is the mean, σ_t is the variance, β_t is Skewness of the IMF and N is the number of samples in IMF.

3. *IMF Spectral Representation*

The important feature of the intensity of EMD is the ability to execute a ghastly investigation of the signs. The ghastly highlights accomplished from IMFs provide an unexpected sign about the EEG signals. Generally when utilising EMD, this otherworldly investigation is carried out by the estimation of momentary IF. Subsequently, when the EEG signals are presented to EMD, mono-part flags are not obtained. Moreover, the count of Power spectral density (PSD). is turned for highlight extraction judgements. The separation intensity of the PSD highlights is examined by their relating plots for three IMFs. The PSD is estimated as follows:

$$P(w) = \sum_{-\infty}^{\infty} r_y[n] e^{-jwn}, \tag{13.14}$$

where $r_y[n] = E(y[m]y * [m])$, which is the autocorrelation of $y[n]$.

13.3.3.2 Kernel Density Estimation (KDE)

KDE estimates the density of decayed IMFs using kernel-smoothing method. The probability of KDE function is given by

$$\delta(x) = \frac{1}{nh} \sum_{i=1}^{n} K\left(\frac{\tilde{x} - \tilde{x}_i}{h}\right), \tag{13.15}$$

where K is the non-negative capacity that coordinates to 1 and h is the portion smoother parameter. The upsides of KDE exceptionally prescribe its adjustment to data streams as the consequent gauge provides a factual model portrayed by the stream data. From the learning profited by EMD and KDE, the succeeding highlights are removed from the disintegrated IMF and can be utilised for characterisation.

1. *Spectral Entropy*

The spectral entropy SE of the EEG signal can be determined from the power spectral density as

$$SE = -\sum_{w=1}^{n} P(w) log_2 \mid P(w) \mid, \tag{13.16}$$

where $P(w)$ is the amplitude of wth frequency bin in the spectrum.

2. *Energy*

   The energy E of the EEG signal can be determined from the power spectral density as

$$E = \sum_{w=1}^{n} P(w)^2. \tag{13.17}$$

3. *Spectral Centroid*

The centroid frequencies SP of the IMFs are extracted from EEG signals. The centroid frequency is a distinct feature used for EEG signals characterisation. This is given as

$$SP = \frac{\sum_w wP(w)}{\sum_w P(w)}. \tag{13.18}$$

After extraction of all features in the EEG signal, comprising spatial and temporal features of individual IMF, its feature vector is obtained by concatenation as

$$F = [\mu_t \, \sigma_t \, \beta_t \, \delta(x) \; SE \; E \; SP], \tag{13.19}$$

where F is the element obtained by combination method. Similarly, the feature vectors are obtained for the given noise free input EEG signal from various IMFs. These results are used as an input for the classification process.

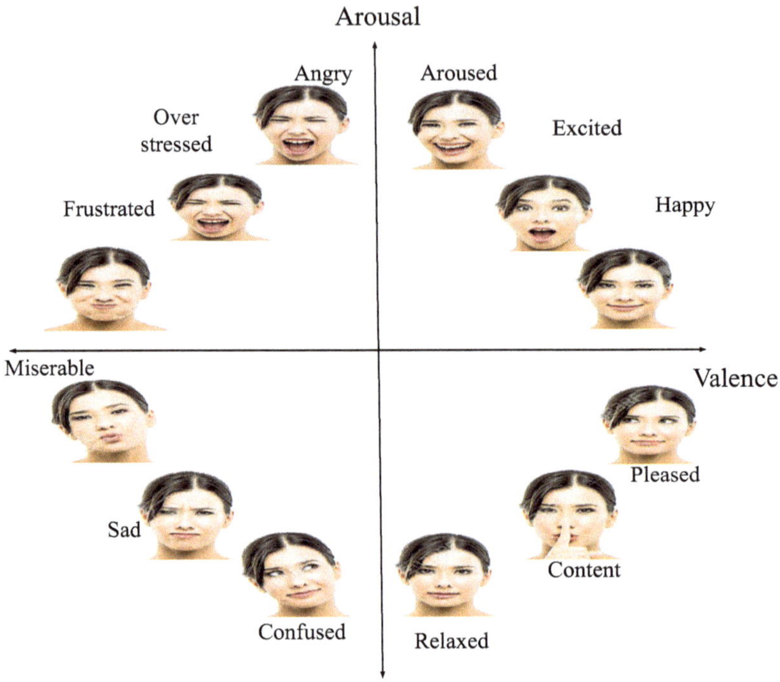

Fig. 13.2 Emotion model

13.3.4 Emotion Classification by SVM

Highlight extraction is trailed by the order of EEG signals using SVM [31]. Human feelings can be grouped into two measurements depending on the excitement and valence among them from EEG flag. The passionate states are portrayed into two classes, which are "certain" and "negative" for valance and "high" and "low" for excitement. The motion model is shown in Fig. 13.2 [32].

SVM build an optimal separating hyper plane in a high-dimensional space to characterise new emotions utilizing statistics learning hypothesis. Given, a preparation informational index $x_1 \ldots x_N$, containing N training labelled samples and coefficients $\delta_1 \ldots \delta_N$, learned in the preparation step. To develop the ideal hyperplane, preparation tests are changed into a higher-dimensional component space by a mapping capacity. A conceivable isolating hyper-plane is determined by

$$w.\phi(x) + b = 0. \tag{13.20}$$

The SVM decision function is estimated as follows:

$$f(x) = \text{sign}\left[\sum_{i=1}^{L} \delta_i K(x_i, F) + b\right], \tag{13.21}$$

where L is the number of support vectors, b is a bias and is occupied as constant function and $K(x_{i,,F})$ is the kernel function which is given by

$$K(x_i, F) = \phi(x_i)^T \phi(F). \tag{13.22}$$

A weight SVM classifier is utilised in this study by considering the conceivable unevenness of tests. Its purpose is to disseminate a substantial discipline factor to the minority class, though a little discipline factor to the greater part class. Therefore, the feelings distinguished are perceived successfully as positive feeling (satisfaction, joy) or negative feeling (trouble, outrage), and the precision results are referenced.

13.4 Implementation and Simulation Results

In this section, an extensive report is produced from the proposed simulation results and the dataset is used to estimate the proposed ER framework execution. Moreover, a succinct examination of the proposed work with existing works is introduced.

13.4.1 Dataset Description and Simulation Setup

In this exploratory study, an online EEG dataset is used [29]. From this dataset, 80% of EEG flag is used for preparing, and the remaining 20% of EEG flag is used for testing segment. This dataset contains an EEG dataset which is the record of number of subjects (human) tuned in to voice chronicles that propose an enthusiastic inclination. The dataset includes 15 unique tracks, where each track proposes a feeling of both positive valence (bliss, joy) and negative valence (bitterness, outrage). Moreover, the dataset combined 32 quantities of sessions and 31 quantities of subjects. At the point when the subject initially starts to feel loaded up with the proposed feeling, they show this by squeezing the finger weight sensor. These dataset signals are favoured from persistent multichannel EEG recording after antiques visual review. These outcomes are executed in the MATLAB stage, and the re-enactment setup with framework design and the underlying parameter settings of the proposed ER framework are given in Table 13.1.

Table 13.1 Initial parameter setting of the proposed emotion recognition system

Parameter	Specification
Number of channels	32
Time duration of each signal	23.6 s
Sampling rate	173.61 Hz
Length of each sample	4097

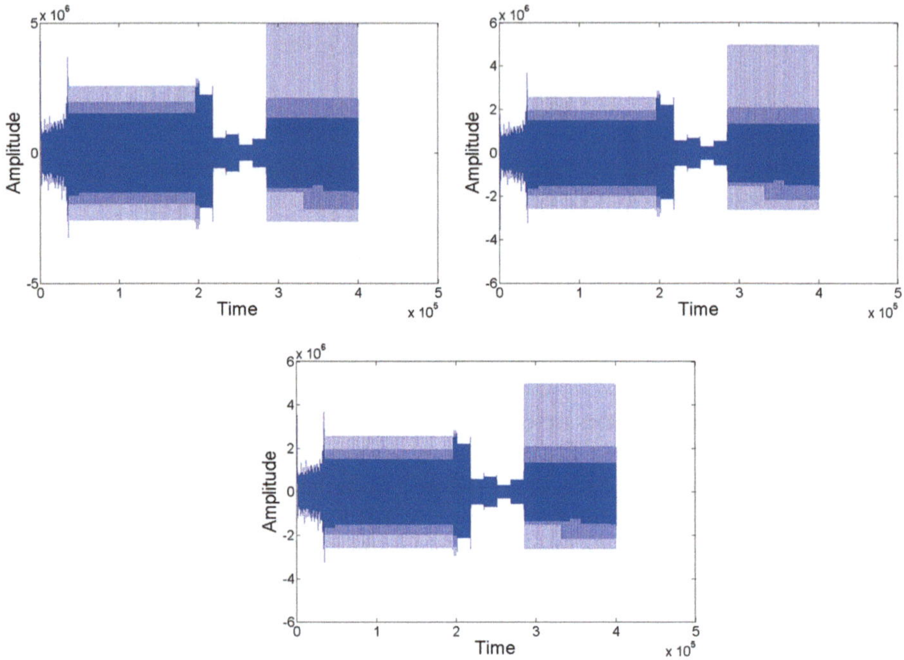

Fig. 13.3 Sample input EEG signals

13.4.2 Results of the Proposed System

The proposed ER framework starts its procedure with a lot of EEG signals obtained from the depicted dataset. Figure 13.3 shows the graphical delineation of the sample input EEG signals.

This EEG flag is pre-handled to evacuate commotions and undesirable signs from the first EEG flag, which should be possible for utilising quick ICA in the proposed ER framework. ICA isolates the clamour from the first flag, and the commotion-free flag was utilised for further handling. Figure 13.4 shows the graphical depiction of the pre-processed EEG signal.

To separate the ideal highlights from the flag in the proposed framework, the melded highlight extraction strategy is used in the proposed framework, which is the combination of two calculations i.e. EMD and KDE. Here, the signal gets decomposed into number of the IMFs which is a lot of narrow-band symmetric waveforms and the spectral and temporal signals of IMFs. The entire EEG band of 256 Hz is tending to extract features instead of decomposing it to diverse frequency bands. By this method, an exceedingly summed up flag is accomplished along these lines creating great information highlights for SVM. Figures 13.5 and 13.6 demonstrate the unearthly and worldly flags of IMFs separately.

From the spectral and temporal IMFs, the optimal features can be extracted and a feature vector is formed before training and testing in the classifier. The proposed framework utilises SVM classifier for characterising various feelings from EEG

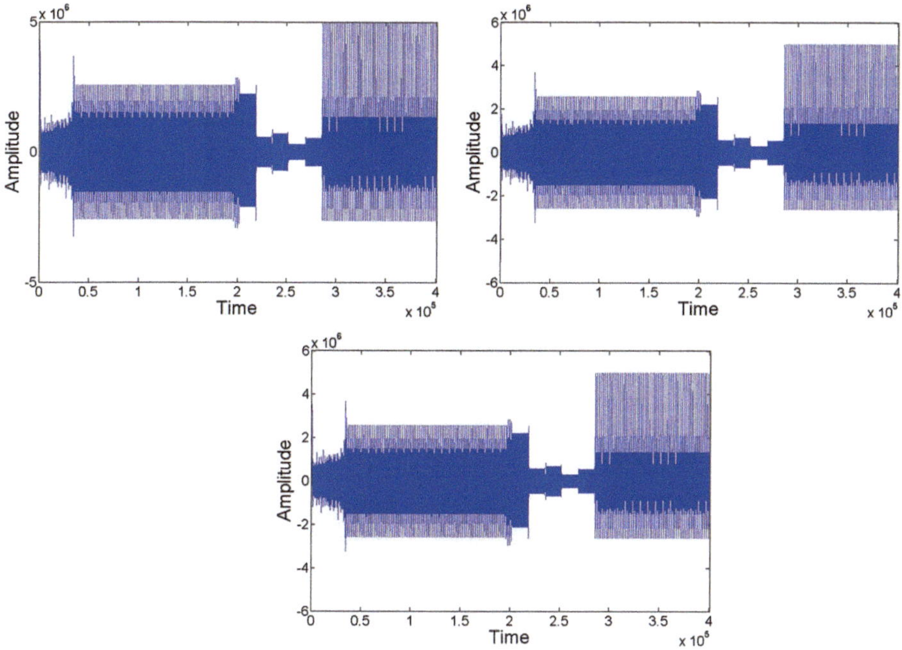

Fig. 13.4 Pre-processed EEG signal

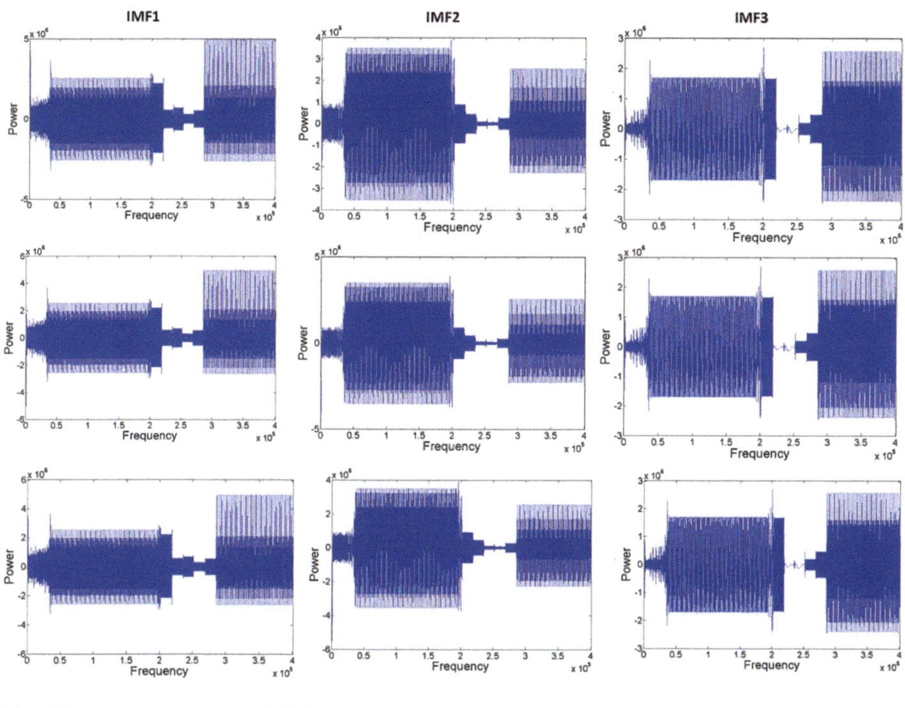

Fig. 13.5 Spectral signals of IMFs

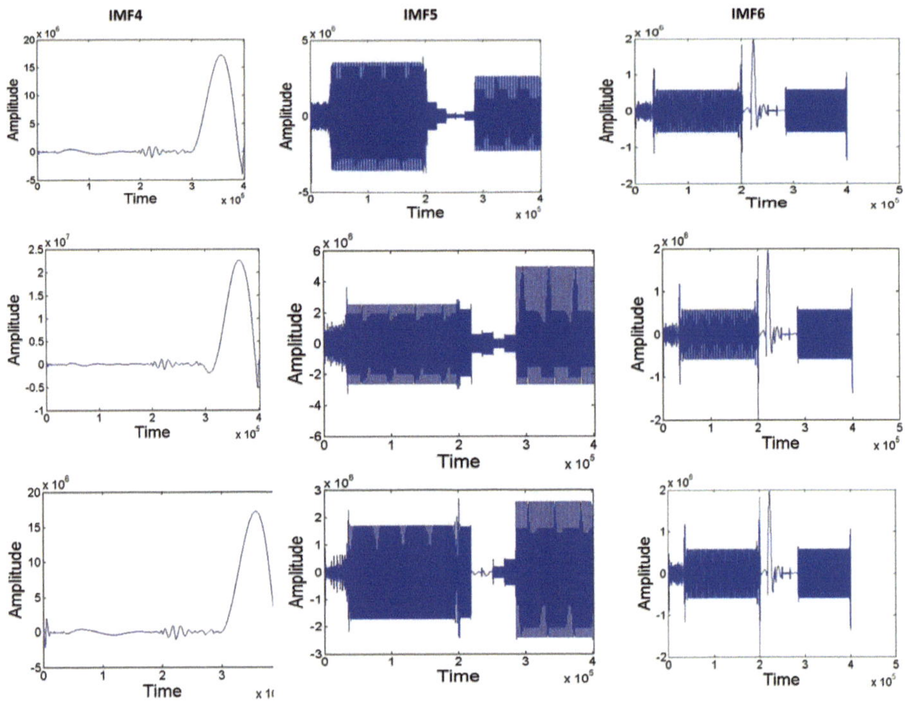

Fig. 13.6 Temporal signals of IMFs

signals. The SVM classifier is at first prepared by the information of recently characterised highlight vectors. In SVM, hyper-planes are resolved to make choice limits between various classes. The SVM distinguishes the ideal isolating hyper-plane to augment the separation from either class to the hyper-plane. The counts are finished with help of a part capacity and predisposition. After characterisation, it is seen as the feeling acknowledgement in EEG dataset for positive feeling (bliss, joy) or negative feeling (trouble, outrage). The precision parameter is evaluated to discover the execution. The precision is characterised as the proportion of number of the effectively arranged flags by the connected EEG information flag and is given as

$$\text{Accuracy} = \frac{(TP + TN)}{(TP + FP + TN + FN)} * 100, \tag{13.23}$$

where TP determines truth-positive, TN shows truth-negative, FP demonstrates false-positive and FN indicates false-negative. The whole procedure gave a computational time of 16.276103 s. The proposed feeling acknowledgement framework execution in examination with the ordinary acknowledgement framework is given in Table 13.2.

Table 13.2 displays a correlation of various feeling acknowledgement frameworks before different component extraction systems and different classifiers.

Table 13.2 Proposed emotion recognition system performance comparison

Approaches	Type of classifier used	No. of channels	Accuracy (%)	Computational time (s)
Sample entropy	SVM	5	80.43	22.4354
HOC + FD + 6 statistics	SVM	4	83.73	19.878
KNN + SVM	KNN	10	86.75	18.76
EMD + KDE (fused proposed)	SVM	32	92.9991	16.276103

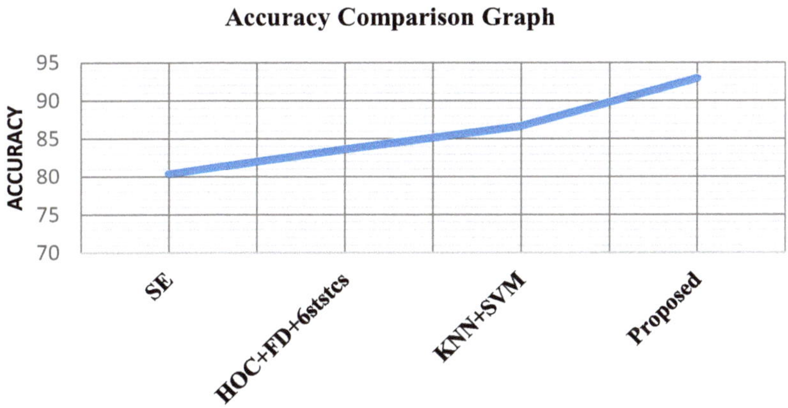

Fig. 13.7 Accuracy performance evaluation

It is unmistakable that the proposed framework accomplishes the greatest precision of 92.9991% with 32 channels, utilising the intertwined highlight extraction technique which is more effective than the current strategies. The achieved precision is fundamentally 8.24% more prominent than the regular ER framework, which utilises entropy, vitality-based element extraction and KNN classifier for order.

The customary ER framework is not ready to achieve a greatest exactness with a decreased number of channels. However, the proposed framework can have the capability to produce a high performance even with an increased number of channels. Figure 13.7 demonstrates the enhancement for a precision of the proposed system with different existing techniques, such as sample entropy (SE), higher order crossings (HOC) + fractal dimension (FD) + 6 insights, KNN (K-nearest neighbours) + SVM. Figure 13.8 portrays the computational time correlation with current strategies.

From above outcomes and correlations, the proposed framework is realised to be productive even with substantial number of channels and achieves a greatest exactness for feeling order. The prevalence of the proposed ER framework over customary feeling acknowledgement frameworks was demonstrated from the information in Table 13.2.

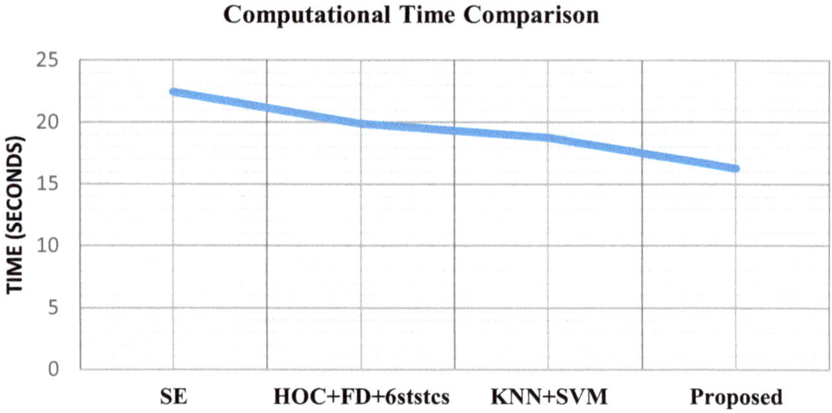

Fig. 13.8 Computational time performance evaluation

13.5 Conclusion

In this study, a proficient ER framework is introduced dependent on SVM classifier. A technique for intertwined highlight extraction is proposed to viably decay and recover the ideal highlights from the first EEG flag. The combination of EMD and KDE makes the framework successful by obtaining the unearthly and worldly flags of IMFs deteriorated by EMD. The proposed framework, furthermore, reduces the computational unpredictability by extricating just the vitality and entropy as ideal highlights. SVM classifier utilised for order has given incredible theory, and its execution is dynamic. Various classifiers such as KNN and naive Bayes are hard to fabricate, and its runtime execution is poor. The re-enactment results demonstrate that the proposed framework accomplishes 92.991% precision, which is higher than the customary feeling acknowledgement frameworks. This uncovers the execution and quality of ER frameworks that can be improved by the use of combination procedures. This can stimulate the use of the proposed ER system in cutting-edge applications.

In future study, this strategy can be considered for human feeling characterisation and open or close eye flag recognition utilising EEG signals. For sifting of non-direct flags, non-straight versatile channels can be utilised at each stage. Moreover, we hope to find the probability to do the equipment execution of the proposed highlights for viable use by the subjects. Furthermore, additional systems, for example, the element choice and decrease instruments, will be executed to discard the overabundance of information and to select the most relevant highlights.

Acknowledgements I thank my co-author Dharmapal Dronacharya Doye for guiding me to complete this research and also am very thankful to my institution Shri Guru Gobind Singhji Institute of Engineering and Technology, Vishnupuri, Nanded, Maharashtra, India, for giving me full support to complete this work.

Conflict of Interest Sujata Bhimrao Wankhade and Dharmapal Dronacharya Doye state that there are no conflicts of interest. Patients' rights and animal protection statements: This research article does not contain any studies with human or animal subjects.

Ethical Statements Animal and human subjects were not used in this study.

Funding This research did not receive any specific grant from funding agencies in the public, commercial or not-for-profit sectors.

References

1. Zou PXW, Sunindijo RY (2013) Skills for managing safety risk, implementing safety task, and developing positive safety climate in construction project. Autom Constr 34:92–100
2. Nissan E (2009) Computational models of the emotions: from models of the emotions of the individual to modelling the emerging irrational behaviour of crowds. AI & Soc 24(4):403–414
3. Mayer RE (2003) The promise of multimedia learning: using the same instructional design methods across different media. Learn Instr 13(2):125–139
4. Liberzon I, Taylor SF, Amdur R, Jung TD, Chamberlain KR, Minoshima S, Koeppe RA, Fig LM (2000) Brain activation in PTSD in response to trauma-related stimuli. Biol Psychiatry 45 (7):817–826
5. Nesse RM Evolutionary explanations of emotions. Hum Nat 1:261–289
6. Chanel G, JulienKronegg DG, Pun T (2006) Emotion assessment: arousal evaluation using EEG's and peripheral physiological signals. Multimedia content representation, classification and security:530–537
7. Bechara A (2004) Disturbances of emotion regulation after focal brain lesions. Int Rev Neurobiol 62:159–193
8. Rosso OA, Blanco S, Yordanova J, VasilKolev AF, Schürmann M, Başar E et al (2001) J Neurosci Methods 105:65–75
9. Petrantonakis PC, Hadjileontiadis LJ (2010) Emotion recognition from EEG using higher order crossings. IEEE Trans Inf Technol Biomed 14(2):186–197
10. Nolte G, OuBai LW, Mari Z, Vorbach S, Hallett M (2010) Identifying true brain interaction from EEG data using the imaginary part of coherency. Clin Neurophysiol 115:2292–2307
11. Perkins KL (2006) Cell-attached voltage-clamp and current-clamp recording and stimulation techniques in brain slices. J Neurosci Methods 154(1):1–18
12. Liu Y, Sourina O, Nguyen MK (2011) Real-time EEG-based emotion recognition and its applications. In: Transactions on computational science XII. Springer, Berlin/Heidelberg, pp 256–277
13. Phelps EA, LeDoux JE (2005) Contributions of the amygdala to emotion processing: from animal models to human behavior. Neuron 48(2):175–187
14. Makeig S, Gramann K, Jung T-P, Sejnowski TJ, Poizner H (2009) Linking brain, mind and behavior. Int J Psychophysiol 73(2):95–100
15. Choi JH, Jung HK, Kim T (2006) A new action potential detector using the MTEO and its effects on spike sorting systems at low signal-to-noise ratios. IEEE Trans Biomed Eng 53 (4):738–746
16. Delorme A, Sejnowski T, Makeig S (2009) Enhanced detection of artifacts in EEG data using higher-order statistics and independent component analysis. NeuroImage 34(4):1443–1449
17. Soleymani M, Asghari-Esfeden S, Fu Y, Pantic M (2016) Analysis of EEG signals and facial expressions for continuous emotion detection. IEEE Trans Affect Comput 7(1):17–28
18. Alarcao SM, Fonseca MJ (2017) Emotions recognition using EEG signals: a survey. IEEE Trans Affect Comput

19. Atkinson J, Campos D (2016) Improving BCI-based emotion recognition by combining EEG feature selection and kernel classifiers. Expert Syst Appl 47:35–41
20. Bhatti AM, Majid M, Anwar SM, Khan B (2016) Human emotion recognition and analysis in response to audio music using brain signals. Comput Hum Behav 65:267–275
21. Lin Y-P, Wang C-H, Jung T-P, Wu T-L, Jeng S-K, Duann J-R, Chen J-H (2010) EEG-based emotion recognition in music listening. IEEE Trans Biomed Eng 57(7):1798–1806
22. Zhang Q, Lee M (2009) Analysis of positive and negative emotions in natural scene using brain activity and GIST. Neurocomputing 72(4):1302–1306
23. Garrett D, Peterson DA, Anderson CW, Thaut MH (2003) Comparison of linear, nonlinear, and feature selection methods for EEG signal classification. IEEE Trans Neural Syst Rehabil Eng 11 (2):141–144
24. Shahabi H, Moghimi S (2016) Toward automatic detection of brain responses to emotional music through analysis of EEG effective connectivity. Comput Hum Behav 58:231–239
25. Rahnuma KS et al (2011) EEG analysis for understanding stress based on affective model basis function. In: Consumer Electronics (ISCE), 2011 IEEE 15th international symposium on. IEEE
26. Jie X, Cao R, Li L (2014) Emotion recognition based on the sample entropy of EEG. Bio-Med Mater Eng 24:1185–1192
27. Hosseini SA, Khalilzadeh MA (2010) Emotional stress recognition system using EEG and psychophysiological signals: using new labelling process of EEG signals in emotional stress state. Biomed Eng Comput Sci 18:1–6
28. Yuan L, Zhou Z, Yuan Y, Wu S (2018) An improved FastICA method for fetal ECG extraction. Comput Math Methods Med
29. http://headit.ucsd.edu/studies/3316f70e-35ff-11e3-a2a9-0050563f2612
30. Dybała J, Zimroz R (2014) Rolling bearing diagnosing method based on empirical mode decomposition of machine vibration signal. Appl Acoust 77:195–203
31. Turker M, Koc-San D (2015) Building extraction from high-resolution optical spaceborne images using the integration of support vector machine (SVM) classification, Hough transformation and perceptual grouping. Int J Appl Earth Obs Geoinf 34:58–69
32. Munawar MN, Sarno R, Asfani DA, Igasaki T, Nugraha BT (2016) Significant pre-processing method in EEG-Based emotions classification. J Theor Appl Inf Technol 87(2):176–190

Part VI

Artificial Intelligence and Computer Aided Diagnosis

Computer-Aided Diagnosis of Life-Threatening Diseases

14

Pramod Kumar, Sameer Ambekar, Subarna Roy, and Pavan Kunchur

Abstract

According to WHO, the incidence of life-threatening diseases like cancer, diabetes, and Alzheimer's disease is escalating globally. In the past few decades, traditional methods have been used to diagnose such diseases. These traditional methods often have limitations such as lack of accuracy, expense, and time-consuming procedures. Computer-aided diagnosis (CAD) aims to overcome these limitations by personalizing healthcare issues. Machine learning is a promising CAD method, offering effective solutions for these diseases. It is being used for early detection of cancer, diabetic retinopathy, as well as Alzheimer's disease, and also to identify diseases in plants. Machine learning can increase efficiency, making the process more cost effective, with quicker delivery of results. There are several CAD algorithms (ANN, SVM, etc.) that can be used to train the disease dataset, and eventually make significant predictions. It has also been proven that CAD algorithms have potential to diagnose and early detection of life-threatening diseases.

Keywords

Computer-aided diagnosis · Machine learning · Deep learning · Disease prediction · Healthcare

Authors Pramod Kumar and Sameer Ambekar have been equally contributed to this chapter.

P. Kumar (✉) · S. Ambekar · S. Roy
Department of Health Research, Biomedical Informatics Centre, ICMR-National Institute of Traditional Medicine, Belagavi, Karnataka, India

P. Kunchur
Department of Computer Science and Engineering, KLS Gogte Institute of Technology, Belagavi, Karnataka, India

S. Paul (ed.), *Application of Biomedical Engineering in Neuroscience*,
https://doi.org/10.1007/978-981-13-7142-4_14

14.1 Introduction

The prolific spread of fatal diseases is responsible for much concern across the world. Some severe diseases are curable and treatable, but many are not. There is a profound need for proper, early diagnosis. It has been proven that early diagnosis often leads to significant positive outcomes. Any disease requires accurate and efficient treatment, and traditional methods are often ineffective. Life-threatening diseases are on the increase, yet there is often a shortage of doctors to treat patients who have these diseases. Often, traditional methods cannot provide early, accurate diagnosis or provide an in-depth overview of a particular disease, nor can they meet the demand for analysis of complex and multi-dimensional data. However, deep learning and artificial intelligence methods can address these problems. These methods are expensive but have fewer requirements, and they can be deployed at a cloud server; as such they can be used to diagnose diseases remotely, i.e., from any part of the world. With these added advantages, anyone can now afford accurate diagnosis. The availability of a vast body of biomedical data contributes immensely in training for the machine-learning model or deep learning model. Deep learning gains valuable knowledge and insight from any given complex, high-dimensional biomedical data. Although previous healthcare instruments and methods often produced a plethora of information, they did not analyze, interpret, or dispense the data. Often imagery data of a lung or kidney might reach a few gigabytes in size. The abundance of biomedical data being generated, for example from fitness bands or even complex magnetic resonance imaging (MRI) machines, requires efficient analytical techniques to yield efficient results.

Computer-aided detection can assist doctors in interpreting medical images such as X-ray, MRI, and ultrasound [1]. The use of CAD has increased in recent years. Deep learning is a broader type of machine learning, and contains supervised, unsupervised, and semi-supervised learning. Studies have proved that early diagnosis is beneficial in many respects [2].

CAD is widely used in the USA to detect breast cancer in mammograms in clinics and medical centers. In the USA, out of 38 million mammographic examinations annually, almost 80% have been studied using CAD. The use of CAD was initiated in the early 1980s at Kurt Rossman Labs at the University of Chicago [3]. Increasingly larger numbers of patients with breast cancer have benefitted from the contributions of automated analysis of breast cancer using algorithms and a vast dataset. The CAD method accurately identifies signs of breast cancer, and in some cases also predicts prognosis [4]. Indeed, the scanned reports and information available in the scan heads of CAD are more efficient and advanced in the diagnosis of any disease, not only the more-feared, life-threatening ones.

CAD has two functions: (1) to detect lesions and (2) to reduce false-positive lesions. Skin cancer detection is one instance that involves looking at an area of the skin that might have lesions other than the particular lesion being examined. In later stages of the process, the lesions that are of no use to the model are therefore discarded. Previously, deep learning was used in the detection of diseases such as breast cancer, lung cancer, and Alzheimer's disease; in breast cancer it was used in the detection, prediction, and classification of lung nodules.

Explaining the technical and logical details of the system that computes the results of deep learning in disease detection is rather complicated. The accuracy of the training data provided as input to the system is of utmost importance; inaccurate training data will inevitably deliver inaccurate information. In some cases, there may be legal and ethical issues regarding the use of medical data to develop a product that is used commercially. In addition, issues may be raised when there is no supervision of the radiologist using the system. Deep learning has been introduced into the field of segmentation and registration, and is showing promising results. Convolution neural networks are being used in the segmentation of lungs, tumors, and other structures of the brain [5].

Radiologists are presently often overwhelmed with too many requests to interpret imaging tests, and deep-learning methods can now be used to acquire more accurate results more efficiently.

Since their introduction, computers and their applications have helped radiologists immensely. Radiologists themselves usually examine reports, and they can now verify their findings using CAD tools. CAD has also been certified by various organizations in the field of healthcare.

CAD also stands for computer-aided diagnosis, which refers to software that analyzes the data – with its features – and predicts the likelihood of the disease. Although not all diagnostic methods can use CAD, many technologies have been developed for the advancement of algorithms, computing power, and libraries, so that CAD can be used more effectively.

In some cases, the CAD approach has proved to be more successful than existing traditional methods being used. Software giants like Google, Microsoft, and IBM are often interested in CAD, and results are being acquired by healthcare companies such as Deepmind, known as Google Deepmind. Deepmind has a number of potentially pivotal projects like diabetic retinopathy detection using artificial intelligence (AI); apart from detection of diabetic retinopathy, AI is also being used in the diagnosis of Alzheimer's disease as well as in lung and other cancers.

CAD processes digital images that are procured from the diagnostic area, and highlights suspicious areas to offer additional support for the decision taken by the professional. CAD uses the following diagnostic processes: machine learning, deep learning, computer vision, and artificial intelligence.

CAD is now widely used in the field of healthcare, where it can be an option to provide treatment to patients, using a doctor's guidance or radiologist's findings. It offers additional evidence to the medical professional who makes medical treatment decisions. In some cases, the goal of CAD is to detect the early signs of a disease, for instance in the case of diabetes, where the early signs of diabetic retinopathy cannot be easily detected by a professional.

14.2 Breast Cancer Detection Using CAD

Cancer is an enormous health problem across the world and it has been estimated that the number of cancer cases could reach 27 million by 2030. Breast cancer is considered the second most common cancer worldwide. The mortality rate of breast cancer is very high [6]. Although the incidence of breast cancer cases in India is not as significant as elsewhere, the number is rising gradually. It is the most common cancer among urban women and the second most common cancer among rural women. It has been observed that this type of cancer is often diagnosed at a relatively advanced stage [7]. As mentioned, early detection could lead to subsequent clinical treatment that would be advantageous to the patient and make it easier for the doctor to treat the patient. In breast cancer, grading staging is usually done manually through visual inspection of the reports by pathologists. It may happen that the image that is being inspected by the pathologist might have a lot of information to be extracted, but this is only partially being done.

Evidence of a tumor found in the body doesn't necessarily mean that it is cancerous. The tumor could well be benign, depending on its dimensions and nature. The traditional approach consists of physically removing a small amount of tissue from the tumor, and then sending it for a biopsy. Thereafter, the report is usually sent to the professional, who examines it and decides whether the tumor is malignant or benign, i.e., cancerous or non-cancerous.

It has been predicted that the number of breast cancer patients could soon exceed 1,797,900 in India [8]. Advances in the field of computer science and data science have led to improvements in the process of diagnosis, provided that the microscopic details of the tumor are known. For instance, the tumor's thickness, radius, area, and other dimensions can be fed into the software. The software is built by making use of machine learning, a suitable dataset, preferably with a sufficient number of examples or algorithms.

One instance of an existing resource is University of California Irvine (UCI)'s machine learning repository from which the dataset can be procured for hands-on analysis. The dataset is called the Wisconsin Diagnostic Dataset, and consists of 569 instances where the patient's tumor details are listed according to 32 attributes.

14.2.1 Breast Cancer Detection Using Machine Learning

The process outlined below can be followed to detect breast cancer using machine learning:

1. A dataset is procured, using online web databases or any other source, as shown in Fig. 14.1.
2. The dataset is examined for any missing values, or values that may have been incorrect.
3. Such values can be fixed at either zero or the mean or median of the corresponding column or row.

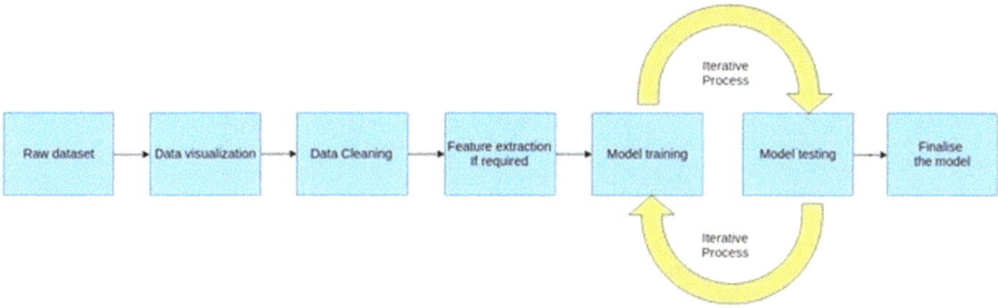

Fig. 14.1 Detection process using comma-separated values (CSV) data

4. Once the dataset has the proper values and there are no missing values, it is ready for the next stage.
5. The next stage consists of dividing the dataset into train dataset and test dataset. A percentage of the dataset is used to train and the rest is used to test the outcomes of predicted and actual values.
6. The train dataset is fed into various machine learning algorithms, where the algorithm trains itself from the examples that are fed into it.
7. Once the training is complete, various algorithms can be used to validate its performance, using a variety of tests.
8. By applying scikit-learn's test metrics, the accuracy and other metrics, such as the confusion matrix, recall, and precision, can be computed.
9. Once the test results are available for every algorithm, they can be compared, and the algorithm with the highest accuracy can be chosen.

The steps involved in the data visualization process of the Wisconsin dataset are depicted in Fig. 14.2. In this figure, the observation made in most cases is that as the value of the radius_mean, texture_mean, perimeter_mean, area_mean, smoothness_mean, compactness_mean, concativity_mean, concave_points_mean, symmetry_mean, or fractal_dimension_mean increases, the possibility of the tumor being cancerous increases. However, this does not apply in all cases. Although an increase in value marks the possibility of the tumor being cancerous, in the case of fractal_dimension_mean, the cases of malignant and benign tumors are spread out throughout the area for all values.

Visualization data analysis can be performed using heat maps; this helps to identify how the parameters in the data are related to each other by the values of covariance. If there are two variables that are highly correlated, if one value increases, the other variable would also increase, which means that one of the variables should be removed so that there would be fewer variables; this would simplify the computing of the data. One method used is to eliminate some variables by correlation. The data visualization process is one of the most important stages of building machine learning, since it gives a comprehensive view of the data. Plots such as the violin plot, heat map, and scatter plot can also be useful.

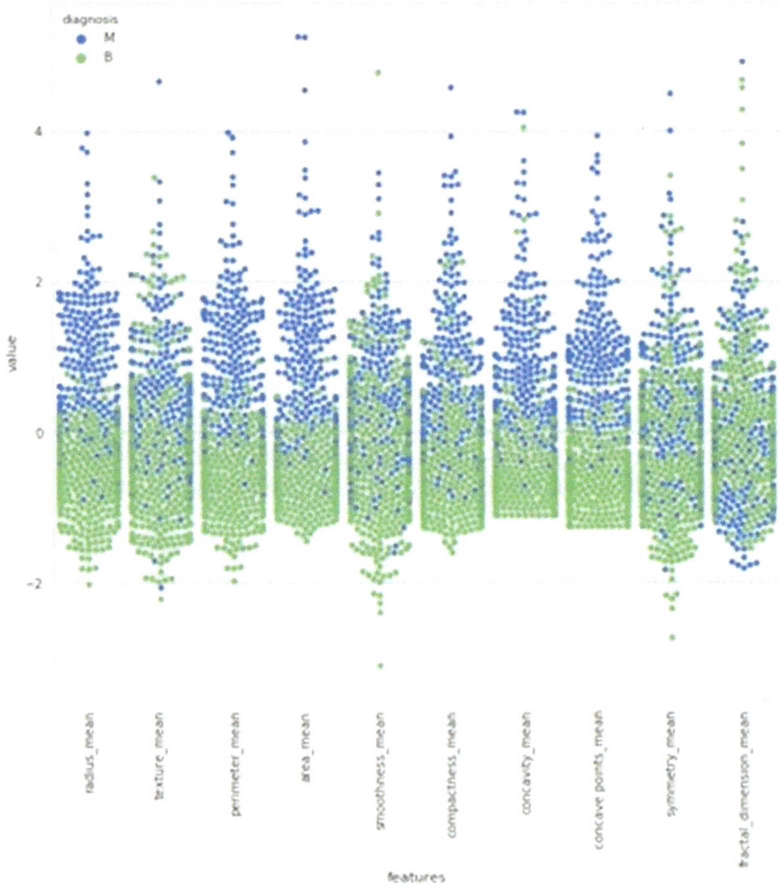

Fig. 14.2 Data visualization using libraries of Python

The dataset usually consists of images; several datasets are available in the public domain and can be used for CAD analysis. The described dataset contains images that can be divided into two groups: malignant and benign. The dataset described here contains more than 9000 images collected from 82 patients who were categorized into groups; furthermore, on the basis of magnification size, it can be used for CAD analysis. Computer vision concept can be applied to the mentioned dataset, and with the deep learning method, images can be processed and analyzed to detect the presence of cancerous tumors.

14.2.2 Breast Cancer Detection Using Deep Learning for Images

The classification of an image process is often very difficult and challenging for a physician as a result of complicated structures and rich geometry. The given image shows both geometry and complication of structure (Fig. 14.3).

Methods of deep learning such as the convolution neural network can be used to observe and analyze the images more effectively, and it is not as time-consuming a process; it also offers good accuracy.

CNN is a class of neural networks that can be applied to visual imagery. CNN can be used for tasks such as natural language processing and other cognitive tasks [10]. It includes an input layer, various hidden layers, and an output layer. The hidden layer can consist of convolution layers, RELU layers, an activation function, and normalization layers [11].

The CNN architecture consists of various distinct layers that distinguish the output layer from the input layer. The convolution layer is said to be the building

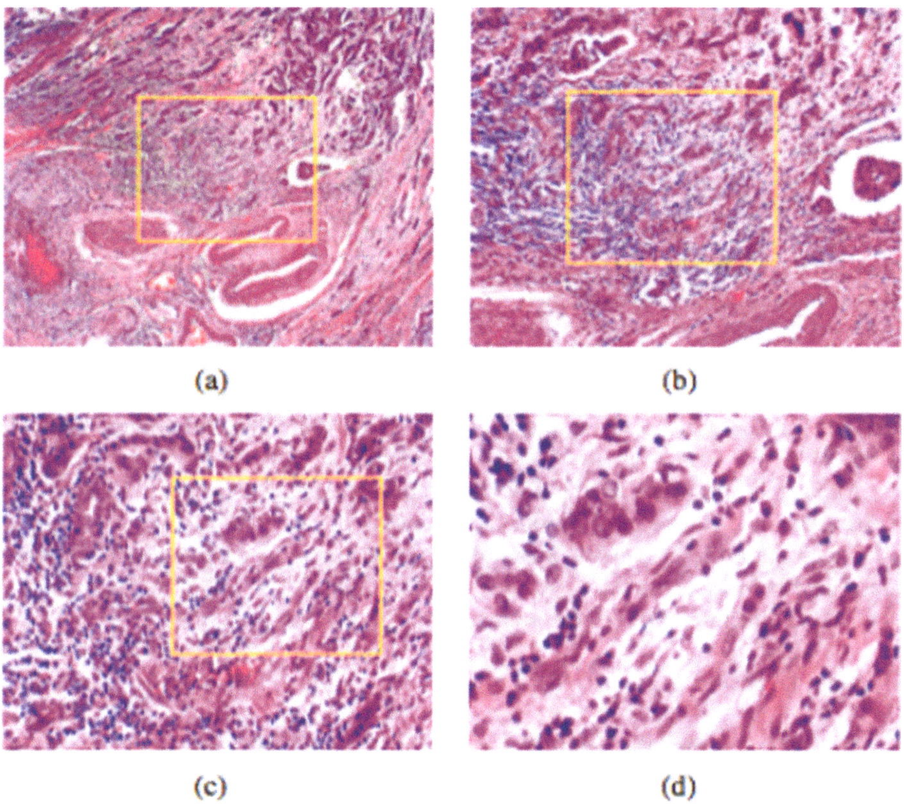

Fig. 14.3 Slide images of a malignant tumor (breast) (stained with HE) in different magnification factors: (**a**) 40×, (**b**) 100×, (**c**) 200×, and (**d**) 400×. Highlighted rectangle (manually added for illustrative purposes only) [9]

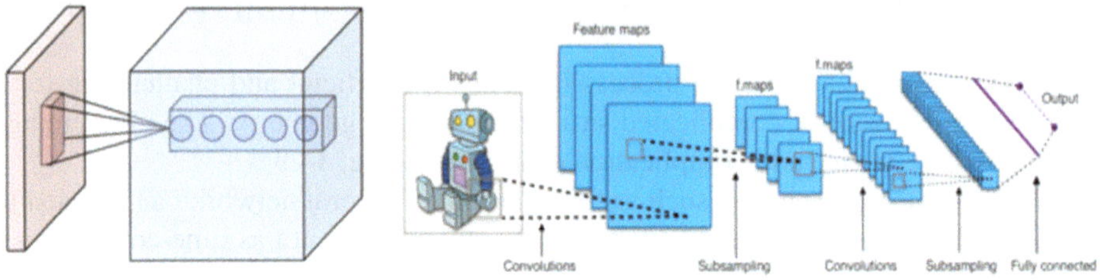

Fig. 14.4 A typical CNN

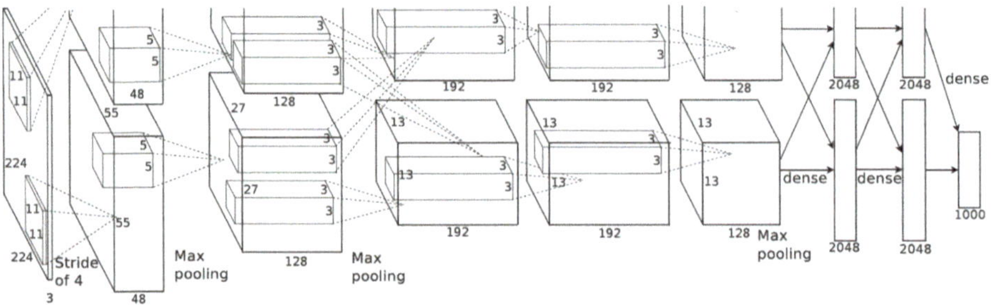

Fig. 14.5 Architecture proposed by Alex Krizhesvsky for the Imagenet challenge

block of CNN. This layer consists of filters of a specified stride that convolve through the width and height of the input value and compute the dot product between the filter and input layer, which, in turn, produces a two-dimensional output (Fig. 14.4).

A pooling layer is also inserted in the hidden layers. The pooling layer picks out the highest amongst the set of inputs given to it, and sends it to the next layer. The pooling can be used to resize the layer, i.e., down-sampling the volume spatially (Fig. 14.1).

Alexnet is the name of the CNN architecture designed by Alex Krizhesvsky (Fig. 14.5), and can be used for our requirements.

The method described by Spanhol FA et al. used to detect breast cancer is described below.

The architecture in the figure contains the following layers:

1. Input layer:

This is used to feed the input. Feature scaling can be done in this layer. In our case we feed it with an image of dimensions 32 × 32 or 64 × 64 pixels and use RGB as color channels.

2. Convolution layer:

This comprises one or many convolution layers to detect features (such as edges) that are present in the images. A filter consists of a small size matrix that traverses over the input matrix (input image) and computes the dot product. Three convolution layers can be used in this model. The size of the kernel is 5×5 and the stride is 1. The first and second convolution layers will each learn 32 filters. 64 layers will be learnt from the last layer.

3. Pooling layer:

The function of this layer is to down-sample the layer that is given as the input to this layer. One pooling layer is present in between every two convolution layers. They are 3×3 in size and the stride is 2.

4. ReLU layer:

This is one of the activation functions used in CNN architecture. The function returns 0 if it receives any negative input, or else it returns the same value as the input. It can be defined as $f(x) = \max(0x)$ where "x" is the input value. The default value of ReLU is 0. Three layers of ReLU have been added in this model.

5. Inner product layers:

They produce the output as a single vector.

The following strategy can be used: instead of using the large image immediately, small patches that can be extracted randomly are used. These extracted patches can then be used for the subsequent stages. In order to classify, patches are combined to make up the large image. Since the model received input image in the form of patches for training, in order to test these patches, partial images can be sent instead of the whole image and can be passed through the model (Fig. 14.6).

14.3 Use of CAD to Detect Early Signs of Diabetic Retinopathy

It is evident that early diagnosis of any disease could be crucial and have a beneficial effect in the subsequent treatment of a patient. One such area where CAD can be applied due to its computational ability in detection of early signs is diabetic retinopathy, the leading cause of blindness in the working-age population of the developed world. It is said that India will emerge as the diabetic capital of the world. By 2030, it is expected that the number of patients affected by diabetes mellitus will reach 79.4 million, the largest number in any nation in the world. Two-thirds of Type 2 and almost all Type 1 diabetics are expected to develop diabetic retinopathy.

Currently, detection of diabetic retinopathy is time consuming and consists of a manual process that requires a trained clinician to examine and evaluate the fundus images of the retina. Since the data are in the form of images, observing and

Fig. 14.6 Learning from two different GPUs [12]

quantifying associations can be difficult due to the number of colors, textures, and patterns present in the image.

Deep learning has already shown promising results in detecting skin cancer and also plant diseases through analysis of given images. Apart from this, studies have shown that deep learning can identify whether a person is a smoker or a non-smoker and can also predict approximate systolic blood pressure through analysis of retinal fundus images. This shows that retinal fundus images alone could be a treasure trove of information, and a gateway to a new area of research, which could provide presently unearthed, surprising, and unexpected results. The abundance of retinal images is an added advantage of this project. There are around 35,000 retinal fundus images available for analysis in the process. This proposed model harnesses the power of deep learning in the field of image processing and classifies accordingly by training it with the available images. It can also detect any abnormal features in the retinal fundus images of the patient.

There are several scientific articles as well as much evidence to indicate that when sets of retinal images were analyzed, early signs of diabetes retinopathy could be detected. Google is one of the giants in the field of detecting diabetes retinopathy in India. In collaboration with Indian eye hospitals, the company has initiated mass screening. Once the mass screening is done, analysis of the retinal images will begin by training the deep-learning model, and then diabetic retinopathy can be predicted by merely taking the patient's retinal image.

14.3.1 Dataset

There is a dataset available in the public domain that can be utilized. This consists of high-resolution retinal images [13]; in every image left and right views are provided for every subject. Since the amount by which retinopathy has affected patients can vary, the images have also been graded on a scale of 0–4. Automation has indeed become a requirement for the enhancement of healthcare efficiency, affordability, and accessibility.

14.3.2 A Method that Could be Utilized

This method consists of the following:

1. A dataset is procured.
2. Exploratory data analysis is done, giving an overview of the data.
3. The data is pre-processed: images are scaled down to a smaller size, which is beneficial for subsequent stages.
4. Additional image-related operations are performed, such as mirroring and alignment of all the images. It is important that the model is trained by images of similar sizes and orientation; this eases the training process and contributes to the accuracy of the model.
5. The Neural Network Architecture model is built by using frameworks such as keras, tensorflow, and theano.
6. Once the building of the CNN has taken place, the process of classification of the images begins.
7. Accuracy is computed in order to measure the performance of various algorithms and also the time taken to execute the process.

14.3.2.1 Method and Structure of CNN

The structure of CNN was described by Harry Pratt et al. [14]: the architecture consists of ten convolution filters with a ReLU layer present in between, and also a maxpool layer between the convolution layers. The size of convolution filters is 3×3, and a large number of convolution filters have been added so that the model can learn about deep features. The max pooling kernel size is 3×3 and the stride is 2×2. A flattened output can be received at the end of CNN.

As an example, a one-dimensional image is a flattened model sent for training. This ensures edge detection, where what is needed is picked out. To avoid overfitting, weighted class weights are used in each class. Approximately 120 epochs can be used to train about 78,000 images. SVM is the most accurate method, and CNNs can be used to classify cases of diabetic retinopathy efficiently (Fig. 14.7).

A dataset like the Wisconsin one may also consist of textual or CSV data. An example is the PIMA Indian Diabetes dataset, which can be retrieved from the UCI machine learning repository. The PIMA Indian dataset has certain parameters, and contains only female patients above the age of 20 years. The dataset consists of parameters, as well as the result. The parameters include the number of pregnancies, BMI, insulin level, age, and many others [15].

Once the dataset is retrieved, the model can be built, using machine learning. The same procedure has been followed for breast cancer. However, there are missing

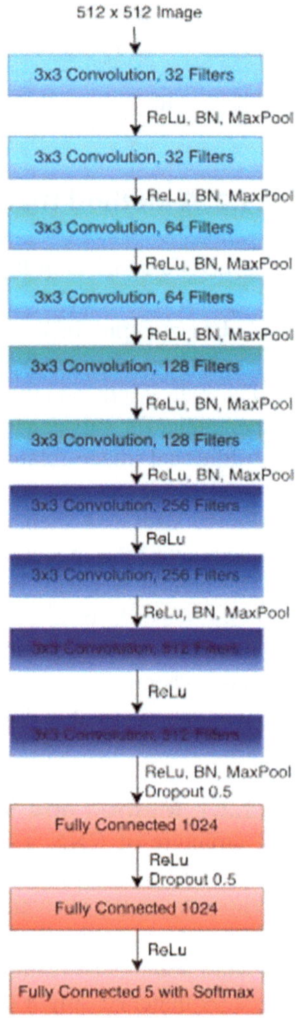

Fig. 14.7 Network architecture [14]

values that need to be dealt with here. Unfortunately, some missing values have been marked as 0, which is irrelevant when we consider parameters where data are present in the form of text.

In the described dataset, there are values where input such as the blood pressure of a patient cannot be zero; if it is considered to be zero, it would not be relevant for the machine-learning model. Hence a data-cleaning process is needed from which values such as NaN or zero will be replaced by either the mean or median of all the values in a particular column.

The procedure is the same as the procedure followed for the detection of breast cancer, using CSV data (Table 14.1).

Table 14.1 Shows a comparison of the results obtained from various journal articles

	Maximum Avg Accuracy
Harry Pratt et al.	0.9470
Majji et.al	0.9327
Sheet et al.	0.8213
Stall et al.	0.9422
Zana et al.	0.9377
Chadhuri et al.	0.8773

14.4 Skin Cancer – Melanoma

Skin cancer, also known as melanoma, is one of the most common diseases worldwide [16]. It is assumed that melanomas emerge because of the development of abnormal cells that have the ability to invade or spread to other parts of the body. There are three types of skin cancer: basal cell carcinoma, squamous cell carcinoma, and melanoma. We will limit ourselves to the study of melanoma and its diagnosis in this chapter, since it is the most common type of cancer, and diagnosis has been simplified with the advancement of new technologies and research.

It has been observed that the primary cause of melanoma is ultraviolet light (UV). The source of UV light may be sunlight or other sources such as tanning devices. Melanoma develops from the pigment in cells known as melanocytes [17]. Melanoma is considered the most dangerous type of skin cancer. In 2012, 232,000 cases of melanoma were reported, while in 2015 the number of cases had grown to 3.1 million in India.

Melanoma is responsible for 75% of all skin cancer deaths [18]. The incidence of malignant melanoma has been increasing on a daily basis [19]. Presently, the average lifetime risk has risen from 1 to 50 in Western countries. It is found to be more prevalent in countries where the population is generally fair-skinned. Melanoma is now the fifth most common cancer amongst males and the sixth most common cancer amongst females. Heredity is a strong indicator for risk of melanoma.

We know that early detection of malignant melanoma is beneficial and is associated with lower mortality. Unlike many other diseases, detection of malignant melanoma does not require the physician to examine the screen or perform surgery; instead the physician can detect melanoma by examining the affected area on the skin. Melanoma can be either malignant (cancerous) or benign (non-cancerous) (Fig. 14.8).

14.4.1 Detection Methods

The technique of deep learning can be used for the detection of skin cancer. As has already been mentioned, deep learning is an efficient tool in the prognosis of various diseases such as breast cancer, lung cancer, and diabetic retinopathy. Deep learning

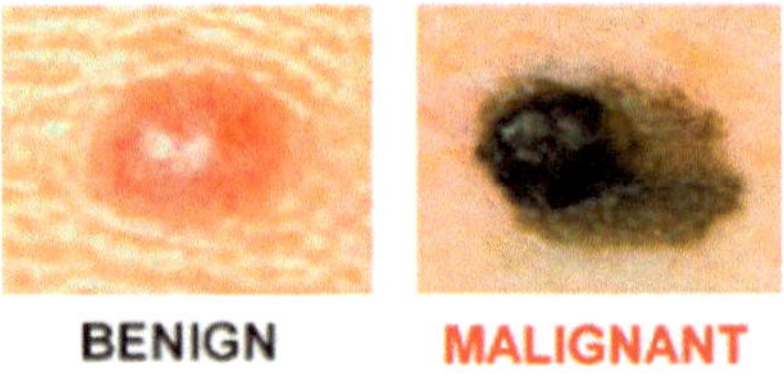

Fig. 14.8 Typical skin cancer visuals [20]

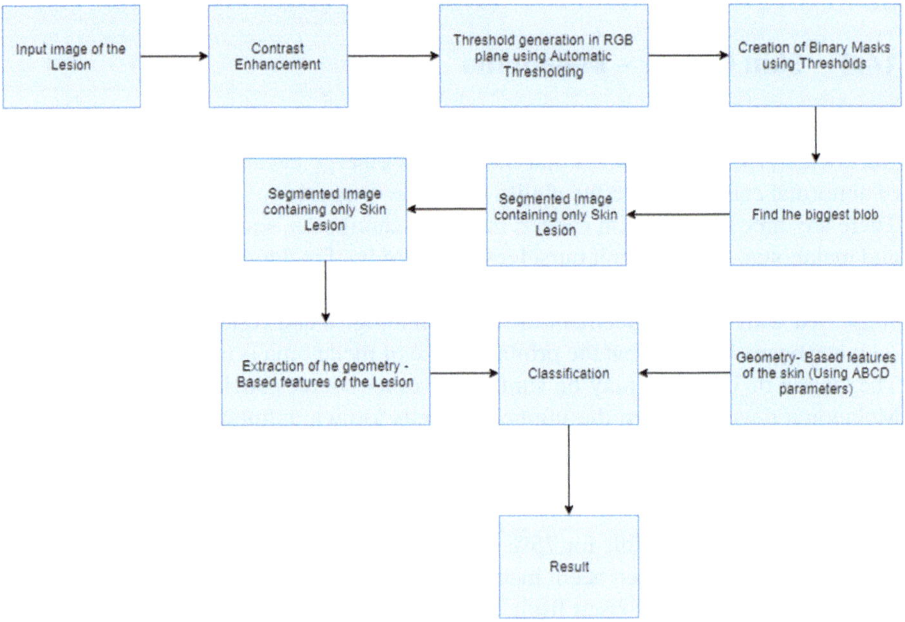

Fig. 14.9 Block diagram of method [18]

can also be used to analyze whether the patch that appears on the skin, in the case of melanoma, is malignant or benign.

There are many techniques that have been used for the detection of melanoma. One of the methods, by Jain S and Pise N et al., is illustrated in Fig. 14.9.

14.4.1.1 Procurement of the Dataset

The vast dataset on www.isic-archive.com is one of the most important contributors to the success of this process. There are many labelled images of melanoma that can be procured and utilized in building the deep-learning model.

14.4.1.2 Image Processing

Since the number of images of various kinds and varying image details is so vast, pre-processing is essential. We can use techniques such as cropping the main area from the image, image resizing, and adjusting the contrast and brightness, as required. Image processing libraries, such as Python, or tools from Matlab, can be used.

14.4.1.3 Image Segmentation

Otsu is used to do the image thresholding in each lane. This is done in order to increase the segmentation accuracy, followed by edge detection, which helps the model to differentiate between different areas. We need to extract the lesion from the surrounding skin, since that is the area of focus and what the training needs. However, to extract the skin lesion, we look out for the biggest blob, because the possibility exists that smaller blobs may be encountered that do not constitute the skin lesion.

14.4.1.4 Feature Extraction

After receiving the image from the previous stage, significant features such as area, perimeter, greatest diameter, circularity index, irregularity index A, irregularity index B, irregularity index C, and irregularity index D should be extracted.

14.4.1.5 Classification

Subsequently, by using the ABCD classification method and certain other rules, classification of the skin cancer takes place.

14.5 Alzheimer's Disease

This is a common disease where the patient experiences difficulty remembering recent information. Alzheimer's disease is described as incurable and is considered a brain disorder. Early diagnosis helps in preventing brain tissue damage. It usually affects those in the age group of 60–65 years. It is estimated that by 2050 around 0.64 billion people will be affected. Alzheimer's disease affects the part of the brain which is responsible for learning.

Diagnosis of Alzheimer's disease involves a process where the physician examines the MRI scans of the patient and makes a positive or negative assessment. However, there are a few areas in MRI scans that may be overlooked by the physician. Deep-learning methods can be used to overcome this problem, since they can perform advanced analysis of any image fed to them and can observe even small grains in an image. However, for deep learning to function properly, a suitable dataset that consists of labelled MRI scan images is required, and based on this learning, the deep learning model can be used to predict Alzheimer's disease from the new scan report. Memory and learning are not the only functions that are affected by Alzheimer's disease. Alzheimer's disease destroys the part of brain that is responsible for breathing and heart functionality, which ultimately leads to the death of the patient. Presently, detection of Alzheimer's disease is not accurate until it reaches the moderate level. Its diagnosis is based on clinical history as well as MRI data.

14.5.1 Method

According to Islam J., Zhang Y., et al., deep learning is a method that consists of four basic operations – convolution, batch normalization, rectified linear unit, and pooling. While these operations are performed, all layers are connected to each other. During the process, MRI data patches can usually be created from three physical planes of imaging: axial, coronal, and sagittal, and these are used as input for the model. In data augmentation, the number of samples in the dataset can be increased. Sensitive training is used in order to handle any imbalance (Fig. 14.10).

14.6 Lung Cancer

Lung cancer is the leading cause of death of all types of cancer. In the USA, approximately 225,000 people are diagnosed with lung cancer every year. It is known that early detection can decrease the mortality rate.

CAD is used as a "second opinion" by doctors of the of the scan reports [22]. The most common type of lung cancer is non-small-cell lung cancer (NSCLC), which accounts for 80–85% of all cases. According to the American Lung Cancer Association, there were approximately 1.6 million cases in 2012 alone, and around 1 million deaths were resulted from lung cancer. The main reason why lung cancer is so deadly is that its detection or identification is problematic in the early stages. About 84% of patients are diagnosed only after the cancer has spread beyond the lungs. Lung cancer is treatable and even curable — if it can be identified at an early stage. Early detection of lung cancer is possible through regular CT scans, and in this way the need for surgery can be minimized.

Methods Involved in the Diagnosis of Lung Cancer

- Imaging tests: an abnormal mass is seen through X-ray, MRI, CT, and Pet scans. These scans provide much detail regarding the disease.
- Sputum cytology: If production of phlegm during coughing exists, microscopic examination can determine the presence of cancer cells.

If the tumor cells are cancerous, a biopsy report can be requested. Tissue should be extracted by making use of the following procedures:

- Bronchoscopy: While the patient is sedated, a lighted tube is passed down the throat and into the lungs, providing a closer look into the area of infection.
- Mediastinoscopy: The physician makes an incision at the base of the neck. Samples are taken from the lymph nodes. This surgery can be performed only after anesthesia has been administered.
- Needle: A needle is inserted through the chest wall.

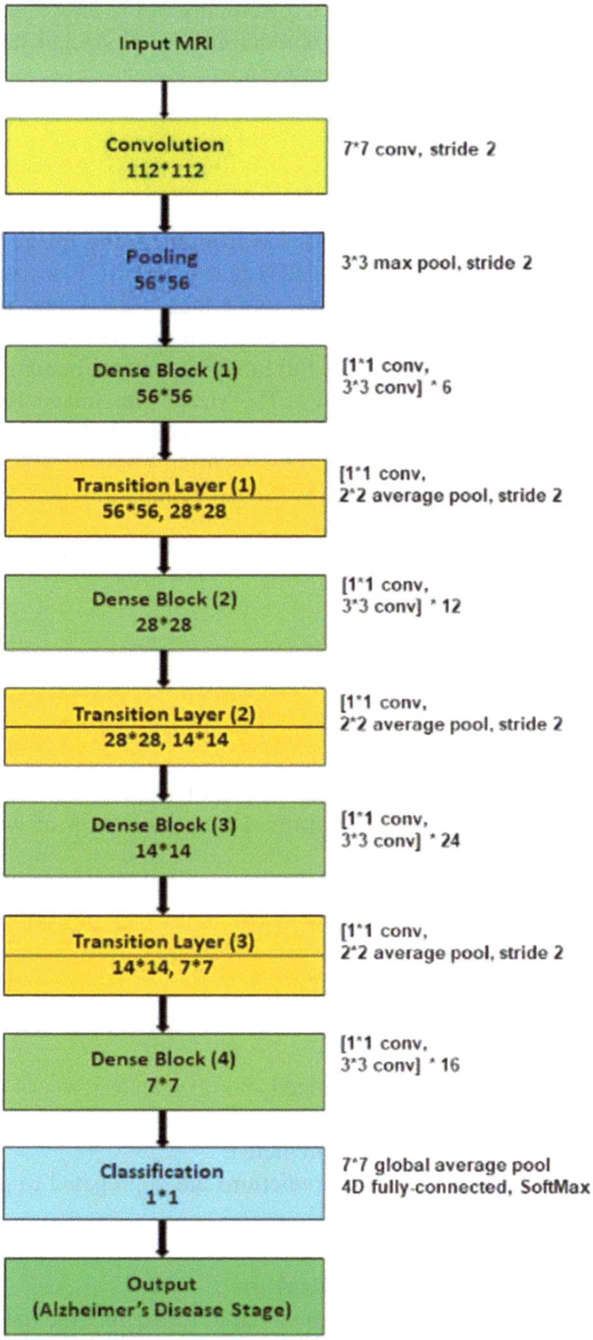

Fig. 14.10 Network architecture [21]

Once tissues are collected from any of the above methods, samples are sent to a pathologist for analysis. If required, the physician may prescribe further tests to confirm the presence of lung cancer.

14.6.1 Dataset

It is possible to retrieve datasets from the Lung Image Database Consortium image collection (LIDC – IDRI) [23] that contain CT scans with marked up, annotated lesions. These data contain more than 1000 scans from high-risk patients. The dataset size is around 124 GB and includes many images. All the images are marked according to levels: Level 1 to Level 2 represent benign cases, while Level 4 to Level 5 indicate malignant cases. The size of an image is around $512 \times 512 \times 400$ volumetric pixels. Every image is associated with a binary label. LUNA16 dataset is in use and is available on the internet. This is combined with the LIDC dataset to give us information about diameter, slobulation, and spiculation.

14.6.2 Malignancy

- Calcification,
- Sphericity.

After including this with our existing dataset, we found an additional 900 CT scans. Attributes such as diameter, lobulation, speculation, and malignancy were the focus of attention. Malignancy provides an indication of whether a tumor is cancerous or not.

14.6.3 Method

The method consists of four steps as described by Daniel Hammack et al. [24].

1. The CT scan is normalized.
2. Regions where nodules are present are investigated.
3. Nodule attributes are predicted.
4. The nodule attribute predictions are aggregated to global patient level diagnosis forecasts.

14.6.3.1 Data Normalization

In this process, steps are involved to restructure the data elements in order to reduce redundancy and improve data integrity. Furthermore, resizing of the CT scan takes place to about 1 mm^3 volume, which allows the same model to be applied to scans that differ from each other.

14.6.3.2 Detecting Candidate Nodules

In this step, the aim is to detect regions where useful information for diagnosis can be found. A pipeline may help detect regions with no abnormality in them.

To detect abnormalities, a model can be built that identifies the areas that are abnormal. This model can be trained with the same dataset.

In every scan, the number of abnormal regions should fall within a range of 1–50. If more than 50 are found, the size of the region could be 64 mm^3. This step decreases the computation time and the number of resources used.

14.6.3.3 Predicting Nodule Attributes

This solution predicts attributes such as diameter, lobulation, speculation, and malignancy. This is one of the most important steps in building the model.

Test augmentation can be used to improve the accuracy of the model. It was found that nodule attribute training was improved satisfactorily when training on multiple objectives occurred (Fig. 14.11).

14.6.3.4 Forecasting Diagnosis from Nodule Attributes

The last step of the pipeline predicts cancer when the nodule attribute predictions are present. At this step, we know the number of predicted attributes and also the number of nodules per scan. The algorithm runs over the clustering algorithm on the nodule locations.

14.6.3.5 Neural Model and Architecture

The input of all neural network models consists of 64 mm^3 regions of the CT scan. There are five convolution blocks, then global max pooling and a non-negative regression layer.

Three-dimensional convolutions of the size $3 \times 3 \times 3$ are used, with pooling $2 \times 2 \times 2$, and stride 2. The size of the train set was 75% and the rest was used to test the model (Fig. 14.12).

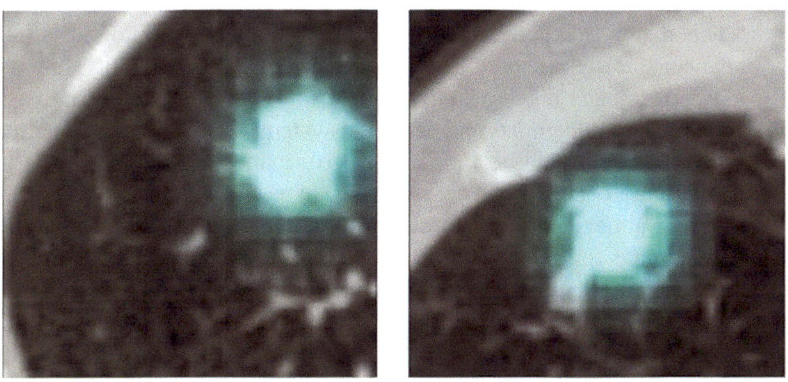

Fig. 14.11 Saliency map [24]

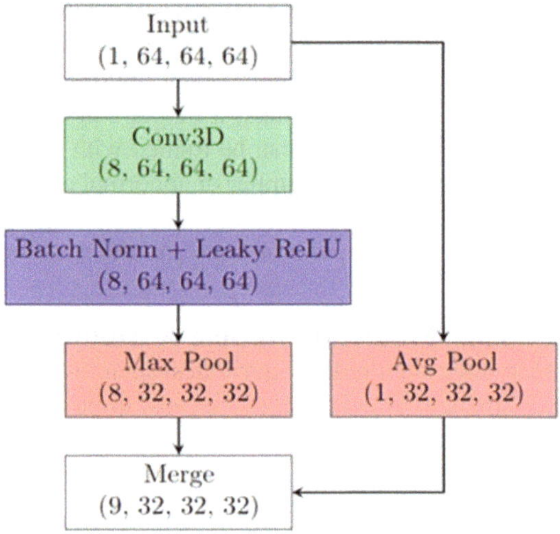

Fig. 14.12 Architecture of the network used

Since there are only 1200 positive samples, data augmentation can be used so that the area of the training size can be increased.

14.7 Discussion

In CAD technology, a number of major areas can be considered when performing CAD: training and test data sets; reference standards; mark-labelling criteria; stand-alone performance assessment metrics and methodologies; reader performance assessment metrics and methodologies, as well as study sample size estimation. In all of the above areas, we discussed the current state of knowledge, identified practical techniques and methodologies to be followed, provided recommendations that might be useful in real-world situations, and identified some reporting and study design requirements that might be critical. We also tried to describe possible future research. The methods, approach and strategies discussed could be helpful for further enhancement of structured guidelines, performance, scalability, and reliability of CAD performance results. In this chapter, most of the discussion was about commonly used lesion diagnosis and detection systems, as well as the principles and potential approach that could be a guide for performance assessment of other CAD systems. An accurate and proper assessment of CAD could affect the user by facilitating better understanding regarding the limitations and effectiveness thereof. This could stimulate further development and research in CAD technologies, and eliminate existing problems that result from inappropriate use. This could lead to an improvement in utility efficacy of CAD in clinical practice.

References

1. Freer TW, Ulissey MJ (2001) Screening mammography with computer-aided detection: prospective study of 12,860 patients in a community breast center. Radiology 220(3):781–786
2. Fabian TC, Patton JH Jr, Croce MA, Minard G, Kudsk KA, Pritchard FE (1996) Blunt carotid injury. Importance of early diagnosis and anticoagulant therapy. Ann Surg 223(5):513
3. Doi K (2007) Computer-aided diagnosis in medical imaging: historical review, current status and future potential. Comput Med Imaging Graph 31(4–5):198–211
4. Warren Burhenne LJ, Wood SA, D'Orsi CJ, Feig SA, Kopans DB, O'Shaughnessy KF, Sickles EA, Tabar L, Vyborny CJ, Castellino RA (2000) Potential contribution of computer-aided detection to the sensitivity of screening mammography. Radiology 215(2):554–562
5. Lee JG, Jun S, Cho YW, Lee H, Kim GB, Seo JB, Kim N (2017) Deep learning in medical imaging: general overview. Korean J Radiol 18(4):570–584
6. Spanhol FA, Oliveira LS, Petitjean C, Heutte L. Breast cancer histopathological image classification using convolutional neural networks. In: Neural Networks (IJCNN), 2016 international joint conference on 2016 July 24, pp 2560–2567. IEEE
7. Agarwal G, Ramakant P (2008) Breast cancer care in India: the current scenario and the challenges for the future. Breast Care 3(1):21–27
8. Malvia S, Bagadi S, Dubey U, Saxena S (2017) Epidemiology of breast cancer in Indian women: breast cancer epidemiology. Asia Pac J Clin Oncol 13:289. https://doi.org/10.1111/ajco.12661
9. Spanhol FA, Oliveira LS, Petitjean C, Heutte L (2016) A dataset for breast cancer histopathological image classification. IEEE Trans Biomed Eng 63(7):1455–1462
10. Lawrence S, Giles CL, Tsoi AC, Back AD (1997) Face recognition: a convolutional neural-network approach. IEEE Trans Neural Netw 8(1):98–113
11. Simonyan K, Zisserman A (2014) Very deep convolutional networks for large-scale image recognition. arXiv preprint arXiv:1409.1556
12. Spanhol FA, Oliveira LS, Petitjean C, Heutte L (2016) Breast cancer histopathological image classification using convolutional neural networks. In: Neural Networks (IJCNN), 2016 international joint conference on 2016 July 24, pp 2560–2567). IEEE
13. https://www.kaggle.com/c/diabetic-retinopathy-detection/data
14. Pratt H, Coenen F, Broadbent DM, Harding SP, Zheng Y (2016) Convolutional neural networks for diabetic retinopathy. Procedia Comput Sci 90:200–205
15. Kaggle.com (2018) Pima Indians Diabetes Database. [online] Available at: https://www.kaggle.com/uciml/pima-indians-diabetes-database/home. Accessed 28 Dec 2018]
16. Esteva A, Kuprel B, Novoa RA, Ko J, Swetter SM, Blau HM, Thrun S (2017) Dermatologist-level classification of skin cancer with deep neural networks. Nature 542(7639):115
17. Miller AJ, Mihm MC Jr (2006) Melanoma. N Engl J Med 355(1):51–65
18. Jain S, Pise N (2015) Computer aided melanoma skin cancer detection using image processing. Procedia Comput Sci 48:735–740
19. Rastrelli M, Tropea S, Rossi CR, Alaibac M (2014) Melanoma: epidemiology, risk factors, pathogenesis, diagnosis and classification. In Vivo 28(6):1005–1011
20. Skincancer.org (2018) Melanoma-SkinCancer.org. [online] Available at: https://www.skincancer.org/skin-cancer-information/melanoma. Accessed 27 Dec 2018
21. Islam J, Zhang Y. (2018) Early diagnosis of Alzheimer's disease: a neuroimaging study with deep learning architectures. In: Proceedings of the IEEE conference on computer vision and pattern recognition workshops 2018, pp 1881–1883
22. Sun W, Zheng B, Qian W (2016) Computer aided lung cancer diagnosis with deep learning algorithms. InMedical imaging 2016: computer-aided diagnosis 2016 Mar 24, Vol 9785, p 97850Z). International Society for Optics and Photonics

23. LIDC-IDRI – The Cancer Imaging Archive (TCIA), Public Access – Cancer Imaging Archive Wiki [Internet]. Wiki.cancerimagingarchive.net. [cited 26 December 2018]. Available from: https://wiki.cancerimagingarchive.net/display/Public/LIDC-IDRI
24. Hammack D (2018) Forecasting Lung Cancer Diagnoses with Deep Learning. [Internet]. Daniel Hammack; [cited 27 December 2018]. Available from: https://raw.githubusercontent.com/dhammack/DSB2017/master/dsb_2017_daniel_hammack.pdf

Implementation of Optogenetics Technique for Neuron Photostimulation: A Physical Approach

15

Saurav Bharadwaj, Sushmita Mena, and Dwarkadas Pralhaddas Kothari

Abstract

Optogenetics is an opto-triggered neuron-switching technique in specific neuron clusters of the nervous system. Specifically, the chapter redefined the implantation of physical optode–electrode assembly on primate for simultaneous photo-switching and data recording from the complexly arranged neuron clusters. Specially, a number of modern commercial commutators are listed on the basis of flexible rotator optical joint, number of LEDs, magnetic base and low torque. However, the chapter highlights certain standard protocols used in the implantation of optode and electrode on the primate skull. For efficient optical stimulation and signal detection, it revised certain significant contributions on implementation of multimode fibres, photonic CMOS integrated, super flexible optofluidic ultrathin channel, iridium oxide electrodes and pulse width modulation control on different power levels and duty cycles. Specifically, the authors plotted different irradiance curves of mammalian brain tissues in linear scale at three wavelengths 473 nm, 561 nm and 630 nm. Successively, it is a collection of different wireless optogenetics systems of certain strength in each model as lightweight, super flexibility, optofluidic light delivery, small size, minimal heat emission and wide range output power control. Practically, a number of biomedical imaging techniques, that is, computer tomography, magnetic resonance imaging, micro-positron emission technology, micro-ultrasound, opto-micro-electrocorticography and electroretinography is implemented for studying the physiology and behaviour changes of primate under simultaneous photostimulation.

S. Bharadwaj (✉)
Indian Institute of Information Technology Guwahati, Guwahati, Assam, India

S. Mena
Assam Down Town University, Guwahati, Assam, India

D. P. Kothari
Indian Institute of Technology Delhi, New Delhi, India

S. Paul (ed.), *Application of Biomedical Engineering in Neuroscience*,
https://doi.org/10.1007/978-981-13-7142-4_15

Keywords
Carousel commutators · Tungsten electrodes · Stereotaxic frame · Irradiance · Transcranial magnetic stimulation

15.1 Introduction

Optogenetics is an optically switched cell-targeting method that modulates neurons of a specific cluster in respond to photoluminescence. Two decades ago, researchers *Boris Zemelman* and *Gero Miesenbock* for the first time named the technique as *optogenetics* after numerous experimentation to prove that the light controls rhodopsin-sensitive neurons [1]. Successively in the year 2013, these two neuroscientists proved a new theory that neurons can independently be activated and control the inotropic channels in response to light. Later, the research was followed by biophysicist, *Peter Hegemann*, who demonstrated the algal response to optogenetic photostimulation [2]. Presently, neuroscientists, Karl Deisseroth, Peter Hegemann and Ed Boyden, who are the pioneers in the field of optogenetics, have demonstrated the variation of microbial opsin, cell targeting, optical projection on the mammalian nervous system and behaviour of primates on photostimulation [3–6].

Specifically speaking, the chapter focuses on optogenetics signal and system, construction and implantation of the optical fibre probe and LED driver system. Considering a number of practical conditions of the fibre, authors illustrated three irradiation curves at three different frequencies 473 nm, 561 nm and 630 nm in two power levels of 7 mW and 9 mW. The curves show the variation of irradiance with respect to the probe tip distance on tissues for photostimulation.

15.2 Related Literatures

Optogenetics is an optostimulation technique to control the ion flow in channels in response to light. In the year 2013, scientists Ernst Bamberg, Edward Boyden, Karl Deisseroth, Peter Hegemann, Gero Miesenbock and Georg Nagel were awarded the *Grete Lundbeck European Brain Research Prize* for the invention of optogenetics technique [7].

However, the story of optogenetics was translucent three decades ago. But, from the observation of photosensitive cells in response to different wavelengths of light, Francis Crick, in the year 1999, stated in his literature *The Impact of Molecular Biology on Neuroscience* that it is possible to fire a cluster of neurons using light as a stimulator [8].

Successively in the year 2002, Zemelman demonstrated the functional analysis of neural circuits on the photoreceptor of neurons in *Drosophila* [1]. In the following years, researchers from different laboratories conducted studies on current flow in protein encodings across cellular membrane in photostimulation and projection and intensity control in specific regions of the brain. Limitlessly, researchers from Deisseroth Lab, Stanford University, California, and Ed Boyden, Mc Govern Institute, Massachusetts Institute of Technology, Cambridge, are studying the behavioural, psychological and physiological disorders and treatments on a mammalian body [3–5]. Extensively, a number of complex environment experimental setups are built to study the decision making, reflexes in response to external stimuli and response to primary sources including olfactory, audio and vision signal to the brain.

15.3 System Overview of the Technique

Optogenetics focus on optically stimulating a cluster of neurons in injection of an external DNA to modify the original sequence. The experimental setup is peculiar as an optode–electrode pair is implanted in the cortex of the mouse. Generally, optodes are pointed to the specific regions of neuron clusters stimulated with different light wavelengths on optical switching in on–off states. Technically, LED and the drive system, compactly named as LED module, is responsible for an accurate switching of the LEDs. The LED driver provides precise constant current pulse and regulated output current. A single channel controls the output of a LED and multichannel control the output of multiple LEDs [9, 10]. A multichannel LED driver is highly functional and software-defined hardware as it accepts digital input to trigger user-defined output pattern. In the primate, the output patch cable is hanged over the head and is connected to a commutator for free movement of the mouse without twisting the cable. Commercially, there are a number of commutator systems including dual LED, cascaded dual LED and 16 channels, cascaded dual LED and fluid swivel, carousel and motorized carousel commutators.

However, for effective neural stimulation, a number of electrodes, probes and microelectrode arrays are being manufactured for short- or long-term signal recording [5]. Therefore, penetrating probes are implanted in cortical layers for chronic recording. μ-LEDs are used in the device to prevent heat production and damage of tissues. They are designed with photopolymerized lens to measure the electrical signals and neuron spikes using tungsten electrodes at various frequencies. Practically, μ-lens-coupled LED photodetectors are adjusted at an effective distance of 5 mm to avoid congestion of optode–electrode assembly [3, 11] (Fig. 15.1 and Table 15.1).

15.4 Implantation of Optodes and Electrodes on Primate

Literature of *Sparata* states certain standard protocols in construction and implantation of optical fibres in the cortex of a mouse as primate under study using optogenetics technique [11]. In this section, the construction of optical fibre and patch cord assembly, implantation of fibre in cortex layers of primate and behavioural study of the mouse are being illustrated using the technique (Fig. 15.2).

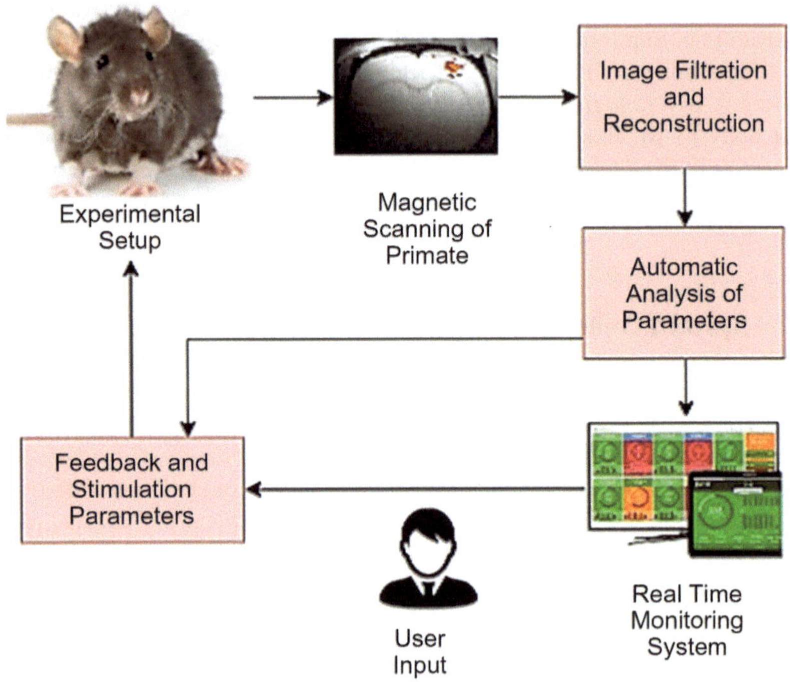

Fig. 15.1 Experimental setup of optogenetics technique [12]

Table 15.1 Classification of commutators

Commutators	Specifications
Dual LED	Rotary LEDs in optical rotary joint
Dual LED 16 channel	Rotary 2 LED modules on magnetic base
Carousel	Low-torque movement of LED module
Dual LED fluid swivel carousel	Magnetic arm transfers torque to commutator
Motorized carousel	Four digital headstages with two LEDs

Source: (Plexon Optogenetics Commercial Products)

Construction of Implantable Optical Fibre

- Optical fibre is spooled with a diamond knife, 25 mm fibre core is stripped off from the 200 μm core fibre to separate the coating of the fibre; 10 mm of unstripped fibre is left with 35 mm of the fibre. Hot epoxy is filled in a 1 ml syringe and applied on a bunted needle of 25 gauge.
- Tightly fix the ferrule in a vice and insert the fibre in it leaving 5 mm of stripped fibre.
- In case of resistance in the ferrule, replace it with a smooth one. Heat the epoxy bead for 10 min to form a complete curve with an indication of black or dark purple colour.
- Place the ferrule at the convex end, score the fibre with a diamond knife and polish the ferrule to a smooth surface.

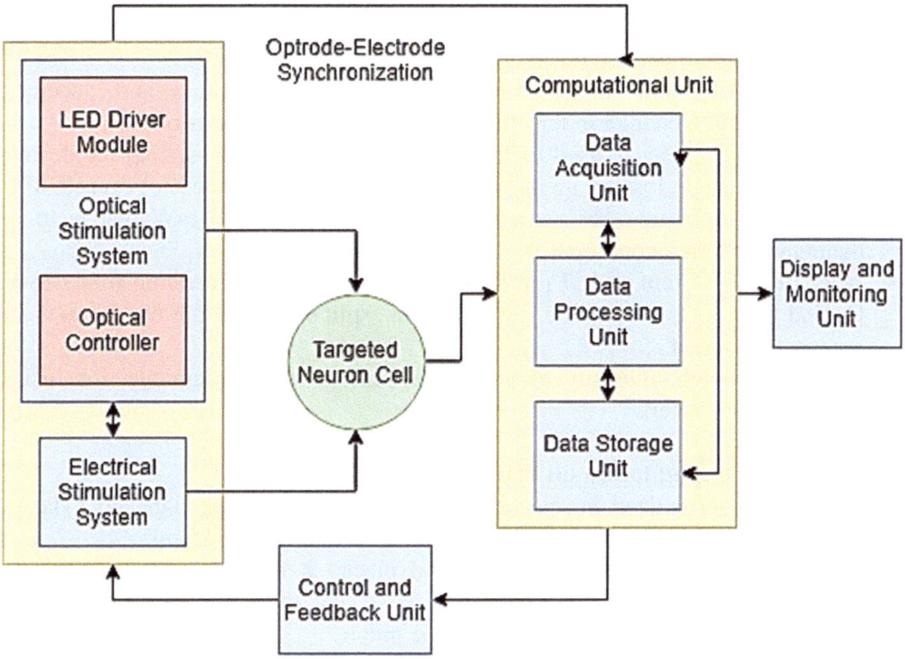

Fig. 15.2 Optode–electrode and computational unit [3, 12]

- A better polish of ferrule is done by placing it perpendicular on polish paper for concentrated light output.
- Cut a polyethylene tube at a length of 5 m used as optical fibre protector that is placed over implantable fibre.
- Store the optical fibre in a sealed container.

Patch Code Construction In Vivo Experiments

- Multimode fibre of core diameter 50 μm and furcation tubing of the desired length. A length of 50–100 cm and 200–300 cm patch cord is used for implantable optical fibres and behavioural experiments, respectively.
- An optical fibre is threaded through the furcation tube and strip 25 mm at both ends. The multimode ferrule assembly is fixed in the vice.
- Stripped fibre is inserted into the ferrule to ensure a correct bore diameter and epoxy is applied to the ferrule.
- Heat the epoxy for 2–3 min as it turns brownish black and attaches rigidly the optical fibres to the ferrule. Polish the ends of the ferrule in the polishing disc.
- Check for excessive epoxy or cracks on the fibre core. Insert 127 μm ferrule in the vice with the convex side facing down and the fibre should fit it smoothly without obstruction.
- Place a 2 cm piece of heat shrink tube over the interface of furcation tubing and the ferrule to prevent light emission. In order to increase structural rigidity, a bead of 5 min epoxy is placed in between the ferrule and furcation tube.

Measurement of Light Output

- Connect the fibre cable connector to the coupler of the laser.
- Turn on the LED and wait for 15 min to investigate the brightness and concentric circles; a minute change in the coupling will change the intensity of light.
- Optical power meter is used to measure the light output on setting the desired wavelength. Light loss for an ideal 50 μm core patch cable does not exceed 30%.
- Reduce the light intensity to a minimum threshold of 10 mW that is to be maintained in the experiment.
- A length of 3–4 mm of 127 μm ferrule is inserted into the ceramic split sleeve. Optical fibre is inserted in the gap of ceramic split sleeve so that two ferrules are compactly framed.
- Measure the light intensity of the output through the implantable fibre using optical power meter.

Optical Fibre Implantation on Primate

- Mouse is anaesthetized in injection of ketamine (100 mg per kg) and xylazine (10 mg per kg). The anaesthetic process takes a duration of 15 min.
- Hairs of the mice is being plucked and placed it in a stereotaxic frame for immovability of the primate during surgery. Check the reflex arc of the primate in response to external stimuli such as toe pinch.
- Injection 0.3 ml of 0.9 (weight/volume) sodium chloride to prevent dehydration.
- Apply ophthalmic ointment in the eyes of the mouse to prevent drying.
- Coat the surgical area with Betadine and alcohol to minimize the risk of infection.
- Identify the bregma and lambda by swapping the skull with cotton swab wetted in 70% (volume/volume) of ethanol solution.
- Insert the fibre by drilling the bregma maintaining an accurate dorsal–ventral coordinates on the mouse placed in the stereotax.
- Stereotaxic atlas is used to determine the brain region of interest. Once the skull is removed, use a 26-gauge syringe tip to break through the dura without damaging the cortex.
- Two holes are drilled in the skull to insert anchoring screws that is 5 mm away from the implanted optical fibre site.
- Stereotaxic arm is lowered to implant fibre in the brain region of interest in a very slow rate of 2 mm per minute.
- Cement the implant fibre to fix it rigidly using a 1 ml syringe and let it dry for 15 min.
- A number of major issues of heath and medication need to be taken care of the surgical mouse.
- After the surgery, mouse need to be kept inside a clean cage on a warm blanket until it recovers from anaesthesia.
- Monitor the weight of mouse regularly and observe the sense of pain and distress.
- Once the mouse adjusts in the habitat, behavioural experiments are ready to be conducted.

15.5 Optical Stimulation and Signal Detection

Optogenetics is the integrated technique of photocontrolled genetics. Literatures reveal a number of methods to stimulate and analyse quantitatively on the targeted living cells.

- Thin multimode fibres are designed to decrease the size of the fibre. It reduces the intensity of light and prevent damages due to phototoxicity and photobleaching, and allowance of high signal-to-noise ratio.
- Photonic CMOS integrated circuit is fabricated on silicon nitride layer. Neurons are stimulated by straight and bent wavelength of 450 nm [13].
- Optofluidic neural probes of ultrathin channels for soft mircofluidic drug delivery. Cellular-scale inorganic LED arrays stimulate the neuronal activities [14].
- Iridium thin oxide electrodes of thickness 75 μm are coupled to opto-μ-ECoG arrays for neuronal stimulation [5].
- Pulse width modulated optical stimulation is done at different powers including 50 μW, 75 μW, 100 μW, 150 μW and 200 μW in control of duty cycles as 75%, 50%, 38% and 25% to regulate the probe tip optical power.

In detection of synaptic signal, quantitative phase imaging shows neuron growth, dynamics, substrate interaction and tomography. Additionally, digital micromirrors are solid-state devices that are used in photoprojection in a single or multi-excitation of light. These devices are μ-optic-electromechanical (μ-optic-EM) system that consists of a thousand of small size mirrors arranged in arrays [15]. In binary on state, light from the source is individually reflected to the lens and turns bright the pixel of the screen. Alternatively in off state, light beam is diverted elsewhere and makes the pixel dark. The intensity or the phase of the focussed beam of light is modulated by the technique of spatial light modulation. In advance, multiphoton intrapulse interference modulation technique is a light phase characterization and correction of an ultrashot light pulse to measure the intensity of light in an extreme short duration. The technique of dispersion phase spectroscopy is used to study the mass transport in cellular and subcellular levels. Microelectrode arrays are implanted to enhance the computer–brain interface to connect neurons to electronic circuit for in vivo experiments. The technique studies the potential due to ion current flow in the neuron channels. These electrodes have the ability to simultaneously record the real-time relative signals of different sites from the array of electrodes implanted in adjacent locations. In integration with confocal microscopy, it allows the constriction of 3D images of network activity and synapse produced in neurons (Figs. 15.3 and 15.4) [16].

Authors studied the irradiance curves of mammalian brain tissues in linear scale using Brain Tissue Light Transmission Calculator from Optogenetics Resource Centre, Stanford University. It considers certain practical conditions specifically optical fibre numerical aperture, fibre core radius and light power from fibre tip to estimate accurate curves. Therefore, irradiance increases in magnitude depending on the power delivered on the cortex and is directly proportional to the optical power

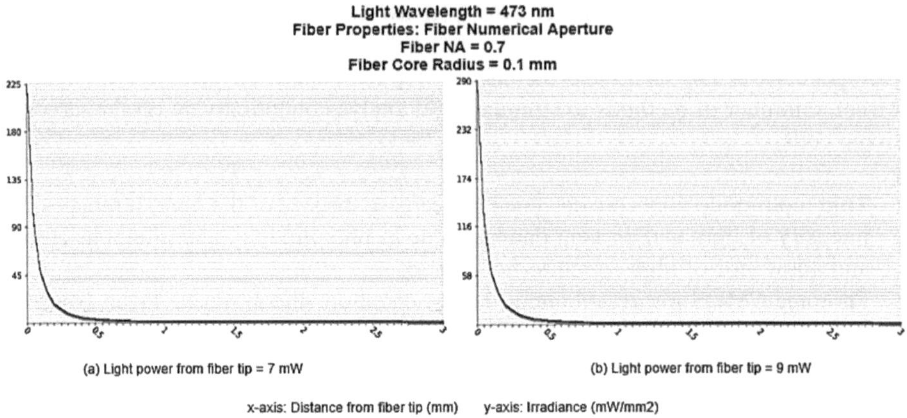

Fig. 15.3 Irradiance curve of mammalian brain at 473 nm

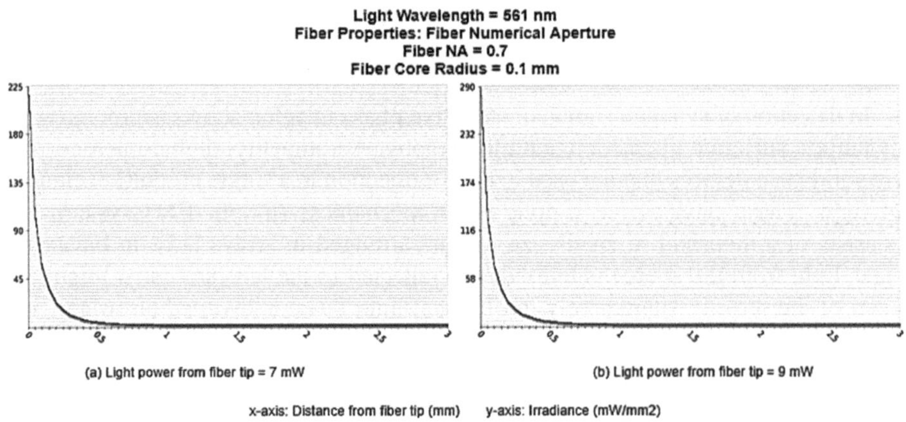

Fig. 15.4 Irradiance curve of mammalian brain at 561 nm

from the probe. It is not much affected by a slight change in light wavelength within a range of 473–630 nm. An increase in the fibre core changes the gradient of the irradiance curve in transforming the smoothness of the curve to steps (Fig. 15.5).

15.6 Wireless Optogenetics Systems

Balasubramaniam reviewed on different standard wireless optogenetic systems that are experimentally proved and are highly acceptable. The section highlights the contributions of these wireless systems and its location in the brain [17–23] (Fig. 15.6).

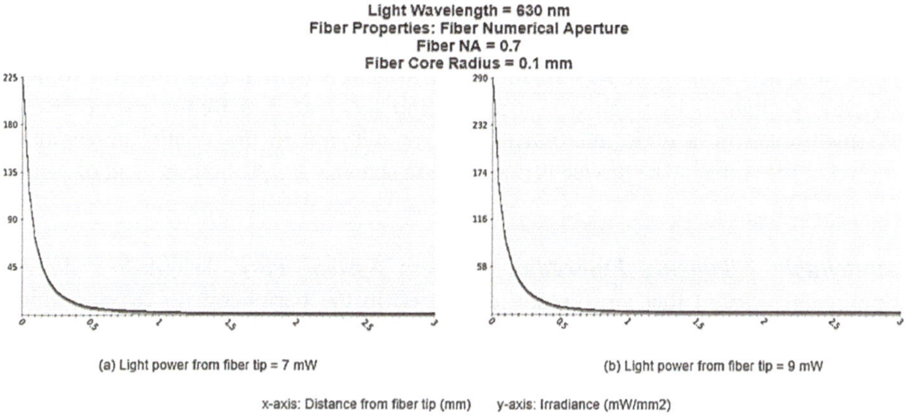

Fig. 15.5 Irradiance curve of mammalian brain at 630 nm

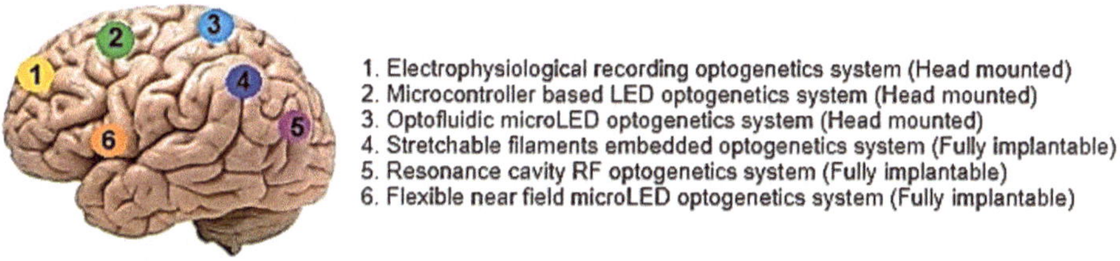

Fig. 15.6 Wireless optogenetic systems and implant locations [17]

Electrophysiological Recording Wireless System Gagnon-Turcotte designed a special wireless communication named as wireless headstage with real-time spike detection and delta compression for combined optogenetics and multichannel electrophysiological recordings. An optode–electrode assembly is being implanted in a transgenic rodent that has 32 recording and stimulating channel pairs to stimulate the regions of somatosensory cortex and hippocampus. Additionally, it consists of a radio transmitter operated at a frequency of 2.4 GHz and consumes currents of 11.3 mA and 13.3 mA at transmitter and receiver models. It is fabricated on a printed circuit board of dimensions 17 × 18 × 10 mm^3 and weight 2.8 g that achieve a value as high as 97.28% at a low SNR of 5 dB [24].

Microcontroller-Based LED Wireless System Hashimoto designed wireless LED stimulator for the chronic stimulator. The wireless system communicates over infrared waves and controlled by a PIC microcontroller. It drives three LEDs independently on IR transmitter and specify pulse for a duration of (>1 ms), pulse width (>1 ms) and frequency (<500 Hz). Additionally, the microcontroller is programmed to control the duration of irradiation, frequency and pulse width of the light pulse. Muscle movement of the primate is displayed using electromyography technique in 30 frames per second. Experimentally, the stimulator is programmed to generate pulses for 50 ms for 2 s at a standard frequency of 10 Hz [25].

Optofluidic μ-LED Wireless System Jeong designed a flexible biocompatible wireless optogenetic circuit that are directly interfered in the peripheral nerves and spinal epidural space without skeletal fixation. The system is built to communicate in radio frequency channel between a transmitter and receiver at a frequency of 20 Hz. Photostimulation is done at a wavelength of 470 nm in the sciatic and epidural region of the rodent. The device is miniature in size $0.7 \times 3.8 \times 6\ \mathrm{mm}^3$ and of weight 16 mg [14].

Stretchable Filaments Embedded Wireless System *Park* designed a flexible optogenetic circuit that are directly interfered in the peripheral nerves and spinal epidural space. The specific technique used is the antenna that is widely stretched over the elastomeric circuitry substrate. Communication link between the transmitter and receiver antenna is setup by radio frequencies operated at a frequency of 20 Hz [26].

Resonance Cavity RF Wireless System Montgomery and his team designed the smallest wireless optogenetic system of weight 20 mg and volume 10 mm^3 operated on RF power source and controlled system. It can precisely stimulate in the peripheral nervous system with minimal heat emission. The module is implemented in three different regions of neurons are premolar cortex of the brain, peripheral nerve ending the hind jaw and dorsal horn the spinal cord [27].

Flexible Near Field μ-LED Wireless System Shin designed a fully implantable optogenetics neuromodulator that eliminates large headstages and batteries in wireless communication. The system operates in an integrated circuit of a miniature antenna and minimizes smashing of the hardware in the cage. The μ-CT images of the respiratory signals and MRI images of the mouse are observed to study the activities of the brain and nervous system. Additionally, the system is compatible with various light wavelengths (red, green yellow, blue and ultraviolet) and wide range output power control (0–100 $\mathrm{mW/mm}^2$) [28].

15.7 Optogenetic Control on Potassium Channel

Prior to the development of light-gated ion channels, optogenetic tools were limited to provide a transient inhibitory impact on neural cells. However, manipulation of ion channels with light enabled long-lasting output in accurate control of neuronal activities. Practically, archaerhodopsin, halorhodopsin and channelrhodopsins are defined as optogenetic tools for fast on–off kinetics [2]. Along with it, certain *Haloarchaea* mediate very powerful, safe and multiple colour silencing of neurons. Strictly speaking, gene archaerhodopsin-31 of *Halorubrum sodomense* optimizes complete silencing of neurons in the awaked brain on viral insertion in the cortical region and photostimulated with yellow light. Additionally, archaerhodopsin-31 is effective in the functioning of neuron silencing than its other counterparts like channelrhodopsins or natural spike firing for its pH excursions and eventually

Table 15.2 Biomedical imaging techniques

Techniques	Specifications
Computed tomography (CT)	Combined -X-rays and produces detail image
Magnetic resonance imaging (MRI)	Strong and radio waves to produce image
μ-positron emission technology/CT (μ-PET/CT)	Respiration of the primate can be observed
μ-photoacoustic tomography (PAT)	Detect ultrasound reflected by EM waves of brain
μ-ultrasound	Produce 2D and 3D images
Opto-μ-electrocorticography (μ-ECoG)	Neural modulation and recording of spikes
Voltage and current clamp recording	Measures membrane potential
Electroretinography	Intravitreal contraction in eyes of the mouse

reduced by self-limiting mechanisms [29]. Thus, light-driven proton pumps have high-performance potential in independently silencing of neural cells. Successively, the activity of depolarization showed by channelrhodopsin (ChR2) extracted from an algal protein motivates researchers to investigate other strains of microorganisms in order to complement the activity of ChR2 to show more inactivation of neurons by promoting hyperpolarization. Halorhodopsins from two strains of archaea, *Halobacterium salinarum* and *Natronomonas pharaonis* are explored for complementing with the activity of ChR2 [30]. However, among the halobacteriums, *Haloarcula salinarum* is effective in inhibition of neuron activity on high wavelength red light photostimulation. It can promote fast silencing on injection in the cortical regions of the retina in the mouse. Moreover, it can non-invasively mediate transcranial optical inhibition of neurons in the brain of awake mice. This non-invasive technique of inhibiting neural cells has established a powerful application in the field of neuroscience [31, 32] (Table 15.2).

15.8 Techniques to Study the Behaviour of Primate

Researchers from global laboratories performed numerous experiments on the surgical mouse in response to optogenetic stimulation. Hence, a number of biomedical imaging techniques are being implemented to capture the brain synapses in respond to reflex actions. Successively, a major issue is precise and accurate imaging of the electrophysiological and behavioural change of the primate. The electrophysiological analysis is to detect the membrane potential and measure the current and voltage across it in a variation of light wavelength. However, multiphoton excitation microscopy technique is an important tool for investigating neural signals at different time intervals. Specifically, two-photon laser-scanning microscopy has shorter excitation wavelength in comparison to the emission wavelength. It visualizes deep inside a single neuron with high-quality light detection and simultaneous photoabsorption. Investing the brain-electronics interface in neural signal analysis, the technique of electrocorticography (ECoG) is beneficial over electroencephalography (EEG).

However, a number of PEDOT-laminated iridium tin oxide electrodes is being implanted in the surface of the cortex on a grid. It is preferable to analyse deep brain signals in ECoG technique for better spatial resolution and high SNRs. In advance, functional magnetic resonance imaging (fMRI) is another technique to analyse brain functions based on the blood flow in the cerebral nervous system and the presence of oxygenated blood. A major advantage of the technique is that non-radiation that reduces the risk of side effects. However, it has a very low spatial resolution and signal quality deteriorate as the primate moves in the cage. Another technique is transcranial magnetic stimulation (TMS) that captures the electrical signals from the specific regions of the brain in a change of externally applied magnetic field. It offers a low spatial resolution and a non-target-specific and cell excitation technique. In controlled optogenetic stimulation, optical power should be kept in a range of 50 μW, 75 μW, 100 μW, 150 μW and 200 μW. A duty cycle of 75%, 50%, 38% and 25% for PWM LED intensity modulation need to be maintained. Practically, every technique has certain limitations and are overcome by one or the other techniques.

15.9 Conclusion

Optogenetics is in the initial phase of development. Generally, the technique is used to crack neural codes and measure the rate of mean firing and timing of action potential, control action of the neural circuit, analysing neural diseases models and therapeutic treatments. However, a majority of the key features of the technique are still not identified for insufficient mathematical parameters and models. Technically, the aspheric lens is coupled to a solid-state laser for stable light delivery in targeted regions of the brain. An array of LEDs is implanted at a various depth and still possible to deliver only one-third of the injected light as an output. Since optical loss is a serious issue, therefore, an accurate PWM intensity control of the photonic integrated circuits minimizes the absorption and increases the brightness of LED module. Practically, the network of the brain is so complex that it is still not possible to extract complete information of a single neuron. Presently, the neurons are studied in clusters by cell-target methods and electric, magnetic and chemical stimulation. However, optogenetics increases the precision of cell targeting with a minimal side effect. In the future, the optogenetics technique will be able to map the nervous system and affected regions from neurological and psychiatric diseases. The laboratories around the globe are focussed on solving the diseases related to neurodegradation, memory loss and disorders, and diseases. Optogenetics study is extended to understand the synaptic network complexity and billion node communication in every time instant.

References

1. Zemelman BV, Lee GA, Ng M, Miesenbock G (2002) Selective photostimulation of genetically chARGed neurons. Neuron 33(1):15–22
2. Sierra YAB, Rost BR, Pofahl M, Fernandes AM, Kopton RA, Moser S, Holtkamp D, Masala N, Beed P, Tukker JJ, Oldani S, Wolfgang B, Kohl P, Baier H, Schneider-Warme F, Hegemann P, Beck H (2018) Reinhard Seifert10 and Dietmar Schmitz, Potassium channel-based optogenetic silencing. Nat Commun 9(1):4611
3. Kim CK, Adhikari A, Deisseroth K (2017) Integration of optogenetics with complementary methodologies in systems neuroscience. Nat Rev Neurosci 18(4):222
4. Zhang F, Wang L-P, Brauner M, Liewald JF, Kay K, Watzke N, Wood PG, Bamberg E, Nagel G, Gottschalk A, Deisseroth K (2007) Multimodal fast optical interrogation of neural circuitry. Nature 446(7136):633
5. Yong Kwon K, Sirowatka B, Weber A, Li W (2013) Opto-ECoG array: a hybrid neural interface with transparent ECoG electrode array and integrated LEDs for optogenetics. IEEE Trans Biomed Circ Syst 7(5):593–600
6. Yan B, Nirenberg S (2018) An embedded real-time processing platform for optogenetic neuroprosthetic applications. IEEE Trans Neural Syst Rehabil Eng 26(1):233–243
7. Reiner A, Isaco EY (2013) The Brain Prize 2013: the optogenetics revolution. Trends Neurosci 36(10):557–560
8. Crick F, Frs O (1999) The impact of molecular biology on neuroscience. Philos Trans R Soc Lond Ser B Biol Sci 354(1392):2021–2025
9. Lee ST, Williams PA, Braine CE, Lin D-T, John SWM, Irazoqui PP (2015) A miniature, fiber-coupled, wireless, deep-brain optogenetic stimulator. IEEE Trans Neural Syst Rehabil Eng 23 (4):655–664
10. Pashaie R, Falk R (2013) Single optical fiber probe for fluorescence detection and optogenetic stimulation. IEEE Trans Biomed Eng 60(2):268–280
11. Sparta DR, Stamatakis AM, Phillips JL, Hovels N, van Zessen R, Stuber GD (2012) Construction of implantable optical fibers for long-term optogenetic manipulation of neural circuits. Nat Protoc 7(1):12
12. Pashaie R, Baumgartner R, Richner TJ, Brodnick SK, Azimipour M, Eliceiri KW, Williams JC (2015) Closed-loop optogenetic brain interface. IEEE Trans Biomed Eng 62(10):2327–2337
13. Gagnon-Turcotte G, Khiarak MN, Ethier C, De Koninck Y, Gosselin B (2018) A 0.13-μm CMOS SoC for simultaneous multichannel optogenetics and neural recording. IEEE J Solid-State Circ 53(11):3087–3100
14. Jeong JW, McCall JG, Shin G, Zhang Y, Al-Hasani R, Kim M, Li S, Sim JY, Jang KI, Shi Y, Hong DY (2015) Wireless optofluidic systems for programmable in vivo pharmacology and optogenetics. Cell 162(3):662–674
15. Cao H, Gu L, Mohanty SK, Chiao J-C (2013) An integrated μ-LED optrode for optogenetic stimulation and electrical recording. IEEE Trans Biomed Eng 60(1):225–229
16. Bi X, Xie T, Fan B, Khan W, Guo Y, Li W (2016) A flexible, micro-lens-coupled LED stimulator for optical neuromodulation. IEEE Trans Biomed Circ Syst 10(5):972–978
17. Balasubramaniam S, Wirdatmadja SA, Barros MT, Koucheryavy Y, Stachowiak M, Jornet JM (2018) Wireless communications for optogenetics-based brain stimulation present technology and future challenges. IEEE Commun Mag 56(7):218–224
18. SeungWook O, Harris JA, Ng L, Winslow B, Cain N, Mihalas S, Wang Q, Lau C, Kuan L, Henry AM, Mortrud MT, Ouellette B, Nguyen TN, Sorensen SA, Slaughterbeck CR, Wakeman W, Li Y, Feng D, Ho A, Nicholas E, Hirokawa KE, Bohn P, Joines KM, Peng H, Hawrylycz MJ, Phillips JW, Hohmann JG, Wohnoutka P, Gerfen CR, Koch C, Bernard A, Dang C, Jones AR, Zeng H (2014) A mesoscale connectome of the mouse brain. Nature 508 (7495):207

19. Al-Atabany W, McGovern B, Mehran K, Berlinguer-Palmini R, Degenaar P (2013) A processing platform for optoelectronic/optogenetic retinal prosthesis. IEEE Trans Biomed Eng 60(3):781–791
20. Chen CH, McCullagh EA, Pun SH, Mak PU, Vai MI, In Mak P, Klug A, Lei TC (2017) An integrated circuit for simultaneous extracellular electrophysiology recording and optogenetic neural manipulation. IEEE Trans Biomed Eng 64(3):557–568
21. Gerhardt KP, Olson EJ, Castillo-Hair SM, Hartsough LA, Landry BP, Ekness F, Yokoo R, Gomez EJ, Ramakrishnan P, Suh J, Savage DF, Tabor JJ (2016) An open-hardware platform for optogenetics and photobiology. Sci Rep 6:35363
22. Wietek J, Rodriguez-Rozada S, Tutas J, Tenedini F, Grimm C, Oertner TG, Soba P, Hegemann P, Simon Wiegert J (2017) Anion conducting channelrhodopsins with tuned spectra and modified kinetics engineered for optogenetic manipulation of behaviour. Sci Rep 7(1):14957
23. Quadrato G, Nguyen T, Macosko EZ, Sherwood JL, Yang SM, Berger DR, Maria N, Scholvin J, Goldman M, Kinney JP, Boyden ES, Lichtman JW, Williams ZM, McCarroll SA, Arlotta P (2017) Cell diversity and network dynamics in photosensitive human brain organoids. Nature 545(7652):48
24. Gagnon-Turcotte G, LeChasseur Y, Bories C, Messaddeq Y, De Koninck Y, Gosselin B (2017) A wireless headstage for combined optogenetics and multichannel electrophysiological recording. IEEE Trans Biomed Circ Syst 11:1), 1–1),14
25. Hashimoto M, Hata A, Miyata T, Hirase H (2014) Programmable wireless light-emitting diode stimulator for chronic stimulation of optogenetic molecules in freely moving mice. Neurophotonics 1(1):011002
26. Park SI, Brenner DS, Shin G, Morgan CD, Copits BA, Chung HU, Pullen MY, Noh KN, Davidson S, Oh SJ, Yoon J, Jang K-I, Samineni VK, Norman M, Grajales-Reyes JG, Vogt SK, Sundaram SS, Wilson KM, Ha JS, Xu R, Pan T, Kim T-i, Huang Y, Montana MC, Golden JP, Bruchas MR, Gereau RW, Rogers JA (2015) Soft, stretchable, fully implantable miniaturized optoelectronic systems for wireless optogenetics. Nat Biotechnol 33(12):1280
27. Montgomery KL, Yeh AJ, Ho JS, Tsao V, Iyer SM, Grosenick L, Ferenczi EA, Tanabe Y, Deisseroth K, Delp SL, Poon AS (2015) Wirelessly powered, fully internal optogenetics for brain, spinal and peripheral circuits in mice. Nat Method 12(10):969
28. Shin G, Gomez AM, Al-Hasani R, Ra Jeong Y, Kim J, Xie Z, Banks A, Lee SM, Han SY, Yoo CJ, Lee J-L, Lee SH, Kurniawan J, Tureb J, Guo Z, Yoon J, Park S-I, Bang SY, Nam Y, Walicki MC, Samineni VK, Mickle AD, Lee K, Heo SY, McCall JG, Pan T, Wang L, Feng X, Kim T-i, Kim JK, Li Y, Huang Y, Gereau RW, Ha JS, Bruchas MR, Rogers JA (2017) Flexible near-field wireless optoelectronics as subdermal implants for broad applications in optogenetics. Neuron 93(3):509–521
29. Chow BY, Han X, Dobry AS, Qian X, Chuong AS, Li M, Henninger MA, Belfort GM, Lin Y, Monahan PE, Boyden ES (2010) High-performance genetically targetable optical neural silencing by light-driven proton pumps. Nature 463(7277):98
30. Zhang F, Wang LP, Brauner M, Liewald JF, Kay K, Watzke N, Wood PG, Bamberg E, Nagel G, Gottschalk A, Deisseroth K (2007) Multimodal fast optical interrogation of neural circuitry. Nature 446(7136):633
31. Chuong AS, Miri ML, Busskamp V, Matthews GA, Acker LC, Srensen AT, Young A, Klapoetke NC, Henninger MA, Kodandaramaiah SB, Ogawa M (2014) Noninvasive optical inhibition with a red-shifted microbial rhodopsin. Nat Neurosci 17(8):1123
32. Alberio L, Locarno A, Saponaro A, Romano E, Bercier V, Albadri S, Simeoni F, Moleri S, Pelucchi S, Porro A, Marcello E, Barsotti N, Kukovetz K, Boender AJ, Contestabile A, Luo S, Moutal A, Ji Y, Romani G, Beltrame M, Del Bene F, Di Luca M, Khanna R, Colecraft HM, Pasqualetti M, Thie G, Tonini R, Moroni A (2018) A light-gated potassium channel for sustained neuronal inhibition. Nat Method 15(11):969

Part VII

Nanomaterials in Therapeutics

Nanoparticle: Significance as Smart Material in Therapeutic Strategies in Drug Delivery in Biological Systems

16

Kamal Dhungel and Jyoti Narayan

Abstract

Nanoparticles have gained tremendous potential as smart materials in various therapeutic strategies in biological and non-biological systems. Due to the small size (less than 100 nm), nanoparticles penetrate into even smaller capillaries, which are taken up within cells, allowing an efficient drug accumulation at the targeted sites in the body. The use of biodegradable materials for nanoparticle preparation allows sustained drug release at the targeted site over a period of days or even weeks after injection. Peptides, proteins, nanogels, and antisense drugs have been synthesized in an effort to combat central nervous system (CNS) diseases. Molecules such as dalargin and loperamide are loaded onto nanoparticlees with the aim to drug delivery. Nanoparticles such as pegylated-poly(hexadecylcyanoacrylate) (PEG-PHDCA) have been investigated for the treatment of several CNS diseases. Liposomes are synthetic and spherical molecules, consisting of single amphiphilic lipid bilayers, which can entrap therapeutic molecules, including drugs, vaccines, nucleic acids, and proteins. Solid lipid nanoparticles are spherical, stable nanocarriers that possess a solid hydrophobic lipid core matrix stabilized by aqueous surfactant. Micelle is utilized for the delivery of curcumin for targeting glioma and treating Alzheimer's disease as a nanocarrier. Gold, silica, and carbon nanotubes/nanoparticles have been used to deliver specific drug across the blood-brain barrier. Gold nanoparticles functionalized with peptides are utilized for the treatment of Alzheimer's disease and functionalized with L-DOPA have been reported for the treatment of Parkinson's disease. Chemically functionalized multi-walled carbon nanotubes and polymer-coated carbon nanotubes have been applied for the delivery of drugs for brain cancer therapy. Various shapes such as nanosphere, nanostar, nanorods, and nanocage are utilized in brain tumor diagnosis. In the noninvasive approach,

K. Dhungel · J. Narayan (✉)
Department of Basic Sciences and Social Sciences (Chemistry Division), School of Technology, North Eastern Hill University, Shillong, Meghalaya, India

S. Paul (ed.), *Application of Biomedical Engineering in Neuroscience*,
https://doi.org/10.1007/978-981-13-7142-4_16

polymeric nanoparticles, especially PBCA nanoparticles coated with polysorbate 80, have recently received much attention from neuroscientists as an attractive and innovative carrier for brain targeting.

Keywords
Blood-brain barrier · Biocompatible nanoparticle · Functionalized nanoparticle · Therapeutic strategies

16.1 Introduction

There are some diseases such as multiple sclerosis, brain cancer, Parkinson's disease, Alzheimer's disease, and stroke becoming a great concern in elderly population. These diseases are caused by genetic and environmental factors, pathologies in process involving protein aggregation that lead to neurodegeneration or dys regulation of the immune process, or abnormalities regarding the development and function of the brain [1]. Invasive and noninvasive are the pathways in the transport of therapeutic agents to the brain [2]. In invasive route, the drug is administered directly inside the brain, thus providing a sufficient dosage without causing toxicity. The non-invasive administration strategies are based on anatomical structures of the brain capillaries and extra cellular environment of the cell. They are based on the directional transfer of fluids across the brain, the main routes, including the nasal drug molecules that are delivered to the brain [3]. Net charge, polarity, molecular weight, solubility and affinity for hydro or lipid moieties have played a major role in the drug delivery to the brain. Inorganic nanoparticles (INPs) are also utilized to transport the drugs through the blood-brain barrier (BBB).

The nasal route of systemic drug delivery is preferred, in which the drug is delivered to the brain; this increases the bioavailability and reduces the degradation of the drug [4]. The blood-brain barrier (BBB) acts as a local gateway against the circulating toxins and cells through a selective permeability system [5]. The pathways for crossing the BBB are through paracellular transport, between endothelial cells; through transcellular transport, involving passive or active mechanisms; across the luminal side of the endothelial cells; through the cytoplasm; and, subsequently, across the abluminal side, into the brain interstitium [6]. Nanoparticles [7], liposomes [8], dendrimers [9], carbon nanotubes [10], and micelles [11] are utilized to deliver the drug to the brain.

Nanoparticles have small size, can penetrate into small capillaries, are taken up within cells, allowing an efficient drug accumulation at the targeted sites in the body [7]. Biodegradable materials are utilized in the preparation of nanoparticles. These biodegradable materials are suitable for drug release at the targeted site in a period of days or even weeks after injection.

The drug molecules are adsorbed, attached, entrapped, or encapsulated into the nanoparticle. Nanoparticle-mediated drug delivery provides better bioavailability in the brain. Non-ionized, low molecular weight, lipophilic molecules can freely diffuse through the neural tissues. The presence of lipophilic groups increases drug delivery to the brain by passive diffusion.

These types of techniques have been successful for targeted delivery of anticancer drugs to cancer cells and the transport of higher amount of drug to the brain. Encapsulated drug molecules are delivered to the brain. Organic nanoparticles such as liposomes, polymers, emulsions, solid lipid nanoparticles, dendrimer and inorganic nanoparticles like silica, carbon, and gold are the carrier materials. Inorganic carrier nanoparticles are also utilized in the potential to cross the blood-brain barrier [12].

It has been observed that chemical conjugates modified with biological molecules, like pristine polymers and carbon nanotubes (CNT) have resulted in the enhancement of the stability of the nanoparticles [13].

16.2 The Blood-Brain Barrier

It was observed that intravenously injected dyes can stain all organs but it could not stain the brain in animals. However, the dyes injected directly into the brain could stain only the brain and not the other parts of the body [14]. These results showed that there exists a barrier between the brain and other parts of the body. This barrier is known as the blood-brain barrier (BBB), and its existence was confirmed by electron microscopic studies [12]. Delivery of drugs to the central nervous system has always being a challenging task due to the presence of blood brain barrier. Succeeding over the resistance of this barrier system, can lead to the positive treatment of the diseases related to the central nervous system. The blood-brain barrier inhibits the transport of ions and molecules from blood to the brain.

Blood-brain barrier have enzymatic machinery that restricts the entry, stability, and bioavailability of drugs in the cerebral region. The Blood-brain barrier thus maintains structural integrity restriction by restraining the entry of pathogens and toxic molecules, thereby maintaining brain homeostasis [15]. The blood brain barrier is characterized by its unique structure that differs remarkably from the cells present in other parts of the body. The protective basal lamina lies in the endothelial cells due to the presence of lesser paracellular diffusion of hydrophilic molecules. The presence of high number of mitochondrial cells and intercellular tight junctions, lead to high metabolic activities, bearing higher number of active transporter [15]. The blood brain barrier inhibits the transport of ions and molecules from blood to the brain.

16.3 Nanomaterial-Mediated Drug Delivery

Proteins, peptides (vasopressin), and antisense (nusinersen) drugs were utilized in the treatment of central nervous system (CNS) disease [16]. Some of these drugs are found to be ineffective due to their unfavorable invivo properties. These drugs have poor stability in the biological fluids, rapid enzymatic degradations, inadequate release profiles and unfavourable pharmaco-kinetic properties. These are the main reasons that results in the reduction of their efficacy fail to achieve in the clinical trial [17]. It has been observed that drugs may associate with NPs by being adsorbed, covalently dissolved, encapsulated, or bound covalently [18]. The incorporation of various functionalities into the nanoparticle system is known as nanoparticle functionalization and has proven advantageous in enhancing drug delivery to the brain [19].

In the process of therapeutic agent to cross the blood-brain barrier, it should be controllable and safe, there should not be any negative effect on the integrity of the blood-brain barrier. The drug should be biocompatible and selective. In target-specific drug delivery, the drug should be delivered at the affected site of the brain only, where the drug load should be acceptable, so that therapeutic concentration is obtained and maintained for sufficient time at the targeted site in the brain [20].

16.4 Coated Nanoparticle

Dalargin [21] and loperamide [22] have been encapsulated in biocompatible nanoparticles with the aim to delivery at the targeted sites of the brain. Dalargin and loperamide were adsorbed onto the surface of poly(butylcyanoacrylate) (PBCA) nanoparticles, coated with polysorbate 80 (PS-80). This drug when was experimented on mice, resulted in an analgesic effect and thus wasfound to be useful. On the other hand PBCA nanoparticles coated with PS-80 displayd some toxic effects towards the BBB [23]. Even the mixture of dalargin and PBCA did not show any allergic effect [24]. There is an advantage that coated nanoparticles can help to get effective results when delivered as therapeutic agents. The nanoparticles can bind to the inner endothelial lining of the brain capillaries and thus, provides a drug concentration gradient which results in the improvement of passive diffusion or the brain endothelial cell uptake of nanoparticles. This can occur through endocytosis or transcytosis pronounced analgesic effect, that was found to be successful [24]. Poly (butylcyanoacrylate) nanoparticles coated with PS-80, displayed some toxic effect towards the BBB [25].

Various mechanisms were utilized to explain the enhancement of drug transport through the BBB. Firstly, the nanoparticles bind to the inner endothelial lining of the brain capillaries, and there by, provides a drug concentration gradient, which improves the passive diffusion. Secondly, brain endothelial cell uptake of nanoparticles may occur through endocytosis or transcytosis. Apolipoproteins (APO) are involved in the brain penetration of PBCA nanoparticles overcoated with PS-80 [26].

PS-80 nanoparticles are utilized to transport molecules to the brain [27]. It has been observed that the administration of 8-chloro-4-hydroxy-1-oxo-1, 2-dihydropyridazino (4, 5-b) quinoline-5-oxide choline salt (MRZ 2/576) loaded onto PS-80 overcoated nanoparticles prolonged the duration of the anticonvulsive activity too [28]. Doxorubicin adsorbed on the PS-80 coated nanoparticles when were administered in rats for the treatment of experimental glioblastoma, were found to have higher survival time [29]. Although the acute toxicity of doxorubicin was reduced if it was associated with PS-80-coated nanoparticles only [30]. In noninvasive approach, polymeric nanoparticles, such as PBCA coated with polysorbate 80, have received attention from neuroscientists as an attractive and innovative carrier for brain targeting [31]. PBCA nanoparticles are defined as a submicron drug-carrier system and they are polymeric in nature. These nanoparticles were based on non-biodegradable polymeric systems.

16.5 Pegylated Nanoparticle

Nanoparticles coated with polyethylene glycol (PEG) is a commonly used approach to improve the efficiency of drugs and gene delivery to target cells and tissues. This is called as PEGylation. Pegylated-poly(hexadecylcyanoacrylate) (PEG-PHDCA) nanoparticles have been utilized in the treatment of brain tumor [32], encephalomyelitis (EAE) [33], and prion diseases [34]. PEG-PHDCA nanoparticles have been shown to penetrate in the brain, to a greater extent than all the other nanoparticles includings, PS-80 nanoparticles [32]. In EAE rats, PEG-PHDCA nanoparticles could reach the brain by two mechanisms: passive diffusion, due to the increase of blood-brain barrier permeability and transport by nanoparticles-containing macrophages, which infiltrated the inflammatory tissues. This study claims that PEG-PHDCA nanoparticles had appropriate characteristics for penetration into CNS under pathological conditions, especially, in neuroinflammatory diseases [35]. After intravenous administration in rats bearing intracerebral well-established gliosarcoma, PEG-PHDCA nanoparticles have accumulated preferentially in the tumoral tissue. PEG-PHDCA nanoparticles concentrated much more in the gliosarcoma than did their non-pegylated counterparts (PHDCA nanoparticles). Based on the test of sucrose permeability, PEG-PHDCA nanoparticles did not display any toxicity toward the BBB [35]. Firstly, in gliosarcoma, the preferential accumulation of PEG-PHDCA nanoparticles was attributable to their reduced plasma clearance. Secondly, in the normal brain, the applied pharmacokinetic model suggested an affinity of the PEG-PHDCA nanoparticles for the endothelial cells of the BBB, allowing their translocation [35].

16.6 Nanogel

Nanogels are made from ionic polyethylene imine (PEI) and nonionic PEG chains (PEG-cl-PEI). Emulsion evaporation method was utilized to synthesize nanogels [36]. Basically, the biologically active molecule (retinoic acid, indomethacin and valproic acid) are bonded to the nanogels by the electrostatic interactions, van der waal forces, hydrophobic or covalent bonding. The ionic group of these molecules interact with nanogel. Since PEI chains have a tendency to collapse, they result in the decrease of the volume and size of the nanoparticles. However, because of the steric stabilization of PEG chains, the collapsed nanogel forms a stable dispersion with a mean particle size of 80 nm. Nanogels have been tested as a potential carrier for oligonucleotide delivery to the brain, by using polarized monolayers of bovine mammary epithelial cells (BMEC) also [37].

16.7 Lipid-Based NPs

Lipid-based NPs are typically stable, nontoxic carriers that are well suited to brain drug delivery applications. Among these, the most common types are liposomes and solid lipid NPs (SLNs), both of which are discussed below.

16.7.1 Liposome

Liposomes are synthetic molecules composed of fatty acid. They are spherical vesicles in nature, consisting of single amphiphilic lipid bilayers. Drugs, vaccines, nucleic acids, proteins and therapeutic molecules can be encapsulated in the liposome. The advantage of liposome is that it has the ability to incorporate and transport large amount of drug and have the possibility to encapsulate its surface with various ligands. The safety and effectiveness of drug delivery can be improved by using liposome. Liposomes are utilized to deliver anticancer drugs, such as methotrexate [38], 5-fluorouracil [39], paclitaxel [40], doxorubicin [41] and erlotinib [42]. Thus, liposomes can be coated with various molecules. Liposomes have been utilized as carrier for gene therapy also. They consist of one or more lipid bilayers known as lamellae bounding an internal aqueous space. Cholesterol can increase the stability of liposomal formulation [43]. Liposomes can be neutral, anionic, or cationic; depends on the lipids used in the formulation. For cationic liposomes, one of the most commonly used lipids is 1, 2-dioleoyl-3-trimethylammonium-propane (DOTAP), mixed with dioleoyl-phosphatidylethanolamine (DOPE) [44]. The positively charged cationic liposomes may enable them to interact more effectively, with the negatively charged surface of the brain microvessel endothelial cells (BMEC). This results in their incapability of accumulating in the brain in greater amounts, but also creates potential issues including decreased stability and increased cytotoxicity in vivo (liposomes generally possess low toxicity) [45].

16.7.2 Solid Lipid Nanoparticles

Solid lipid nanoparticles are spherical, stable nanocarriers, possessing a solid hydrophobic lipid core matrix stabilized by aqueous surfactant [46]. The core is typically composed of biocompatible lipids such as triglycerides, fatty acids and waxes, which have the ability to solubilize lipophilic molecules. The stabilizing surfactants are composed of biological membrane lipids such as phospholipids, sphingomyelins, bile salts and cholesterol [47]. Drug can be dissolved or dispersed in solid lipid nanoparticles. The advantage of solid lipid nanoparticles is its biocompatibility, significant drug entrapment efficiency, increased drug stability and ability to provide controlled drug release over a time scale of several weeks [48]. Solid lipid nanoparticles can be surface functionalized to limit reticuloendothelial system (RES) uptakes and improve specific targeting to the brain. Surface charge plays a major role in the SLN-mediated brain uptake and toxicity. The positively charged solid lipid nanoparticle demonstrated maximal brain uptake (14-fold) higher brain drug accumulation compared to negatively charged solid lipid nanoparticles at 4 hours post administration [49].

16.8 Micelles

Micelles are amphiphilic molecules, with particle size ranging between 5 to 50 nanometre in diameter. They are formed in aquous solution with appropriate concentration and at specified temperature. Micelles have potential to deliver hydrophobic (poorly water-soluble) molecules. They can provide sustained and controlled release of the drugs at the targeted site, also can maintain the chemical and physical stability of the drugs, improving drug bioavailability [50].

Several other polymers, such as poly(styrene)-poly(acrylic acid), poly(ethylene glycol)-b-poly(lactic acid), distearyl-sn-glycero-3-phosphoethanolamine-N-methoxy and poly(ethylene glycol) also provide cellular uptake and thus facilated the delivery of encapsulated polymeric drug to the specific site of the brain [51]. Micelles are utilized in the delivery of curcumin for targeting glioma and treating Alzheimer's disease [52].

16.9 Inorganic Nanomaterial

Gold, silica and carbon nanotube have been used to deliver specific drugs across the blood-brain barrier [53]. Inorganic nanoparticles offer advantages over polymeric nanoparticles in terms of control over size and shape and simplicity of preparation and functionalization. Most importantly, inorganic nanoparticles are easier to track by microscopy techniques, e.g., magnetic resonance imaging (MRI), field emission scanning electron microscopy (FESEM), high resolution transmission electron microscopy (HR-TEM), or analytical techniques, inductively coupled plasma-mass spectroscopy (ICP-MS). However, inorganic nanoparticles have disadvantages too as they might not be degraded or eliminated through the kidneys, thus resulting in undesireable toxicity (e.g., carbon nanotubes and fullerenes may lead to lipid peroxidation and oxygen radical formation) [54].

16.9.1 Gold Nanoparticles

Gold nanoparticles (AuNPs) are biocompatible, user-friendly and widely therapeuticaly applicabile, making them an ideal candidate for spontaneous BBB penetration. It was observed that hydrophilic nanoparticles were able to penetrate through the BBB. Since BBB is lipophilic in nature; hydrophilic AuNPs have higher probability to spontaneously penetrate through the BBB [55]. Gold nanoparticles are functionalized with amyloid-specific peptides, were utilized for the treatment of Alzheimer's disease [56]. Whereas, multi-branched nanoflower-gold nanoparticles functionalized with L-3, 4-dihydroxyphenylalanine (L-DOPA) were utilized for the treatment of Parkinson's disease [57]. Insulin-coated gold nanoparticles having diameter around 20 nm were found to be the most widespread particles accumulating within the brain. Gold nanoparticles modified with transcription peptide with a 5 nm core size containing doxorubicin, an anticancer drug and gadolinium chelates, as imaging contrast agents, have been administered for theranostic application in glioblastoma [58]. These nanocarriers have the potential to penetrate the blood-brain barrier and efficiently deliver anticancer drugs thus enhancing brain tumor imaging. Citrate and polyethylene glycol-coated gold nanoparticles are found to disrupt cortical vascular changes, which are associated with disruption in the blood-brain barrier. Nanoparticles with 5 nm or smaller diameters could be useful in the diagnostics of early stages of blood-brain barrier dysfunction for drug delivery [59]. Gold nanoparticle can be used as computerized tomography imaging contrast agent. Various shapes such as nanosphere, nanostar, nanorods, and nanocage were utilized in brain tumor diagnosis [60]. The size of these gold nanostructures ranged from 20 to 120 nanometer.

16.9.2 Silica Nanoparticle

Functionalized fluorescent silica nanoparticles are utilized as nanocarrier for drug delivery to the brain [61]. If lactoferrin is attached on the surface of the polyethylene glycol-coated silica nanoparticles (25 nanometer); the process of receptor-mediated transcytosis of these nanocarriers has been noted to get enhanced. Silica nanoparticles can deliver nootropics (Smart drugs, cognitive enhancers, supplements and the substances that can enhance memory and creativity), such as piracetam (2-oxo-1-pyrolidine acetamide), pentoxifylline (3,7-dimethyl-1-(5-oxohexyl) purine-2, 6-dione), and pyridoxines (4, 5-Bis (hydroxymethyl)-2-methyl pyridine-3-ol) that are designated to enhance the permeability of the blood-brain barrier [62]. Silica nanoparticles are more efficient nanocarriers as a drug in comparison to the encapsulated drugs. The latter was not detected in the brains of the mice [63]. Polylactic acid-coated mesoporous silica nanoparticles, conjugated with a ligand peptide of low-density lipoprotein receptor, to enhance the transcytosis process across the blood-brain barrier, have been employed in the delivery of resveratrol, a therapeutic agent for excess reactive oxygen species and reactive nitrogen species removal [59].

16.9.3 Carbon Nanotube

Carbon nanotube are cylindrical molecules consisting of graphite sheets with nano-scaled diameter. Carbon nanotube can be single-walled or multi-walled, with open ends or closed with fullerene caps [64]. They have gained a great interest as nanocarrier systems due to the possibility of functionalization with specific chemical compounds, thus, modifying their physical and biological properties. Chemically functionalized multi-walled carbon nanotubes and polymer-coated carbon nanotubes have been applied for the delivery of drugs for brain cancer therapy [64]. Both in vitro and in vivo experiments indicated the penetration of the blood-brain barrier and enhanced uptake in tumors [65].

The permeation of amino-functionalized multi-walled carbon nanotubes through the blood-brain barrier has been studied in vitro, using a co-culture model comprising primary porcine brain endothelial cells and primary rat astrocytes, and in vivo, through the systemic administration in mice [65]. The results of the study could pave the way for carbon nanotube application in the delivery of drugs and biologics to the brain, causing no toxic effects on the cells.

16.10 Prodrugs

Prodrugs are basically bioreversible derivative of the drugs that undergo chemical or enzymatic biotransformation to convert into an active drug to result in pharmacological action in the body [66]. Prodrug approach is helpful to overcome drug formulation and delivery problems, such as poor aqueous solubility, chemical instability, insufficient oval absorption, rapid presystemic metabolism, inadequate brain penetration, toxicity and local irritation [67]. Prodrugs are chemical compounds that form pharmacological drug after metabolism. They are synthesized to overcome problems, such as low bioavailability, degradation by protective mechanism after reaching the targeted site and widespread systematic exposure, resulting in significant off-target effects [68].

16.11 Conclusion

Various types of nanoparticles are being utilized in the treatment of brain diseases. These nanoparticles, such as liposome, PBCA, silica nanoparticles, carbon nanotubes, gold nanoparticle, micelles and Pegylated nanoparticle, are able to cross the blood-brain barrier. Therefore, these nanoparticles with advanced functionalization can play a significant role in future drug delivery, into the brain in coming years.

References

1. Teleanu DM, Chircov C, Grumezescu AM, Volceanov A, Teleanu RI (2018) Blood-brain delivery methods using nanotechnology. Pharmaceutics 10:269
2. Beduneau A, Saulnier P, Benoit JP (2007) Active targeting of brain tumors using nanocarriers. Biomaterials 28:4947–4967
3. Mo X, Liu E, Huang Y (2019) The intra-brain distribution of brain targeting delivery systems. In: Brain targeted drug delivery system. Academic Press, Cambridge, MA, 409–438
4. Crowe TP, Greenlee MHWE, Kanthasamy AG, Hsu WH (2018) Mechanism of intranasal drug delivery directly to the brain. Life Sci 195:44–52
5. Alexander JJ (2018) Blood-brain barrier (bbb) and the complement landscape. Mol Immunol 102:26–31
6. Hersh DS, Wadajkar AS, Roberts N, Perez JG, Connolly NP, Frenkel V, Winkles JA, Woodworth GF, Kim A (2016) Evolving drug delivery strategies to overcome the blood brain barrier. J Curr Pharm Res 22:1177–1193
7. Masserini M (2013) Nanoparticles for brain drug delivery. ISRN Biochem 2013:1–18
8. Peng Y, Zhao Y, Chen Y, Yang Z, Zhang L, Xiao W, Yang J, Guo L, Wu Y (2018) Dual-targeting for brain-specific liposomes drug delivery system: synthesis and preliminary evaluation. Bioorg Med Chem 26:4677–4686
9. Moscariello P, David YWN, Jansen M, Weil T, Luhmann HJ, Hedrich J (2018) Brain delivery of multifunctional dendrimer protein bioconjugates. Adv Sci 5:1700897
10. Guo Q, Shen XT, Li YY, Xu SQ (2017) Carbon nanotubes-based drug delivery to cancer and brain. J Huazhong Univ Sci Technol 37(5):635–641
11. Gothwal A, Khan I, Kesharwani P, Chourasia MK, Gupta U (2018) Chapter 11- Micelle-based drug delivery for brain tumors. In: Nanotechnology-based targeted drug delivery systems for brain tumors. Academic press, United states, Cambridge, Massachusetts, pp 307–326
12. Mudshinge SR, Deore AB, Patil S, Bhalgat CM (2011) Nanoparticles: Emerging carriers for drug delivery, Saudi Pharmaceutical Journal 19: 129-14
13. John AA, Priyadharshni A, Muthu S, Vellayappan V, Balaji A, Mohandas H, Jaganathan SK, (2015) Carbon nanotubes and graphene as emerging candidates in neuroregeneration and neurodrug delivery International Journal of Nanomedicine 10: 4267–4277
14. Ehrlich P, (1885) The oxygen demand of the organism: a color analysis study. Berlin: Hirschwald
15. Weiss N, Miller F, Cazaubon S, Couraub P (2009) The blood brain barrier in brain homeostasis and neurological diseases, Biochimica et biophysica acta, United states, Salmon Tower Building, New York City 1788: 842-857
16. Khorkova O, Wahlestedt C (2017) Oligonucleotide therapies for disorders of the nervous system, Nature Biotechnology 35 (3): 249-263
17. Olivier JC (2005) Drug transport to brain with targeted nanoparticles, Neuro Rx 2(1): 108-19
18. Wohlfart S, Gelperina S, Kreuter J (2012) Transport of drugs across the blood-brain barrier by nanoparticles. J. Controlled Release 161(2): 264-73
19. Joao M, Bárbara G, Susana M, Sarmento MB Nanoparticle functionalization for brain targeting drug delivery and diagnostic, Handbook of Nanoparticles 941–959 Springer (United states, Salmon Tower Building, New York)
20. Banks WA (2009) Characteristics of compounds that cross the blood-brain barrier. BMC Neurol. 9 (Suppl 1): S3
21. Kreuter J, Alyautdin RN, Kharkevich DA, Ivanov AA (1995) Passage of peptides through the blood-brain barrier with colloidal polymer particles (nanoparticles) Brain Res. 13: 674 (1) 171-4
22. Alyautdin RN, Petrov VE, Langer K, Berthold A, Kharkevich DA, Kreuter J (1997) Delivery of loperamide across the blood–brain barrier with poly-sorbate 80-coated polybutylcyanoacrylate nanoparticles Pharm. Res. 14: 325–328
23. Troster SD, Muller U, Kreuter J (1990) Modification of the body distribution of poly-(methylmethacrylate) nanoparticles in rats by coating with surfactants. Int. J. Pharma. 61: 85-100

24. Schröder U, Sabel BA (1996) Nanoparticles, a drug carrier system to pass the blood brain barrier, permit central analgesic effects of i.v.dalargin injections, Brain Research 710: (1–2) 121-124
25. Olivier JC, Fenart L, Chauvet R, Pariat C, Cecchelli R, Couet W (1999) Indirect evidence that drug brain targeting using polysorbate 80-coated polybutylcyanoacrylate nanoparticle is related to toxicity, Pharmaceutical research 16: 12 1836-42
26. Kreuter J, Shamenkov D, Petrov V, Ramge P, Cychutek K, Brandt CK, Alyautdin R (2002) Apolipoprotein-mediated transport of nanoparticle-bound drugs across the blood-brain barrier. Journal of drug targeting 10: (4) 317-325
27. Grabrucker AM, Ruozi B, Belletti D, Pederzoli F, Forni F, Vandelli MA, Tosi G (2016) Nanoparticle transport across the blood brain Barrier. Tissue Barriers 4: (1) 1153568
28. Friese A, Seiller E, Quack G, Lorenz B, Kreuter JÈ (2000) Increase of the duration of the anticonvulsive activity of a novel NMDA receptor antagonist using poly(butylcyanoacrylate) nanoparticles as a parenteral controlled release system. European Journal of Pharmaceutics and Biopharmaceutics 49: 103-109
29. Steiniger SCJ, Kreuter J, Khalansky AS, Sidan IN, Bobruskin AI, Smirnova ZS, Severin SE, Reiner UHL, Kock M, Geiger KD, Gperina SE (2004) Chemotherapy of Glioblastoma in Rats Using Doxorubicin-Loaded Nanoparticles. Int J Cancer 109: 59–767
30. Gelperina SE, Khalansky AS, Skidan IN, Smirnova ZS, Bobruskin AI, Severin SE, Turowski B, Zanella FE, Kreuter J, (2002) Toxicological studies of doxorubicin bound to polysorbate 80-coated poly(butyl cyanoacrylate) nanoparticles in healthy rats and rats with intracranial glioblastoma. Toxicology Letters 126: (2) 131-41
31. Garcia EG, Andrieux K, Gilb S, Couvreur P (2005) Colloidal carriers and blood–brain barrier (BBB) translocation: A way to deliver drugs to the brain? International Journal of Pharmaceutics 298: 274–292
32. Karanath H, Murthy RSR (2008) Nanotechnology in brain targeting. International journal of pharmaceutical sciences and nanotechnology 1: (1) 9-24
33. Calvo P, Gouritin B, Villarroya H, Eclancher FË, Giannavola C, Klein C (2002) Quantification and localization of PEGylatedpolycyanoacrylate nanoparticles in brain and spinal cord during experimental allergic encephalomyelitis in the rat. European Journal of Neuroscience 15:1317-1326
34. Calvo P, Gouritin B, Brigger I, Lasmezas C, Deslys JP, Williams A, Andreux JP, Dormont D, Couvreur P, (2001) PEGylatedpolycyanoacrylate nanoparticles as vector for drug delivery in prion diseases. Journal of Neuroscience 111: (2) 151-155
35. Garcia-Garciaa E, Gil S, Andrieux K, Desmaële D, Nicolas V, Taran F, Georgin D, Andreux JP, Roux F, Couvreur P (2005) A relevant in vitro rat model for the evaluation of blood-brain barrier translocation of nanoparticles. Cell. Mol. Life Sci. 62: 1400–1408
36. Serguei V, Vinogrado V, Kohli E, Zeman AD (2005) Cross linked Polymeric Nanogel Formulations of 5-Triphosphates of Nucleoside Analogs: Role of the Cellular Membrane in Drug Release. Mol. Pharm. 2(6): 449–461
37. Serguei V, Batrakova E, Kabanov AV (2004) Nanogels for Oligonucleotide Delivery to the Brain. Bioconjug Chem.15: (1) 50–60
38. Wang X, Liu P, Yang W, Li L, Li P, Liu Z, Zhuo ZX, Gao Y (2014) Microbubbles coupled to methotrexate-loaded liposomes for ultrasound-mediated delivery of methotrexate across the blood–brain barrier. International Journal of Nanomedicine 9: 4899–4909
39. Lakkadwala S, Singh J, (2018) Dual functionalized 5-Fluorouracil liposomes as highly efficient nanomedicine for glioblastoma treatment as assessed in an in vitro brain tumor model. Journal of Pharmaceutical Sciences 107: (11) 2902-2913
40. Liu Y, Ran R, Chen J, Kuang Q, Tang J, Mei L, Zhang Q, Gao H, Zhang Z, He Q (2014) Paclitaxel loaded liposomes decorated with a multifunctional tandem peptide for glioma targeting. Biomaterials 35: 4835-4847
41. Ashley JD, Quinlan CJ, Schroeder VA, Suckow MA, Pizzuti VJ, Kiziltepe T, Bilgicer B (2016) Dual Carfilzomib and Doxorubicin–Loaded Liposomal Nanoparticles for Synergistic efficacy in multiple myeloma. Mol Cancer Ther 15: (7) 1452-9

42. Hui XZ, Shi TKH (2018) Development of nanoliposomal formulation of erlotinib for lung cancer and invitro/in vivo antitumoral evaluation. Drug design, development and therapy 12: 1–8
43. Webb MS, Harasym TO, Masin D, Bally MB, Mayer LD (1995) Sphingomyelin-cholesterol liposomes significantly enhance the pharmacokinetic and therapeutic properties of vincristine in murine and human tumour models. Br. J. Cancer 72: (4) 896-904
44. Ciani L, Ristori S, Salvati A, Calamai L, Martini G, (2004) DOTAP/DOPE and DC-Chol/DOPE lipoplexes for gene delivery: zeta potential measurements and electron spin resonance spectra. Biochim Biophys Acta 1664: (1) 70-79
45. Pedroso de Lima MC, Neves S, Filipe A, Duzgunes N, Simoes S (2003) Cationic liposomes for gene delivery: from biophysics to biological applications. Curr Med Chem 10: (14) 1221-31
46. Müller RH, Mäder K, Gohla S (2000) Solid lipid nanoparticles (SLN) for controlled drug delivery-a review of the state of the art. Eur J Pharm Biopharm 50: (1) 161-77
47. Manjunath K, Venkateswarlu V (2005) Pharmacokinetics, tissue distribution and bioavailability of clozapine solid lipid nanoparticles after intravenous and intraduodenal administration. J Controlled Release, 107: (2) 215-28
48. Mishra B, Patel BB, Tiwari S (2010) Colloidal nanocarriers: a review on formulation technology, types and applications toward targeted drug delivery. Nanomed 6: (1) 9-24
49. Reddy LH, Sharma RK, Chuttani K, Mishra AK, Murthy RR (2004) Etoposide-incorporated tripalmitin nanoparticles with different surface charge: Formulation, characterization, radiolabeling, and biodistribution studies. The American Association Pharmaceutical Scientist Journal 6: (3) 55-64
50. Xu W, Ling P, Zhang T, (2013) Polymeric micelles, a promising drug delivery system to enhance bioavailability of poorly water-soluble drugs. Journal of Drug Delivery 2013
51. Ahmad Z, Shah A, Siddiq M, Kraatz HB (2014) Polymeric micelles as drug delivery vehicles. RSC Adv 4: 17028
52. Keshari P, Sonar Y, Mahajan H (2019) Curcumin loaded TPGS micelles for nose to brain drug delivery: in vitro and in vivo studies. Advanced Performance Materials 34 (7) 423-432
53. Pokale A (2007) Inorganic and organic hybrid nanocapsules based anticancer drug delivery MSC clinical research article 18: 21-26
54. Lai F, Fadda AM, Sinic C (2013) Liposomes for brain delivery, Expert Opin Drug Deliv 10: (7)
55. Sela H, Cohen H, Elia P, Zach R, Karpas Z, Zeiri Y (2015) Spontaneous penetration of gold nanoparticles through the blood brain barrier (BBB). J Nanobiotechnol 13: (71) 1-9
56. Ruff J, Hüwel S, Kogan MJ, Simon U, Galla HJ (2017) The effects of gold nanoparticles functionalized with ß-amyloid specific peptides on an in vitro model of blood-brain barrier Nanomedicine 13: (5) 1645–1652
57. Gonzalez-Carter DA, Ong ZY, McGilvery CM, Dunlop IE, Dexter DT, Porter AE (2019) L-dopa functionalized, multi-branched gold nanoparticles as brain-targeted nano-vehicles Nanomedicine 15: (1) 1–11
58. Chen Y, Dai Q, Morshed R, Fan X, Wegscheid ML, Wainwright DA, Han Y, Zhang L, Auffinger B, Tobias AL, Rinc6n E, Thaci B, Ahmed AU, Warnke P, Chuan H, Lesniak MS (2014) Blood-Brain Barrier permeable gold nanoparticles: an efficient delivery platform for enhanced malignant glioma therapy and imaging. Small 29: 10 (24) 5137–5150
59. Subramani K, Mehta M (2018) Chapter 19—Nanodiagnostics in microbiology and dentistry. In: Subramani K, Ahmed W (eds) Emerging nanotechnologies in dentistry, 2nd edn. William Andrew Publishing, Norwich, pp 391–419
60. Liu Y, Yuan H, Fales AM, Register JK, Dinh TV (2015) Multifunctional gold nanostars for molecular imaging and cancer therapy. Frontiers in Chemistry 3:51
61. Tamba BI, Streinu V, Foltea G, Neagu AN, Dodi G, Zlei M, Tijani A, Sefanescu C (2018) Tailored surface silica nanoparticles for blood-brain barrier penetration: Preparation and in vivo investigation. Arab J Chem 11:1–990
62. Jampilek J, Zaruba K, Oravec M, Kunes M, Babula P, Ulbrich P, Brezaniova I, Opatrilova R, Triska J, Suchy P (2015) Preparation of silica nanoparticles loaded with nootropics and their in vivo permeation through blood-brain barrier. BioMed Res. Int. 812673

63. Mendiratta S, Hussein M, Nasser HA, Ali AAA (2019) Multidisciplinary role of mesoporous silica nanoparticles in brain regeneration and cancers: from crossing the blood–brain barrier to treatment. Particle Particle System Characterization 1900195: 1-22.
64. Kafa H, Wang JTW, Rubio N, Klippstein R, Costa PM, Hassan HAFM, Sosabowski JK, Bansal SS, Preston JE, Abbott NJ, Al-Jamal KT (2016) Translocation of LRP1 targeted carbon nanotubes of different diameters across the blood–brain barrier in vitro and in vivo. J Control Release 225:217–229
65. Kafa H, Wang JTW, Rubio N, Venner K, Anderson G, Pach E, Ballesteros B, Preston JE, Abbott NJ, Al-Jamal KT (2015) The interaction of carbon nanotubes with an in vitro blood-brain barrier model and mouse brain in vivo. Biomaterials 53:437–452
66. Rautio J, Laine K, Gynther M, Savolainen J (2008) Prodrug approaches for CNS delivery. AAPS J 10(1):92–102
67. Singh Y, Palombo M, Sinko PJ (2008) Recent trends in targeted anticancer prodrug and conjugate design. Curr Med Chem 15(18):1802–1826
68. Rais R, Jančařík A, Tenora L, Nedelcovych M, Alt J, Englert J, Rojas C, Le A, Elgogary A, Tan J, Monincová L, Pate K, Adams R, Ferraris D, Powell J, Majer P, Slusher BS (2016) Discovery of 6-Diazo-5-oxo–norleucine (DON) prodrugs with enhanced CSF delivery in monkeys: a potential treatment for glioblastoma. J Med Chem 59(18):8621–8633

17 Biomedical Application of Nanoparticles for Channel Protein Modulation to Control Neural Disorder with Special Reference to Seizure

Pankaj Kalita and Manash Barthakur

Abstract

Channel proteins are the regulators of entry and exit of different molecules and ion to and fro from the cells. Regulation of entry and exit of molecules through different channel proteins can control different disorders. Therapeutic agents are used to target the channel protein to regulate ionic entry. Engineered channel proteins are developed to modify the channel protein movement. Epilepsy, which is marked by repeated seizures, is one of the serious neural disorder prevailing worldwide. Epileptic disorder is an electrophysiological alteration in the neuronal level and these electrophysiological changes are regulated by inward and outward movement of sodium, calcium, potassium, chloride ions, etc. Ions move through different channel proteins, and their movements are regulated by different channel proteins. These channel proteins are charge dependent and can be modulated by charged molecules. Nanoparticles are charged molecules and can be used to modulate channel proteins. Besides, nanoparticles have more exceptional properties than its raw materials which are helpful in the drug delivery approach. In the present article, it is targeted to focus and highlight the structural and functional approach of channel proteins and application of nanoparticles to control channel protein regulation which can help control different neural disorders including seizure.

Keywords

Channel protein · Therapeutic agent · Seizure · Nanoparticle

P. Kalita
Department of Biophysics, Pub Kamrup College, Baihata Chariali, Kamrup, Assam, India

M. Barthakur (✉)
Department of Zoology, Pub Kamrup College, Baihata Chariali, Kamrup, Assam, India

S. Paul (ed.), *Application of Biomedical Engineering in Neuroscience*,
https://doi.org/10.1007/978-981-13-7142-4_17

17.1 Introduction

Neural disorder is any kind of disorder of the brain and spinal cord or its supporting organ systems that disturbs the normal brain activity. It may appear as anatomical changes of the brain or other neural organ systems or may be reflected as physiological dysfunctions of the brain's activity marked by biochemical or molecular changes in the nervous system. The anatomical deformations of the brain are generally congenital deformation or deformation after birth due to different ailments such as cancer, stroke, accident, infection or other brain injury. Anatomical deformations include Alzheimer's disease and Parkinsonism, etc.. Shrinkage of brain and some area of the central nervous system is observed in Alzheimer's disease. Anatomical deformation may lead to physiological and biochemical changes of the neuron and other supportive cells. Protein misfolding in the brain tissue also causes functional disorder of the nervous tissue. Seizure is a kind of neural disorder caused by interrupted brain electrical signalling.

17.2 Brain's Insights of Seizure and Its Types

Neurons are the anatomical and physiological unit of the nervous system having electrical activity. Normal brain cells communicate by the mechanism of charge alterations between the nerve cells of different body parts. These cells pass the signals through the neuronal membrane in the form of an electrical passage in a controlled manner constituting a nerve impulse. This electrical communication takes place through a solution containing different solutes like proteins and inorganic salts dissolved in water. The nerve impulse transmission from one position to the other position of the nerve is a fair and controlled physiology. Anomaly in nerve impulse conduction may result in an unordered firing causing the neural disorder: seizure which lasts only a few seconds to a few minutes, namely 1 to 2 min. Single seizure may persist even in a normal individual. But repetitive seizures lead to the occurrence of the neurological disorder called epilepsy. In an epileptic patient, nerve cells send powerful and unordered electrical signals (up to four times more than their normal rate), disrupting the brain's normal activity. Continuous seizures in an individual lead to epilepsy, needing antiepileptic drugs (AEDs). Any part in the brain may act as the seizure centre, and accordingly, the symptoms of seizure may vary depending on the seizure centre in the brain (Fig. 17.1).

Although seizure may be attributed due to various factors, major cases are caused by an unknown origin (Fig. 17.2). Any factor which can cause brain damage or disrupt the normal electrical or chemical activity may result into epilepsy.

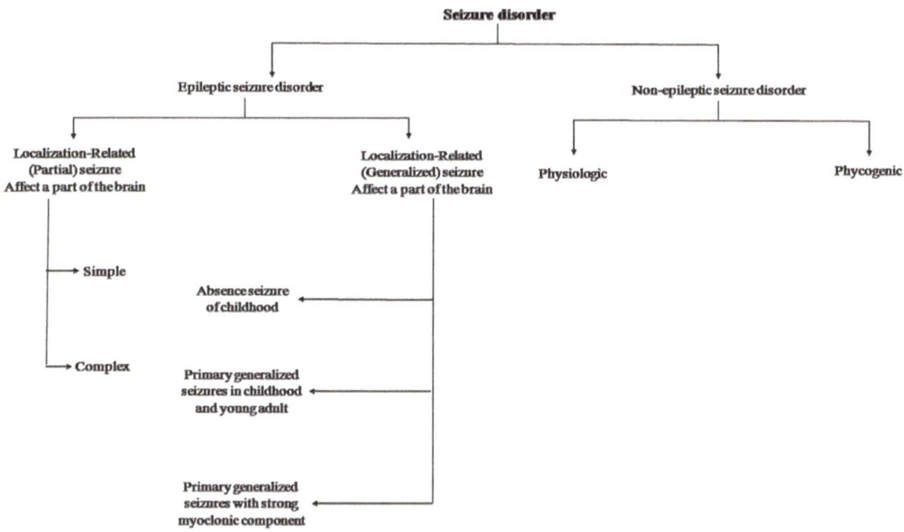

Fig. 17.1 Different types of seizures. (Source: PhD thesis of Pankaj Kalita (first author). Title of the thesis "Identification of Phytochemicals in Controlling Seizure and Associated Biochemical and Histological Changes in Experimental Animal Model," submitted at Assam University, Assam, India)

Fig. 17.2 Contribution of different factors towards epilepsy burden. (Source: PhD thesis of Pankaj Kalita (first author). Thesis title "Identification of Phytochemicals in Controlling Seizure and Associated Biochemical and Histological Changes in Experimental Animal Model" submitted at Assam University, Assam, India)

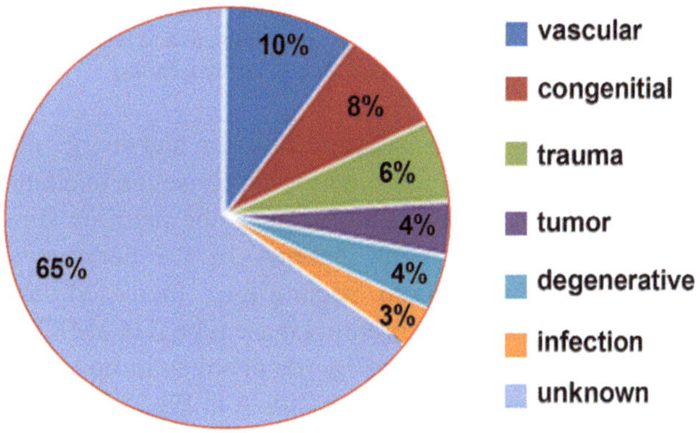

17.3 Mechanism of Seizure

As already mentioned, disturbances in neuronal firing and imbalance between neuronal excitation and inhibition lead to the generation of a seizure. Various underlying molecular, cellular and biochemical changes like neurotransmitter

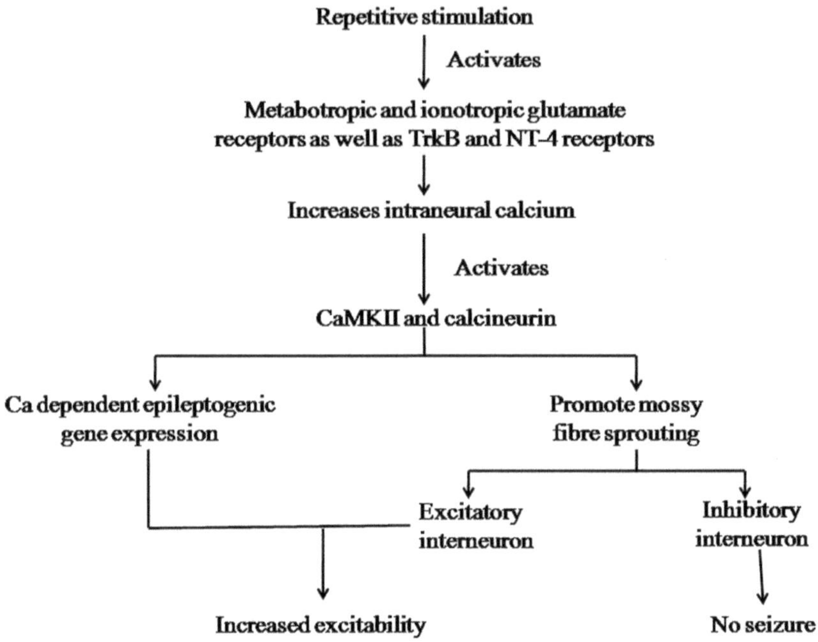

Fig. 17.3 Diagrammatic representation of mechanistic pathway for seizure generation. (Source: PhD thesis of Pankaj Kalita, (first author). Title of the thesis"Identification of Phytochemicals in Controlling Seizure and Associated Biochemical and Histological Changes in Experimental Animal Model" submitted at Assam University, Assam, India)

level, neurotransmitter receptor function and energy metabolism or alterations in ion channel modulation and functioning cause the excitability of the neurons and result in seizures. Repeated stimulation of the neurons stimulates the metabotropic and ionotropic glutamatergic channels of the nervous system which results in the increase of intraneuronal calcium ions. Increased calcium activates the calcium–/calmodulin-dependent protein kinase type II (CaMKII) and calcineurin which in turn causes genetic and epileptogenic changes in epileptogenic gene expression which ultimately result in seizure as shown in Fig. 17.3.

17.4 Channel Protein and Neuronal Function

Neuronal physiology depends upon the types of electrolytes present on both sides of the neurilemma. Nerve impulses are generated after inward movement of sodium ions and outward movement of potassium ions. Not only the sodium and potassium ions but also calcium and chloride ions play an important role in the generation of nerve impulse. Electrolytes cross the plasma membrane through different processes like active transport and passive transport. Different channel proteins are involved in the transportation of different ions. Dysregulation of different types of channel proteins causes occurrence of different ailments termed as channelopathies. Autoimmune diseases, toxic substances contamination or mutations in the channel protein-encoding genes attribute to the evidence of channelopathies.

Most of the neural disorders are channel dependent. Different channel proteins present in the neural membrane regulate the entry and exist of different ions. The entry and exit of different ions from extracellular medium to intracellular fluid and vice versa is important for functioning of the neurons and astrocytes. Cations like sodium, potassium, calcium and anions like chloride and bicarbonate ion transportation are the regulating factor for proper brain activities. Among different ions, calcium ion plays an important role in cell survival, neuroprotection and death signal initiation. Movement of different ions takes place through some channel and pores present in the neuronal membrane. Some channels are voltage dependent, and opening and closing of the channel is completely regulated by the potential difference between two planes of the neurons. Some channels are high-voltage-dependent and some are low-voltage-dependent channels. Opening or closing of channel is also mediated by activation of some receptors and these are called receptor-activated channel protein.

17.5 Types of Calcium Channels

There are different types of calcium channels observed in many cells. Various Ca^{2+} channel types are attributed by different workers [1, 2]. Accordingly, the calcium channels are classified as types mentioned below.

1. **L**-type (referring to long lasting) calcium channel is a high-voltage-activated calcium channel present in the plasma membrane of all excitable cells, including the neurons. L type calcium channels are mostly prevalent in spinal cord and dorsal root ganglia.
2. **P**-type calcium channel is also a high-voltage-activated calcium channel, which is a voltage-dependent channel. The $\alpha 1$ subunit determines most of the channel's properties. These channels play a role in neurotransmitter release at the presynaptic terminal and in neuronal integration in the neuronal system.
3. **N**-type calcium channel is another type of high-voltage-activated calcium channel. They also play a role in neurotransmitter release. This type of calcium channel is the target for the development of drugs to relieve chronic and neuropathic pain, hypertension, autism spectrum disorder, osteoarthritis and other medical diagnoses.
4. **R**-type calcium channel is intermediate-voltage-activated calcium channel, which is also voltage-dependent channel. R-type channels are mostly expressed in the cortex, hippocampus, striatum, amygdale and interpeduncular nucleus.
5. **Q**-type calcium channel activity is also regulated by the alpha subunit which requires higher threshold of activation and lower kinetics.
6. **T**-type of calcium channel is the low-voltage-activated calcium channel which opens during membrane depolarization.

In addition to these types of calcium channel proteins, there are other types of calcium-permeable channel proteins in many cellular systems. These are discussed below.

17.6 Receptor-Activated Calcium Channel

These types of non-voltage-gated channels, also called receptor-operated calcium channels (ROCCs), which are activated by agonists acting on a range of G-protein-coupled receptors and store-operated calcium channels (SOCCs), and also activated by depletion of the calcium load in the sarcoplasmic reticulum.

17.7 IP3-Receptor-Operated Calcium Channel

Inositol triphosphate is present in the plasma membrane and the receptor is activated by the depletion of calcium ions in the intracellular storage and causes entry of calcium ions through store-operated calcium channel.

17.8 Ryanodine Receptor

Ryanodine receptors are modulated by the plasma membrane coupled voltage sensors and activated by calcium ions. These receptors are modulated by a sequential manner from its primordial ryanodine receptors. The intracellular Ca^{2+} activity governs the activity consequent ryanodine receptor channel activity.

17.9 Two-Pore Channel

Two-pore channels are second-messenger-regulated channel protein which control the entry of calcium ions into the cytosol. Functional loss of two-pore channel protein causes substrate accumulation and creates congestion; on the other hand, increase in the function of two-pore channel protein causes enlargement of lysosome [3].

17.10 Store-Operated Channels

Store-operated calcium channels are present in the plasma membrane. The entry of calcium ions starts after activation of store-operated calcium channel after depletion of the stored calcium level in the endoplasmic reticulum.

17.11 Channel Proteins and Seizure

Neurons communicate between them through some specialized proteins embedded in the plasma membrane (termed the channel proteins). Among the other channel proteins, the calcium signalling plays a key role in excessive firing of the neurons. The calcium spike propagates to the neuronal soma triggering further action potential which finally results into a burst in action potential. Low-threshold calcium spike generates burst firing. When cell or neuron is in hyperpolarized condition, it responds to low-threshold calcium spike and burst firing. In hyperactivated neurons, the level of calcium conductance is reduced to a certain level and results in the membrane polarization abnormally and finally burst. Characteristic of a low-threshold calcium spike is the latency of the spike affected by the power of the initial depolarizing current. The calcium latency is reduced with an increase in depolarizing current and quickly activates the exponential activation of growth of calcium spike. The size of the transient current has a correlation with the amplitude of low-threshold calcium spike. Membrane hyperpolarization and depolarizing input regulates the amplitude of the low-threshold calcium spike. Neurons also show high-threshold calcium spike after modification of depolarizing input. Depolarizing input after amplification following sodium conductance enhances membrane excitability in the tonic mode and causes high-threshold calcium spike. The voltage-dependent conductance plays a role in balancing the high-threshold Ca^{2+} spike firing and the short-term Ca^{2+} influx.

Electrical signals are received from a neuron and transferred to the next soma via the axonal terminals. Dendritic spike transmits signal at relaxed rate than the axonal action potential. The sodium and potassium voltage-gated ion channels are caused to open by the dendritic channels. A higher level of local threshold voltage is required for initiation of dendritic spike than the action potential initiated in the axon and therefore spike initiation requires a strong input.

17.12 Problems and Prospects in Epileptic Medications

Antiepileptic medications have not resulted in potential AEDs with superior efficacy because most of these AEDs have some unwanted side effects. In some pyknoleptic patients, they develop some resistance to these drugs, thus indicating an importance of a new drug. Drug toxicity is an important issue of AEDs. Since the discovery of phenobarbitone, different AEDs like phenytoin, ethosuximide, carbamazepine, sodium valproate and different derivatives of benzodiazepines have been regarded as antiepileptic drugs prevailing in the market among the clinicians. Although a large number of antiepileptic drugs are available, their usage becomes limited due to drug resistance, their toxicity and their interactions with others. Vigabatrin can block the GABA transaminase activity and increasing GABA level, thus controlling the seizure; but it has severe retinotoxicity. Similarly, levetiracetam is another highly effective antiepileptic drug, but it has also some side effects like increased anxiety, mood change and severe headache [4].

17.13 Calcium Channel Blocker

Entry and exit of calcium ions through different protein present in the plasma membrane can be controlled or blocked by application of some exogenous molecules, termed as calcium channel blocker. Calcium channel blockers are the chemicals showing neuroprotective effect which can regulate the intracellular calcium level. These channel-targeted chemicals can be used as therapeutic agent to control burst firing of neurons [5]. In active neuron, there is a close relationship between calcium transient decay with calcium amplitude and calcium wave [6]. In ion pores present in the cell membrane, negatively charged gold nanoparticle blocks the ion pores after binding to the sulphur group of the cystine loop of the ion channel. The calcium level is important for neuronal survival and a transient calcium influx may trigger subsequent calcium-independent events leading to increased neuronal survival [6].

17.14 Seizure Prevention and Modulation of Channel Protein

Channel proteins are charged molecules which can be modulated by different micromolecules and macromolecules. Therapeutic chemicals can change the functioning of channel proteins and regulate the entry and exit of different ions. Concept of application of nanosized particles to change the channel protein functioning is relatively a new idea and can be targeted to control functioning of neuronal activities and can regulate neural disorder. Natural processes like mutation and methylation are the basis that can change the functioning of channel protein and lead to alteration of brain function and diseases. The functional alteration of channel proteins can be corrected with the help of nanoparticles. Surface functionalized nanoparticles can be prepared after the incorporation of a wide variety of targeted ligands on the nanoparticle surfaces, allowing them to be used in sensing of biomolecules and cells, diagnosis of diseases and intracellular delivery. For example, nanoparticles functionalized with ligands exhibiting differential affinity towards proteins and cell surface molecules have been employed for their identification. Surface functionalized positively charged nanoparticles can be prepared after incorporation of ligands exhibit higher internalization into cells compared to the neutral and negatively charged particles [7]. The process of surface functionalization of gold nanoparticle (AuNP) with targeted ligand is depicted in Fig. 17.4. The surface of AuNP is negatively charged and the cell surface is also negatively charged, thereby preventing the interaction of AuNP and cell surface in normal conditions. But when the AuNP surface is modified with a ligand which has the efficiency converting the AuNP surface into positively charged surface, then the AuNP and the ligand can be internalized by the cell enhancing the expression of the ligand inside the cell.

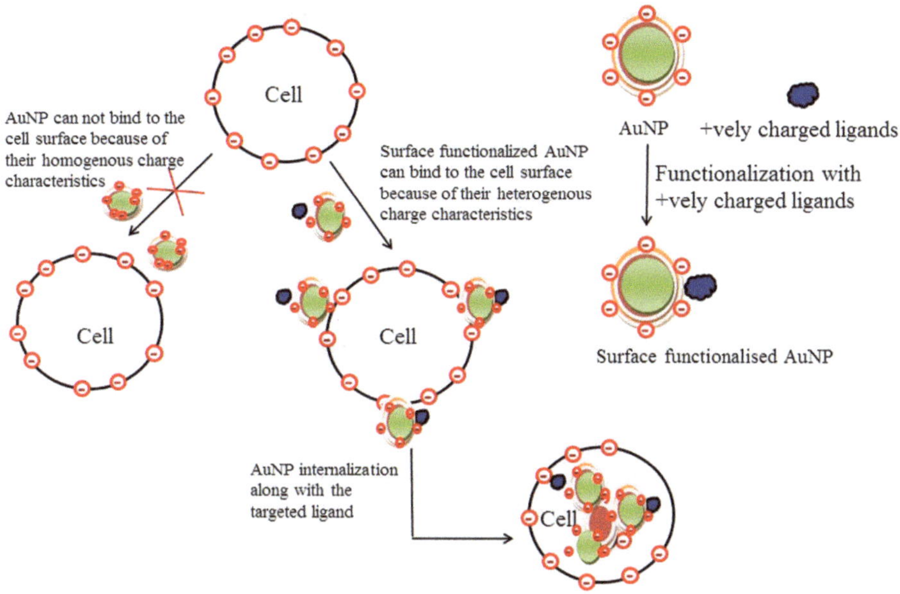

Fig. 17.4 Schematic illustration of surface charge-dependent nanoparticle internalization by cells. (Illustration is prepared by the authors based on the concept of earlier publications [7])

17.15 Problems and Prospects in Epileptic Medications

Since the discovery of phenobarbitone, different AEDs like phenytoin, ethosuximide, carbamazepine, sodium valproate and different derivatives of benzodiazepines have been used as antiepileptic drugs to control seizure and, as a whole, epilepsy. But the drug discovery approach for antiepileptic drug formulation have not succeeded in potential AEDs with superior efficacy development. Although a number of AEDs are available, seizures remain uncontrolled in more than 20% of the patients. It has shown some severe side effects. In some pyknoleptic patients, they develop some resistance to these drugs. Drug toxicity is another important issue towards AEDs research. It was reported that vigabatrin blocks the GABA transaminase activity and increasing GABA level, thus controlling the seizure; but it has severe retinotoxicity. Similarly, levetiracetam is another highly effective antiepileptic drug, but it has also some side effects like increased anxiety, mood change and severe headache [4]. Therefore, a drug with lower and/or no unwanted effects is a most urgent and needy product formulation research area in the AEDs research.

17.16 Application of Nanotechnology in Neuroscience

Nanotechnology has wide scope of applicability in different applied field. In biological science and medical science, nanoparticles can be used for diagnosis, therapy and drug delivery. Nanoparticles have unique properties having surface charge characteristics with different plasmon band and, hence, can be applied in different field of biological science. Surface charge characteristics of nanoparticle can be applied to control protein misfolding disorder in the brain of Alzheimer's patients.

17.17 Surface Charge of Different Nanoparticles (Fig. 17.5)

Surface charge of nanoparticles reflects the electric potential of a particle that is influenced by its composition and the medium in which it is dispersed. The surface charge of the nanomaterial is expressed as zeta potential. Zeta potential is also termed as electrokinetic potential in colloidal dispersion. Surface charge of nanomaterials also changes after loading a drug or other biological materials. Nanomaterials having a surface charge of ± 30 mv show stability in suspension, indicating that surface charge prevents aggregation. Stern potential is another term that indicates electric surface potential of a particle. Stern potential is the potential generated by the electric charges opposite the charges of the nanoparticles. Beyond the stern layer both positive and negative charged ions are present.

17.18 Application of Gold Nanoparticles to Control Neurological Disorder

Among different nanosized particles, gold nanoparticles have a wide scope of applicability in biomedical field due to its lesser toxicity. Nanoparticles are charged particles that can be applied to control or regulate neural function. Although, gold nanoparticles regulate different charges of the excitable cells, but at the intracellular level they also have adverse effect. Elevation of intracellular calcium is observed in a resting neuron in presence of gold nanoparticles.

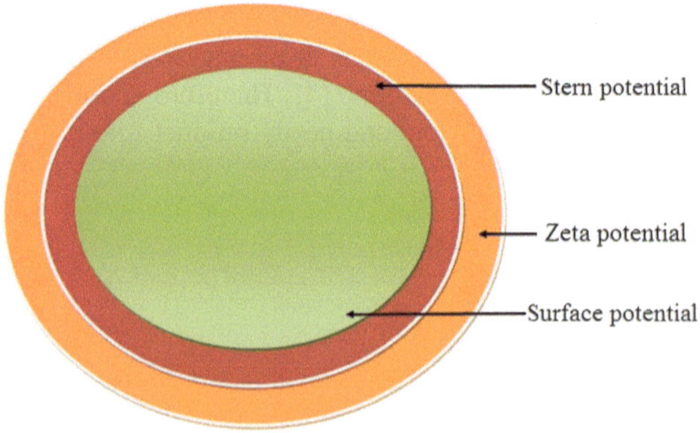

Fig. 17.5 Electric potential of a nanoparticle. (Source: Wikipedia, Retrieved on 21 March 2019)

Use of AuNPs in neuroscience research has the potentiality to find new strategies for diseases that are not currently treatable. Unique properties of gold nanoparticles like chemical and physical stability, relatively low toxicity and wide range of possible surface functionalized opportunities bring the possibility of use of gold nanoparticle in neuroscience research. For example, functionalization with specific legends allows cellular and molecular specificity, which enables the interaction with target cells and tissues in controlled ways. The functionalized AuNPs can bind to the voltage-dependant Na channels, TRPV1 (transient receptor potential vanilloid member 1) channels and P2X3 receptor ion channel in dorsal root ganglion neurons [8].

Large biological molecules can be compared with some nanoparticles. In this context, AuNPs have already been used in several biomedical applications, such as biosensing, bioimaging, drug delivery, therapy and tissue engineering. Most of the commonly used particle-reducing ligands for the synthesis of AuNPs are toxic to the cellular system [9].

17.19 Interaction of Gold Nanoparticles with Calcium Channel

Different charged particles can interact with protein molecules present in the plasma membrane and can alter the functional activity of the neurons. Nanosized negatively charged gold particles can block ion pores after binding to the sulphur group of the cystine loop of the of the calcium ion channel.

Recent study suggests that surface charge of the nanoparticles is mainly responsible for the translocation of nanoparticles into the cells. After exposure to AuNP, most of the particles are translocated to the interior of the cell. Surface charge can modify the membrane stability and lead calcium entry to the interior of the cells. There is a controversy about the elevation of intracellular calcium level in gold-nanoparticle-treated neuronal systems. Study on the biological effect of gold nanoparticles particularly on membrane transport indicates that anionic gold nanoparticle is translocated across the cell membrane. Surface charge of gold nanoparticles regulates not only the translocation through the biological membrane but also other cytotoxic effects such as the anionic, cationic and neutral surface charge of gold nanoparticles, which show different cytotoxic effects. Besides, the size of the nanoparticles is also responsible for toxic effect on different cells.

Gold nanoparticles also have impact on cells' self-protecting mechanism. Calcium homeostasis is an important cell-protecting mechanism, and AuNP-treated neuron calcium homeostasis mechanism is lost.

Intracellular calcium homeostasis depends upon the activity of different types of membrane proteins. Membrane proteins like voltage-gated channel protein present in different biological membranes regulate the entry of calcium from cytosol to mitochondria and endoplasmic reticulum. Studies on different types of voltage-gated calcium channels reported that L-type calcium ion channels are not gated by changes of membrane potential. These types of channels have the characteristics that an additional calcium ion inactivates further calcium entry. So, calcium ion influx elevates intracellular calcium ions and inhibits further calcium ion entry into the neuron.

17.20 Conclusion

Epilepsy and seizure are electrophysiological disorders of brain and its supporting cells, with abnormal depolarization or burst firing of neurons. Normal depolarization process is regulated by entry of sodium ions into the interior of the neurons and exit of potassium ions into the intercellular space. Besides, calcium ions also pay an important role. Studies on different channel proteins and their characteristics indicate that most of the channel proteins are charge dependent and potential differences between two sides of the neuronal membrane regulate the opening and closing of the channel proteins. Functional tuning of different channel proteins can alter the movement of ions and molecules through different channel proteins and can regulate the functions of neurons. For tuning of channel proteins, different macromolecules, micromolecules and nanomaterials can be used.

Previous studies relating to the correlation between intracellular changes of calcium ion and depolarization suggest that depolarization induces increase of calcium ions and similar effect is also observed when neurons are stimulated by positive-surface-charged AuNP. Scientists also observed that AuNP-induced calcium increase can be compared with the depolarization induced by 40 millimolar potassium chloride. Though positively charged AuNP can induce intracellular calcium level in neurons up to a certain level but these charged nanoparticles unable to increase intracellular calcium level when the action of L-type calcium channels are inhibited by different inhibiting agents. Studies on the impact of positively charged gold nanoparticles on intracellular calcium level in absence of extracellular calcium report that, in the absence of extracellular calcium, intracellular calcium wave becomes blunt, suggesting other mechanisms involved in intracellular calcium release. Though L-type calcium channel regulates intracellular calcium level from extracellular media, some other calcium-elevating mechanism may be involved to release calcium ion from intracellular storage. Most of the earlier works suggest the possibility of calcium release from mitochondrial source in AuNP-treated neurons, but some other scientists suggest the release of calcium from ER in AuNP-treated neurons. AuNP-induced increase in calcium was small, confirming a role for calcium release from ER stores. AuNPs elevate calcium by stimulating plasma membrane calcium influx and endoplasmic reticulum calcium release. The transient nature of the cytosolic calcium ions signal is probably due to voltage- and cytosolic-calcium-dependent inactivation of calcium ion channels upon sustained cell depolarization. The calcium-buffering system and the membrane-intrinsic calcium transport proteins are capable of removing the calcium from the cytosol into the extracellular milieu. Gold nanoparticles also can be able to alter the function of acetylcholine-activated potassium channel in different excitable cells including cardiac muscle cells.

References

1. McCleskey EW, Fox AP, Feldman D, Tsien RW (1986) Different types of calcium channels. J Exp Biol 124:177–190

2. Zamponi GW, Striessnig J, Koschak A, Dolphin AC (2015) The physiology, pathology, and pharmacology of voltage-gated Calcium channels and their future therapeutic potential. Pharmacol Rev 67:821–870
3. Zhu MX, Evans AM, Ma J, Parrington J, Galione A (2010) Two-pore channels for integrative Ca signaling. Commun Integr Biol 3(1):12–17
4. Izumi Y, Ishikawa M, Benz AM, Izumi M, Zorumski CF, Thio LL (2004) Acute vigabatrin retino-toxicity in albino rats depends on light but not GABA. Epilepsia 45(9):1043–1048
5. Eisenberg MJ, Brox A, Alain N, Bestawros MD (2004) Calcium channel blockers: an update. Am J Med 116(1):35–43
6. Barthakur M (2018) Impact of gold nanoparticles on electrophysiology and intracellular calcium in cultured neuron
7. Mout R, Moyano DF, Rana S, Rotello VM (2012) Surface functionalization of nanoparticles for nanomedicine. Chem Soc Rev 41(7):2539–2544. https://doi.org/10.1039/c2cs15294k
8. Paviolo C, Stoddart PR (2017) Gold nanoparticles for modulating neuronal behavior. Nanomaterials (Basel) 7(4):92. https://doi.org/10.3390/nano7040092
9. Lovisolo D, Gilardino A, Ruffinatti FA (2014) When neurons encounter nanoobjects: spotlight on calcium signalling. Int J Environ Res Public Health 11(9):9621–9637. https://doi.org/10.3390/ijerph110909621

Biomaterials and Its Medical Applications 18

Saili Dharadhar and Anuradha Majumdar

Abstract

The human body undergoes wear and tear with age, and it is subjected to various diseases and disorders throughout the lifespan. Sometimes, administration of medical agents does not suffice for complete recovery from the ailments. External agents or materials are required to support the normal functioning of the body. The materials that are engineered to interact with biological systems for the purpose of diagnosis or treatment of diseases and ailments are known as biomaterials. Any resource to be engineered and used as a biomaterial must possess the qualities of biocompatibility, inertness, mechanical stability, and ease of fabrication. The site and application of the biomaterial may demand specific properties. Biomaterials for dental and orthopedic applications should possess substantial mechanical strength and prolonged rates of biodegradation, while those for visceral organ and dermal applications must be flexible with faster rates of degradation. Biomaterials may be of natural or synthetic origin. The chapter discusses various applications of biomaterials in the medical and pharmaceutical industry. It highlights the types of biomaterials and discusses their properties specific to each application.

Keywords

Biomaterials · Medical applications · Biocompatibility · Implants

S. Dharadhar
Novartis Healthcare Pvt. Ltd., Hyderabad, India

A. Majumdar (✉)
Department of Pharmacology and Toxicology, Bombay College of Pharmacy, Mumbai, India
e-mail: anuradha.majumdar@bcp.edu.in

S. Paul (ed.), *Application of Biomedical Engineering in Neuroscience*,
https://doi.org/10.1007/978-981-13-7142-4_18

Abbreviations

3D	three-dimensional
Al_2O_3	aluminum oxide
ALCL	anaplastic large cell lymphoma
BMS	bare-metal stents
CaP	calcium phosphate
CB	cardiovascular biomaterials
DES	drug-eluting stent
ECM	extracellular matrix
e-PTFE	elongated-PTFE
IL	interleukin
IOLs	intraocular lenses
HA	hydroxyapatite
NSS	nitrogenated stainless steel
PE	polyethylene
PEG	polyethylene glycol
PET	polyethylene terephthalate
PGA	polyglycolic acid
PLA	polylactic acid
PLGA	poly (DL-lactide-co-glycolide)
PLGA-Fn	PLGA immersed in fibronectin
PLLA	poly-L-lactic acids
PLLA-Fn	PLLA immersed in fibronectin
P (LL-co-CL)	poly-D,L-lactide-co-epsilon-caprolactone
PMMA	polymethyl methacrylate
PP	polypropylene
PTFE	polytetrafluoroethylene
PVDF	polyvinylidene fluoride
T/L	tendons and ligaments
TJR	total joint replacement
TNF	tumor necrosis factor
ZrO_2	zirconia

18.1 Introduction to Biomaterials

Biomaterials are natural or synthetic materials engineered to interact with biological systems for the purpose of diagnosis or treatment of diseases and ailments. The most common approach of categorizing biomaterials is based on the type of the materials used. They can be classified into metals, ceramics, polymers, and composite substances. They can also be categorized into natural or synthetic, depending on their source.

Table 18.1 Overview of biomaterials

Type	Advantages	Disadvantages	Applications
Metals and its alloys (Gold, titanium, platinum, steel, cobalt chromium)	Good mechanical strength	Corrosive	Orthopedic implants, pins, screws, and plates
	Easy of fabrication and sterilization	Aseptic loosening	
		Excessive elastic modulus	
Ceramics and carbon compounds (Calcium phosphate [CaP] salts, aluminum and titanium oxides)	Good mechanical strength	Challenging fabrication	Cochlear implants
	Biocompatibility	Excessive elastic modulus	Dental implants
	Corrosion resistance		Orthopedic implants
Polymers (Polymethyl methacrylate [PMMA], polycaprolactone, polylactic acid [PLA], polyurethanes)	Biodegradable	Leachability in body fluids	Orthopedic and dental implants
	Biocompatible	Difficulty in sterilization	Tissue engineering scaffolds
	Ease of fabrication		Prostheses
	Suitable mechanical strength		Drug delivery systems
	Ease of availability		
Composites (Dental filling composites, carbon fiber-reinforced methyl methacrylate + ultrahigh molecular weight polyethylene [PE])	Excellent mechanical properties	High cost	Porous orthopedic implants
	Corrosive resistant	Manufacturing difficulties	Rubber catheters and gloves
			Dental fillings

For any biomaterial to be clinically successful, it must possess the properties of biocompatibility, inertness, mechanical stability, and ease of fabrication [1]. Biocompatibility, a dynamic term, refers to the performance of any material in a specific environment with apt host response. The material is considered biocompatible if it does not elicit clinically significant immune response in the human body after implantation. Table 18.1 provides the list of biomaterials with their advantages and disadvantages [1].

The most common technique used for fabricating biomaterials into implant designs is the three-dimensional (3D) printing technology. It involves a technique of producing 3D entities with the help of a digital file, wherein computer-aided design software is used for designing the objects [1].

18.2 Applications

Biomaterials find applications in various fields including the medicine, pharmaceutical industry, food industry, fashion industry, and household appliances. In the medical field, biomaterials are used for fabricating dental fixtures, implants, prosthesis, and tissue scaffolds. Pharma industry utilizes biomaterials for manufacturing customized implants for drug delivery. They are also used in manufacturing dosage forms like tablets and capsules [1].

The selection of biomaterial depends on its application and site of utilization. Biomaterials used for dental and orthopedic applications should possess substantial mechanical strength and prolonged rates of biodegradation, while those for visceral organ and dermal applications must be flexible with higher rates of degradation. Below are the various applications of biomaterials and their anticipated properties [1].

18.2.1 Breast Implants

Breast implants are intended to modify the appearance of women's breast in terms of the size and shape. Breast implants may be used for reconstructive plastic surgery or breast augmentation surgery. Reconstructive plastic surgery aims to reinstate the natural breast mound or resolve chest wall deformities and inherent defects, whereas breast augmentation surgery is used for cosmetic enhancement of the appearance of the breast.

Three types of breast implant devices are utilized, namely, silicone gel, saline solution, and composite fillers. In the case of the saline implants, saline solution is filled into a shell made of silicon during the surgery as opposed to silicone implants, wherein a gel with considerable viscosity is prefilled into the shell. The latter type includes fillers such as polypropylene (PP) string and soy oil. Europe and the United States have imposed a ban on this type of implants due to the health risks associated with its fillers.

In 1992, a moratorium was issued by the FDA on gel implants. This instigated higher investigation on the safety of silicone gel implants. Assessments made by scientific panels concluded that there was no connection between silicone gel implants and systemic diseases. Upon return of gel implants to the market, the FDA limited gel implants for use only in women older than 22 years old, unless used in the reconstruction [2].

When silicone was injected into women's breasts, it frequently migrated from the site leading to many complications. In 1963, Dow Corning Corporation came up with breast implant composed of a shell to contain the silicone gel. This led to the availability of the saline implants. However, they had a higher tendency to ripple and had higher failure rates (estimated more than 75%) [2].

In the mid-2000s, the FDA introduced cohesive gel implants. Recently, the structured (bi-lumen) ideal implant has been introduced and is available for use [2].

Complications associated with silicone implants limit their use. Around 50% of patients experience capsular contracture, the formation of a fibrotic capsule around the implant along with concomitant foreign body reaction post-breast augmentation and/or reconstruction. It may cause implant distortion and displacement [3].

The development of capsular contracture is significantly associated with the implant surface texture. Since cells react to the structures that they border, the topography of the surrounding surfaces influences cell proliferation, polarization, migration, alignment, adhesion, morphology, and attachment at the nano and micro levels. Studies have reported that the surface texture can result in the formation of multiplanar conformations of collagen fibers. These form more flexible capsules and demonstrate lower probability to contract. Breast implants with smooth surface had significantly higher association with capsular contracture compared to those with textured surfaces [3].

Textured implants can aid in reducing capsular contracture. Implants with pore diameters of 250–350 μm and 600–800 μm with depth 40–100 μm led to normal breast firmness compared to smaller pore diameters of 70–150 μm with depth of 10–760 μm and 150–200 μm. Macrotextured implants resulted in significantly reduced risk of fibrous capsular contraction. However, textured implants are associated with other complications such as development of late seromas and double capsules in contrast to smooth implants. Additionally, textured surface capsules have a significantly higher risk of developing implant-associated anaplastic large cell lymphoma (ALCL) [3].

The modern nano/micro-engineering methods have allowed the fabrication and analysis of natural topography of tissues. The fabrication and functional characterization of novel biocompatible polydimethylsiloxane surfaces with numerous nano/micro topographies are liable to foreign body reactions [3]. Ultrasound, mammogram, and MRI may be necessary for evaluating rupture concerns as well as implant displacement mentioned prior on physical exam for gel implants [2].

In March of 2017, the FDA confirmed a relationship between implants and the development of ALCL. Information released by the American Society of Plastic Surgeons shortly after the FDA announcement stated that "This is a T-cell lymphoma that is not a breast cancer."

Studies suggest that textured implants are the only common thread with this phenomenon that is estimated to occur 1:30,000 worldwide but smooth-surfaced implants cannot be excluded from this. ALCL has been seen with both saline and gel implants [2].

18.2.2 Cardiovascular System

Cardiovascular Biomaterials

Some cardiovascular diseases significantly affect the quality of life and cannot be treated by drugs alone. Supportive implants such as heart valves and vascular grafts need to be implanted in the human body to ameliorate the condition. The materials used as implants in the cardiovascular system that interact with biological components are known as cardiovascular biomaterials (CB). Based on the current demand and investment, CB are dominant in the field of biomaterials. CB are of three types, metals, polymers, and biological materials. Application of CB depends on the properties of the materials used. For example, although polymers have materialized as a versatile choice for numerous applications in the cardiovascular field, blood compatibility remains the major drawback. New methods such as surface modifications are being investigated to overcome this issue. Comparative properties of biomaterials for cardiovascular use are given in Table 18.2 [4].

18.2.2.1 Blood Vessel Prostheses/Vascular Grafts

Blood vessel prosthesis is used for the repair of injured or diseased blood vessels, whereas a vascular bypass, also known as a vascular graft, is a clinical procedure, involving surgery, to reconnect blood vessels by redirecting the flow of the blood from one area to another. It is usually performed for bypassing a diseased blood vessel and maintaining the blood flow between two healthy areas. It may be performed in ischemic conditions caused by atherosclerosis, as an aid in organ transplantation, or during hemodialysis. Although the most preferred material for vascular bypass is one's own vein (autograft) or donor vein (allograft), other materials include polyethylene terephthalate (Dacron) and polytetrafluoroethylene, viz., PTFE (Teflon).

Synthetic biomaterials are the first choice for vascular graft design owing to the ease and flexibility of modifying their mechanical properties. Commonly employed nonbiodegradable biomaterials used as vascular grafts include expanded Teflon, Dacron, and polyurethane. One of the factors to be considered during bypass is the rate of thrombosis and blockage of the blood vessel. Patency rate is the likelihood that a blood vessel will remain open. Polyurethane has higher capacity to mimic native vasculature. However, the grafts composed of this material have relatively poor patency rates. Application of chemical modifications and coatings to functionalize the surface of the polymer is one of the techniques for improving the thromboresistance and endothelialization of the polymers. PTFE is successfully used as vascular grafts, when the devices are implanted in high-flow, large-diameter arteries such as the aorta. However, it faced issues when implanted below aortic bifurcations. Another form of PTFE called elongated-PTFE (e-PTFE) was explored. This form of PTFE was indicated for use in smaller arteries that have lower flow rates leading to low thrombogenicity, lower rates of hemostasis and restenosis, less calcification, and biochemically inert properties [4].

Cells can modulate vascular remodelling on scaffolds of biodegradable polymers. Polyglycolic acid (PGA), polycaprolactone, polyhydroxyalkanoates, and polyethylene glycol (PEG) are some of the polymers used to fabricate cell-seeded scaffolds. Proteolytic functionalization of these scaffolds can be used to achieve controlled release of therapeutic agents from the polymer matrix. Although degradable biomaterials aid to enhance the production of extracellular matrix (ECM) and increase infiltration of vascular cells into the graft site, they face significant issues of compliance mismatch, challenges of cell sourcing, and prolonged duration of cell culture [5].

Table 18.2 Comparative properties of biomaterials for cardiovascular use

Properties	Stiffness	Strength	Corrosion resistance	Blood compatibility	Hardness	Rigidity	Blood flow dynamics
Metals							
Stainless steel	Best	Better	Good	Good	NA	NA	NA
CoCr alloys	Better	Good	Better	Better	NA	NA	NA
Titanium alloys	Good	Best	Best	Best	NA	NA	NA
Polymers							
Polyamides	NA	Medium	NA	Good	Medium	Medium	NA
Polyolefin	NA	Good	NA	Better	High	High	NA
Polyester	NA	Good	NA	Moderate	High	High	NA
Polytetrafluoroethylene (PTFE)	NA	High	NA	Low	High	High	NA
Polyurethanes	NA	Better	NA	Good	Medium	Medium	NA
Biological materials							
Allograft	NA	Moderate	Moderate	Best	NA	NA	Better
Xenograft	NA	Moderate	Moderate	Better	NA	NA	Better

Protein polymers such as collagen and fibrin fibers and gels are another type of polymers that possess the ability to simulate native proteins and can embrace the properties of the arterial walls. This provides a new supportive system for the development of vascular grafts. These biopolymers significantly aid in the development of artificial blood vessels by binding to proteins that direct cell fate. Mechanical strength and integrity of the graft are provided by the cross-linked collagen fibers, whereas elastic fibrin simulates the role of elastin in the human vasculature. Modulation of extracellular matrix (ECM) production and maintaining mechanical integrity are possible by techniques such as preconditioning treatments, smooth muscle cell seeding, and culture techniques.

Recombinant protein polymers such as elastin-mimetic protein polymers are obtained in significant quantities by exploiting the biosynthetic machinery of microorganisms. The polymers are prepared from sequences of primary amino acids and can self-assemble into a 3D folded structure. Decellularized allogeneic and xenogeneic tubular tissues are studied as vascular conduits including the umbilical vein and contain structurally organized and intact ECM. Although such grafts have been successful to some extent, their progress is hurdled by the incompetence to modify matrix content and architecture, progressive biodegradation, and the associated risk of viral transmission from animal tissues [5].

18.2.2.2 Heart Valves

Artificial heart valve is a machine implanted in the heart of patients with valvular heart disease, when one of the four heart valves malfunctions.

Artificial heart valves are classified into mechanical and biological. Mechanical heart valves are manufactured from several materials such as metals, polymers, and ceramics. Some of the examples include stainless steel, pyrolytic carbon, titanium, and silicone. Biological valves are composed of synthetic constituents (e.g., PTFE, Dacron) and elements of biological origin. The transcatheter aortic valve implantation is fabricated from metals, ceramics, and polymers (titanium, stainless steel, Dacron) and biological material (pig heart valve) [6].

Stainless steel is widely used in heart valves, especially in making struts for supporting leaflets to avoid corrosion and provide mechanical strength to the valves. A new alloy prepared from cobalt-chromium-tungsten-nickel, also known as L-605, is used for making heart valves. Among polymers, PP, PTFE, and segmented polyurethanes are utilized for making heart valve structures. PTFE is particularly used in implantable prosthetic heart valve rings [4].

18.2.2.3 Stents

A stent is a tube inserted into the lumen of biological blood vessels aimed to keep the passage open. It may be metallic or plastic in nature. Stents are of different types including coronary stents, vascular stents, urethral stents, prostatic stents, esophageal stents, and biliary stents [4].

Based on functional and physical characteristics, stents can be classified into three types, namely, bare-metal stents (BMS), drug-eluting stent (DES), and bioabsorbable stents [4].

Although alloys of cobalt have been used for medical purposes since 1937, they are recently being investigated for the production of stents. They are preferred for fabricating coronary stents because the thinner dimensions required for coronary interventionists can be easily attained using these alloys. Additionally, properties of being nonferromagnetic in nature and possessing higher density than stainless steel have made cobalt-based alloys a feasible option for the production of coronary stents [4]. Besides cobalt, titanium alloys have a wide application in stent manufacturing. The nickel-titanium alloy, nitinol, which has the shape memory effect is an interesting feature and is widely employed for the production of self-expanding memory stents. The BMS possess good mechanical characteristics. However, they could not be successful due to the major drawback of developing stent thrombosis, which necessitates sustained antiplatelet treatment and the stent to vessel size discrepancies. Additionally, metallic stents negatively affect the vessel geometry and hinder side branches [4].

To overcome these hurdles, DES were developed [4]. The DES are further classified into polymer-free stents and polymer (carrier) containing metallic stents for controlled release of the drug. DES comprise of three parts: stent platform, coating, and the drug. Amazon Pax (MINVASYS), BioFreedom (Biosensors Inc.), and Optima (CID S.r.I.) are some examples of polymer-free DES. Another version of DES contains bioabsorbable polymers as a carrier matrix for drugs. Although DES were considered as a headway in the development of stents, they are associated with subacute and late thrombosis. Additionally, the polymer engaged as the vehicle for drug delivery may result in irritation, vessel hypersensitivity, endothelial dysfunction, and chronic inflammation at the stent site [4].

Bioabsorbable stents were then developed to overcome the problems faced by the former types of stents. Bioabsorbable stents stay at the site for a limited duration and enhance healing and repair of the blood vessel. They are aimed to aid arterial remodelling which may require 6–12 months. A wide range of biodegradable polymer blends has been investigated with a reasonable degradation life of 12–24 months to avoid the need of prolonged medication and also reduce the risk of late stent thrombosis. The materials for biodegradable stents and their degradation products should be biocompatible. In addition, the materials should not move from the site before complete bioabsorption, and the radial force of the stent must be sufficient for scaffolding effect throughout the arterial remodelling period. Due to these requirements, magnesium and iron have been investigated for this application [4].

Among polymers, polyesters are preferred for the production of bioabsorbable stents. Poly (D, L-lactide/glycolide) copolymers, poly-L-lactic acids (PLLA), and PGA are some of the commonly used bioabsorbable polymers. The Igaki-Tamai stent manufactured from PLLA was the first absorbable stent implanted in humans and showed low complication rates for stent thrombosis. The BVS everolimus-eluting stent is another type of bioabsorbable stent. It is coated with poly-D, L-lactide, has a metallic base, and is used to carry the antiproliferative drug everolimus. The clinical records expressed lack of stent thrombosis and ensure total vascular function restoration [4].

Table 18.3 Biomaterials for cochlear implants

Biomaterials		Applications
Metals	Titanium	Case; encapsulation
	Gold	Coil; encapsulation
	Platinum	Electrode
	Iridium	Electrode
	Zirconium	Case
Nonmetals	Ceramic	Case; feedthroughs
	Silicone rubber	Carrier; encapsulation
	Glass	Feedthroughs
	Parylene	Insulation coating
	Teflon	Insulation coating

18.2.3 Cochlear Replacements

A cochlear implant is a neuroprosthetic device that converts sound into electrical signals, which stimulates the auditory nerve. It is surgically implanted in people with sensorineural loss of hearing. It is composed of two parts. The sound processor, which is the outer component, worn behind the ear, transmits signals across the skin to the implant. It consists of microphones, electronics including DSP chips, battery, and a coil. The inner component receives signals and has an array of electrodes placed into the cochlea that stimulate the cochlear nerve. The safety of a cochlear implant is attributed to the biocompatibility of the material used, sterility, its potential to induce structural damage to tissues, and ability to cause neural damage via exposure to energy [7].

Table 18.3 provides a list of biomaterials frequently used in cochlear implants. Although some biomaterials are generally considered to be biocompatible, the specific components equipped in cochlear implants require special attention to the materials used.

Titanium, when used as electrodes, can induce more neural damage than platinum-iridium electrodes at the same stimulation exposure levels [8]. Frequently used electrodes in cochlear implants consist of platinum-iridium alloys, possessing 90:10 ratio of platinum to iridium. This composition provides good electrical conductivity, mechanical strength, and resistance to failure by fatigue. The properties of the alloy can be modulated by varying the ratio [7]. Another important requirement of the implant is sterility. However, for successful sterilization, the material should possess the ability to withstand exposure to harsh chemicals and high temperatures during the process. Ethylene oxide is the commonly used sterilizing agent for cochlear implants. In addition, for sterilization process to be effective, it is necessary that the external surface of the implant should be free of pockets, crevices, and small spaces, which may harbor bacterial growth [7, 9]. In addition to being biocompatible and sterile, the implant should not induce any structural or neural damage postimplantation. Hence, the design of the implant must ensure its stability on implantation [7]. An optimally shaped top, rounded corners and soft silicone rubber encapsulation can aid in reducing internal tissue trauma. Severe tissue necrosis may result in a need for reimplanting the prosthetic [7]. Another factor that must be considered for cochlear implants is the exposure to safe levels of energy. Cochlear nerve stimulation necessitates exposure to sufficient amount of electrical energy, and parameter used to quantify this energy is charge density. The limit of safe charge density is less than 15–65 $\mu C/cm^2$/phase. However, higher values are also considered safe for nerve tissue stimulation [10].

18.2.4 Dental Implants for Tooth Fixation

A dental implant (also called as endosseous implant/fixture) plays the role of an interface between the jawbone and skull and supports a dental prosthesis such as a bridge, crown, denture, and facial prosthesis or functions as an orthodontic anchor. Osseointegration is a process wherein an intimate bond is formed between the metal and the bone. It is the basis for modern dental implants. First, the implant fixture is allowed to osseointegrate, followed by addition of the dental prosthetic. Osseointegration requires a variable amount of healing time before attachment of the dental prosthetic to the implant or positioning of an abutment to hold the prosthetic [11].

Biological response to any biomaterial depends on its biocompatibility, which is governed by the basic bulk and surface properties of the material. In an implant procedure, strength of the implant material and the type of bone in which the implant is to be placed are of prime importance. The other factors to be considered include the implant design, abutment availability, abutment choices, surface finish, and biomechanical consideration of restorative treatments [11].

Commonly used biomaterials for this purpose include metals and alloys (titanium and titanium-6 aluminum-4 vanadium [Ti-6AI-4V], cobalt-chromium-molybdenum-based alloy, iron-chromium-nickel-based alloys), ceramics (aluminum, titanium and zirconium oxide, bioactive and biodegradable ceramics), and polymers and composites (PMMA, PE, PTFE, silicone rubber, polysulfone) [11].

Due to the unique combination of chemical, physical, and biological properties of titanium and its alloys, they are widely utilized in dentistry as prosthetic appliances. Frequently used titanium are of two types, the commercially pure titanium (Ti-160) and the alloy of Ti-6%Al-4% Va (Ti-318). The favorable properties of titanium include its low density of 4.5g/cm and relatively high flexure strength comparable to that of the cast forms of stainless steel and cobalt alloys. The modulus of elasticity of Ti-6Al-4V is closer to that of the bone compared to any other metallic biomaterial. Although titanium is considered to be biocompatible, there are reports of generalized health problems attributed to titanium. Additionally, titanium dioxide that covers the surface of most of the titanium-based biomaterials has the potential to induce adverse tissue responses in humans. Overall, chronic exposure to low-level metal may induce metal sensitization and contribute to undesirable effects. Some clinical studies also report the presence of allergic reactions attributed to titanium dental implants in rare circumstances [11].

Recently the focus is shifted on the application of tooth-colored implant materials that not only add to the aesthetic appeal of dental implants but are also highly biocompatible and have the ability to withstand the forces present in the oral cavity. Ceramic implants, which then came into picture, have low capacity to withstand tensile or shear stress induced by occlusal loads. However, they can tolerate high amount of compressive stress. Ceramics are plasma sprayed or coated onto the metallic surfaces, which are more thermodynamically stable, hydrophilic, and non-conductive of heat and electricity and thereby produce high-strength integration with bone. The commonly used standard material for ceramic dental implants is aluminum oxide (Al_2O_3), attributed to its inertness and no evidence of adverse reactions. However, dental implants of this material are withdrawn from the market due to its poor survival rate. Alumina has higher surface wettability compared to metallic implants. Zirconia (ZrO_2) is another ceramic used for dental implants and possesses high degree of inertness. It is a suitable material for utilization in high load situations due to its toughening mechanism and is thus useful for crack deflection, zone shielding, contact shielding, and crack bridging. These ceramic implants are not bioactive, i.e., they do not promote the formation of bone [11].

The bone is composed of 60–70% CaP. Hence, CaPs find wide applications in grafting and bone augmentation. CaPs are nonimmunogenic and biocompatible with host tissues. The two most commonly used CaPs are hydroxyapatite (HA) and tricalcium phosphate. They promote and achieve direct bond of implant to hard tissues and are hence used as bone graft material to serve as template for new bone formation. They also promote vertically directed bone growth as well as relatively strong bond to bone. Studies have reported accumulation of osteocytes adjacent to HA granules after 4 weeks of implantation, which indicates the possibility of osteogenesis. Utilization of these ceramics depends on their physical and chemical properties, when used as coatings. Higher crystallinity of HA coating indicates better resistance to clinical dissolution [11].

Another bioactive biomaterial that has the ability to stimulate bone formation is bioglass (SiO_2-CaO-Na_2O-P_2O_5-MgO). It can form a carbonated HA layer in vivo due to their calcium and phosphorus content. However, its brittle nature renders it unsuitable for use when the implant may face the possibility of being subjected to stress [11].

The titanium foam, a new material in dental implants, can make dental implants less invasive. It is prepared by blending titanium powder and a polymer along with foaming agents that expand the polymer on heating. The polymer is then removed through a high-temperature heat treatment, and the metallic particles are consolidated so that the porous structure obtains mechanical strength. The friction generated between the implant and the bone by the rough surface enables bone growth into the pores and helps to fix the implant in place.

An important consideration during implant material selection is its affinity toward bacteria and plaque. Some bacteria including *Porphyromonas gingivalis*, *Streptococcus sanguis*, short rods, and cocci have lower adherence to zirconia compared to titanium surface. The adhesion of streptococcus to glass ceramics and zirconia is considerably similar [11].

Polymers possess inferior strengths and lower elastic moduli and greater elongations to fracture compared to other categories of biomaterials. They are electrical and thermal insulators and are relatively resistant to biodegradation. Polymers can be used in solid and porous forms for tissue attachment, replacement, and augmentation and as coatings for force transfer to tissue regions. When intended to be used as implants, steam or ethylene oxide cannot be used for sterilization. Moreover, polymeric biomaterials possess electrostatic surfaces and attract dust and other particulate matters if exposed to semi-clean oral environments.

Vitreous carbon does not induce significant response from the host tissues and is one of the most frequently employed biomaterials for implant. Under physiological conditions, carbon is inert compared to metallic biomaterials. Since its modulus of elasticity is comparable to dentine and bone, its rate of deformation is similar to that of tissues and thus aids biomechanical force transmission. The challenge faced with carbon is its susceptibility to fracture under tensile stress. Another variety of carbon is pyrolytic carbon or low-temperature isotropic carbon that possesses the ability to absorb energy on impact and is almost four times greater than that of glassy or vitreous carbon. It is manufactured by the pyrolysis of a gaseous hydrocarbon and depositing carbon onto a preformed substrate such as polycrystalline graphite [11].

18.2.5 Drug Delivery Systems

For any formulation to be efficacious, the active component has to be bioavailable at the target site in the human body in optimum quantities. In the pharmaceutical industry, drug delivery systems depend on the active component, the delivery device or dosage form, and the release of the active ingredient at the required site of action. The utilization of conventional dosage forms including tablets and capsules is limited by the need of higher doses and associated higher toxicity profiles. To overcome these barriers, novel drug delivery systems are developed. Novel drug delivery systems comprise of the active ingredient and excipients with properties that modulate the release of active ingredients and provide optimum benefit at the target site with minimal side effects. Some of these systems include liposomes, microspheres, nanoformulations, and osmotic drug delivery systems. They can be used for oral administration or parenteral use. One of the systems specific to transdermal applications is the mucoadhesive drug delivery system [12].

In addition to these novel drug delivery systems, the conventional dosage forms are modified to overcome the drawbacks and achieve optimum release of active ingredients. These techniques include sugar coatings, film coatings (enteric coatings or sustained release coatings), and compression coatings [12].

Table 18.4 provides a list of novel drug delivery systems and the biomaterials used for their fabrication, and Fig. 18.1 depicts the different dosage forms.

Table 18.4 Novel drug delivery systems

Drug delivery system	Characteristics	Biomaterials
Liposomes	Drug encapsulated in a phospholipid bilayer to enhance bioavailability	Phospholipids: phosphatidylcholine, phosphatidylethanolamine
		Steroids: cholesterol
Nanoformulations	Nano-sized formulations	Nanoparticles; titanium oxide, zinc oxide, carbon, magnetite, polystyrene, iron oxide
	Types:	Nanotubes and nanowires: carbon black, titanium dioxide, silicon, boron nitride
	Nanoparticles	Quantum dots: cadmium sulfide, cadmium selenide, cadmium telluride
	Nanotubes	
	Nanowires	
	Quantum dots	
	Solid lipid nanoparticles	
Microspheres	Free-flowing powders (proteins/ synthetic polymers); biodegradable; particle size <200 μm	Synthetic polymers: PMMA, glyceryl methacrylate, epoxy polymers, glycolides, lactides and copolymers, poly alkyl cyanoacrylates, poly anhydrides
		Natural polymers: albumin, collagen, chitosan, gelatin, agarose, carrageenan, starch
Osmotically controlled drug delivery systems	Osmotic pressure is the driving force for controlled drug release	Osmotic agents: sorbitol, sodium chloride, potassium chloride, xylitol, fructose, citric acid, dextrose
	Formulation (tablet) having an outer semipermeable membrane and ≥ one small laser-drilled hole. Osmosis causes water absorption through the semipermeable membrane; osmotic pressure pushes the active drug through the opening in the formulation to release the drug in controlled manner	Semipermeable membrane: polymers: cellulose acetates, cellulose propionate, cellulose acetate butyrate, ethyl cellulose
		Wicking agents: colloidal silicon dioxide, kaolin, titanium dioxide, polyvinylpyrrolidone, sodium lauryl sulfate
Mucoadhesive systems	Provide mucoadhesion and significantly increased residence time of sustained release systems in the mucosa of ocular, nasal, vaginal, and buccal pathways	Mucoadhesive agents: polymers containing hydrogen bonding groups: polyacrylic acid, polyvinyl alcohol, sodium carboxymethyl cellulose, sodium alginate, hydroxyethyl cellulose, hydroxypropyl methylcellulose, hydroxypropyl cellulose, agarose, gelatin, chitosan

(continued)

Table 18.4 (continued)

Drug delivery system	Characteristics	Biomaterials
	Dosage forms: tablets, patches, gels, and solutions	Penetration enhancers: PEG, benzalkonium chloride, dextran sulfate, menthol, sodium EDTA, polysorbate 80
Tablet/capsule coatings	Sugar coating: taste masking	Sugar coating: sugar syrup
	Film coating: aesthetic appeal and taste masking	Film coating: polymer solution
	Enteric coating: prevent degradation of drugs in gastric pH of the stomach and release of active agents in the intestine	--Enteric coatings: methyl acrylate-methacrylic acid copolymers, cellulose acetate succinate, cellulose acetate phthalate, hydroxypropyl methylcellulose phthalate, polyvinyl acetate phthalate, hydroxypropyl methylcellulose acetate succinate, methyl methacrylate-methacrylic acid copolymers, shellac, zein, sodium alginate
	Sustained release coatings: Achieve drug release at predetermined rate and thus maintain constant drug concentration for specific time period	
	Compression coating: the entire surface of the core is surrounded by a coat to achieve controlled release of the core materials	

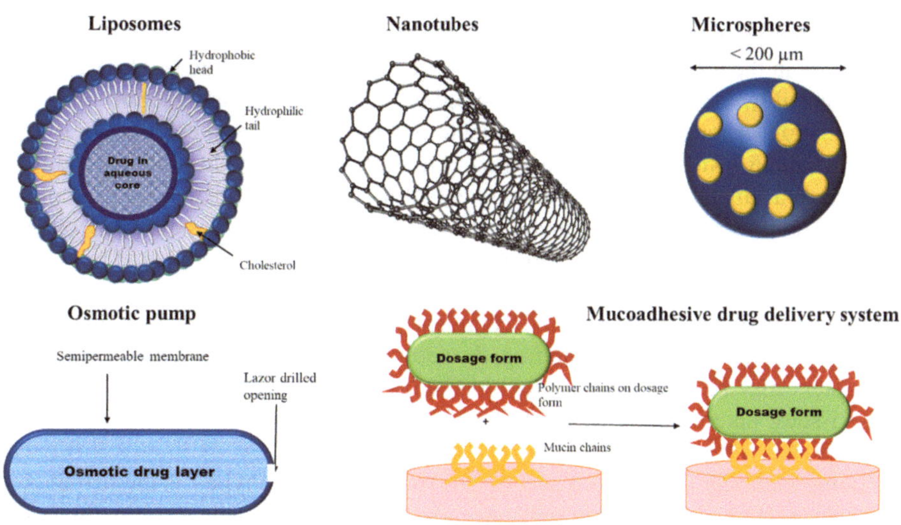

Fig. 18.1 Novel drug delivery systems

18.2.6 Musculoskeletal System

18.2.6.1 Artificial Ligaments and Tendons

Tendons and ligaments (T/L) are sheets of connective tissue that provide support and stability to the musculoskeletal system. Injuries to T/L due to sports or other physical activities result in the need for implantation of an artificial T/L.

For utilization in the musculoskeletal system, the biomaterial used must have structural and functional characteristics similar to that of the native T/L. Currently, combinatorial approaches are on a surge. They include techniques such as merging of two materials or functionalization of the material surface [13]. The desirable properties of any biomaterial utilized for T/L include enough mechanical strength to bear load immediate after implantation and a degradation rate similar to that of developing cellular and tissue ingrowth. The biomaterial is expected to degrade and be substituted by the regenerating T/L matrix. This property has highlighted the use of polymers for T/L tissue engineering [13].

Owing to the property of hydrolytic degradation, polyhydroxyesters attain higher interest for use in T/L implants. The degradation rates of these materials are modified to yield optimum properties. High hydrophilicity is detrimental to cell adhesion. The two commonly used polymers include poly (L-lactic acid) immersed in fibronectin (PLLA-Fn) and poly (DL-lactide-co-glycolide) immersed in fibronectin (PLGA-Fn). The preference for PLLA-Fn is higher due to their slower rate of degradation.

Although single materials are used frequently, combination approach may aid to mimic the multi-protein matrix of T/L. This can be achieved by developing a single scaffold inclusive of fibers with varied diameters and mechanical properties. Synthetic polymers possess the advantage of being easily electrospun into fibers with diameters in the nanometer range similar to that of natural collagen fibrils. However, the major drawback with synthetic nanofibers is their rapid degradation rate. The combination of microfibers and nanofibers is aimed to ensure mechanical stiffness and resistance to degradation, attributed to microfibers, and hydrophilicity and high surface area suitable for cellular attachment attributed to nanofibers. One example of such combination approach is the co-fabrication of PLGA scaffolds with collagen type I to produce porous microsponges [13, 14]. Although synthetic polymeric biomaterials possess benefits of being easily fabricated to different shapes and sizes and flexible mechanical and chemical properties, they may lack functional groups essential for cellular binding. In addition, acidic by-products or unnatural polyesters may be released into the bloodstream during degradation. To encounter these issues, natural, protein-based fibers are being explored as alternatives for T/L tissue engineering.

Being naturally present as a part of T/L tissues, collagen type I is the first choice to explore. However, collagen gels have elastic moduli and tensile strengths in the range of 10–30 kPa and 5–10 kPa, respectively, and are significantly weak. Hence, collagen-based scaffolds can be used to study the mechanisms of differentiation and regeneration of tendons but do not play a major role in T/L replacement [13, 15]. Silk fibroin is a widely used natural biomaterial for T/L tissue engineering. Unlike collagen, silk not only possesses remarkable tensile strength and toughness but also has surface amino acids for cell adhesion. It has additional advantages of remaining structurally unchanged in aqueous solutions with a slow degradation rate in vivo and ease of fabrication into films, gels, braided fibers, or nanofibers. These features have made silk fibroin one of the best choices for support of cellular and tissue ingrowth during T/L tissue engineering [16].

Due to the inherent biocompatibility of natural polymeric macromolecules, they are preferred biomaterials for T/L tissue engineering. Although collagen type I gels face the issue of low tensile strength, silk fibroin is an excellent alternative [13].

18.2.6.2 Bone Cement

Bone cement forms an elastic zone by filling the empty space between the prosthesis and the bone. It consists of two components, a powder (pre-polymerized PMMA/PMMA copolymer beads/amorphous powder, initiator) and a liquid (MMA monomer, inhibitor). Free radical polymerization is initiated on addition of the initiator to the accelerator, and there is significant change in the viscosity of the cement as it turns into a hard material [17].

It is necessary that a suitable bioresorbable material should fill the spaces to allow growth of the new bone into the defects. In the absence of proper bioresorbable material, fibrous tissue ingrowth may prevent the formation of the bone within the defects.

Bioceramics made of CaP are extensively used for bone regeneration due to their good biocompatibility, osteoconduction, and osseointegration [18]. Hip implants are one of the prime applications of CaPs. CaP is coated to the stem's surface that promotes bonding between bones and osteoconduction of the implant. This technique is known as interface bioactive bone cement. CaP is also used in reconstruction surgery as filler of bone defects. Hyaluronic acid is the efficient biomaterial attributed to its ability to physicochemically bind to the bone and high mechanical strength. CaP can also be used to promote bone growth in the osteonecrotic sites in the body. Bone grafts are also used as nonhardening injectable CaPs in the form of pastes. They consist of a blend of CaP in the form of powder, granules or particles, and a hydrogel. Self-setting CaP pastes are used for minimally invasive surgery [18].

18.2.6.3 Joint Replacements

Joint replacement surgery, also known as replacement arthroplasty, is a procedure of replacing a dysfunctional joint with an orthopedic prosthesis. It may be performed on the shoulders, hips, knees, ankles, or fingers. It is indicated for some diseases such as osteoarthritis and rheumatoid arthritis, when other therapies are unable to suffice the need [19]. Total joint replacement (TJR) aims to treat arthritis, usually when it impedes daily activities due to the intense pain.

Essential properties of the biomaterials used for joint replacement include robustness and biocompatibility. The life expectancy of biomaterials used as implants is limited due to the wear and the reaction of the host to wear by-products. The implants remain biocompatible and well tolerated if they have mechanical stability within the bone and remain unaffected by microorganisms. Excessive wear of the biomaterials induce inflammation, leading to periprosthetic osteolysis, damage to the support, loosening, and failure of the implant [20].

Commonly used biomaterials are metals and their alloys, polymers, ceramics, and composites [21]. Metal alloys including cobalt-chromium alloys, titanium alloys, and stainless steel are the most frequently used materials. Titanium-6 aluminum-4 vanadium (Ti6Al4V) and cobalt-chromium-molybdenum (CoCr or CoCrMo) are the first choice, followed by other alloys like iron-chromium-nickel (stainless steel, AISI 316 L), titanium-aluminum-niobium (Ti6Al7Nb), commercially pure titanium (cpTi), and tantalum. Polymers that play a role in TJR include ultrahigh molecular weight polyethylene (PE), highly cross-linked PE, and PMMA. The least frequently employed biomaterials are the ceramics (alumina and zirconia) [20].

Among metals, titanium alloy has the advantage of having lower modulus of elasticity. However, it has poor wear resistance. In addition, titanium particles induce proinflammatory cytokine (IL-6, TNF-α, and MCP-1) production and lead to a minor inflammatory reaction. It differs from the reaction with alloys of cobalt, wherein significant cell necrosis is observed [22]. The cobalt-chromium alloy CoCr undergoes corrosion and repassivation, releases metal ions, and induces immunological response immediately postimplantation [23]. CoCr implants have the potential to cause cytotoxicity, genotoxicity, and hypersensitivity [21]. Another metal stainless steel is composed of iron, chromium, nickel, and copper. Bulk stainless steel induces transient inflammation but does not root cytotoxicity [21]. When used in the form of particulates and ions, stainless steel releases ions depending on the composition and surface finish of the alloy. Its corrosion products affect the survival of osteoblasts and produce macrophages-dependent inflammation [21]. Among 316 L stainless steel and nitrogenated stainless steel (NSS) alloys, 316 L stainless steel particles have higher potential to induce toxicity compared with NSS particles. However, both the alloys can induced significant rise in the expression of IL-1β and TNF-α.

Among polymers, PE has wide applications. However, PE induces biological reaction on phagocytosis or cell contact. Activation of nuclear transcription factors by PE causes release of cytokine and chemokine, inducing the systemic recruitment of proinflammatory macrophages that differentiate into osteoclasts. Osteoclast accumulation near the implant leads to osteolysis [20].

Alumina is one of the commonly used bulk ceramics. On implantation, a fibrous membrane is formed around the implant. The degree of osseointegration with alumina depends on if it is implanted under nonloaded or loaded conditions, the latter proving to be supportive of osseointegration. Currently used orthopedic ceramics for joint replacement are mixtures of alumina (AL_2O_3) and zirconia (ZrO_2), wherein alumina forms primary continuous phase (70–95%) and zirconia is the secondary phase (30%–5%) [20]. Overall, ceramic particles do not induce genotoxicity, and the nanometer-ranged alumina particles have minimal impact on cell survival and insignificant influence on cytokines production [20].

Table 18.5 Generations of nerve conduits

Generation	Functions/characteristics
First generation	FDA approved
	Guide nerve growth
	Provide support
	Provide barrier function
Second generation	FDA approved
	Semipermeable/porous
	Biodegradable
Third generation	Loaded with support cells
	Provide release of nerve growth factors
	Contain luminal fillers
	Provide electrical conductivity

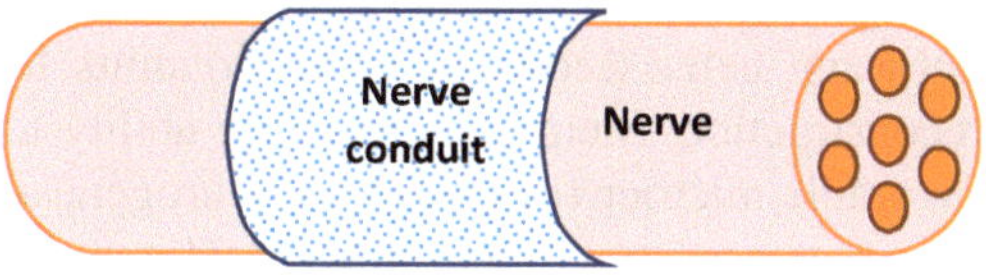

Fig. 18.2 Nerve conduit

18.2.7 Nerve Conduits

The peripheral nervous system is more prone to mechanical injury as is not protected by the vertebral column and skull like the central nervous system. Serious injuries leading to denervation of muscles and tissues may cause loss of motor and sensory functions and negatively affect the quality of life of these patients. Autologous nerve grafting is the standard therapy when direct suturing of the two ends of the damaged nerve is not possible. Allografts and xenografts are also used to bridge the gap, but they require adequate immunosuppression. Although autografts offer the best opportunity for nerve reconstruction, the limited availability of donor tissue and requirement for another surgical site limit their use. Also, an acceptable level of function is acquired in only 40–50% of the patients. Currently, research on nervous tissue engineering focuses on developing bioartificial nerve conduits. Nerve guidance conduits are implants that promote axonal regrowth and aid nerve regeneration. The choice of biomaterials used depends on the nerve environment. The critical factors to consider include biodegradability, mechanical integrity, regeneration, implantation, and sterilization. Nerve conduits have evolved over time and can be classified into three generations. Table 18.5 depicts the generations of nerve conduits [24], and Fig. 18.2 gives diagrammatic image of a typical nerve conduit.

The first-generation nerve conduits were prepared from nonresorbable silicone or PTFE (ePTFE, Gore-Tex®) [25]. Silicone tubes were successful when utilized to bridge very short nerve gaps (2–3 mm). However, compression syndrome and encapsulation of the implants by fibrous tissue urged for the requirement of secondary surgeries for removal. Second-generation conduits are manufactured from biocompatible and resorbable materials. They are aimed to promote remyelination of axons and eliminate the need for a secondary removal surgery. These conduits are mainly hollow in nature and composed of a biocompatible material. Commonly used materials include hydrogels that may be biological or synthetic. Synthetic hydrogels include collagen-terpolymer, poly (lactic-co-glycolic acid) category, poly (glycerol sebacate), and PEG hydrogels. Biological hydrogels include polysialic acid, collagen type I/III, silkworm silk fibroin, spider silk fiber, chitosan, alginate, aragonite, and hyaluronic acid hydrogels [24]. Table 18.6 enlists characteristics of frequently used second-generation biomaterials for nerve conduits.

Although second-generation nerve conduits aid nerve regeneration, they are not proved to be superior to autologous nerve grafts [26]. The third-generation nerve conduits, not approved by the FDA, are currently the center of research in this field. They are aimed to provide controlled release/delivery of cells or factors that aid in the growth of neurons. Such cells are called glial cells, namely, Schwann cells, astrocytes, and olfactory ensheathing cells [27]. Additionally, stem cells can also aid in repairing and regeneration as they can differentiate into neurons or glial cells. Schwann cells seeded in nerve conduits have been successfully used for nerve reconstruction. Stem cells have the ability to secrete neurotrophic factors, create a favorable microenvironment for neurogenesis, and increase Schwann cell proliferation in peripheral nerve repair. The nerve conduits can also be loaded with neurotrophic substances, in the lumen or in the wall. These substances when released at the site of injury at a controlled rate can promote nerve growth. Some of the neurotrophic factors include transforming growth factor β superfamily; neurotrophins 3, 4, and 5; neuregulin-1; nerve growth factor; and ciliary, brain-derived, and glial cells-derived neurotrophic factors [24].

18.2.8 Ophthalmic

Intraocular Lenses (IOLs) for Eye Surgery

Implantation of artificial IOLs is commonly found during the cataract surgery [28].

Cataract surgery includes removing the opacified contents in the capsular bag of the cataractous lens and placing the IOL inside the capsular bag to restore the normal refractive power [29]. Biocompatibility is an important requirement of an IOL. The blood-aqueous barrier is broken after a cataract surgery, and proteins and cells are released into the anterior chamber. Inflammatory cells can deposit up to 1 year after surgery. Fibroblast-like cells show maximum deposition by 1 month, whereas giant cell deposition is maximum at 3 months. The duration and intensity of each cellular response vary as per the biomaterial used. However, being of a very low intensity, it is clinically insignificant [28].

IOL may be composed of acrylic or silicone material. Acrylic materials may be foldable or non-foldable, depending on their flexibility. Foldable lenses can be composed of hydrophobic or hydrophilic acrylics, whereas non-foldable lenses are made of PMMA. Silicone lenses are foldable. Flexibility of the material enables easy insertion or injection into the eye via small incisions [28].

The rigid PMMA lenses that required creating a large incision in the eye were used primitively [29]. However, with the development of ultrasound

Table 18.6 Characteristics of second generation biomaterials

Biomaterial	Characteristics
PGA	First clinically available bioabsorbable conduit (NeuroTube by Synovis Micro Companies Alliance)
	Advantages
	More flexible than silicone
	Oxygen diffusion possible due to porosity
	Achieved superior results when compared to end-to-end repair for gaps <4 mm
	Can regenerate small motor nerve defects successfully
	Drawbacks
	Degradation before completion of nerve regeneration process
	Toxic effects of lactic acid degradation by-product
Type I collagen	Five commercially available FDA-approved collagen type I nerve conduits: NeuraGen, NeuroMatrix, NeuroFlex, NeuraWrap, and NeuroMend
	Advantages
	Capable of splinting small nerve defects up to 20 mm
	Ability to reduce the severity of symptoms associated with neuropathic pain and alter the regrowth of transected nerves
	Drawbacks
	Variable degradation time (4–48 months)
	Degradation before completion of the nerve regeneration process
Caprolactone	A poly-D,L-lactide-co-epsilon-caprolactone conduit (P(LL-co-CL)) consists of lactic acid and caprolactone monomers. Neurolac® is the only currently FDA-approved synthetic caprolactone conduit
	Advantages
	Fewer toxic degradation side products produced
	Long degradation time (up to 16 months)
	No difference in the outcomes by single P(LL-co-CL) conduits vs the filling of P(LL-co-CL) conduits with muscle tissue
	No variation in outcomes with varying degrees of porosity
	Drawbacks
	Controversial efficacy
	Requirement of more controlled clinical trials
N-Fibroin	Silk protein, produced initially as a soluble protein in the glands of silk worms and later arranged into fibrous structures during the spinning process
	Advantages
	Superior biocompatibility and low immunogenicity
	Degradable with excellent mechanical stability
	Chemically modified silk nanofibers with gold nanoparticles provide enhanced cellular interactions such as adhesion, differentiation, and proliferation; gold nanoparticles immobilize specific molecules of the nanofiber without generating significant cytotoxicity
	Superior results of silk fibroin conduits containing nerve growth factors demonstrated
	Drawback
	Not approved by the FDA or other administrations

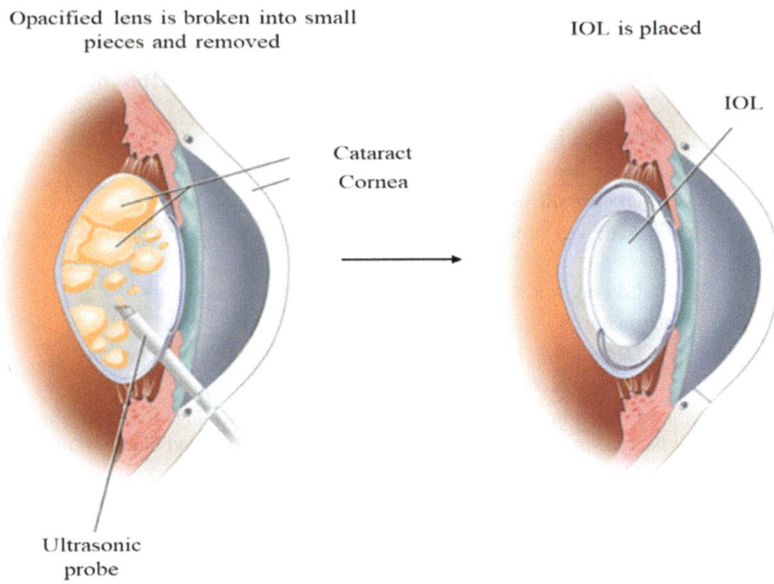

Fig. 18.3 IOL in the human eye

phacoemulsification and advent of the foldable acrylic lenses, PMMA use has reduced significantly. The design of the currently available foldable acrylic lens is fabricated from a different acrylic copolymer, with a different refractive index, water content, glass transition temperature, mechanical properties, etc. Hydrophilic acrylic lenses can be easily inserted through very small incisions of 1.8 mm in microincision surgery due to their flexibility and high water content [28]. Silicone lenses are fabricated of a polyorganosiloxane backbone and have lower refractive indices than those of acrylic lenses. They are comparatively thicker than their acrylic counterparts [30]. Figure 18.3 depicts the placement of an IOL.

18.2.9 Skin Tissue Substitutes

The skin, the largest organ, protects the human body from external environment. In case of trauma, the stem cells present in the epidermis provide the ability of self-renewal and healing. However, if the injury is deep, this process of regeneration is not sufficient, and additional therapy is required for cure. Any loss of more than 4 cm diameter of full-thickness skin requires grafting for its treatment [31].

Depending on the biomaterial used, skin substitutes can be classified as biological (allogeneic, autologous, and xenogeneic) or synthetic (biodegradable and nonbiodegradable). Synthetic substitutes are composed of acellular elements. Their primary function is to check fluid loss and inhibit microbial contamination. The frequently

Table 18.7 Synthetic and acellular skin substitutes

Type	Product	Company	Constituents
Acellular	Alloderm®	LifeCell Inc.	Human acellular lyophilized dermis
	SureDerm	HANS BIOMED Corporation	Human acellular lyophilized dermis
	OASIS Wound Matrix	Cook Biotech Inc.	Porcine acellular lyophilized small intestine submucosa
	Biobrane®	Mylan Bertek Pharmaceuticals	Ultrathin silicone as epidermal analog film and 3D nylon filament as dermal analog with type I collagen peptides
	Integra® DRT	Integra® Lifesciences Corp.	Dermal analog–bovine collagen and chondroitin-6-sulfate GAG; epidermal analog–silicone polymer polysiloxane
Synthetic	Hybrid nanofibrous PLGA/chitosan membrane	Tianjin University	PLGA/chitosan hybrid electrospun nanofibrous membrane
	Biodegradable polyurethane microfibers	University of Delaware	Biodegradable polyurethane microfibers

utilized synthetic skin substitutes include Alloderm™, Biobrane®, TransCyte™, and Integra®. On the other hand, the biological substitutes, also called tissue engineered skin, are cultured cell suspensions/sheets, which may be used alone or in conjunction with a dermal matrix. Examples of frequently employed biological substitutes with allogeneic cells are Apligraf™, Dermagraft™, and OrCel™, and with autologous cells is Epicel™ [31]. Table 18.7 provides the list of synthetic and acellular substitutes [31].

18.2.10 Surgical Sutures and Surgical Mesh and Staples

Surgical suture is a device used to hold body tissues together after a surgery or an injury. Sutures consist of a needle with a surgical thread. Surgical sutures are of two types, absorbable and nonabsorbable, depending on their fate in the body.

Originally, absorbable sutures were prepared from materials of biological origin, such as catgut and silk. These materials absorbed bodily fluids that frequently led to infection. Currently the absorbable sutures are prepared from synthetic materials such as PGA, PLA, polydioxanone, and caprolactone. In the body, these polymers either undergo hydrolysis or are enzymatically degraded within 10 days to 8 weeks. Table 18.8 gives retention period and absorption rate of commonly used absorbable materials [32].

Nonabsorbable sutures are prepared from silk, PP, polyester, or nylon. Orthopedic and cardiac surgeries frequently involve stainless steel wires, with or without coatings. These types of sutures are preferred in stressful internal environment, such as the cardiac tissue, which involves constant movement, and the bladder, which has

Table 18.8 Absorbable biomaterials for surgical sutures

Name	Composition	Retention period	Absorption rate
Polyglactin 910	Poly (glycolide-co-L-lactide)	28–35 days	56–70 days
Polyglecaprone 25	Glycolide and ε-caprolactone copolymer	21–28 days	90–120 days
Polydioxanone	Poly (p-dioxanone) polyester	Up to 90 days	180–210 days

adverse chemical conditions. They induce less immune response and hence cause minimal scarring. They can be removed after a certain time or left permanently [32].

Surgical mesh can be used as a support for organs and tissues during surgery. It is commonly used for hernia repair surgery. It is also used for pelvic organ prolapse. Biomaterials used for hernia include PP, polyethylene terephthalate (PET), and PTFE. PP causes discomfort in patients postimplantation, PET faces the issue of degradation after few years of implantation, and PTFE loses stability due to its inability to integrate with the surrounding tissues [33]. A new type of mesh, the polyvinylidene fluoride (PVDF), or nanofibrous mesh is currently under investigation. It is more resistant to disintegration and hydrolysis and does not get stiffer with age, unlike PET and PP, respectively. Another significant benefit of PVDF is the development of capillaries post 12 weeks, which is crucial for wound healing. The sooner the neovascularization, the faster can the tissue be repaired, which reduces the probability of extrusion of the mesh [34]. Biocompatibility of these materials depends on the porosity of the mesh. Porosity is defined as the ratio of the pore to the total area. Bacteria enter and proliferate in the mesh sizes below 10 μm. Macrophages and neutrophils being larger in size are unable to eliminate the bacteria, rendering such mesh prone to infections. Fibroblasts, blood vessels, and collagen fibers can pass through pore size larger than 75 μm, which aids tissue regeneration. Overall, larger pore sizes are more suitable to promote tissue development.

Surgical staples have the same function as sutures and are used to close skin wounds. They have the advantage of inducing lower local inflammatory response, higher reduction in width of the wound, and faster resealing time over surgical sutures. Although surgical staples are frequently prepared from titanium, stainless steel is also used. Titanium induces lower immune response and does not significantly interfere with MRI scanners. Among bioabsorbable staples, PGA is frequently used.

References

1. Tappa K, Jammalamadaka U (2018) Novel biomaterials used in medical 3D printing techniques. J Funct Biomater 9(1)
2. Wong CS, Schaffner AD (2018) Breast, implants. StatPearls Publishing, Treasure Island (FL)

3. Kang SH, Sutthiwanjampa C, Heo CY, Kim WS, Lee SH, Park H (2018) Current approaches including novel nano/microtechniques to reduce silicone implant-induced contracture with adverse immune responses. Int J Mol Sci 19(4)
4. Jaganathan SK, Supriyanto E, Murugesan S, Balaji A, Asokan MK (2014) Biomaterials in cardiovascular research: applications and clinical implications. Biomed Res Int 2014:459465
5. Ravi S, Chaikof EL (2010) Biomaterials for vascular tissue engineering. Regen Med 5(1):107–120
6. Kostrzewa B, Rybak Z (2013) History, present and future of biomaterials used for artificial heart valves. Polim Med 43(3):183–189
7. Zeng FG, Rebscher S, Harrison W, Sun X, Feng H (2008) Cochlear implants: system design, integration, and evaluation. IEEE Rev Biomed Eng 1:115–142
8. Harnack D, Winter C, Meissner W, Reum T, Kupsch A, Morgenstern R (2004) The effects of electrode material, charge density and stimulation duration on the safety of high-frequency stimulation of the subthalamic nucleus in rats. J Neurosci Methods 138(1-2):207–216
9. Mendes GC, Brandao TR, Silva CL (2007) Ethylene oxide sterilization of medical devices: a review. Am J Infect Control 35(9):574–581
10. Shannon RV (1992) A model of safe levels for electrical stimulation. IEEE Trans Biomed Eng 39(4):424–426
11. Ananth H, Kundapur V, Mohammed HS, Anand M, Amarnath GS, Mankar S (2015) A review on biomaterials in dental implantology. Int J Biomed Sci 11(3):113–120
12. Bhagwat R, Vaidhya IS (2013) Novel drug delivery systems: an overview. Int J Pharm Sci Res 4(3):970–982
13. Kuo CK, Marturano JE, Tuan RS (2010) Novel strategies in tendon and ligament tissue engineering: advanced biomaterials and regeneration motifs. Sports Med Arthrosc Rehabil Ther Technol 2:20
14. Cooper JA, Lu HH, Ko FK, Freeman JW, Laurencin CT (2005) Fiber-based tissue-engineered scaffold for ligament replacement: design considerations and in vitro evaluation. Biomaterials 26(13):1523–1532
15. Roeder BA, Kokini K, Sturgis JE, Robinson JP, Voytik-Harbin SL (2002) Tensile mechanical properties of three-dimensional type I collagen extracellular matrices with varied microstructure. J Biomech Eng 124(2):214–222
16. Greenwald D, Shumway S, Albear P, Gottlieb L (1994) Mechanical comparison of 10 suture materials before and after in vivo incubation. J Surg Res 56(4):372–377
17. Havelin LI, Espehaug B, Vollset SE, Engesaeter LB (1995) The effect of the type of cement on early revision of Charnley total hip prostheses. A review of eight thousand five hundred and seventy-nine primary arthroplasties from the Norwegian Arthroplasty Register. J Bone Joint Surg Am 77(10):1543–1550
18. Eliaz N, Metoki N (2017) Calcium phosphate bioceramics: a review of their history, structure, properties, coating technologies and biomedical applications. Materials (Basel, Switzerland) 10:4
19. Joint Replacement Surgery and You. In Arthritis MaSDo. April 2009. http://www.niams.nih.gov/#
20. Gibon E, Cordova LA, Lu L, Lin TH, Yao Z, Hamadouche M et al (2017b) The biological response to orthopedic implants for joint replacement. II: Polyethylene, ceramics, PMMA, and the foreign body reaction. J Biomed Mater Res B Appl Biomater 105(6):1685–1691
21. Gibon E, Amanatullah DF, Loi F, Pajarinen J, Nabeshima A, Yao Z et al (2017a) The biological response to orthopaedic implants for joint replacement: Part I: Metals. J Biomed Mater Res B Appl Biomater 105(7):2162–2173
22. Kim JA, Ihn HJ, Park JY, Lim J, Hong JM, Kim SH et al (2015) Inhibitory effects of triptolide on titanium particle-induced osteolysis and receptor activator of nuclear factor-kappaB ligand-mediated osteoclast differentiation. Int Orthop 39(1):173–182
23. Willert HG, Buchhorn GH, Fayyazi A, Flury R, Windler M, Koster G et al (2005) Metal-on-metal bearings and hypersensitivity in patients with artificial hip joints. A clinical and histomorphological study. J Bone Joint Surg Am 87(1):28–36

24. Gaudin R, Knipfer C, Henningsen A, Smeets R, Heiland M, Hadlock T (2016) Approaches to peripheral nerve repair: generations of biomaterial conduits yielding to replacing autologous nerve grafts in craniomaxillofacial surgery. Biomed Res Int 2016:3856262
25. Konofaos P, Ver Halen JP (2013) Nerve repair by means of tubulization: past, present, future. J Reconstr Microsurg 29(3):149–164
26. Bozkurt A, Lassner F, O'Dey D, Deumens R, Bocker A, Schwendt T et al (2012) The role of microstructured and interconnected pore channels in a collagen-based nerve guide on axonal regeneration in peripheral nerves. Biomaterials 33(5):1363–1375
27. Zhang BG, Quigley AF, Myers DE, Wallace GG, Kapsa RM, Choong PF (2014) Recent advances in nerve tissue engineering. Int J Artif Organs 37(4):277–291
28. Nguyen J, Werner L (1995) Intraocular lenses for cataract surgery. In: Kolb H, Fernandez E, Nelson R (eds) Webvision: the organization of the retina and visual system. University of Utah Health Sciences Center, Salt Lake City (UT). Copyright: (c) 2018 Webvision
29. Werner L (2008) Biocompatibility of intraocular lens materials. Curr Opin Ophthalmol 19(1):41–49
30. Maddula S, Werner L, Ness PJ, Davis D, Zaugg B, Stringham J et al (2011) Pathology of 157 human cadaver eyes with round-edged or modern square-edged silicone intraocular lenses: analyses of capsule bag opacification. J Cataract Refract Surg 37(4):740–748
31. Vig K, Chaudhari A, Tripathi S, Dixit S, Sahu R, Pillai S et al (2017) Advances in skin regeneration using tissue engineering. Int J Mol Sci 18(4)
32. Gierek M, Kusnierz K, Lampe P, Ochala G, Kurek J, Hekner B et al (2018) Absorbable sutures in general surgery – review, available materials, and optimum choices. Polski Przeglad Chirurgiczny 90(2):34–37
33. Todros S, Pavan PG, Natali AN (2015) Biomechanical properties of synthetic surgical meshes for pelvic prolapse repair. J Mech Behav Biomed Mater 55:271–285
34. Ding J, Deng M, Song XC, Chen C, Lai KL, Wang GS et al (2016) Nanofibrous biomimetic mesh can be used for pelvic reconstructive surgery: a randomized study. J Mech Behav Biomed Mater 61:26–35

Bioheat Physics for Hyperthermia Therapy 19

Gurmeet Singh, Neeraj Kumar, and Pramod Kumar Avti

Abstract

The chapter describes the physical and mathematical aspects of hyperthermia. The different types of hyperthermia application modalities, such as local, regional, and whole-body hyperthermia, along with the mathematical models that govern the transfer of heat in biological domains are described. The mathematical models include Penne's bioheat model, Mitchell and Myers model, Keller and Seiler model, Chen and Holmes model, Weinbaum and Jiji model, etc. Primarily these models consider the mechanism of tissue-tissue and tissue-blood heat transfer based on the diffusion and convection of heat. In addition to this the different types of hyperthermia applicators are described in this chapter. These applicators include electromagnetic heating, ultrasound heating, heating with laser, and magnetic nanoparticles. The recently evolving methodology of magnetic nanoparticle induced hyperthermia is further described in detail. Thus, this chapter provides the basic as well as application-based details of hyperthermia, and its application methodologies.

Keywords

Bioheat transfer models · Hyperthermia and applicators · Magnetic nanoparticle hyperthermia

G. Singh · N. Kumar (✉)
Department of Mechanical Engineering, Thapar Institute of Engineering & Technology, Patiala, India
e-mail: neerajkumar@thapar.edu

P. K. Avti (✉)
Department of Biophysics, Postgraduate Institute of Medical Education and Research (PGIMER), Chandigarh, India
e-mail: avti.pramodkumar@pgimer.edu.in; pramod.avti@gmail.com

S. Paul (ed.), *Application of Biomedical Engineering in Neuroscience*,
https://doi.org/10.1007/978-981-13-7142-4_19

19.1 Introduction

Cancer has become an epidemic in today's world. The conventional treatments available like surgery, radiation therapy, and chemotherapy are highly invasive. There are several painful and harmful effects of the abovementioned treatment strategies, which can be permanent. Therefore, to have a minimally invasive treatment methodology that does not have such aftereffects has always been the need of the society. Currently several efforts are being made worldwide for such noninvasive treatment methodologies for cancer therapy.

Hyperthermia, the oldest treatment methodology for cancer treatment, has minimum aftereffects [1]. Many centuries ago, hyperthermia was used to cure breast lumps by Romans, Greeks, and Egyptians [2]. The clinical use of hyperthermia began in India around 3000 B.C. by Indian Ayurvedic physicians [3]. In 1867, a German physician Busch reported that certain bacteria such as erysipelas can regress tumor [4]. However, the cause of erysipelas, a streptococcus organism, to regress tumor was not known until 1881, when William Bradley Coley, an American bone surgeon and cancer researcher, concluded that having a severe infection could cause cancer to regress. Sustained elevated temperature (fever, 39–40 °C) is critical for tumor regression [5]. In the last few decades, there has been much advancement in the therapeutic procedure of hyperthermia. Hyperthermia is used either as a singular therapy or as an adjuvant therapy with radiation and drugs [6, 7]. Efficiency of such combinations has been demonstrated by various clinical trials [8–11]. This treatment procedure has fewer complications than more costly and risky surgical treatment [12].

One of the critical components of hyperthermia therapy is the elevation of tumor tissue temperature in controlled environment. This procedure is performed with various hyperthermia applicators. The elevation of temperature depends on various thermophysical parameters of tissues like thermal conductivity, specific heat, density, and vascular system of tissues. Furthermore, the heating pattern and heating capacity of these applicators also play a critical role in this thermotherapy. Forecasting of temperature distribution during this therapy will help in optimizing this procedure. Bioheat transfer physics has a critical role in prediction of temperature in tissue. Thus, during the last few decades, an increased interest in the bioheat transfer phenomena has been observed, especially for its therapeutic and diagnostic applications. This is the outcome of the collaborations among physiologists, clinicians, and engineers in the field of bioheat transfer. These collaborative works in the area of bioheat field have resulted in improvement and optimization of biological systems [13]. The detailed understanding of bioheat transfer is critical for designing the thermotherapy applicators.

19.2 Hyperthermia

Based on the location, depth, and tumor stages, clinically three main methods of hyperthermia are practiced, such as local, regional, and whole-body strategy, to dissipate heat to the superficial, deep-seated, or metastatic tumors.

19.2.1 Local Hyperthermia

Tumors of the size range $\leq$3 cm–6 cm within the accessible range or located in the superficial regions of the esophagus, rectum, breast, etc. are usually treated with this strategy [14]. Applicators of either superficial, intracavital, interstitial, or intraluminal with energy sources of radio waves, microwaves, or ultrasound are applied to heat the tumor regions of interest. During the treatment, normal tissue surrounding the tumor is usually cooled with water boluses to maintain the normal temperature of 37 °C and minimize or reduce the side effects of heating which may result in blistering or tissue/skin burns. Natural cavities or orifices in the human body, in cases of rectal, esophageal, and prostate cancers, are usually accessed for hyperthermia by way of intraluminal or intracavital applicators for the heat delivery. In case of tumors present in the skin or just beneath the skin, they are usually accessed by directly placing the applicators or antennas near the tumor region of interest. However, interstitial tumors are treated with minimal invasive procedures by placing the microwave (MW) antennas, radiofrequency (RF) electrodes, ultrasound (US) transducers, heat sources, and optical fibers into the tumor regions [15]. In cases of lung, liver, bone, and kidney focal tumors, radiofrequency ablation procedure is used to raise the temperature up to 50 °C for >5 min to cause vascular stasis, coagulation, and necrosis [16].

19.2.2 Regional Hyperthermia

In this treatment, advanced tumors in the regions of the abdomen, pelvis, and soft tissue sarcomas are generally treated. For treating the tumors in the abovementioned regions, three main approaches are in practice [17].

A. External applicators used for deep-seated tumors are applied in the form of circular rings around the patient as in case of dipole antenna pairs of coherent arrays. The phase and amplitude (70–150 MHz) in such antenna are usually controlled in response to the specific absorption rate (SAR) where temperatures in the tumor regions could be achieved up to 42 °C. Advancement in the heating methods and planning and monitoring systems is in progress, taking advantage of noninvasive measurements of tissue temperature and perfusion using magnetic resonance tomography.
B. The second approach is applied treat to arms, legs, and organs such as lungs and livers, having tumors. This treatment procedure is carried out by passing the extracorporeal heated blood through major arteries and veins.
C. The third strategy includes the treatment modality used for the tumors affecting the abdominal cavity called as continuous hyperthermic peritoneal perfusion (CHPP) or hyperthermic intraperitoneal chemotherapy. Here primary tumor resection is carried out to reduce the peritoneal metastases as seen in some cases of liver metastases.

Table 19.1 Different types of hyperthermia and the energy sources used for various clinical applications

S. no	Type of energy source	Frequency	Hyperthermia type	Region	Applications used
1.	Microwave (MW)	433–2450 MHz	Local	Superficial	Soft tissue sarcomas
	Radiofrequency (RF)	100–150 MHz		Intracavital	Head and neck cancers
	Ultrasound (US)	0.1–10 MHz		Intraluminal	Esophageal cancer
	Hot sources	NA		Intracranial	Rectal cancer
	Nanoparticles			Interstitial	Malignant gliomas
	Hot water perfusion				Breast cancer
					Locally advanced or recurrent
	Ferromagnetic implants				NA
	Resistive wire implants				
2.	Microwave (MW)	NA	Regional	Abdominal	Peritoneal carcinomatosis
	Radiofrequency (RF)			Pelvic	Mesothelioma
	Ultrasound (US)	NA		Limbs	Bladder cancer
	Hot sources				Ovarian cancer
					Soft tissue sarcoma
					Prostate cancer
					Rectal cancer
					Cervical cancer
3.	Hot water blankets	NA	Whole-body	Disseminated cancer	Malignant melanoma
	Infrared			Metastatic cancer	Ovarian cancer
	Thermal chambers				Recurrent soft tissue sarcoma

19.2.3 Whole-Body Hyperthermia

In this treatment strategy, thermal chambers are designed, such as hot water blankets or infrared radiators, to achieve a temperature of 42 °C so as to treat the metastases of soft tissue sarcomas or ovarian cancer or melanomas. The design of hyperthermia chambers also restores the electrolyte fluids and temperature loss considering the fact about the high-temperature stress to organs such as the brain, heart, lungs, and liver [14] (Table 19.1).

19.3 Bioheat Transfer Model

One of the remarkable features of the human thermoregulatory system is the maintenance of core temperature around 37 °C across the range of environmental conditions and during temperature-related stress [13]. The blood circulation through our vascular system has the key responsibility to thermoregulate our body temperature by convective heat transfer. Thus, it works as a sink or source of heat depending upon the local tissue temperature. Therefore, the efficiency of hyperthermia therapy is greatly influenced by various aspects of heat transfer mechanisms, i.e., conduction (primarily at tissue level), convection (through blood vasculature), and heat loss to environment primarily due to convection. Several theoretical and experimental studies have been performed to understand heat transfer between blood and tissue [18–20]. The dominant mode of heat transfer (removal) in tissues is through blood circulation. Exact evaluation of blood circulation through the vasculature is still a challenging issue that makes theoretical prediction of temperature distribution in tissues a nontrivial task. Thus, mathematical modeling of thermal behavior in tissues is a topic of great interest for the all the stakeholders, i.e., physiologists, physicians, mathematicians, and engineers.

There are primarily two approaches to model the interaction of blood with tissue for thermal physics.

Continuum Model In this model instead of quantifying spatial distribution of blood flow, the lumped/averaged effect of blood flow in the region of interest is considered. This makes mathematical modeling of thermo-physics much simpler in comparison with considering the individual effect of every blood vessel. This model works well for the region where blood circulation is primarily through capillaries or microvasculature. Based on this model, one of the simplest and widely accepted mathematical models called *Pennes bioheat model* [21] was developed. However, a shortfall of this model is that it cannot be appropriately applied in regions where blood recirculation is through macro or major blood vessels.

Vascular Model This model includes the effect of major blood vessels on heat transfer phenomena through tissues. Again, due to complexity in the vascular system, very few critical blood vessels along with their major branches can be considered for heat transfer phenomena. Furthermore, the phenomena of blood flow and heat transfer combined make this physics complex and nonlinear. In most of the cases, solution can be obtained only through computational and numerical solutions of the combined fluid flow and heat transfer governing equations.

The following are some of the key bioheat transfer models:

19.3.1 Pennes Bioheat Model

The famous Pennes bioheat transfer model [19], proposed by H. H. Pennes, is based on the series of experiments that he had conducted to measure temperatures on human volunteers' peripheral appendages such as foramens. He derived a thermal energy conservation equation based on the experimentation which later became well known as the Pennes bioheat transfer equation (PBHTE). The general form of PBHTE is given as

$$\rho c_t \frac{\partial T}{\partial t} = \nabla k_t \nabla T + q_m - q_{blood} + q_g \tag{19.1}$$

where T is temperature in the tissue, which is space as well as time dependent, ρ (kg. m^{-3}) the tissue density, c_t (J.kg^{-1}.°C^{-1}) the tissue-specific heat, and k_t (W.m^{-1}.° C^{-1}) the thermal conductivity. The terms q_m, q_{blood}, and q_g having units W.m^{-3} are the tissue metabolic heat generation, convective heat transfer by blood circulation through tissue, and heat generation/deposition by an external applicator or source in the region of interest, respectively. The last term in equation (1), q_g, accounts for the elevation of tissue temperature above the therapeutic limit during application of hyperthermia applicator. The higher the intensity of this term (q_g), the higher is the elevation of temperature above the core body temperature (37 °C). This term is also popularly known as the specific absorption rate (SAR). SAR is defined as the heat energy deposited in the tissue or tumor per second per unit mass or volume of tissue, by the hyperthermia applicator. In all practical applications, the value of SAR is not uniform throughout the region of hyperthermia application. Thus, suitability of the specific applicator primarily depends upon optimum deposition of the energy/heat at the region of interest in the tissue. This ensures that without any damaging effect to the healthy tissue, the required elevation of temperature up to the therapeutic range is achieved. Another critical term of equation (1) on the right-hand side denotes the heat carried away (q_{blood}) by network of blood vessels called capillaries in tissue. The expanded form of this term is given by

$$q_{blood} = \omega \rho_b c_b (T - T_a) \tag{19.2}$$

where ω (m^3.m^{-3}.s^{-1}) is called the blood perfusion rate (volume of blood perfused per unit volume of tissue per second), ρ_b (kg.m^{-3}) the blood density, c_b(J.kg^{-1}.° C^{-1}) the blood-specific heat, and T_a (°C) the local arterial temperature (constant core body temperature) for simplification in the analysis is considered. Due to self-regulated thermal system, the heat carried away by blood perfusion may get enhanced by enhancement in the blood recirculation rate through the tissue. Thus, efficacy of hyperthermia applicator also depends on how the heat deposition rate varies to maintain the required therapeutic temperature under the action of varying blood perfusion. However, due to the assumptions that local arterial temperature remains constant, and simplifications primarily in the term q_{blood}, the effect of blood perfusion is overestimated. Another limitation of this model is its non-applicability in the vicinity of major blood vessels. Despite these limitations of the Pennes bioheat equation, its predictions result in reasonable agreement between the theoretically evaluated and experimentally measured temperature profiles in perfused tissues subjected to various hyperthermia applicators [13].

19.3.2 The Mitchell and Myers Model

This model uses the concurrent heat exchange mechanism to describe the thermal energy balance [22]. According to this energy conservation law, the energy pattern arising from the concurrent heat exchange can be divided into three components based on the following six assumptions:

(a) The arterial and venous temperatures vary in the flow direction with distance.
(b) The thermal conductance is independent with respect to the distance of extremities between an artery and a vein, the artery and its environment, and the vein and its environment.
(c) The mass flow rates of the arterial and venous flows are equal and constant with distance.
(d) The thermal energy generated due to metabolism is small relative to the heat transfer terms.
(e) The limb is in a steady state; temperatures and flow rates are not changing with time.
(f) The thermal properties of the blood and tissue are constant.

Mitchell and Myers investigated the effect of countercurrent heat exchange between an artery and a vein and the heat exchange between the vessels and its surrounding normal tissue.

The arterial flow equation is given as

$$mc\frac{dt_a}{dx} + UA'(t_a - t_v) + (UA')_a(t_a - t_\infty) = 0 \tag{19.3}$$

The venous flow equation is given as

$$-mc\frac{dt_v}{dx} + UA'(t_v - t_a) + U_A{}'(t_v - t_\infty) = 0 \tag{19.4}$$

The temperature and boundary conditions are given as

$$x = 0 : t_a = t_0;x = L : t_a = t_v \tag{19.5}$$

where U is the thermal conductance, A' is the heat transfer area/length, and (Δt) the difference in temperature causing the heat flow, and temperature t with the subscripts a and v refer to the artery and vein, respectively. As the blood flow rate reduces, the countercurrent effect becomes significant.

19.3.3 The Keller and Seiler Model

This model focuses on the heat transfer mechanism of subcutaneous region below the skin [23]. In this model the region is divided into two compartments: (a) an isothermal center and (b) a peripheral region, where temperature varies in the normal direction to the skin surface. According to this model, Keller and Seiler not only assumed that in x-direction the heat conduction is small or negligible as compared with the blood convection effect but also included (a) concurrent heat exchange parameter [20] and (b) surrounding tissue energy conservation equation coupled with artery and vein equations:

$$k\frac{d^2T}{dx^2} + (ha - cg)(T_a - T) + ha(T_v - T) + q_m = 0 \tag{19.6}$$

$$\left[\dot{m}_{a0} - \int_0^x \dot{g}dx\right] c\frac{dT_a}{dx} + ha(T_a - T) = 0 \tag{19.7}$$

$$\left[(\dot{m}_a)_0 - \int_0^x \dot{g}dx\right] c\frac{dT_v}{dx} + (ha + c\dot{g})(T - T_v) = 0 \tag{19.8}$$

with the following boundary conditions:

$$x = 0, T = T_a = T_b; x = \delta, T = T_v = T_s \tag{19.9}$$

where k is the tissue thermal conductivity, x is the length variable (normal to surface direction), h is the average heat transfer coefficient (vessel to surrounding tissues), a is the average area of heat transfer/unit volume, c is the thermal capacity, $\dot{g}$ is the perfusion rate of capillary, $\dot{m}$ is the flow rate of blood, δ is the tissue thickness, and subscripts a, b, v, and s the artery, isothermal core, vein, and skin, respectively. In addition, Keller and Seiler found that the increasing heat transfer in subcutaneous region was induced by the capillary perfusion rate and by the decreasing atrial precooling.

19.3.4 The Chen and Holmes Model

Chen and Holmes in 1980 have proposed their model for bioheat transfer equation (BHTE) [17]. Their thermal model is based on physical and physiological behavior of tissue. The mathematical equation for model is expressed as

$$\rho c_p \frac{\partial T}{\partial t} = \nabla(k_t + k_p)\nabla T + q_m - q_{blood} + q_g - \rho_b c_b u \nabla T \tag{19.10}$$

In comparison with Pennes bioheat equation, few additional terms are incorporated in this equation. The term $\rho_b c_b u \nabla T$ is convective heat loss due to termal interaction between tissues and blood vessels. The term $\nabla k_p \nabla T$ denotes tissue conductive heat transfer rise due to tissue blood perfusion, where k_p is perfusion conductivity which is the function of blood perfusion rate in tissue. Thus, this model incorporates heat transfer through main supplying vasculature ($-\rho_b c_b u \nabla T$) along with heat transfer through capillary structures in tissue (q_{blood}). However evaluation of the last term ($-\rho_b c_b u \nabla T$) of Eq. (19.10) needs detailed knowledge of the vascular anatomy and flow patterns. This enhances the complicacy of the solution.

19.3.5 The Weinbaum and Jiji Model

A mathematical model for BHTE was proposed by Weinbaum and Jiji [18] based on countercurrent heat exchange between a thermally significant artery and vein pair. The model says that there is a net rise in energy transfer due to "incomplete countercurrent heat exchange" between arteries and veins (diameters of 50–500 μm) [24]. This enhanced energy transfer is accommodated by enhancement in tissue conductivity in the axial direction of the vessels. The mathematical form of Weinbaum and Jiji bioheat transfer model is given as

$$\rho c_p \frac{\partial T}{\partial t} = \nabla \left(k_{eff}\right) \nabla T + q_m \tag{19.11}$$

where k_{eff} is enhanced effective conductivity due to blood flow. The term k_{eff} α (blood flow rate)^2 within the vessels. In a 1D situation, where temperature gradient and blood vessels are in the same direction, this term is expanded as [25]

$$k_{eff} = k\left[1 + \frac{n\left\{\pi r_b^2 (\rho c)_b V \cos\gamma\right\}^2}{\sigma k^2}\right]$$

where n is the number density of blood vessels, r_b is the blood vessel radius, $(\rho c)_b$ is the product of density and blood-specific heat, V is the average velocity of blood in vessel, σ is a shape factor for the thermal conduction resistance among the adjacent countercurrent vessels, and γ is the angle of the pair of countercurrent vessels' axes relative to temperature gradient.

However, in practical situations, if the above equation is considered, then detailed understanding of the countercurrent vessel sizes and orientations and blood flow velocities is important. This enhances the complexity of this equation and solution procedure. The assumption of perfect countercurrent heat exchange, i.e., that most of the heat leaving the artery is recaptured by its countercurrent vein, is which this model is strongly criticized for by Wissler [26].

Other New model for Bioheat Equation Several other models were proposed to overcome shortcoming of Pennes, Chen and Holmes, and Weinbaum and Jiji models. These models were proposed with addition of correction factor in the Pennes perfusion term [27–29]. Weinbaum and coworkers [29] modified the Pennes source term and the modified model is given as

$$\rho c_p \frac{\partial T}{\partial t} = \nabla\left(k_t + k_p\right)\nabla T + q_m - \epsilon\omega\rho_b c_b (T - T_a) + q_g \tag{19.12}$$

The blood perfusion term is modified with correction coefficient term ϵ. The value of this coefficient varies from 0.6 to 0.8 under normal physiological condition [13]. Thus, we can say overprediction heat loss in blood perfusion term is 20 to 40%, or this much amount of rewarming takes place in tissue due to countercurrent veins.

Thus, we can see that various models have been developed for bioheat transfer equation. However, for all the tissues, none of these models can be universally applied. Suitability of specific models depends upon the vascular structure in tissue. The Pennes bioheat model could be applied appropriately to the tissue regions that are away from major blood vessel. Weinbaum and Jiji (WJ) model could be appropriately applied to the regions where most of the vessels are paired (such as muscle). Chen and Holmes (CH) model could be used for small vessel regions where the vessel's majority portion is not paired.

The microvascular structure of tumor is quite different and unstructured in comparison with healthy tissue. Thus, suitability of a specific model will always associate with itself an error factor depending upon the physiological condition of tumor tissue. However, the Pennes model is still the simplest and practical model for efficient predictions of profiles of transient temperature of certain hyperthermia conditions [30].

19.4 Hyperthermia Applicators

The treatment objective of hyperthermia is to raise the temperature of a tumor higher than 43 °C for a period of 30–60 min while keeping the surrounding healthy tissue below 43 °C. It is observed that the temperature rise produces cytotoxic response mediated by either heat induction, radiation, or drugs [12]. Thus, the success of this treatment methodology strongly depends upon the achieved temperature rise and the heating duration in tumor region. The size and location of tumor have a significant impact on the type of heating and applicator design. Various heating devices utilized for deposition of heat in tissues are based on electromagnetic heat absorption (microwave or radiofrequency), electric conductive heating, high-intensity-focused ultrasound heating, laser ablation, and magnetic particle heating. Each hyperthermia-inducing methodology therapeutically has its own advantages and limitations. The following section will briefly discuss the different methodologies/devices used for hyperthermia.

19.4.1 Direct Heating to Tumor Tissue

Direct heating to tumor is done by suitable applicators or tubes with hot fluid flow or electrically heated probes. Thermal damage to normal tissues is partially prevented by thermally insulating suitable length of the applicator. However, these applicators are suitable for tumors located only in the skin or just below it. Another limitation of these applicators is that they are not commercially available as dedicated systems. Direct heating is also used for whole-body hyperthermia (WBH). Thermal chambers, infrared radiators, and hot water blankets are used for WBH. This therapy is generally used for metastatic cancer [31]. Two types of thermotherapies, radiant hyperthermia and extracorporeal hyperthermia, are utilized for WBH therapy. In radiant WBH, the use of hot wax, hot water blankets, thermal chambers, or inductive coils supplies the heat externally to the whole body. The core body temperature of patients is elevated to 41.5–42.0 °C for 60 min. This therapy is usually used combined with systemic chemotherapy. Extracorporeal WBH is performed by external heated blood perfusion through extracorporeal loop. External blood circuit is created using the femoral artery outside the body. Taking advantage of hot air or water bath, the circulating blood is heated and injected into the major vein. By changing the hot reinfused blood flow volume and the rate, the required body temperature could be controlled [32]. These therapies are however considered difficult to execute.

19.5 Electromagnetic Heating

Heating of tumors by electromagnetic fields is categorized into low-frequency heating and high-frequency heating. This heating of biological tissues strongly depends on frequency.

19.5.1 Low-Frequency Electromagnetic (Radiofrequency) Heating

At low-frequency waves, the effect of electromagnetic field is negligible. The standard radiofrequency (RF) generator used in a catheter produces a sinusoidal wave at frequency of 200 to 1000 KHz. A current is induced between two electrodes that are attached to the tissue. This is because the charges in tissue are in motion. This induced current in the tissue leads to resistive or ohmic heating (I^2R loss). The electric field distribution and currents depend upon tissue's geometrical parameters and their conductivities. The positioning of electrode is not a trivial task in this heating, because the electrode's position determines the electric field distribution. Current density in tissue is inversely proportional to the square of the distance from the electrode. Thus, resistive heating decreases with the distance from the electrode to the fourth power. Therefore, maximum heating typically occurs in tissues that are surrounding the catheter electrodes but decreases in-between them. This causes variations of heating in tumor tissue volume. Various improved RF hyperthermia systems now propose to reduce the heterogeneity of the RF heating [33, 34]. If invasive approach is to be followed for the treatment, then multiple electrode implantations with thermal sensors can produce a controlled temperature. Sufficient power deposition is also achieved by electrode multiplexing.

19.5.2 High-Frequency Electromagnetic (Microwave) Heating

High-frequency electromagnetic heating is referred to as microwave hyperthermia, where wavelength of the applied field is in orders of centimeters and the frequency is in orders of GHz. Heating in high-frequency electromagnetic heating is dielectric instead of ohmic. The propagating electromagnetic wave in tissue raises the energy of dielectric molecules. The alternating electric field energizes the neighboring dipole molecules, which collide with each other, and the electromagnetic energy converts into thermal energy. This results in the generation of heat in the tissue. However, the main limitation of high-frequency electromagnetic heating (microwave heating) is the depth of penetration of heat. Thus, high-frequency electromagnetic heating (434–915 MHz) is beneficial only for superficial tumors, whereas low-frequency heating (radiofrequency) is suitable for deep-seated tumor. Many RF-based systems have evolved to reduce the heterogeneity of the RF heating [33, 34]. These applicators use electrically insulating materials around the electrodes.

19.5.3 Ultrasound Heating

When mechanical oscillations (vibrations) propagate in continuous matter medium, the local oscillation's phase difference induces a stress. The propagation of mechanical waves leads to stress/relaxation cycles in the medium, which transforms the oscillation's mechanical coherent energy into random movement. Thus, some part of acoustic energy gets absorbed by tissue which leads to a local temperature rise. However, in ultrasound heating, there is risk of overheating the surrounding bone-tissue interface because of the high ultrasound absorption in bone. Ultrasound heating has the inherent properties which if used can aid in overcoming some of the difficulties observed while using other electromagnetic heating methods. In tissues, the energy loss attenuation coefficient of ultrasound is proportional to the frequency. The penetration depth of the ultrasound could be controlled by varying the frequency approximately from 1 cm to 10 cm or use of a focused beam in the range of 0.5–10 MHz. With the advent of new technological developments in fabrication of ultrasound transducers, combined with high-order wavelengths in the range of millimeters, the abovementioned frequency range helps in the formulation of specific applicator dimensions and configuration. This would help in having either electronically phased arrays or beams of focused transducers to increase the depth or heating in a controlled manner. Ultrasound heating method is considered as a noninvasive hyperthermia-inducing method in which therapeutic temperature levels can be achieved with high-intensity-focused ultrasounds (HIFU). In these systems the phased array transmitter usually transmits small volumetric acoustics which has the ability to destroy lesions located deep within tissues by the process of thermocoagulation. This avoids inadvertent targeting of surrounding tissues [35]. In this method, the focal peak intensities lying between 300 and 2000 W/cm^{-2} with 0.1–10 s per lesion of sonication times are typically used.

19.5.4 Laser Heating

Laser photocoagulation is a use of a laser beam or other intense light source to coagulate and destroy small areas of tissue. This is a minimally invasive thermotherapy in which laser energy is deposited into a target tissue volume through one or more implanted optical fibers. Laser is used in medicine for incision and explosive ablation of tumors and other tissues. It is also used for blood vessel coagulation in various tissues. Laser beam power ranges from milliwatts to several watts. Usually the laser energy is focused on a small tissue area of a radius less than 300 μm, resulting in a very high heat flux. The wavelength of the laser determines the overall absorption of energy in the tissues or blood. Longer wavelength laser can penetrate deeper into a tissue than the lasers with short wavelength.

19.5.5 Magnetic Nanoparticle Heating

Another hyperthermia approach, currently being explored for targeted therapeutic heating of tumors, is the microparticle or nanoparticle heating. This method is considered as one of the minimally invasive techniques of inducing hyperthermia because the magnetic nanoparticles can be easily injected either systemically or intratumorally into the body. Magnetic nanoparticle (MNP) hyperthermia procedure consists of distribution of magnetic nanoparticles within tumor tissue or tumor vasculature followed by application of an external alternative magnetic field. This agitates the particles and thus produces heat [13, 36]. The magnetic nanoparticles function as heat sources and generate heat due to hysteresis loss, Néel relaxation, Brownian motion, or eddy currents. Therapeutic temperature distribution can be achieved by manipulating the particle distribution in the tumor and tuning the magnetic field parameters. To induce localized hyperthermia, MNPs are either injected intravenously or intratumorally. In intravenous injection, MNPs are injected into the veins that carry blood to the tumor, whereas in intratumoral injection, they are injected directly into the tumor.

Generally, single-domain MNPs whose hydrodynamic diameters are less than 100 nm are used in hyperthermia. Heat generation in particles of such small size occurs only due to relaxation mechanisms, viz., Néelian and Brownian relaxation mechanisms [37]. Brownian relaxation occurs when an MNP under the influence of an AMF experiences magnetic dipole reversal with respect to its surrounding fluid. This leads to heat generation due to shearing between the MNP and its surrounding fluid. Néelian relaxation is observed when magnetic dipole reversal of the MNP occurs, with respect to its crystal structure. Out of the two, whichever relaxation mechanism takes minimum time is the more dominating one for heat generation.

The parameters that have been observed to affect the temperature distribution during hyperthermia treatment are the hydrodynamic diameter of the particle, the MNP dose to be injected, amplitude and frequency of the applied field, and spatial distribution patterns of MNPs. Along with the effects of other parameters, MNPs' spatial distribution patterns on temperature distribution arising due to different

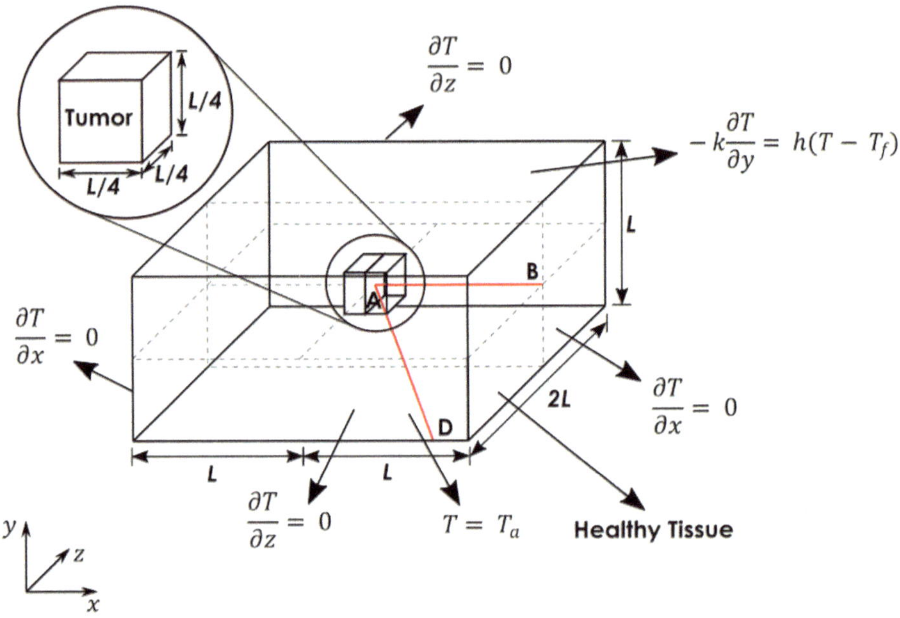

Fig. 19.1 3D model assumed for analysis [37]

intratumoral injection strategies are reported by Singh et al. [38]. The three-dimensional model considered to simulate MNP hyperthermia is shown in Fig. 19.1. This physical model consists of a central tumor region of a cube with sides L/4, surrounded by a domain of healthy tissues, with sizes of 2L, 2L, and L, along the three axial directions. The model assumes its lower surface to be at the core body temperature of 37 °C. The upper surface of the model loses heat by way of convection to the surrounding air. The heat loss is considered negligible along all other surfaces.

The possible MNPs' distribution patterns spatially considered are shown in Fig. 19.2. Iron oxide (magnetite) nanoparticles are distributed within the tumor. A typical dose of 5 mg Fe/cm^3 of tumor is considered for this study. Nonuniform distribution of MNPs is approximated by Gaussian distribution.

Temperature distribution along line AB (passing from the tumor center in the direction of x-axis (Fig. 19.1)) and line AD (from the tumor center along the diagonal (Fig. 19.1)) is plotted in Fig. 19.3, since they cover maximum tumor region. The temperature at point A (tumor center) is highest (57 °C) for the case of single injection site, whereas for multiple injection sites and uniform MNP distribution, it is 51 °C and 44 °C, respectively. This is due to the greater accumulation of MNPs at point A and in its vicinity for single injection site. Figure 19.3a, shows that along AB the temperature reduces up to 37 °C for all the spatial distribution patterns. However, more than the therapeutic temperature of 42 °C is attained from the tumor up to its boundary. Along AD (Fig. 19.3b), temperatures for uniform distribution and single

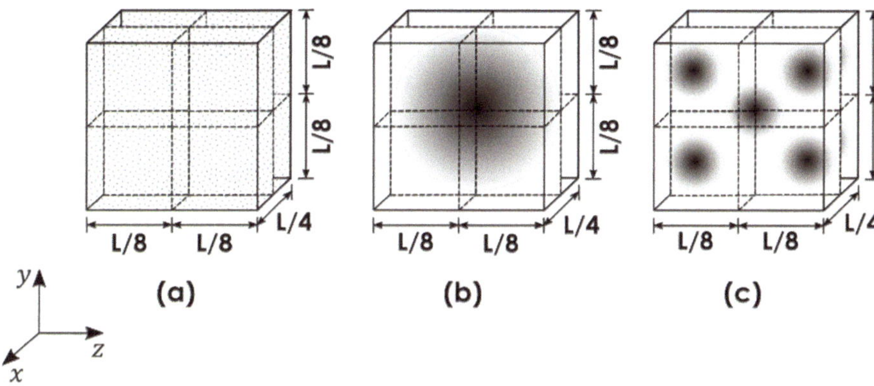

Fig. 19.2 MNPs' spatial distribution patterns injected within the tumor: (**a**) uniform distribution, (**b**) nonuniform distribution at tumor center injection, (**c**) nonuniform distribution around the tumor center [37]

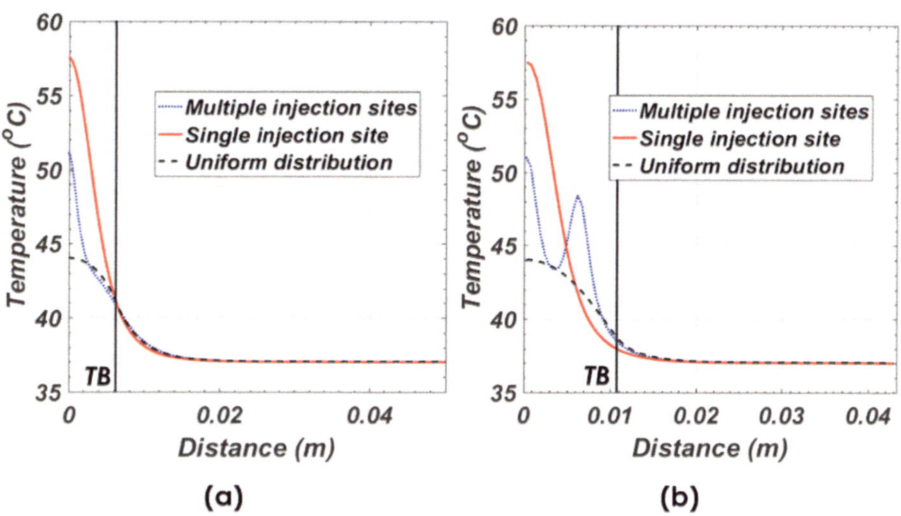

Fig. 19.3 Plots of temperature distribution along (**a**) AB and (**b**) AD. TB indicates tumor boundary [38]

injection site fall below the therapeutic limit before reaching the tumor boundary. For multiple injection sites, an initial fall followed by a rise, in temperature, is observed. This rise, due to the presence of MNPs at additional injection sites, maintains the temperature above the therapeutic limit in the tumor, in comparison with the other two spatial distribution cases. This enhances the effectiveness of the therapeutic procedure.

In comparison with most of the noninvasive heating approach, this treatment procedure is capable of delivering adequate heat to the tumor with minimal heat penetration into surrounding normal healthy tissue [13].

References

1. Hornback NB (1989) Historical aspects of hyperthermia in cancer therapy. Radiol Clin N Am 27:481–488
2. Van der Zee J (2002) Heating the patient: a promising approach? Ann Oncol 13:1173–1184
3. Singh BB (1991) Hyperthermia: an ancient science in India. Int J Hyperthermia 7:1–6
4. Busch W (1868) Verhandlungen artzlicher gesellschaften. Berl Klein Wochenschr 5:137–138
5. Coley WB (1893) The treatment of malignant tumors by repeated inoculations of erysipelas with a report on ten original cases. Am J Med Sci 105:487–511
6. Anderson RL, Kapp DS (1990) Hyperthermia in cancer therapy: current status. Med J Aust 152:310–315
7. Steevs RA (1992) Hyperthermia in cancer therapy: where we are today and where are we going? Bull N Y Acad Med 68(2):341–350
8. Overgaard J (1989) The current and potential role of hyperthermia in radiotherapy. Int J Radiat Oncol Biol Phys 16:535–549
9. Herman TS, Teicher BA, Jochelson M, Clark J, Svensson G, Coleman CN (1988) Rationale for use of local hyperthermia with radiation therapy and selected anticancer drugs in locally advanced human malignancies. Int J Hyperthermia 4:143–158
10. Sneed PK, Stauffer PR, McDermott MW, Diederich CJ, Lamborn KR, Prados MD, Chang S, Weaver KA, Spry L, Malec MK et al (1998) Survival benefit of hyperthermia in a prospective randomized trial of brachytherapy boost +/- hyperthermia for glioblastoma multiforme. Int J Radiat Oncol Biol Phys 40:287–295
11. Wust P, Hildebrandt B, Sreenivasa G, Rau B, Gellermann J, Riess H, Felix R, Schlag PM (2002) Hyperthermia in combined treatment of cancer. Lancet Oncol 3:487–497
12. Dewhirst MW, Prosnitz L, Thrall D, Prescott D, Clegg S, Charles C, MacFall J et al (1997) Hyperthermia treatment of malignant diseases: current status and a view toward the future. Semin Oncol 24(6):616–625
13. Zhu L (2009) Heat transfer applications in biological systems. In: Kutz M (ed) Biomedical engineering & design handbook, Volume 1: Bioengineering fundamentals. McGraw-Hill, New York, pp 2.33–2.67
14. Sneed PK, Stauffer PR, Li GC, Stege GJJ (2004) Chapter 70: Hyperthermia. In: Leibel SA, Phillips TL (eds) Textbook of radiation oncology, 2nd edn. Saunders, Pennsylvania, pp 1569–1596
15. van der Zee J (2002) Heating the patient: a promising approach? Ann Oncol 13:1173–1184
16. Stauffer PR, Goldberg SN (2004) Introduction: Thermal ablation therapy. Int J Hyperthermia 20 (7):671–677
17. Wust P, Hildebrandt B, Sreenivasa G et al (2002) Hyperthermia in combined treatment of cancer. Lancet Oncol 3:487–497
18. Chato J (1980) Heat transfer to blood vessels. ASME J Biomech Eng 102:110–118
19. Chen MM, Holmes KR (1980) Microvascular contributions to tissue heat transfer. Ann N Y Acad Sci 335:137–150
20. Weinbaum S, Jiji LM, Lemons DE (1984) Theory and experiment for the effect of vascular microstructure on surface tissue heat transfer—Part I: Anatomical foundation and model conceptualization. ASME J Biomech Eng 106:321–330
21. Pennes HH (1948) Analysis of tissue and arterial blood temperatures in the resting human forearm. J Appl Physiol 1:93–122

22. Mitchell JW, Myers GE (1968) An analytical model of the countercurrent heat exchange phenomena. Biophys J 8:897–911
23. Keller KH, Seiler L (1971) An analysis of peripheral heat transfer in man. J Appl Physiol 30:779–786
24. Huang H-W, Horng T-L (2015) Bioheat transfer and thermal heating for tumor treatment. In: Becker SM, Kuznestove AV (eds) Heat transfer and fluid flow in biological processes. Academic Press Elsevier, New York
25. Song WJ, Weinhaum S, Jiji LM (1987) A theoretical model for peripheral heat transfer using the bioheat equation of Weinhaum and Jiji. ASME J Biomech Eng 109:72–78
26. Wissler EH (1987) Comments on the new bioheat equation proposed by Weinbaum and Jiji. J Biomech Eng 109:131–139
27. Baish JW (1994) Formulation of a statistical model of heat transfer in perfused tissue. ASME J Biomech Eng 116:521–527
28. Brinck H, Werner J (1994) Estimation of the thermal effect of blood flow in a branching countercurrent network using a three-dimensional vascular model. ASME J Biomech Eng 116:324–330
29. Weinbaum S, Xu LX, Zhu L, Ekpene A (1997) A new fundamental bioheat equation for muscle tissue: Part I—Blood perfusion term. ASME J Biomech Eng 119:278–288
30. Arkin H, Xu LX, Holmes KR (1994) Recent developments in modeling heat transfer in blood perfused tissues. IEEE Trans Biomed Eng 41(2)
31. Mallory M, Gogineni E, Jones GC, Greer L, Simone CB II (2016) Therapeutic hyperthermia: the old, the new, and the upcoming. Crit Rev Oncol Hematol 97:56–64
32. Wiedemann GJ, d'Oleire F, Knop E, Eleftheriadis S, Bucsky P, Feddersen S et al (1994) Ifosfamide and carboplatin combined with 41.8°C whole-body hyperthermia in patients with refractory sarcoma and malignant teratoma. Cancer Res 54(20):5346–5350
33. Strahan MA, Norman A (1982) A localized current field hyperthermia system for use with 192 Iridium interstitial implants. Med Phys 9:419–424
34. Hartov A, Colacchio TA, Hoopes PJ, Strohbehn JW (1994) The control problem in hyperthermia. Adv Heat Mass Transf Biol Syst:119–126
35. Hynynen K, Jones RM (2016) Image-guided ultrasound phased arrays are a disruptive technology for non-invasive therapy. Phys Med Biol 61(17):R206–R248
36. Gilchrist RK, Medal R, Shorey WD, Hanselman RC, Parrott JC, Taylor CB (1957) Selective inductive heating of lymph nodes. Ann Surg 146:596–606
37. Rosensweig RE (2002) Heating magnetic fluid with alternating magnetic field. J Magn Magn Mater 252:370–374
38. Singh G, Kumar N, Avti PK (2018) Effects of spatial distribution patterns of magnetic nanoparticles on temperature distribution in magnetic hyperthermia. In: EMF-Med 1st world conference on biomedical applications of electromagnetic fields.

Part VIII

Emotion, Stress and Other Neurological Dysfunctions

Anxiety, Stress, and Neurological Dysfunction: From Basic Biology to Present Therapeutic Interventions

20

Ravindra Pramod Deshpande and Phanithi Prakash Babu

Abstract

Emotion is a way to express the inherent feeling. In present Indian society, psychological disturbance is still considered as stigma and often not rendered proper counseling and emotional help. Presently, we have tried to conceptualize the borders of mental and emotional well-being and how deviations thereof cause the psychological disturbances. The psyche of each individual is molded by the immediate surrounding which may define the internal thresholds to bear the stress. Presence of emotional stress or anxiety as a comorbidity with other diseases as stroke and cancer often goes unnoticed and deserves a special attention and treatment. We also shed light on how internal as well as external factors lead to disturbances of hormones and neurotransmitters causing anxiety and stressful conditions and how present drugs and the cognitive behavioral treatment (CBT) help to alleviate the disease. In addition, traditional ways as mindfulness and heartfulness meditation can help in proper conditioning of mind in turn helping to lead a balanced and stress-free life.

Keywords

Emotion · Threshold · CBT · Stress · Neurotransmitter

R. P. Deshpande (✉)
Department of Biotechnology and Bioinformatics, School of Life Sciences, University of Hyderabad, Hyderabad, India

Department of Caner Biology, Wake Forest School of Medicine, Winston Salem, NC, USA
e-mail: prakash@uohyd.ac.in

P. P. Babu
Department of Biotechnology and Bioinformatics, School of Life Sciences, University of Hyderabad, Hyderabad, India

S. Paul (ed.), *Application of Biomedical Engineering in Neuroscience*,
https://doi.org/10.1007/978-981-13-7142-4_20

20.1 Introduction

20.1.1 Introduction on Definition of Emotion, Stress, and Anxiety

Emotion is a way to express the thoughts and feelings. Stress and anxiety are the most common emotions in psychological conditions. Peoples suffering from anxiety and stress have a repetitive pattern of intrusive emotions and thoughts. Stress is a feedback response given by an individual to the feeling of emotion arising from external stimuli [1, 2] and is a known risk factor to be associated with neurological disorders and diseases as Alzheimer's disease and associated dementia [3], Parkinson's disease [4], and multiple sclerosis [5, 6].

20.1.2 Present Social Burden and Morbidity

WHO has estimated that neuropsychiatric complications represent the 6.3% of global burden on the disease at present which is projected to increase by 12% in 2030. High- and middle-income countries are reported to contribute in major proportion and projected to raise more by 2030. Although the high-income countries have high disease burden, the estimated deaths are more pronounced in low-income countries [7, 8]. It may be attributed to unavailability of proper remedies either to diagnose or detect the medical condition or in taking the corrective measures. This assessment provides the present dire need to understand the biology and physiology of the neurological diseases in detail to improve the service to the diseased population.

20.1.3 Measures of Emotion

Emotion is a response given to a particular condition or situation. The reaction can be behavioral or physiological. Emotion can be measured in terms of the feedback given; however, there are no "gold standards" available for its measure. The response given to an emotion can be measured in terms of central or peripheral physiology. EEG, fMRI, and the PET scan are the measures of central physiology, while the autonomic nervous system (ANS) measures the peripheral physiology. The behavioral response is measured with several parameters as observer rating given by whole body behavior, facial and vocal outputs, etc. Stress and anxiety remain the most shared readouts for the behavioral response [9].

20.1.4 Neurobiology of Emotion, Stress, and Anxiety

The regions of the brain that are associated with the processing of emotions are the insula (deals with emotion processing and awareness), amygdala (emotion processing), striatum (deals with regulation of attention and emotion processing), posterior and anterior cingulate cortex (deals with the self-awareness and regulation

Table 20.1 Neuroimaging techniques frequently used in psychological research and diagnosis

Sr no	Neuroimaging technique	Working principle/use	References
1	Single-photon emission computed tomography (SPECT)	Identifies the blood flow and metabolism	[11]
2	Positron emission tomography (PET)	Glucose uptake	[12]
3	Functional MRI (fMRI)	Measures bilateral cerebral blood flow (CBF)	[13]
4	Transcranial magnetic stimulation (TMS)	Used to noninvasively stimulate or activate particular regions in the cerebral cortex	[14]
5	Diffusion tensor imaging (DTI)	Used to access the integrity of white matter tracks and functionality of cortico-cortical connections	[15]

of emotion, attention), and prefrontal cortex (emotion and attention) [10]. The stress and anxiety disorders are often measured by the neurochemical disruptions in the synthesis or secretion of neurotransmitters and hormones. The changes in the mechanism of signaling or changes in the brain structure are also attributed to the anxiety-related behavior and the associated phenotypes. Neuroimaging is often used to visualize the changes happening in the brain of mentally diseased persons (Table 20.1). The general readouts are measured by altered blood flow and the glucose uptake in the specific regions as the amygdala and hippocampus in response to anxiety-provoking stimuli.

20.2 Role of Neurotransmitters in Anxiety Disorder

20.2.1 Glutamate and GABA

Neurotransmitter signaling is often found to be dysregulated in the cases of anxiety and depression. The main components that are involved in causing the mental unrest are glutamate and γ-aminobutyric-acid (GABA). Increased neurotransmission by glutamate or decreased inhibitory signaling mediated by GABA are prevalent in the central nervous system (CNS). The inhibitory signaling driven by GABA originates at the amygdala and plays a pivotal role in normal as well as pathologic conditions. The activity of the GABA is modulated by the presence of allosteric binding sites on its surface. Antianxiety drugs in principle work by binding to the allosteric sites. In anxiety mental state, there are changes in the endogenous modulators of this allosteric site and the deviations in the subunit composition of the GABA receptor [16].

Two types of GABA receptors are known to impact the neuropsychology: ionotropic and metabotropic. Ionotropic receptors are regulated by the chloride ions and mediate rapid inhibition. On the other hand, the metabotropic receptors are regulated by G protein-coupled signaling (calcium or potassium channels) and have slow but prolonged effect [17, 18].

Glutamate is the excitatory neurotransmitter in the CNS and noted to play an important role in anxiety disorders as well as in learning, in memory retention, and in neurodegeneration. Glutamate regulates the release of neurotransmitters from presynaptic neurons. Glutamate mediates its action mainly through ionotropic receptors. The classic example of it is the N-methyl-d-aspartate (NMDA) receptor. The NMDA receptor is made of four subunits in turn introducing ample heterogeneity in regulation of its activity. Glutamate signaling plays an important role in fear conditioning in anxiety condition. Glutamate excitotoxicity is mediated by stress response leading to neuronal damage and/or death. Diminished glutamate release leads to activation of NMDA receptor [19, 20]. Antidepressant drugs work by decreasing the activity of glutaminergic neurons. Similarly the NMDA antagonists are demonstrated to mediate the antidepressant and antianxiety effect by blocking its activity in clinical and preclinical study [21, 22].

20.2.2 Serotonin

Serotonin, also known as 5-hydroxytryptamine (5-HT), is an important neurotransmitter for emotional well-being. In addition to regulating the mood and the associated anxiety, serotonin also helps to normalize the sleep, energy homeostasis, and cognitive functions. Postmortem study showed patients suffering from depression and suicidal deaths have upregulated 5-HT expression in the midbrain [21, 22]. Reduced serotonin receptor is reported in the cases of patients suffering from depression and anxiety [23–25]. SNRIs (serotonin-norepinephrine reuptake inhibitors) are the class of drugs which selectively inhibit the reuptake of serotonin.

20.2.3 Melatonin

Melatonin is a sleep hormone. It is secreted at night and regulates the sleep-wake cycle. It functions through the G protein-coupled receptors and regulates the wide array of physiological responses as sleep, anxiety, circadian rhythms, and immune functions. Melatonin is often recommended as initial treatment for patients suffering from insomnia, disturbed circadian rhythms, and anxiety [26, 27].

20.3 Functional Neurological Stress and Disorders

Neurological disorder manifests at the border of psychiatry, mind, and physical body intertwined with the neural connections. Patients suffering from this disease are often noted to have abnormal neurological symptoms in absence of a neurological disease. These disorders have disabling and distressing consequences in the life of patients [28, 29]. Early life stressful events as child, physical abuse, emotional negligence, and the relevant incidences are postulated to associate with the occurrence with the functional neurological disorders.

To understand the neurological stress and the associated deviation from the normal mental health, it is important first to identify the boundary between the normal and abnormal brain conditioning. The distress or the anxiety originating from the distress can be transient or situation dependent. On the other hand, the anxiety disorders are often characterized by persistent or chronic stress, which to an extent compromises the normal physiological responses [30]. The dysregulated physiological readouts in the condition of stress and anxiety are measured by the few common responses as neurological disorders, dermatological illness, and cardiovascular disorders.

20.3.1 Neurological Stress and Its Impacts

The most common readout of the neurological stress is migraine. The patients suffering from the neurological stress are frequently identified with the migraine, and often its intensity is used as a measure. Stress originating from the early life events is often found to be associated with the cases of temporal lobe epilepsy. The severity of stress is also found to influence the symptoms and clinical progression of Parkinson's disease. The worsened stress response is seen with concomitant rise in the partial body tremor which compromises the normal physiological actions [31].

Stress is known to impact the intensity and magnitude of cognitive perception. As cognitive development has its origin in the temporal lobe and amygdala, it is imperative that the stress has its influence to those particular regions of the brain. Stress is known to have negative impact on cognitive development. Stress induces the release of glucocorticoids in the blood and waves cognitive responses [32, 33]. Stress is also known to induce the level of responsive hormones as cortisol and the associated factors as IL-6 and cAMP responsive element binding protein as well as the brain-derived neurotrophic factor (BDNF) which are mostly upregulated in peoples suffering from anxiety and distress [34]. The levels of stress are hypothesized to have a positive impact on memory and cognitive development when its intensity is below the threshold, which is a subjective measure. Once the stress crosses its threshold, it negatively impacts the memory and the cognitive development.

Stress is reported to create immunosuppressive environment in the body. Stress-mediating hormones as adrenocorticotropic hormone (ACTH) and corticotrophin-releasing hormone (CRH) are influenced by the neuroendocrine system. It mediates the immune impairment once they cross the blood-brain barrier and are available in the body. Stress downregulates the release of growth hormones while promotes the secretion of opioids to fight the stress [35, 36].

Stress is known to affect the function of cardiovascular system on account of positive and negative effect by affecting the function of autonomous nervous system. Persistent stress induces the release of hormones and impacts the physiology of the heart by increase in pulse and contraction of arteries. Stress-induced activation of parasympathetic nervous system decreases the blood pressure, can stop the heartbeats, and increases the risk of compromised blood supply to the brain leading to higher chances of platelet aggregation and ischemic stroke [37, 38].

Stress is also recognized to impact the health of the gastrointestinal (GI) tract by influencing the relevant functions. For instance, stress modifies the digestive functions and intestinal permeability and activity of ion channels which controls the acid secretion [39] and influences appetite, GI tract movement by reducing the water and food intake [40]. Further, stress impacts the GI inflammation and the bacterial count and peptic ulcers. Together, the stress controls the functional aspects of the GI tract by impacting the functions as modifying the GI secretion, microbiome composition, inflammatory responses and the food absorption. Stress, thus, has both beneficial and hazardous effects on the human physiology. The beneficial effects are visible as long as the stress levels are under threshold. Stress response should be dealt with the personalized medicines as each individual may vary with the respective stress-bearing thresholds to have the maximum benefits.

20.3.2 Anxiety and Cancer

Cancer, on account of its occurrence and present mode of treatment, is observed to associate with the stress disorders as anxiety and depression. Nearly 50% of the cancer patients are found to suffer from the psychiatric disorders, partly because of the trauma because of the disease and partly as side effect of the treatment. Particularly, the breast cancer patients are found to be diagnosed with mood disorders, alcohol addiction, nicotine dependence, and mental abnormalities as a result of treatment regime [41, 42]. Similarly, children suffering from cancer are found to suffer from anxiety, depression, and post-traumatic stress disorder as compared with the general population [43, 44]. Growing body of evidences also suggests that the psychologic health of parents impacts the mental health of children, particularly when suffering from cancer [45]. Among the other cancers noted, 15% of the glioma and 20% of meningioma patients are found to have subclinical depression symptoms [46, 47]. There are heterogeneous evidences on association of depression and brain tumors. High-grade glioma patients with depressive symptoms are noted to have poor prognosis [48, 49], while other reports note no significant association of high-grade glial tumors with depressive symptoms. This report however intelligences the lesser survival in low-grade glioma patients [50]. Approximately 30–60% of the brain tumor patients are diagnosed with the anxiety disorders. Glioma and meningioma patients with anxiety are reported to have reduced cognitive functions and poor quality of life [51, 52]. Meningioma patients with severe depression are found with sevenfold enhancement in 5-year mortality [47]. Anaplastic astrocytoma and glioblastoma patients detected with the clinical depression are found to have median survival of 7 months, while those who are not suffering were recorded with 11 months of median survival post diagnosis. This survival in this study was found to be independent of adjuvant treatment, age of patients, or the extent of resection [48]. Most of the studies have retrospectively analyzed the association between brain tumors and the psychological abnormalities, often by extracting the details from medical records. There could be many incidences which are not treated separately and go unnoticed. So, it needs methodologically rigorous study to observe the link and to investigate the underlying mechanisms between the pathology of brain tumors and the clinical depression and anxiety. Further efforts are needed to consider the association of psychological abnormalities separately in brain tumor cases, particularly in meningiomas, and to mitigate with the antidepressant treatment [53].

20.3.3 Anxiety and Ischemic Stroke

Anxiety is reported to affect nearly 25% of stroke patients. Most of the patients are observed to have anxiety symptoms post stroke and often go neglected or unnoticed. It may be due to the fact that stroke often occurs in the elderly population while the anxiety is less common in the similar age group. As stroke patients may have the similar symptoms, there is need to revisit the anxiety thresholds and recalibrate the scale to diagnose anxiety and depression [54]. There are overlapping symptoms between the patients of anxiety and stroke, for example, sleep disturbance, fatigue which are commonly observed in the patients of anxiety, and stroke, which can make the diagnosis complicated [55]. Stroke patients are often observed to have "fear" of stroke recurrence. The risk was further found to be elevated in patients with previous history of anxiety disorder. Comorbidity of psychologic unrest can further aggravate the severity of existing symptoms in stroke patients. Together, there is a need to improve our understanding of anxiety disorder in stroke patients, particularly in post-diagnostic scenario.

20.3.4 Gut-Brain Connection and Stress

Gut microbiota is reviewed to be an important component in modification of central nervous system (CNS) immunobiology. There is a bidirectional cross talk between the neurotransmitters secreted by the CNS on gut microbes. The gut microbes are reported to mediate their part of communication in few ways as by inflammatory mediators, hormones, and enzymes. The classic example includes antibiotics mediated shift in intestinal bacterial population alters the pro/anti-inflammatory cytokines that impacts the brain function. Change in the metabolic reactions in the gut, for example, variation in tryptophan metabolism, is known to influence the behavior and have impact on axonal growth. On the other hand, activation of hypothalamic-pituitary-adrenal (HPA) axis by the brain affects the gut microbiota composition and gut permeability and causes shift in the local immune cell composition [56]. Gut microbes are reported to synthesize molecules as catecholamine, acetylcholine, gamma-aminobutyric acid (GABA), melatonin, serotonin, and histamines [57]. Emotional and psychological stress is reported to actuate the gut microbiota population [58]. Healthy peoples were noted to have reduced lactobacilli in their gut during the part of neurological stress and compared to the time when they were healthy [59]. Similarly, a study in rhesus monkeys has shown that when the neonates are forced with maternal separation during the age of 6–9 months, it causes reduction in the fecal lactobacilli [60]. Patients suffering from anxiety or the mood disorders are recommended with the bacterial nutrient supplement, also called as psychobiotics [59]. Nutritional supplement of certain probiotics is shown to benefit lab rats from the depression. On similar note, the probiotics have been reported to inhibit the stress-induced activation of GABA receptors in the brain [61–64].

20.3.5 Therapeutic Measures to Alleviate Stress

In present time, the major approaches undertaken as therapeutic measure include cognitive behavioral therapy (CBT), meditation activities, and drugs given to modify the course of hormonal balance.

20.3.5.1 Cognitive Behavioral Therapy (CBT)

CBT is the most widely used method to alleviate the stressful conditioned mind. It is the evidence-based psychological intervention prescribed to improve the mental health of the patients [65]. CBT treatment is based on principles of cognitive and behavioral psychology. The behavioral component includes the sleep control and body relaxation. Together with the cognitive and the educational component, the treatment is reported to yield the desired results [66]. Principally, CBT is given as treatment for the primary and comorbid sleeplessness in adults as well as in younger population [67]. The effect of CBT is also studied on immunological and psychological front [68]. In addition, CBT is also found to be beneficial in older adult patients suffering from mental worry and the psychological stress. Patients who underwent the CBT were observed with reduced stress and anxiety and improved mental health.

20.3.5.2 Antidepressant and Antianxiety Medications

Antianxiety and antidepressant medications work in principle by influencing the activity or release of neurotransmitters. These medications are known to take 4–6 weeks to stabilize its effects. Often initially, the patients are observed to note the worsened symptoms before the medications have the noticeable effect.

The medications are categorized into few classes on account of its mode of action: selective serotonin reuptake inhibitors (SSRIs), tricyclics, monoamine oxidase inhibitors (MAOIs), and the beta-blockers. SSRIs influence mood, sleep, sexual desire, and memory by increasing the serotonin uptake. The most common drugs prescribed under this category are paroxetine, fluoxetine, sertraline, and escitalopram. The most common adverse effects of these medications are drowsiness, dizziness, dry mouth, and muscle weakness. Tricyclics work on similar note as SSRIs. They are generally started with the low dose and increased gradually in response to the condition of patient. Common examples of this category are imipramine and clomipramine. MAOI class of drugs are prescribed when the patient has symptoms of phobia and panic disorder. The approved drugs under this category are selegiline, tranylcypromine, isocarboxazid, and phenelzine. These are relatively older class of drugs and have been observed to be associated with more adverse side effects as increased blood pressure leading to life-threatening consequences. Beta-blockers are used to treat the anxiety situations. The commonly prescribed medicine under this category is propranolol [69–71].

20.3.5.3 Meditation and Yoga to Alleviate Stress

Yoga is ancient part of Indian culture and spiritual practice. In literal terms, "yoga" means reunion of self with the ultimate being. Yoga is now seen in two ways: a form of physical exercise and, in holistic terms, maintenance of mind and body harmony. Additionally, yoga exercise has been found to improve the cardiovascular performance and glucose tolerance [72]. Similar to yoga, mindfulness meditation is an integral part of Eastern philosophy and suggests to focus attention in a unbiased way, by keeping mind in the present conditions with a view of acceptance [73]. Mindfulness meditation is found to be beneficial in psychological conditions of anxiety [74], depression [74], and obsession [75]. Recently, yogic transmission-based heartfulness mediation is being practiced in India. This type of meditation is based on the unique element of transmission, a form of divine energy for spiritual and material benefit. This type of meditation is observed to be beneficial for emotional and mental well-being [76].

20.4 Conclusions

We would like to highlight the underlying biology of neurological stress, emotion, and anxiety. The departure of emotional well-being in certain cases is caused by the deficiency of nutritional factors, hormonal imbalance, and predisposition to mental stress which on accumulation leads to the psychological abnormalities. The exclusive feature of psychological deviations is unique threshold baseline of tolerance, which may vary in every individual. The equal multifarious way is to recognize the need of medical and counseling help. The incidence of psychological condition makes the biology complex when associated with the metal disorders and diseases as Alzheimer's, ischemic stroke, and cancer. The most common way used to address the clinical help are through certain drugs which act as inhibitor to uptake or activity of certain neurotransmitters or hormones and the CBT treatment. In addition, mindfulness and the heartfulness meditation ways are used as adjuvant therapies and proven to yield fair results.

References

1. Le Moal M (2007) Historical approach and evolution of the stress concept: a personal account. Psychoneuroendocrinology 32(1):30
2. Selye H (1973) The evolution of the stress concept. Am Sci 61(6):692–699
3. Johansson L, Guo X, Waern M, Ostling S, Gustafson D, Bengtsson C et al 2010 Midlife psychological stress and risk of dementia: a 35-year longitudinal population study. Brain 133(Pt 8):2217–2224

4. Sugama S, Sekiyama K, Kodama T, Takamatsu Y, Takenouchi T, Hashimoto M et al (2016) Chronic restraint stress triggers dopaminergic and noradrenergic neurodegeneration: possible role of chronic stress in the onset of Parkinson's disease. Brain Behav Immun 51:39–46
5. Artemiadis AK, Anagnostouli MC, Alexopoulos EC (2011) Stress as a risk factor for multiple sclerosis onset or relapse: a systematic review. Neuroepidemiology 36(2):109–120
6. Lovera J, Reza T (2013) Stress in multiple sclerosis: review of new developments and future directions. Curr Neurol Neurosci Rep 13(11):013–0398
7. Alonso J, Liu Z, Evans-Lacko S, Sadikova E, Sampson N, Chatterji S et al (2018) Treatment gap for anxiety disorders is global: results of the World Mental Health Surveys in 21 countries. Depress Anxiety 35(3):195–208
8. Lopez AD, Murray CC (1998 Nov) The global burden of disease, 1990-2020. Nat Med 4(11):1241–1243. https://doi.org/10.1038/3218
9. Mauss IB, Robinson MD (2009) Measures of emotion: a review. Cogn Emot 23(2):209–237
10. Tang YY, Holzel BK, Posner MI (2015) The neuroscience of mindfulness meditation. Nat Rev Neurosci 16(4):213–225
11. Lee YS, Hwang J, Kim SJ, Sung YH, Kim J, Sim ME et al (2006) Decreased blood flow of temporal regions of the brain in subjects with panic disorder. J Psychiatr Res 40(6):528–534
12. Kent JM, Mathew SJ, Gorman JM (2002) Molecular targets in the treatment of anxiety. Biol Psychiatry 52(10):1008–1030
13. Engel K, Bandelow B, Gruber O, Wedekind D (2009) Neuroimaging in anxiety disorders. J Neural Transm 116(6):703–716
14. Paus T, Castro-Alamancos MA, Petrides M (2001) Cortico-cortical connectivity of the human mid-dorsolateral frontal cortex and its modulation by repetitive transcranial magnetic stimulation. Eur J Neurosci 14(8):1405–1411
15. Lim KO, Hedehus M, Moseley M, de Crespigny A, Sullivan EV, Pfefferbaum A (1999) Compromised white matter tract integrity in schizophrenia inferred from diffusion tensor imaging. Arch Gen Psychiatry 56(4):367–374
16. Martin EI, Ressler KJ, Binder E, Nemeroff CB (2009) The neurobiology of anxiety disorders: brain imaging, genetics, and psychoneuroendocrinology. Psychiatr Clin North Am 32(3): 549–575
17. Bowery NG (2010) Historical perspective and emergence of the GABAB receptor. Adv Pharmacol 58:1–18
18. Sieghart W (2006) Structure, pharmacology, and function of GABAA receptor subtypes. Adv Pharmacol 54:231–263
19. Bergink V, van Megen HJ, Westenberg HG (2004) Glutamate and anxiety. Eur Neuropsychopharmacol 14(3):175–183
20. Nishi M, Hinds H, Lu HP, Kawata M, Hayashi Y (2001) Motoneuron-specific expression of NR3B, a novel NMDA-type glutamate receptor subunit that works in a dominant-negative manner. J Neurosci 21(23):RC185
21. Fineberg NA, Brown A, Reghunandanan S, Pampaloni I (2012) Evidence-based pharmacotherapy of obsessive-compulsive disorder. Int J Neuropsychopharmacol 15(8):1173–1191
22. Gardoni F, Di Luca M (2006) New targets for pharmacological intervention in the glutamatergic synapse. Eur J Pharmacol 545(1):2–10
23. Boldrini M, Underwood MD, Mann JJ, Arango V (2008) Serotonin-1A autoreceptor binding in the dorsal raphe nucleus of depressed suicides. J Psychiatr Res 42(6):433–442
24. Lanzenberger RR, Mitterhauser M, Spindelegger C, Wadsak W, Klein N, Mien LK et al (2007) Reduced serotonin-1A receptor binding in social anxiety disorder. Biol Psychiatry 61(9): 1081–1089
25. Neumeister A, Young T, Stastny J (2004) Implications of genetic research on the role of the serotonin in depression: emphasis on the serotonin type 1A receptor and the serotonin transporter. Psychopharmacology 174(4):512–524
26. Stahl SM, Grady MM, Moret C, Briley M (2005) SNRIs: their pharmacology, clinical efficacy, and tolerability in comparison with other classes of antidepressants. CNS Spectr 10(9):732–747

27. Wilson SJ, Nutt DJ, Alford C, Argyropoulos SV, Baldwin DS, Bateson AN et al (2010) British Association for Psychopharmacology consensus statement on evidence-based treatment of insomnia, parasomnias and circadian rhythm disorders. J Psychopharmacol 24(11):1577–1601
28. Ludwig L, Pasman JA, Nicholson T, Aybek S, David AS, Tuck S et al (2018) Stressful life events and maltreatment in conversion (functional neurological) disorder: systematic review and meta-analysis of case-control studies. Lancet Psychiatry 5(4):307–320
29. Nicholson TR, Aybek S, Craig T, Harris T, Wojcik W, David AS et al (2016) Life events and escape in conversion disorder. Psychol Med 46(12):2617–2626
30. Finlay-Jones R, Brown GW (1981) Types of stressful life event and the onset of anxiety and depressive disorders. Psychol Med 11(4):803–815
31. White MG, Bogdan R, Fisher PM, Munoz KE, Williamson DE, Hariri AR (2012) FKBP5 and emotional neglect interact to predict individual differences in amygdala reactivity. Genes Brain Behav 11(7):869–878
32. Sandi C, Davies HA, Cordero MI, Rodriguez JJ, Popov VI, Stewart MG (2003) Rapid reversal of stress induced loss of synapses in CA3 of rat hippocampus following water maze training. Eur J Neurosci 17(11):2447–2456
33. Scholey A, Gibbs A, Neale C, Perry N, Ossoukhova A, Bilog V et al (2014) Anti-stress effects of lemon balm-containing foods. Nutrients 6(11):4805–4821
34. Song L, Che W, Min-Wei W, Murakami Y, Matsumoto K (2006) Impairment of the spatial learning and memory induced by learned helplessness and chronic mild stress. Pharmacol Biochem Behav 83(2):186–193
35. McCarthy L, Wetzel M, Sliker JK, Eisenstein TK, Rogers TJ (2001) Opioids, opioid receptors, and the immune response. Drug Alcohol Depend 62(2):111–123
36. Reiche EM, Nunes SO, Morimoto HK (2004) Stress, depression, the immune system, and cancer. Lancet Oncol 5(10):617–625
37. Cohen H, Benjamin J, Geva AB, Matar MA, Kaplan Z, Kotler M (2000) Autonomic dysregulation in panic disorder and in post-traumatic stress disorder: application of power spectrum analysis of heart rate variability at rest and in response to recollection of trauma or panic attacks. Psychiatry Res 96(1):1–13
38. Vrijkotte TG, van Doornen LJ, de Geus EJ (2000) Effects of work stress on ambulatory blood pressure, heart rate, and heart rate variability. Hypertension 35(4):880–886
39. DeBerry JJ, Robbins MT, Ness TJ (2015) The amygdala central nucleus is required for acute stress-induced bladder hyperalgesia in a rat visceral pain model. Brain Res 5:77–85
40. Ranjbaran M, Mirzaei P, Lotfi F, Behzadi S, Sahraei H (2013) Reduction of metabolic signs of acute stress in male mice by Papaver rhoaes hydro-alcoholic extract. Pak J Biol Sci 16(19): 1016–1021
41. Kuhnt S, Brahler E, Faller H, Harter M, Keller M, Schulz H et al (2016) Twelve-month and lifetime prevalence of mental disorders in cancer patients. Psychother Psychosom 85 (5):289–296
42. Mehnert A, Brahler E, Faller H, Harter M, Keller M, Schulz H et al (2014) Four-week prevalence of mental disorders in patients with cancer across major tumor entities. J Clin Oncol 32(31):3540–3546
43. Dahlquist LM, Czyzewski DI, Copeland KG, Jones CL, Taub E, Vaughan JK (1993) Parents of children newly diagnosed with cancer: anxiety, coping, and marital distress. J Pediatr Psychol 18(3):365–376
44. Wijnberg-Williams BJ, Kamps WA, Klip EC, Hoekstra-Weebers JE (2006) Psychological adjustment of parents of pediatric cancer patients revisited: five years later. Psychooncology 15(1):1–8
45. Barrera M, Atenafu E, Hancock K (2009) Longitudinal health-related quality of life outcomes and related factors after pediatric SCT. Bone Marrow Transplant 44(4):249–256
46. Bommakanti K, Gaddamanugu P, Alladi S, Purohit AK, Chadalawadi SK, Mekala S et al (2016) Pre-operative and post-operative psychiatric manifestations in patients with supratentorial meningiomas. Clin Neurol Neurosurg 147:24–29

47. Rooney AG, Carson A, Grant R (2011) Depression in cerebral glioma patients: a systematic review of observational studies. J Natl Cancer Inst 103(1):61–76
48. Gathinji M, McGirt MJ, Attenello FJ, Chaichana KL, Than K, Olivi A et al (2009) Association of preoperative depression and survival after resection of malignant brain astrocytoma. Surg Neurol 71(3):299–303
49. Litofsky NS, Farace E, Anderson F, Meyers CA, Huang W, Laws ER (2004) Depression in patients with high-grade glioma: results of the Glioma Outcomes Project. Neurosurgery 54(2): 358–366
50. Mainio A, Hakko H, Timonen M, Niemela A, Koivukangas J, Rasanen P (2005) Depression in relation to survival among neurosurgical patients with a primary brain tumor: a 5-year follow-up study. Neurosurgery 56(6):1234–1241
51. Pringle AM, Taylor R, Whittle IR (1999) Anxiety and depression in patients with an intracranial neoplasm before and after tumour surgery. Br J Neurosurg 13(1):46–51
52. van der Vossen S, Schepers VP, Berkelbach van der Sprenkel JW, Visser-Meily JM, Post MW (2014) Cognitive and emotional problems in patients after cerebral meningioma surgery. J Rehabil Med 46(5):430–437
53. Bunevicius A, Deltuva VP, Tamasauskas A (2017) Association of pre-operative depressive and anxiety symptoms with five-year survival of glioma and meningioma patients: a prospective cohort study. Oncotarget 8(34):57543–57551
54. Campbell Burton CA, Murray J, Holmes J, Astin F, Greenwood D, Knapp P (2013) Frequency of anxiety after stroke: a systematic review and meta-analysis of observational studies. Int J Stroke 8(7):545–559
55. Grenier S, Schuurmans J, Goldfarb M, Preville M, Boyer R, O'Connor K et al (2011) The epidemiology of specific phobia and subthreshold fear subtypes in a community-based sample of older adults. Depress Anxiety 28(6):456–463
56. Petra AI, Panagiotidou S, Hatziagelaki E, Stewart JM, Conti P, Theoharides TC (2015) Gut-microbiota-brain axis and its effect on neuropsychiatric disorders with suspected immune dysregulation. Clin Ther 37(5):984–995
57. Barrett E, Ross RP, O'Toole PW, Fitzgerald GF, Stanton C (2012) Gamma-aminobutyric acid production by culturable bacteria from the human intestine. J Appl Microbiol 113(2):411–417
58. Dinan TG, Cryan JF (2012) Regulation of the stress response by the gut microbiota: implications for psychoneuroendocrinology. Psychoneuroendocrinology 37(9):1369–1378
59. Dinan TG, Cryan JF (2013) Melancholic microbes: a link between gut microbiota and depression? Neurogastroenterol Motil 25(9):713–719
60. Bailey MT, Lubach GR, Coe CL (2004) Prenatal stress alters bacterial colonization of the gut in infant monkeys. J Pediatr Gastroenterol Nutr 38(4):414–421
61. Ait-Belgnaoui A, Durand H, Cartier C, Chaumaz G, Eutamene H, Ferrier L et al (2012) Prevention of gut leakiness by a probiotic treatment leads to attenuated HPA response to an acute psychological stress in rats. Psychoneuroendocrinology 37(11):1885–1895
62. Bercik P, Park AJ, Sinclair D, Khoshdel A, Lu J, Huang X et al (2011) The anxiolytic effect of Bifidobacterium longum NCC3001 involves vagal pathways for gut-brain communication. Neurogastroenterol Motil 23(12):1132–1139
63. Desbonnet L, Garrett L, Clarke G, Bienenstock J, Dinan TG (2008) The probiotic Bifidobacteria infantis: an assessment of potential antidepressant properties in the rat. J Psychiatr Res 43(2): 164–174
64. Hsiao EY, McBride SW, Hsien S, Sharon G, Hyde ER, McCue T et al (2013) Microbiota modulate behavioral and physiological abnormalities associated with neurodevelopmental disorders. Cell 155(7):1451–1463
65. Musiat P, Tarrier N (2014) Collateral outcomes in e-mental health: a systematic review of the evidence for added benefits of computerized cognitive behavior therapy interventions for mental health. Psychol Med 44(15):3137–3150
66. Arico D, Raggi A, Ferri R (2016) Cognitive behavioral therapy for insomnia in breast cancer survivors: a review of the literature. Front Psychol 7:1162

67. Morin CM, Benca R (2012) Chronic insomnia. Lancet 379(9821):1129–1141
68. Wetherell JL, Gatz M, Craske MG (2003) Treatment of generalized anxiety disorder in older adults. J Consult Clin Psychol 71(1):31–40
69. Everitt H, Baldwin DS, Stuart B, Lipinska G, Mayers A, Malizia AL et al (2018) Antidepressants for insomnia in adults. Cochrane Database Syst Rev 5:CD010753
70. Shelton RC (2019) Serotonin and norepinephrine reuptake inhibitors. Handb Exp Pharmacol 6:10
71. Tripathi AC, Upadhyay S, Paliwal S, Saraf SK (2018) Privileged scaffolds as MAO inhibitors: retrospect and prospects. Eur J Med Chem 145:445–497
72. Harinath K, Malhotra AS, Pal K, Prasad R, Kumar R, Kain TC et al (2004) Effects of Hatha yoga and Omkar meditation on cardiorespiratory performance, psychologic profile, and melatonin secretion. J Altern Complement Med 10(2):261–268
73. Lang AJ (2017) Mindfulness in PTSD treatment. Curr Opin Psychol 14:40–43
74. Davis LW, Strasburger AM, Brown LF (2007) Mindfulness: an intervention for anxiety in schizophrenia. J Psychosoc Nurs Ment Health Serv 45(11):23–29
75. Moritz S, Cludius B, Hottenrott B, Schneider BC, Saathoff K, Kuelz AK et al (2015) Mindfulness and relaxation treatment reduce depressive symptoms in individuals with psychosis. Eur Psychiatry 30(6):709–714
76. Thimmapuram J, Pargament R, Sibliss K, Grim R, Risques R, Toorens E (2017) Effect of heartfulness meditation on burnout, emotional wellness, and telomere length in health care professionals. J Community Hosp Intern Med Perspect 7(1):21–27

Physiology of Emotion

21

Rituparna Barooah

Abstract

Emotion is a physiological experience with behavioral expression of feelings in response to any sensory information. The behavioral changes include musculoskeletal, autonomic, and endocrine responses. Emotion is an intriguing aspect of human physiology that has been studied from various viewpoints of philosophy, sociology, psychology, evolutionary biology, etc. Physiologically, activation of specific structures and areas in the brain has been associated with generation of discrete or generalized emotions. Emotions provide motivation, and drive for action as well as aid in decision making. Therefore regulation of emotion results in rational decision making and help improve relationship with self, society, and environment. Psychosomatic and lifestyle noncommunicable disorders are largely result of unregulated emotional behavior over a long period of time.

Keywords

Emotion · Dimension · Behavior · Health · Regulation · Psychosomatic disorder

21.1 Introduction to Emotion

Emotions are an integral component of identity and personality of a human being. It is an intriguing subject studied in various other disciplines as well, including sociology, humanities and arts, evolution and anthropology, philosophy, neurobiology, and psychology [1]. All sensations generate emotions and thoughts, and mind is the flow of subjective emotional experiences and feelings. Mental experiences, therefore, are a result of dynamic admixture of sensations, cognition, emotions, and thoughts that are stored as memory. This frenzied collection of the mental experiences constitute the stream of consciousness. William James' question

R. Barooah (✉)
Department of Physiology, NEIGRIHMS, Shillong, India

S. Paul (ed.), *Application of Biomedical Engineering in Neuroscience*,
https://doi.org/10.1007/978-981-13-7142-4_21

"What is emotion?" way back in 1884 still rings true in the current times. Expressions of emotions differ in different cultures according to the specific norms followed. Darwin, in *The Expression of the Emotions in Man and Animals*, observed few basic emotions that are present across species and cultures, e.g., anger, fear, sorrow, and wonder [2]. For example, the one core emotion shared by all mammals is the infant-mother bond. In fact, the word mammal originates from the Latin word *mamma* meaning breast.

Emotion derives its meaning from the French word *emouvoir* (to excite) and Latin word *emovere*, i.e., "to move." It is a complex psychophysiological experience occurring in the constantly wavering and chaotic mind that is capable of changing the physiology to another level in response to a stimulus. It serves as motivation that fuels an action. The essential elements are a subjective experience, an altered physiology, and an expressed behavioral change [2]. Emotion is an important component in learning, especially survival skills, sexual behavior, and social conditioning. In short, emotion adds color to life.

Seven basic feelings in man were mentioned as early as in the first-century Chinese literature. In 1970, psychologist Paul Ekman described six basic emotions that are inherently present in all humans concerned with the primitive survival instinct. These are automatic, unconscious, and instantaneous, namely, anger, fear, disgust, happiness, sadness, and surprise. Interest and joy were added to the list as independent emotions. Guilt, pride, shame, embarrassment, excitement, and shyness are examples of dependent, secondary emotions. A three-dimensional color wheel model (wheel of emotions) of complex emotions was proposed by Robert Plutchik wherein he grouped eight polar opposite emotions in four groups, i.e., joy-sadness, surprise-longing, anger-fear, and trust-suspicion. A complex emotion can be constructed from the basic, primary ones.

An emotional processing goes through stages of evaluation, and expression, and is classified on the basis of affectivity of the stimulus or valence. Positive emotions are of high valence. Arousal is the ability of the stimulus to produce an immediate survival reaction [3]. An emotional experience is described as pleasant/unpleasant characterized by valence and arousal in the fundamental two-dimensional model [4] (Fig. 21.1).

All said and done, emotions are purely biochemical and physiological events and phenomena that are essential for survival and reproduction, i.e., in preservation of self and species, which are common to all animals who are able to perceive sensations [5]. Thought and memories are nothing but a series of electro-biochemical processes that pass through billions of neural connections responsible for generating emotions at different times, thereby causing fluctuations in motivation and drive (mood and mood swings). Study of emotions has thus brought neurobiology and psychology in a single platform.

Functions of emotions can be clubbed on the basis of their effect on personal, interpersonal, and social human behavior. Emotion plays an important role in homeostasis and stress behavior, habit formation, addiction, motivation, attention, learning, memory, sexual behavior, communication, ability to form and work in team, decision making, reproduction, and recognition and avoidance of danger. The strongest emotions are directed at the most fundamental and basic needs, i.e., for

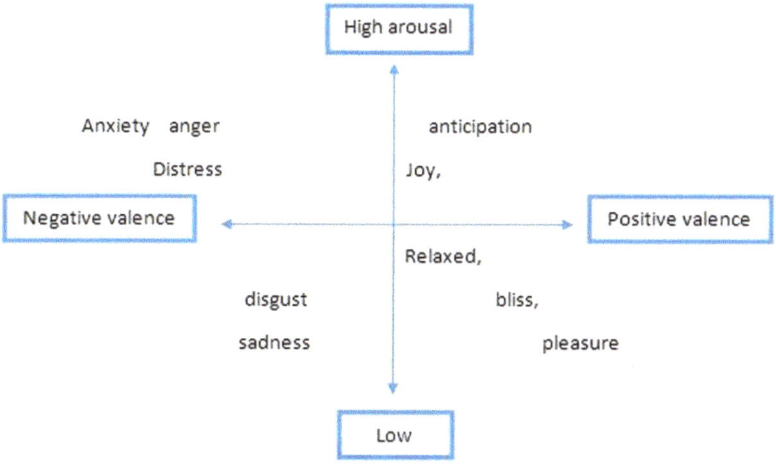

Fig. 21.1 Two-dimensional model of emotion. (Courtesy: author)

preservation of the self (availability of food, perception of threat to survival) and safety and preservation of species – sexual urges and reproduction.

21.2 Theories on Genesis of Emotion

An emotional experience was attributed to a set of behavioral expressions in response to a sensory stimulus by William James [6]. Almost a decade later, Carl Lange in 1885 also proposed a similar theory stating that emotion is determined by specific bodily changes. Different patterns of physical manifestations were thought to represent specific emotions. This theory came to be known as James-Lange theory.

Schachter and Singer's theory of emotion was supportive of the same view that is attributed to simultaneous development of physiological arousal and identification of the emotion. "Embodied cognition" is a concept hypothesizing that emotion and physical manifestations cannot be separate from thought and perception.

On the other hand, Cannon in 1920 observed that the same emotional behavior was exhibited with two different stimuli. With Bard, he established the Cannon-Bard theory following his experiment of "sham rage" on cats [7]. Emotions were demonstrated even in the absence of intact sensory motor apparatus. The Cannon-Bard theory states that the emotional experience results in physical manifestations. For example, seeing a snake generates a feeling of fear resulting in tachycardia. In this case, the feeling of fear had been stored in the memory from some earlier experience either as a negative past happening or as a deep belief. His contention was that autonomic activities were similar in response to emotional as well as somatic stimuli and physical manifestations were not quick enough for generation of emotions [8]. Antonio Damasio, in his somatic marker hypothesis, explained this

Table 21.1 Approaches to the popular theories of emotion

Theories	Approach
James-Lange theory	Physiological
Cannon-Bard	Neurobiological
Schachter-Singer	Cognitive
Richard Lazarus	Cognitive appraisal
Evolutionary theory	Adaptive and motivational
Damasio	Experiential learning

Courtesy: author

phenomenon as a learned behavior on the basis of physical symptoms. Accordingly, he has classified emotion into primary and secondary. It is a primary emotion when the stimulus leads to an emotional experience and physical manifestation almost at the same time. Secondary emotion is a learned behavior following exposure to primary emotion at a later time. There is development of association of either punishment (unpleasant feeling) or reward (pleasant experience). Memory of a previous experience, event, or thought gives rise to the same emotion. Hence, it serves as the stimulus itself, hence the adage "emotions can arise 'de novo.'" Recognition of the stimulus generates feelings followed by a desire for action. Cognitive labeling is a crucial feature of emotion, and so is conation. This finally culminates in expression of physical manifestation of an emotion. Language, executive control, and adaptive and synthesizing qualities of emotion are important aspects in the context of social and interpersonal relationships. In summing up, all sensations are learning experiences and are stored as memory. Pleasant experiences are most often repeated and rehearsed mentally (Table 21.1) culminating in a programmed neural circuit.

21.3 Evolution of Neurobiology

Galen (130 AD–200 AD) investigated into cerebral function, somatic senses, cranial nerve, and autonomic nervous system. He believed the brain to be the seat of intelligence unlike Aristotle in 384 BC–322 BC who had theorized heart to be the seat of intelligence. Cerebral ventricles were considered "seat of emotion" for a long time in 500 AD as proposed by St. Augustine. Da Vinci laid the foundation for development of neural science and mind, thereafter, as studies in neuroanatomy gained momentum, with publication of artwork by Michelangelo and Rembrandt. Research in neuroanatomy received considerable boost during the Renaissance movement [9].

Almost during the same time, neurochemistry was also at a budding stage of research in France. This established the role of molecular biology in genesis of emotion. Darwin's naturalistic theory of emotional expression had by this time established the existence of homologous and common basic set of emotions in all animals and humans as well as physiological basis of emotion. By 1925, Walter B. Cannon from Harvard was a leading physiologist in this field. He introduced a decorticated model of animal experimentation in the field of emotional physiology.

Russian physiologist Bechterect carried Cannon's observation further with experimental lesions in the brain at other levels too. Cannon's experiments were successful in demonstrating psychological influence on somatic manifestation. His experiments culminated in postulating the "thalamic theory of emotions" (Cannon-Bard theory) in collaboration with Philip Bard.

Physiology of emotion was included in brain science during the mid-1920s. Emotions were produced and studied in labs, as animal models came to be used for studying emotions. Physiological and psychological domains became established in a single platform, with physiologists studying brain functions influencing emotion. Mosso's studies of recording physiological parameters inspired Hans Berger to work on brain waves leading to the discovery of electroencephalography.

Thus advanced the advent of experiments regarding brain function during evoked emotions. This is done using behavioral experiments, electrophysiological recordings, functional neuroimaging studies (functional magnetic resonance imaging [fMRI], positron emission tomography [PET]), animal and human lesion studies, etc. Since all experiences are a result of electrical activity in the brain, therefore, it is not impossible to create an entirely virtual world not very different from the real one, in order to study different aspects of emotion. The role of emotion in learning too was explored in many research works. This was corroborated with experiments of Italian physiologist Angelo Mosso in producing graphical representation of physiological parameters in different states of emotion.

Emotions have been studied mainly through the locationist and psychological constructivist approaches [10]. The former postulates designated brain areas for specific and discrete emotions, while the latter endorses a network of psychological activities involving different brain structures in discrete or a group of emotions.

Source(s) of feeling can be direct as well as indirect. Adverse reactions in childhood in the form of threat, deprivation, or poverty of care hamper the development of emotional network both structurally and functionally [11]. Many arise out of visceral efferents indirectly or from neurochemical changes in the brain. A good example is uneasiness felt due to sensory activation of the gut wall in increased gut motility. Pain, often indicative of tissue injury, (either potential or actual) manifests with a great deal of subjective emotional experience [12].

21.4 Physiological Basis of Emotion

The Expression of the Emotions in Man and Animals by Charles Darwin had established existence of a physiological basis of emotions. He investigated not only the behavior in different emotions but also the modes of expressions, such as facial expression, voice, postures, etc. [2]. Affective neuroscience describes the neural physiological basis of emotion, emotion processing, as well as regulation and translation into motor behavior. Discrete areas for punishment and reward were demonstrated through experiments of Milner and James Olds in 1954. In their experiments, electrodes were placed on certain areas of the brain in a self-simulating apparatus. It was observed that the animals would frequently pull the electrodes which were placed on certain areas imparting pleasant feelings.

Therefore, emotional responses are of the following nature:

(a) Neuroendocrine and autonomic (sympathetic) responses.
(b) Skeletomuscular, i.e. physical manifestation of emotion as in facial, laryngeal, and peripheral muscle contraction.
(c) Subjective experience of thoughts and feelings which are further consolidated as memory, depending on the degree of arousal [2].The desire to act and adopt motor movement also is a volitional behavior [13]. Emotion can, therefore be described as energy in motion; and hence the emotional brain is also the seat of motivation and drive, obsession, and addiction.

Expression of emotions is neuroendocrine through activation of the hypothalamic-pituitary-adrenal (HPA) axis (adrenaline, cortisol) and secretion of neurohormones (antidiuretic hormone, oxytocin, prolactin, corticotropin-releasing factor), autonomic (increased heart rate, increased blood flow, increased respiratory rate), and musculoskeletal like pacing up and down, high-pitched voice, smile/frown, etc. Stimulation of punishment areas results in aggressive mechanisms and the sympathetic response of flight, fight, and flee. Integrity of the hypothalamic-pituitary-adrenal axis, the neocortex, or subcortical brain regions and circuits provides the basis for complex forms of emotion. The motor skeletomuscular expression is distinct in voluntary action and involuntary in emotional reaction. For example, a voluntary smile causes contraction of the muscle zygomaticus major and employs the corticospinal pathway, whereas an involuntary smile employs the pathway from the amygdala to the spinal cord via the anterior cingulate gyrus, through the basal ganglia (extrapyramidal) pathway. There is contraction of both zygomaticus major and orbicularis oculi muscles around the eyes. This is a skilled reaction that can be learned with practice, seen in stage acting.

The vertebrate brain developed to that of the human size from the rudimentary reptilian brain (like a mushroom) and was mainly responsible for satisfying the basic urges of survival. In fact, all behavior, actions, and decisions have an emotional and instinctive component generated in the subcortical structures of the primitive emotional brain. Quest for higher needs followed satisfaction of the basic urges. The accumulation of the neocortex over the emotional brain during evolution has lent inhibitory influence as social conditioning to the pursuit of basic urges. The emotions expressed during childhood are learned to be suppressed as chronological age advances as different and varied beliefs are piled up in the neocortex from multiple social interactions.

Paul Broca, a Paris-based neurologist, had named the medial cerebral cortex as the limbic lobe (*limbique* in French meaning loop). He attributed genesis of animalistic emotions to be its function. The rhinencephalon, olfactory tubercle and lobe, and peduncle and pyriform lobe were also included within the brain network areas for emotion as all the primitive and fundamental survival instincts, reproduction, and learning were carried out through the faculty of olfaction in lower animals (Fig. 21.2).

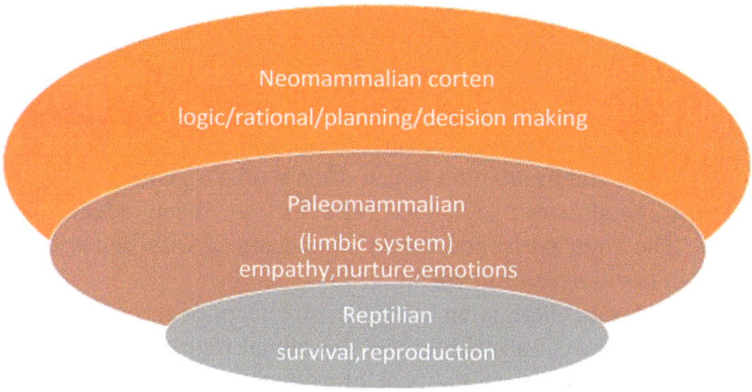

Fig. 21.2 Triune model of the human brain. (Courtesy: author)

The "triune model of the brain" was proposed by Paul MacLean in 1960. It was an anatomical model supporting the generation of emotion and feelings. MacLean's work was in fact an integral explanation of theories of Papez, Cannon-Bard, and Kluver-Bucy. He described the evolution of the reptilian brain (visceral instinctive brain) to be constituting of the striatal cortex and basal ganglia. Paleomammalian or emotional brain evolved with inclusion of newer subcortical structures and three-layered cortices [8]. The primitive emotions, as in fear, were later overpowered by the cortical inhibition due to social conditioning. His new brain system of emotion included the thalamus, hypothalamus, hippocampus, cingulate cortex, amygdala, and prefrontal cortex. Integration of emotion and cognition was hence established. He later coined the term limbic system. Walle Nauta expanded the emotional network to the mesencephalon with inclusion of the habenula.

The neomammalian brain with a six-layered neocortex was the latest brain structure in evolutionary timeline responsible for memory, reasoning and rationality, and planning. The neocortex effectively provided a restraining quality and added rationality to emotional reactions.

Far more organized expression of emotion evolved with conscious rational evaluation as a result of piling up of the further layers of cortical tissue (neocortex) in addition to the amygdala, a pea-sized structure beneath the temporal lobe, which is the seat of unconscious evaluation in the emotional brain [14]. Unconscious evaluation results in immediate, unconscious reaction, often as withdrawal in pain. Another structure that is activated in this process is the hippocampus where consolidation of memory occurs.

Decisions taken by us on a day-to-day basis are usually a blend of both unconscious (emotional) and conscious and rational (cortical) reactions in different proportions. Reasoning, a function of the prefrontal cortex, therefore refines the basic lower emotional reaction. This has been explained in *Descartes' Error* by Antonio Damasio who goes on to quote "emotions are not luxury." Rational evaluation with logic and reasoning (function of the prefrontal cortex) of emotions is necessary to arrive at a productive decision.

More than one brain network area is involved in emotion as well as cognition as also in attention, working memory, and sensory processing. Brain oscillations and high degree of neural connectivity aid in the process of emotional regulation, reasoning, and decision making.

The prefrontal cortex was assigned by MacLean as the area for reward processing, learning, and motivation which often works together with the amygdala. This area also modulates adjustive behavior to meet adequate emotional consequences. The right prefrontal cortex is concerned with behavioral inhibition and the left with positive reinforcement.

The case of Phineas Gage is an example of effects of the prefrontal cortex on emotion processing, decision making, and emotional regulation. He was an intelligent, efficient, and popular personality. Unfortunately, while working in a railroad company, he met with an accident causing an iron rod weighing 6 kg more than 3 cm in length to pierce his left cheek and base of the skull. The rod flew out through the top of his head to land more than 30 meters away. The pointed tip left a clean injury, and he recovered within a few months. But his personality changed as he was recovering. Once a charming personality, his behavior gradually became offensive and inefficient, and finally he passed away at the age 38. Although his faculties were intact, his behavior changed gradually, finally losing the ability to make correct decisions due to damage in the ventromedial prefrontal cortex. His was a case studied by Antonio Damasio in great detail. Egas Moniz also reported of prefrontal leucotomy performed for anxiety which produced similar results of passivity and unsound and minimal emotional reactions.

Ablation of the bilateral temporal lobe for treatment of epilepsy resulted in heightened emotional reactivity, increased exploratory behaviors (oral and otherwise), as well as hypersexuality and aggression. Kluver and Bucy reproduced the same syndrome of hyper-orality and psychic blindness in ablation of the amygdala in monkeys. MacLean later established the temporal lobe as responsible for genesis of emotion. The nucleus accumbens, ventral pallidum, insula, somato-cortical areas, and brainstem were also included as emotional processing areas by 1990.

Assimilation and integration of both limbic, subcortical, and cortical structures is an essential feature of emotion. Neuroimaging studies, since then, have made it possible to map the emotion-processing brain structures [15].

21.5 Cortical and Subcortical Structures Associated with Emotion

Cortical structures are concerned with thinking, control of behavior, foresight, perception and memory, clarity, discrimination, reflection, planning, reasoning, and problem solving.

Subcortical processing of sensory information is reactive: “I act; then I think.” Neuroimaging studies have reported activities in specific structures corresponding to specific emotions.

21.5.1 Cortices (Cerebral)

1. Cingulate cortex – frontal (dorsolateral and orbitofrontal) and anterior: Lesser activity in this area is seen in fMRIs in depressive conditions. More activity is noticed on the left side being more active than the right. The orbitofrontal cortex, in connection with the cingulate cortex, amygdala, and medial striatum, forms a part of the reward circuit, associated with motivation. The lateral part of the orbitofrontal cortex is more concerned with cognitive tasks, judgment, as well as social behavior, anger, and aggression and is associated with emotion of high valence. Frustration is commonly observed with verbal abuse in human clinical conditions involving the area [5]. The pre- and subgenual cingulate cortex is associated with sadness and other visceromotor responses in pain. Both structural and functional changes are noted in the subgenual cortex in clinical depression. Anterior midcingulate cortical activity is noted in motor engagement in emotional behavior. The posterior cingulate cortex is concerned with emotional processing.
2. Dorsomedial prefrontal, medial temporal, and retrosplenial cortex: Regulation of emotional experiences and perception based on memory of similar prior experience is associated with activity in these areas of the cerebral cortex. With the anterior temporal lobe, the ventrolateral prefrontal cortex mediates language-related perception of emotions. Atrophy is noted in semantic dementia with loss of empathy. The dorsolateral prefrontal cortex with the periaqueductal gray regulates autonomic output during behavioral adaptation. Circuits concerned with rage, joy, distress, fear, love, and lust are associated with the periaqueductal gray.
3. Visual cortex: Increased activity in application of visual methods and cues used for generation of unpleasant emotions of fear and disgust. The area is associated with motivated action pattern.
4. Insular cortex: This is an area associated with emotional homeostasis. The anterior insula has reciprocal connections with the amygdala. It processes the emotion of disgust (Damasio thought it to be the one of the most primitive emotions) and discomfort, attention and orientation, and evaluation of emotional memory. The posterior insula receives afferents from the ventral thalamic nucleus and efferents to the somatosensory cortex. The ventral thalamic nucleus is associated with smell and taste and the emotion of disgust, which Damasio described as one of the earliest experiences. The insula is termed as the interceptive relay center as the emotional experience is consciously evaluated in insular neocortical connection.
5. Entorhinal cortex: This is an area in the temporal lobe between the hippocampus and neocortex. The entorhinal hippocampal connection is of importance in the formation of episodic memory and memory consolidation. The presence of the lamina dissecans lends a unique feature due to absence of neuronal cell bodies in this layer. The entorhinal area happens to be the first area to be affected in Alzheimer's disease.

21.5.2 Subcortical Structures

Limbic Structures Amygdala, septum, hypothalamus, habenula, anterior thalamic nuclei, and medial surface of the cerebral hemisphere (Broca's lobe) – dentate gyrus, subiculum, olfactory tract and lobe, and part of the basal ganglia

Diencephalon Forebrain, thalamus, hypothalamus, epithalamiums, subthalamus, hippocampus, and parahippocampal gyrus

The *amygdala* is the key structure in emotional physiology. It is often called the "window" through which an individual perceives his worldview. The amygdala receives information from the thalamus, cortical association areas, medial orbitofrontal cortex, hippocampus, septum, and basal ganglia. The processed information is then transmitted to the hypothalamus via the stria terminalis and regulatory centers in the brainstem for behavioral responses. It is the center of modulation of behavioral responses, fear, automatic regulation (high arousal) of emotion, and emotional memory. This is an area for facial recognition of fear [16]. It's the seat of conditioned fear, genesis of anxiety, and facial recognition of aversion [17]. Sexual behavior for reproduction is also attributed to this structure in all species, by virtue of its proximity with the olfactory bulb. It comprises three nuclear groups [18]:

1. Corticomedial: Receives afferents from the olfactory bulb and efferent connections to the hypothalamus. It is an integrating area for sexual behavioral and endocrine pathway in reproduction.
2. Basolateral: Receives afferents from the sensory association area and hippocampus. Efferent connections project to the ventral striatum, nucleus accumbens, and dorsomedial thalamus.
3. Central: Together with fibers from basolateral and corticomedial nuclei, fibers are projected to the dorsal nucleus of the vagus nerve, facial motor nucleus, and raphe nucleus and locus coeruleus. Lateral division of the nucleus is called the "nociceptive amygdala." Hyperactivity of this area occurs in cases of subjective experience of painful episode and anxiety. The area receives pain afferents from the spinal cord and brainstem through the spinoparabrachial-amygdalar pain pathway [18].

The *hypothalamus* is the emotional transducer where emotional information from the amygdala, insula, orbitofrontal cortex, and other parts of the limbic system is converted to autonomic and endocrine responses. Neuroendocrine and autonomic functions of the hypothalamus aid in homeostasis. The HPA axis and septal area are responsible for the alarm reaction with secretion of adrenaline. Sustained activation of the HPA axis for a considerable length of time results in deleterious effects on other organs and abnormal physiology.

The paraventricular nucleus, by virtue of incoming connection from the amygdala, hippocampus, prefrontal cortex, and locus coeruleus, is responsible for autonomic and hypothalamic output. The preoptic nucleus is associated with sexual behavioral output.

The hypothalamus-amygdala connection integrates the neuroendocrine (activation of the HPG axis), autonomic, and sexual behavior (reproductive neuromodulators in the limbic system). Emotion, sexual behavior, and reproduction are the integrated functions of the limbic system [19].

The *hippocampus* is located just behind the amygdala and is shaped as a seahorse in structure. Through the entorhinal cortex, hippocampal fibers extend to the sensory association cortex and further to mammillary bodies of the hypothalamus through the fornix as part of the Papez circuit. Declarative formation of memory occurs in the hippocampus as well as consolidation of the same and recall of memory.

Thalamus All sensorimotor, cognitive, and emotional information are processed here. The ventral striatum is associated with cognitive processing and along with the nucleus accumbens forms the reward circuit. The corticostriatal-thalamocortical loop (CSTC loop) regulates motor skill in emotion and cognition. The anterior thalamus is a part of the Papez circuit and involved in memory formation.

Striatum The upper dorsal striatum is associated with motor control and cognition. The lower and ventral striatum with the nucleus accumbens is concerned with emotion and behavior in the reward circuit. The striatal fibers synthesize dopamine, a rewarding chemical. The core striatum forms the extrapyramidal motor system and is a part of the CSTC loop.

Thus, emotions emerge from some basic operations working on a set pattern. Designing an effective framework for mapping emotions to brain areas is still being a subject of research.

21.6 Neural Circuits and Loops

The subcortical and cortical structures related to emotions are interconnected, most often in a reciprocal manner (Fig. 21.3).

1. *Papez circuit* (hippocampo-mamillo-thalamocortical circuit): The Papez circuit was described in 1937 by James Papez. Sectional studies on the brain paved the way for the discovery of thalamic connection upward to the neocortex and downward to the hypothalamus, striatal structures, cingulate cortex, and brainstem. The former is responsible for the perception and memory of the emotional experience and the latter for physiological and behavioral expression. The upward output to the neocortex through the cingulate gyrus to the hippocampus serves consolidation of memory. From mammillary bodies of the hypothalamus, fibers further extend to the anterior thalamus and cingulate cortex via the mammillothalamic tract. The anterior cingulate cortex (ACC) is apparently a common interface for conscious learning and memory as well as emotional behavior. This circuit is of considerable importance as it is the site where

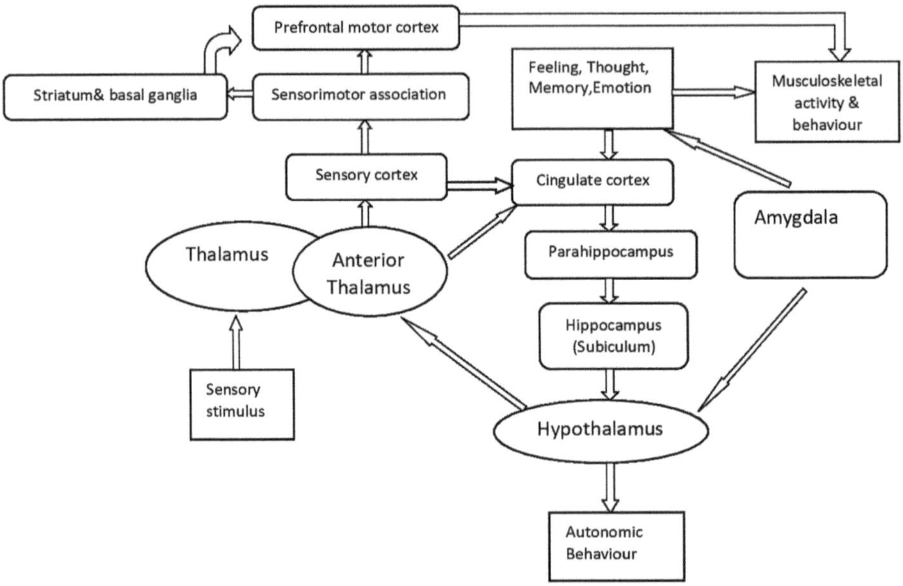

Fig. 21.3 The Papez circuit and physiological expression of emotion. (Courtesy: author)

emotional experience is translated into physical expression. Connection with the pineal body provides impetus to emotions and circadian influence on emotion processing [8].

2. *Associative cortico-striato-thalamo-cortical circuit:* This circuit begins in the lateral and dorsolateral orbitofrontal cortices to the striatum and anteromedial putamen through the internal globus pallidus and substantia nigra (basal ganglia) to the ventral anterior thalamus through the inferior thalamic peduncle and to the prefrontal cortex. The circuit is responsible for processing empathy, working memory, spatial orientation, social interaction, and initiation of action.
3. *Limbic CSTC loop*: The anterior cingulum and orbitofrontal cortex to the ventral striatum including the caudate nucleus, putamen, and nucleus accumbens. Afferent connections arrive from the amygdala, hippocampus, and entorhinal cortex. Efferent connections project to the ventral globus pallidus. This circuit is involved in motivational behavior.

21.7 Neurotransmitter Circuits

The major neurotransmitters that affect the functions of the limbic system belong to the monoaminergic group. These are dopamine, serotonin, noradrenaline, and acetylcholine. They are distributed diffusely creating a “diffuse regulatory system.” Synthesized within the specific brainstem nuclei, these monoaminergic neurotransmitters are responsible for so-called “moods” and for lability of emotions.

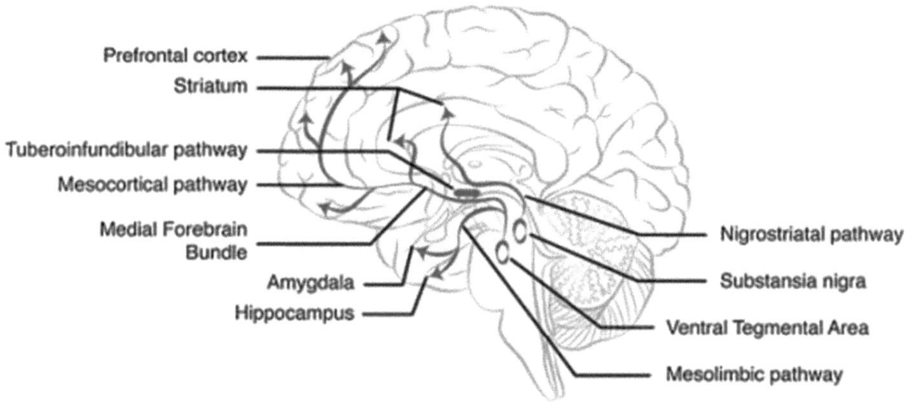

Fig. 21.4 *Dopaminergic pathway*. (Adapted from Lévêque M. Neuroanatomy of Emotions. In: Psychosurgery: New Techniques for Brain Disorders. Springer International Publishing; 2014)

For example the reproductive hormones and neurotransmitters greatly vary through the woman's reproductive cycle. Major neurotransmitters are dopamine synthesized in the substantia nigra and the ventral tegmentum associated with motivation and sleep.

Dopamine is by far the most important neurotransmitter. The four major dopaminergic pathways are the following:

1. Nigrostriatal pathway for motor control
2. Mesolimbic pathway in the reward circuit
3. Mesocortical pathway (to ACC) in concentration and executive functions
4. Tuberoinfundibular pathway in the hypothalamus in inhibition of prolactin hormone

It is also implicated in the reward loop through the CSTC loop (Fig. 21.4).

Serotonin, synthesized in the raphe nucleus, is associated with modulation and expression of pain behavior, mood, and sleep (Fig. 21.5).

Norepinephrine is the sympathetic neurohormone synthesized in the nerve endings as well as locus coeruleus associated with attention, anxiety, vigilance, and alarm reaction (Fig. 21.6).

GABA (gamma-aminobutyric acid) is linked to behavioral apathy in substance abuse and also postpartum depressive behavior. Glutamate, on the other hand, is also concerned with addiction via the anterior cingulate gyrus. Glutamatergic signals in the prefrontal cortex lead to alleviation of the cognitive process in hyperarousal emotional states through corticosterone production.

Acetylcholine, a monoamine synthesized in the nucleus accumbens, is responsible for affective behavior for propelling into action [20].

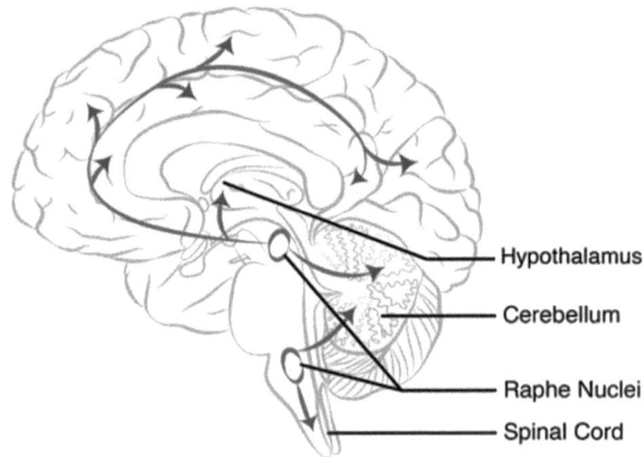

Fig. 21.5 *Serotonergic pathway*. (Adapted from: Lévêque M. Neuroanatomy of Emotions. In: Psychosurgery: New Techniques for Brain Disorders. Springer International Publishing; 2014)

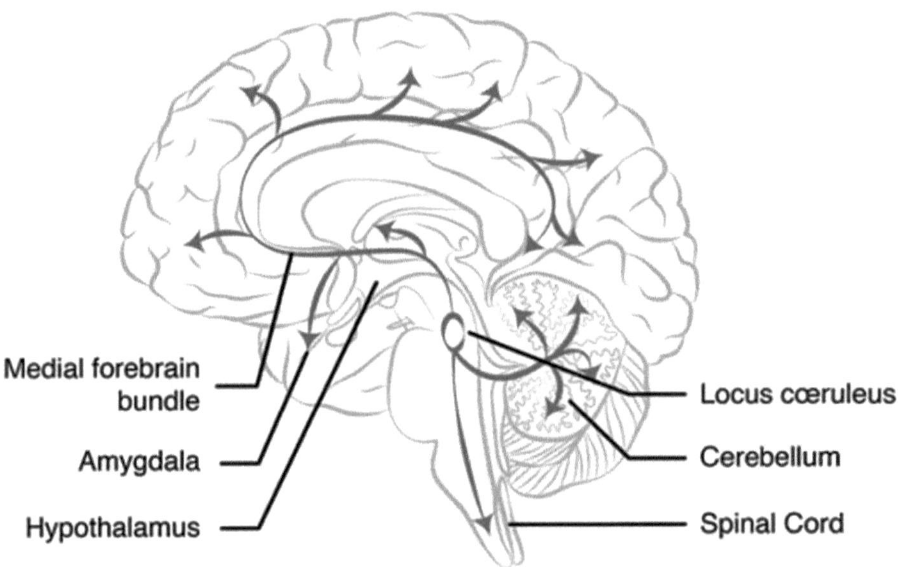

Fig. 21.6 *Noradrenergic system*. (Adapted from Lévêque M. Neuroanatomy of Emotions. In: Psychosurgery: New Techniques for Brain Disorders. Springer International Publishing; 2014)

21.8 Endocrine Influence on Emotion

The affective brain system includes the orbitofrontal cortex, ACC, amygdala, brainstem, and ventral striatum. Hypothalamic hormones with activation of the HPA axis as well as HPG axis regulate these functions. Receptors are abundant within the brain structures. Therefore, any hormonal imbalance results in a social, sexual, and emotional withdrawal/aggressive state.

1. CRH, ACTH, glucocorticoids, and CRF from the hypothalamus (paraventricular nucleus) mediate both neuroendocrine and autonomic outputs. CRF neurons demonstrate neuronal plasticity in sustained activation states. HPA axis activation and excess corticosterone production in hyperarousal states of emotion impair major cognitive processes in the prefrontal cortex through glutamatergic signals [21]. Acute activation in this area shows enhanced working memory. Increased noradrenergic and dopaminergic receptor activity has also been recorded in the area [15]. These molecules are also implicated in addictive emotions and neuroplasticity in stress [22].
2. Oxytocin and vasopressin: Oxytocin is a hormone synthesized in the hypothalamus and secreted in the median eminence. Oxytocin is a nurturing hormone associated with the emotions of trust, altruism, and bonding. Oxytocin, like testosterone, is important for development of social behaviors. Intranasal oxytocin application is an established therapeutic modality in schizophrenia [23]. Vasopressin, by virtue of its receptors in the amygdala, hypothalamus, hippocampus, cingulate gyrus, and brainstem, is associated with emotions of fear, anxiety, and aggression through ACTH of the HPA axis [24]. The role of both the hormones in reproductive and postpartum behavior is well established [25].
3. Prolactin: It is a parental hormone. Elevated levels of prolactin signify association with the emotion of caregiving and child raising in both males and females [26]. Prolactin enhances the physiological effects of growth hormone and TSH. Increased plasma level of prolactin suppresses the release of gonadotropins [27, 28]. Alcohol, caffeine, opioids, nicotine, amphetamine, cocaine, marijuana, and cannabis are exogenous mood elevators They activate the HPA axis via CRH receptors and are commonly used for enhancing performance, motivation to act, or reducing stress [29]. These are substances most commonly abused culminating in ill health.
4. Melatonin: Melatonin is synthesized and secreted from the pineal gland and had been described as "the seat of the soul" by Descartes in his *Passions of the Soul.* The pineal gland and melatonin regulate the circadian rhythm and are also associated with imagination, sensory processing of bodily movements, and memory. Melatonin also affects the electrical activity in the hippocampus and can alter the synaptic transmission in stress [30].
5. Thyroid hormones: Low levels of thyroid hormones are associated with depressive symptoms and blunted emotion [31]. The hypothalamic-pituitary-thyroid axis is activated with the release of TRF in the hypothalamus. It is regulated in the most precise manner by the negative feedback mechanism by the circulating levels of thyroid hormones. Deficiency of any component at any level may produce symptoms like emotional lability, anxiety, depression, irritability, or apathy. Low circulating levels of the thyroid hormones are associated with psychomotor slowing and melancholia, and memory loss is common in hypothyroidism [32].

6. Kisspeptin, GnRH, and GnIH: Kisspeptin is a neuropeptide in the hypothalamus stimulating GnRH release followed by subsequent activation of the HPG axis. It is found to be responsible for readiness for sexual behavior and sexual act in animals. GnRH is an anxiolytic and antidepressant in many psychotic conditions. On the other hand, GnIH has an inhibitory effect on sexual libido. GnRH affinity is recorded in the hippocampus and amygdala besides the olfactory bulb and tubercle, arcuate nucleus, and pyriform cortex, highlighting the importance of the evolutionary survival function. Receptors are located also in the bed nucleus of the stria terminalis (BNST) which is presumed to be an extension of the amygdala and is an influential neuron in the hippocampal periventricular pathway [33].This regulates sustained, social aggression, mating, and bonding of child care [19].
7. Estrogen, testosterone, and sex steroids: Derived from cholesterol, the female reproductive steroid hormones cross the blood-brain barrier. Apart from ovaries, they have been reported to be secreted in the brain also (neurosteroid). The levels vary throughout the reproductive cycle; they are strong neuromodulators and influence eating and reproductive behavior-reward function. Besides neuroregulation, they also regulate the neurotransmitters and neurosteroid (thereby affecting mood and behavior), HPA axis, and immune function [22]. Testosterone is the male hormone responsible for sex drive. It is also associated with aggression, violence, and sexually motivated behavior. Impulsive social and emotional behaviors are commonly noticed. The hormones exert their influence on the parahippocampal area, amygdala, insular cortex, and occipital lobe [34]. High levels of estrogen in females also correspond to high libido and sexual behavior [19].
8. Vitamin D: Deficiency of fat-soluble vitamin D in pregnant women has been seen to be associated with increased incidence of autism and cognitive defect in children. Deficiency in older adults is associated with dementia and depression. Abundant receptors in the brain suggest it to modulate neurotransmitter and nerve growth factor synthesis. Hippocampal growth is retarded with increased risk of psychosis and schizophrenia [35]. Seasonal affective disorder and lowering of mood are common with low levels of vitamin D in blood. This is reversible with supplementation [36].
9. Endorphins: Endorphins and endogenous opioids do not cross the blood-brain area but show abundance of receptors in all areas of the brain especially in the amygdala, hypothalamus, and hippocampus. The extensive neuronal network is responsible for mood elevation, modulation of pain perception, appetite, pleasure, and stress [37].

21.9 Regulation of Emotions: Need and Tools

Descartes had successfully established the influence of mind on body. This paved the way for development of another new science, psychoneuroimmunology.

The neuroendocrine and behavioral changes induced by emotions affect the immune system along with cardiovascular respiratory, gastrointestinal, and metabolic systems. Positive emotions enhance immunity, and negative emotions have been demonstrated to lower the immune status having the body vulnerable to pathogens evidenced by levels of selective biomarkers of natural killer cell activity and lymphocytic proliferation response, which are found lowered in depression, anxiety, and stress.

Action and movement with any intention is greatly influenced by emotion as is preparation for action [13].

The physiological parameters involved in emotional processing are based mostly on assessment of cardiovascular responses or dermal physiology along with brainwave studies [27, 28].

Regulation of emotions is concerned with monitoring, evaluation, and modification of emotional reaction for smooth accomplishment of one's objectives. Self-regulation is the adaptive ability to regulate one's emotions at each level of processing. The process of self-regulation demands awareness and understanding and acceptance of emotions with control of impulsive behavior with use of appropriate regulatory strategies. Customized adaptive strategies for appropriate responses require to be designed for this.

Electrodermal activity, variation in heart rate, and heart rate variabilty (HRV) stand for better emotional regulation and ability to cope with stress. Phasic suppression of HRV occurs in temporary stress, signifying withdrawal of vagal control. HRV could be a scale for measuring emotional regulation. In current times, transcranial magnetic stimulation has emerged as a newer diagnostic tool employed in understanding emotions. Besides transmagnetic stimulation (TMS), functional magnetic resonance imaging (fMRI), positron emission tomography (PET), behavioral genetics, and psychophysiology are other methods for studying emotion. Emotion influences motor control especially during self-regulation which could be assessed by muscle sympathetic nerve activity (MSNA).

21.10 Emotion and Health

Emotional well-being is a major component of human health. Hence, detrimental physiological, behavioral, and cognitive effects of high-valence and high-arousal emotions need conscious regulation for cultivating emotional equanimity. Emotional dysregulations to sensory information over a long period of time often are the cause of psychosomatic disorders [38].

Emotional intelligence provides the scope of understanding, analysis, and interpretation of emotional reactions and responses of others as well for better communication and harmonious performance. A conscious evaluation of a strategy aimed at achieving a wise solution is the basis of emotional intelligence [39]. Emotional intelligence involves simultaneous evaluation both at amygdalar and cortical level. The common etiology of the group of psychosomatic disorders is attributed to aberrant functions and connections within the structures in the limbic system and emotional dysregulation.

Decreased dopamine results in apathy and delay in initiating an action as in Parkinsonism. On the contrary, increased dopamine level causes cognitive impulsiveness and aspiration for higher achievements. Noradrenergic and dopaminergic pathways are greatly responsible for psychological and mental well-being. Bilateral lesions of the cingulate cortex are also a cause of apathy, decreased verbal communication, and blunted emotion even in pain.

Cognitive behavioral therapy, biofeedback, yoga, and meditation are some of the techniques for attaining better emotional regulation, thereby restoring and improving autonomic and cognitive balance [40].

21.11 Historical Perspective

In 1868, Harlow described the case of Phineas Gage – and effects of removal of the prefrontal cortex.

1872 – Publication of *The Expression of the Emotions in Man and Animals* by Charles Darwin.

1884 – William James proposed his theory of emotion.

1885 – Lange proposed a similar theory which came to be known as the James-Lange theory.

1912 – Right hemisphere hypothesis of emotion by Millers.

1931 – The Cannon-Bard theory was postulated.

1937 – The Papez theory was proposed, and the Kluver-Bucy syndrome was described.

1943 – Hess and Bugger described thesis work on single-cell recording in the hypothalamus.

1949 – The limbic model of emotion was described by MacLean.

1956 – Amygdala ablation in monkeys was described by Weiskrantz. Schneirla outlined an approach-withdrawal model of emotion.

1962 – Schachter and Singer described experiments indicating the importance of cognitive factors in determining the nature of emotion experience.

1970 – Pribram and Nauta proposed an early version of the somatic marker hypothesis.

1975 – *Mind and Emotion* by Mandler was published.

1980 – Zajonc put forward his hypothesis on emotion in the absence of cognition.

1983 – Ekman and colleagues proposed that basic emotions can be distinguished autonomically.

1986 – LeDoux proposed multiple amygdala pathways for fear conditioning.

1991 – Damasio outlined his somatic marker theory.

1992 – Panksepp coined the term affective neuroscience.

1994 – Adolphs et al. described unpaired recognition of emotion in facial expression owing to bilateral damage to the human amygdala.

1996 – Cahill et al. revealed importance of the amygdala in consolidation of emotional memories.

1997 – Philips et al. proposed that the insula is a specific neural substrate for perceiving facial expression of disgust.
2000 – Damasio et al. published evidence that different brain regions underline different emotions. Calder et al. described a patient with damage to the insula and basal ganglia who showed impaired recognition and experience of disgust.
2002 – Hariri et al. showed that amygdalar response to emotive stimuli varies as a function of serotonin transporter gene variation. Lambie and Marcel published the theory of conscious emotion experience.

21.12 Conclusion

Emotion is a physiological behavior that has been studied for a long time with much skepticism. Unregulated emotions and repressed feelings with social conditioning have resulted in increased incidence of psychosomatic and lifestyle disorders which are hypothesized to have an epigenetic influence to a great extent. Therefore, much intensive research into understanding of emotions, their regulation, and their molecular changes is the need of the times. Bodily mapping of emotional behaviors using imaging studies would greatly enhance our knowledge on emotions [41].

References

1. Albano C (2008) The puzzle of human emotions: some historical considerations from the 17th to the 19th centuries [Internet]. Dev Med Child Neurol 50:494–497. [cited 2019 Mar 5]
2. Bennett M, Hacker P (2005) Emotion and cortical-subcortical function: conceptual developments. Prog Neurobiol [Internet] 75(1):29–52. 2005 Jan 1 [cited 2019 Mar 24]
3. Trofimova I (2018) Functionality versus dimensionality in psychological taxonomies, and a puzzle of emotional valence [Internet]. Philos Trans R Soc B Biol Sci 373: 20170167 [cited 2019 Mar 24]
4. Trnka R, Balcar K, Kuška M (2011) How many dimensions does emotional experience have? The theory of multi-dimensional emotional experience. ... Spaces From Exp to ... [Internet] 33–40. 2011 [cited 2019 Mar 24]
5. Barrett LF, Mesquita B, Ochsner KN, Gross JJ (2006) The experience of emotion. Annul Rev Psychol [Internet] 58(1):373–403. 2006 Jan [cited 2019 Mar 5]
6. William J (2007) What is an emotion? Mind [Internet] os-IX(34):188–205. 2007 Apr 1 [cited 2019 Mar 21]
7. Cannon WB (2006) The James-Lange theory of emotions: a critical examination and an alternative theory. Am J Psychol [Internet] 39(1/4):106. 2006 Dec [cited 2019 Mar 21]
8. Dalgleish T (2004) The emotional brain. Nat Rev Neurosci [Internet] 5(7):583–9. 2004 [cited 2019 Mar 5]
9. Roxo MR, Franceschini PR, Zubaran C, Kleber FD, Sander JW (2011) The limbic system conception and its historical evolution. Sci World J [Internet] 11:2427–2440. [cited 2019 Mar 21]
10. Lindquist KA, Wager TD, Kober H, Bliss-Moreau E, Barrett LF, Feldman Barrett L (2012) The brain basis of emotion: a meta-analytic review. Behav. Brain Sci [Internet] 35(03):121–143. 2012 Jun 23 [cited 2019 Mar 5]

11. Sheridan MA, McLaughlin KA (2014) Dimensions of early experience and neural development: deprivation and threat [Internet]. Trends Cogn Sci 18:580–585. 2014 [cited 2019 Mar 24]
12. Critchley HD, Garfinkel SN (2017) Interoception and Emotion. Curr Opin Psychol [Internet] 17:7–14. 2017 Oct 1 [cited 2019 Mar 24]
13. Coombes SA, Janelle CM, Duley AR (2005) Emotion and motor control : affective picture processing. J Mot Behav [Internet] 37(6):425–436. [cited 2019 Mar 5]
14. Bliss-Moreau E, Toscano JE, Bauman MD, Mason WA, Amaral DG (2010) Neonatal amygdala or hippocampus lesions influence responsiveness to objects. Dev Psychobiol [Internet] 52(5):487–503. 2010 Apr 21 [cited 2019 Mar 5]
15. Regev L, Baram TZ (2014) Corticotropin-releasing factor in neuroplasticity [Internet]. Front Neuroendocrinol 35:171–916. 2014 [cited 2019 Mar 25]
16. Adolphs R, Tranel D, Damasio H, Damasio A (1994) Impaired recognition of emotion in facial expressions following bilateral damage to the human amygdala. Nature [Internet] 372(6507):669–72. 1994 Dec [cited 2019 Mar 5]
17. Davis M (2002) The role of the amygdala in fear and anxiety. Annu Rev Neurosci [Internet] 15(1):353–375. 2002 Mar 28 [cited 2019 Mar 5]
18. Neugebauer V (2015) Amygdala pain mechanisms. Handb Exp Pharmacol [Internet] 227:261–84. 2015 [cited 2019 Mar 25]
19. Yang L, Comninos AN, Dhillo WS (2018) Intrinsic links among sex, emotion, and reproduction [Internet]. Cell Mol Life Sci 75: 2197–210. 2018 [cited 2019 Mar 25]. Available from: https://doi.org/10.1007/s00018-018-2802-3
20. Schulte MHJ, Kaag AM, Wiers RW, Schmaal L, van den Brink W, Reneman L, et al (2017) Prefrontal Glx and GABA concentrations and impulsivity in cigarette smokers and smoking polysubstance users. Drug Alcohol Depend [Internet] 179:117–23. 2017 [cited 2019 Mar 26]. Available from: https://doi.org/10.1016/j.drugalcdep.2017.06.025
21. Yuen EY, Wei J, Yan Z (2017) Molecular and epigenetic mechanisms for the complex effects of stress on synaptic physiology and cognitive functions. Int J Neuropsychopharmacol [Internet] 20(11):948–55. 2017 [cited 2019 Mar 25]
22. Schmidt PJ, Abate AC, Schiller CE, Rubinow DR, Johnson SL (2016) Reproductive steroid regulation of mood and behavior. Compr Physiol [Internet] 6:1135–60. 2016 [cited 2019 Mar 25]
23. Bachner-Melman R, Ebstein RP (2014) The role of oxytocin and vasopressin in emotional and social behaviours. Handb Clin Neurol [Internet] 124:53–68. 2014 Jan 1 [cited 2019 Mar 26]
24. Stein JL, Hakimi S, Zink CF, Kempf L, Meyer-Lindenberg A (2010) Vasopressin modulates medial prefrontal cortex-amygdala circuitry during emotion processing in humans. J Neurosci [Internet] 30(20):7017–22. 2010 [cited 2019 Mar 26]
25. Carter CS, Grippo AJ, Pournajafi-Nazarloo H, Ruscio MG, Porges SW (2008) Oxytocin, vasopressin and sociality. Prog Brain Res [Internet] 170:331–336. 2008 Jan 1 [cited 2019 Mar 26]
26. Hashemian F, Shafigh F, Roohi E (2016) Regulatory role of prolactin in paternal behaviour in male parents: a narrative review. J Postgrad Med [Internet] 62(3):182. 2016 [cited 2019 Mar 25]
27. Bob P, Fedor-Freybergh P (2008) Melatonin, consciousness, and traumatic stress [Internet]. J Pineal Res 44:341–347. 2008 [cited 2019 Mar 29]
28. Brown-Grant K, Harris GW, Reichlin S (1954) The effect of emotional and physical stress on thyroid activity in the Rabbit [Internet]. J Physiol 26. 1954 [cited 2019 Mar 29].
29. Hage MP, Azar ST (2012) The link between thyroid function and depression [Internet]. J Thyroid Res 2012. Hindawi Publishing Corporation; 2012 [cited 2019 Mar 29]
30. Lebow MA, Chen A (2016) Overshadowed by the amygdala: the bed nucleus of the stria terminalis emerges as key to psychiatric disorders. Mol Psychiatry [Internet] 21(4):450–63. 2016 Apr [cited 2019 Mar 27]
31. Heany SJ, van Honk J, Stein DJ, Brooks SJ (2015) A quantitative and qualitative review of the effects of testosterone on the function and structure of the human social-emotional brain. Metab Brain Dis [Internet] 31(1):157–67. 2015 [cited 2019 Mar 25]
32. Fayyaz M, Anjum S, Samoo Z, Anjum I, Jeffery SS (2018) The role of vitamin D in brain health: a mini literature review. Cureus [Internet] 10(7). 2018 [cited 2019 Mar 27]

33. Penckofer S, Kouba J, Byrn M, Estwing Ferrans C, Ferrans CE (2010) Vitamin D and depression: where is all the sunshine? Issues Ment Health Nurs [Internet] 31(6):385–93. 2010 [cited 2019 Mar 26]
34. Takahashi A (2016) Enkephalin. In: Handbook of hormones [Internet]. Academic, p 55-7A-2; 2016 [cited 2019 Mar 27]. Available from: https://www.sciencedirect.com/science/article/pii/B9780128010280001173
35. Howlett TA, Rees LH (2003) Endogenous opioid peptides and hypothalamo-Pituitary function. Annu Rev Physiol [Internet] 48(1):527–36. 2003 Nov 28 [cited 2019 Mar 27]. Available from: https://www.annualreviews.org/doi/pdf/10.1146/annurev.ph.48.030186.002523
36. Grossman A (1983) Brain opiates and neuroendocrine function. Clin Endocrinol Metab [Internet] 12(3):725–46. 1983 Nov [cited 2019 Mar 27]
37. Roberto M, Spierling SR, Kirson D, Zorrilla EP (2017) Corticotropin-releasing factor (CRF) and addictive behaviors. Int Rev Neurobiol [Internet] 136:5–51. 2017 [cited 2019 Mar 25]
38. Berna G, Ott L, Nandrino J-L (2014) Effects of emotion regulation difficulties on the tonic and phasic cardiac autonomic response. PLoS One [Internet] 9(7):1–10. 2014 [cited 2019 Mar 5]
39. Goshvarpour A, Abbasi A, Goshvarpour A (2017) An accurate emotion recognition system using ECG and GSR signals and matching pursuit method. Biomed J [Internet] 40(6):355–68. 2017 [cited 2019 Mar 5]
40. Bijlani R (2011) Medical physiology. In: Understanding medical physiology, 4th edn. Jaypee Publishers, New Delhi
41. Nummenmaa L, Glerean E, Hari R, Hietanen JK (2013) Bodily maps of emotions. Proc Natl Acad Sci [Internet] 111(2):646–651. 2013 [cited 2019 Mar 30]

Techniques Related to Disease Diagnosis and Therapeutics

22

Saumitra Sen Singh, Sachchida Nand Rai, Hareram Birla, Walia Zahra, Aaina Singh Rathore, Hagera Dilnashin, Chetan Keswani, and Surya Pratap Singh

Abstract

Disease diagnosis is the vital part for the management of any type of disease. Nowadays, human health issues are increasing in a progressive way, which requires accurate and specific techniques which can efficiently diagnose the disease as soon as possible. After the disease diagnosis, therapeutics acts as the second most pillars in the maintenance of human health. Numerous techniques are available these days but are associated with various advantages and disadvantages. Thus, there is a prompt need for the development of techniques related to disease diagnosis and therapeutics which have minimum disadvantages. In this book chapter, we have mentioned few important and useful techniques related to disease diagnosis and therapeutics like ELISA, PCR, EEG, MRI, and CT scan. Here, we have discussed each and every technique in a sequential manner.

Keywords

Disease diagnosis · Technique · ELISA · PCR · EEG · MRI · CT scan

22.1 Enzyme-Linked Immunosorbent Assay (ELISA)

Introduction Enzyme-linked immunosorbent assay (ELISA) is an immunological assay technique used as a diagnostic and analytical tool in various industries for quality control along with detection and quantification of particular antigens including proteins, hormones, glycoproteins, or antibodies in a provided biological sample. The assay is based on a simple concept of immunology involving antigen-antibody combinations. The antigen is fixed on an immobilized surface and allowed to bind

S. S. Singh · S. N. Rai · H. Birla · W. Zahra · A. S. Rathore · H. Dilnashin · C. Keswani
S. P. Singh (✉)
Department of Biochemistry, Institute of Science, Banaras Hindu University, Varanasi, India

S. Paul (ed.), *Application of Biomedical Engineering in Neuroscience*,
https://doi.org/10.1007/978-981-13-7142-4_22

with an enzyme-labeled antibody. This specific binding allows recognition of even sparse quantities of any antigenic molecule or antibody present in the sample. Detection is accomplished when a chromogenic substrate is conjugated with the enzyme and produces a visible colored productor fluorescence. The detection strategy is governed by the principle of highly specific antibody-antigen interaction. Enzyme immunoassay (EIA) and ELISA are frequently used terms in the biomedical fields, industries, regulatory bodies, and quality control assessment organizations. The development of these techniques dates back to the 1960s–1980s when the enzyme labels were used for detection purposes in immunoassays. The ELISA is principally based on radioimmunoassay (RIA) technique and dates back to 1941 [1]. RIA was first described by Berson and Yalow in the 1960s to measure the endogenous plasma insulin level and was awarded with Nobel Prize for it. Similarly, ELISA was also developed during the same time period by two research teams [2, 3]. However, it was made successful in 1971 by the scientists Engvall and Perlmann from Switzerland and developed it with modifications in RIA. With the help of ELISA, various infections including influenza, parainfluenza, and mumps viruses were identified by Bishai and Galli, Leinikki et al., and Ukkonen et al. in 1978, 1979, and 1981, respectively [3–5]. Later in 1980, the technique was modified by Siegle et al. to detect the concentrations of various hormones, peptides, and proteins by using 96-well microtiter plates [35].

Principle ELISA is a diagnostic technique which allows detection of proteins, peptides, hormones, or antibodies in the fluid sample by different combinations of antigen-antibody binding. ELISA uses an enzyme-labeled antigen or antibody which gives a colored product in the presence of a substrate. The enzyme activity is measured colorimetrically by converting the light absorption of the product formed into numeric values. Depending on the antigen-antibody combination and with slight modifications, various kinds of ELISA have been employed such as direct ELISA, indirect ELISA, sandwich ELISA, competitive ELISA, etc.

Type of ELISA Enzymatic immunoassay is classified into two general headings, homogeneous enzymatic immunoassay and heterogeneous enzymatic immunoassay. The homogeneous enzymatic immunoassay method involved irreversible binding of enzyme with the antibody; thus the enzymes become inactivated and also the antigens cannot be separated from the medium, i.e., it has no washing stage. This method is usually taken into account to detect substances in small quantities, like therapeutic drugs. Homogeneous method is expensive and has low sensitivity, and its only merit is its accessibility. On the other hand, the heterogeneous enzymatic immunoassay is the more frequently used method as it is more sensitive than homogenous enzymatic immunoassay method, and washing stage is an essential stage to separate the bound antigen from the free antigen after the antigen-antibody interaction. ELISA is an example of heterogeneous immunoassay technique which detects specific antibodies and soluble antigens [6].

ELISAs can be performed with a number of modifications to the basic procedure: *direct, indirect, sandwich, or competitive*.

(a) **Direct ELISA:** It is the most widely accepted ELISA technique and was developed by Engvall and Perlmann [6] and by Van Weemen and Schuurs [6] in 1971. In this technique, the antigen present in the sample is immobilized in the microtiter well plate and is allowed to bind to the antibody that has been directly conjugated to an enzyme. It is mainly suited for the detection of high molecular weight antigens.
(b) **Indirect ELISA:** This technique was invented as a modification to the direct ELISA by Lindström and Wager in 1978. It is called indirect ELISA because the antigen is determined by secondary antibody instead of primary antibody. In this type of ELISA, the antigen is coated to a multi-well plate and is detected in two stages or layers. Initially, the primary antibody is added which is antigen specific. Further, an enzyme-labeled secondary antibody is added which binds to the primary antibody. The secondary antibody is usually an anti-species antibody and is often polyclonal. This type of ELISA is the most popular assay [6].
(c) **Sandwich ELISA:** The technique was developed in 1977 by Kato et al. This technique is used to identify a specific sample antigen and requires the use of matched antibody pairs, where each antibody is specific for a different, non-overlapping part (epitope) of the antigen molecule. The wells are coated with known amount of primary antibody to confine the antigen. Next, the nonspecific binding sites are blocked using bovine serum albumin, and then the antigen-containing sample is added to the well plate. A specific primary antibody is then added that "sandwiches" the antigen. Thus, when the required protein is fixed between both the antibodies, this method is termed as sandwich ELISA. Sandwich ELISAs have been reported to be 2–5 times more sensitive than all other ELISAs [7].
(d) **Competitive ELISA:** This technique was first used in 1977 by Kato and his co-workers [7]. It is also termed as inhibition ELISA as a result of its mechanism involving competitive binding for the primary antibody between the antigen present in the microtiter well plate and the antigen present in the samples. It is most beneficial in detecting the antigen even with scarce quantities of specific antibody as it is highly sensitive to the varied compositions of complex antigen.

Application of ELISA In general biomedical research, ELISA has been used as a potent diagnostic tool for diverse biological molecules involved in variety of diseases both in animals and humans including diagnosis of HIV infection and pregnancy tests along with measuring soluble receptor levels in serum. Apart from manual ELISA, various customized kits are also commercially available for diagnosis of HIV [8], influenza [9], dengue fever [10–12], Ebola [13], Chagas disease [14], leishmaniasis [15], Lyme disease [16], and West Nile virus [17]. The application of ELISA is not just limited to animal and human diagnostics but is also gaining popularity in plant pathology.

- **Detection of infectious disease:** Both in industries and in medicine, the need for accuracy to detect the sample quality or an infection is of utmost importance, and ELISA has proven to be highly reliable. With the development of new technologies, various highly specific and updated assays are available in the market for the diagnosis and prognosis of infectious diseases, for instance, sexually transmitted diseases (STDs). STD involves various infectious diseases spread through sexual contact, especially prevailing in the developing countries. Till date, numerous kits with highly accurate detection have been made commercially available for different STDs such as syphilis, HIV, chlamydia, and hepatitis. Apart from these, ELISA is effective even in detecting regional or endemic diseases prevailing especially in the tropical and subtropical regions, for example, yellow fever, borreliosis, Chagas disease, and dengue. Unlike other techniques which fail to detect these diseases in early stage, ELISA shows high sensitivity in the efficient diagnosis of such illnesses. There are a number of viral diseases including varicella-zoster virus, Epstein-Barr virus, and various others which result in prenatal infections and have hazardous effects on the fetus if not detected and treated early. The commercially available ELISA kits have proven to be valuable in the clinical diagnosis of infectious agents as well.
- **Cancer diagnosis:** Early detection of cancer provides improved chances of survival for the patient; hence a detection technique which is highly sensitive is an essential requirement. Although it is quite tricky to target cancer biomarkers for detection, ELISA has shown assured results in cancer detection. Slight advancements have always been done to improve any technique including addition of GNPL (gold nanoparticle layers) in ELISA by Zhou et al. to enhance its sensitivity toward the target molecule by lowering its LOD (limit of detection). Herein to test the accuracy of the improved ELISA technique based on GNPL, plasma used was added with a cancer biomarker, carcinoembryonic antigen (CEA), which revealed resourceful application of ELISA in cancer detection. Another improvement to the technique was made by Vazquez-Villegas et al., by incorporating chemically synthesized poly(methacrylate) microspheres in order to have an advanced recognition of microRNA involved in breast cancer [18]. ELISA is a great asset to detect even the slight amount of antigen present in a sample like detection of glycoprotein CA125 in the serum for early diagnosis of ovarian cancer in a patient [19].
- **Pregnancy detection:** A pregnant woman's body produces various hormones involving luteinizing hormone (LH), human chorionic gonadotropin (HCG), thyrotrophin-stimulating hormone (TSH), estriol (E3), and follicle-stimulating hormone (FSH) [20]. ELISA is the cheapest and quickest way to diagnose an early pregnancy using either of the fluids like saliva, blood, or urine [21]. The most common detection is through checking the presence of HCG in the urine using ELISA in the initial month of pregnancy. In addition, ELISA also detects E3 present in saliva after the sixth week of pregnancy. These tests are able to detect the pregnancy in both humans and animals [22]. Along with pregnancy biomolecules, ELISA is also capable of identifying any congenital infection present during pregnancy [23, 24].

- **In vaccine development:** Other than the use of ELISA in the detection of infectious diseases, it also provides application in vaccine development. It can be used to measure the presence of antibody titers in a host intentionally injected with specific antigens to evaluate the immunogenicity of a vaccine [25]. Usually to test immune response, a diverse amount of antigens is introduced to the host and is later narrowed down to the ones which produce most defense reaction with minimal undesirable effects. In vaccine development using ELISA, the selection of suitable positive and negative controls is the most demanding step. Throughout the period of vaccine standardization while using different immunogens, attaining a highly selective diagnostic precision is of paramount importance [26]. However, ELISA is among the most widely used technique for vaccine standardization [27, 28].

22.2 Polymerase Chain Reaction (PCR)

Introduction In the biological research, polymerase chain reaction or PCR is a technique that radically changed the course of molecular science. The discovery of PCR allowed the accomplishment of a great scientific venture like Human Genome Project [31]. It is a powerful technique used for the amplification of a specific target sequence of DNA in order to generate its millions of copies; hence, it is also termed as "molecular phototyping." Along with the amplification, PCR also permits manipulation in the DNA segment for gene alteration during cloning [29, 30] and is being widely used in the diagnosis of various infections and diseases. In 1974, Panet and Khorana first described the principle of DNA amplification in vitro. In the 1980s, Mullis and co-workers developed PCR, and the significance of this work was awarded with a Nobel Prize in 1993. Currently, the technique is being used in various aspects of molecular biology involving clinical as well as research areas for the disease diagnosis, cloning, and sequencing and genomic studies both in plants and animals [31]. Using the basic principle of PCR, multiple advancements have been made to develop various types of PCR such as real-time PCR which is a more precise and faster technique. It is being extensively used in all biomedical areas due to its property of producing quantitative results [32, 33].

Principle PCR is an effortless yet elegant technique allowing exponential amplification of the target DNA sequence in a chain reaction fashion. For the amplification to take place in doubling manner, presence of certain proteins called polymerases is required to initiate the formation of complementary strand to the provided template strand. Both strands can act as the template strands. This requires a presence of DNA building blocks, i.e., the four bases adenine (A), thymine (T), cytosine (C), and guanine (G), to create the resulting DNA strand. Furthermore, to mark the exact sequence of DNA to be amplified, another molecule termed primer is required which is a short sequence of DNA that is detected and amplified initially and serves as the extension point. The segment of DNA to be amplified varies in length and can be as small as the sequence of a particular exon within a specific gene. In simple words, it can be represented as a molecular copier. PCR is a fast method of amplifying a target sequence; it can amplify a DNA segment within a couple of hours. The purity of the template sequence is not essential especially in the case of bacterial colonies. The product obtained after completion of amplification can be used for sequencing and cloning. PCR is an enzymatic assay and requires the presence of a thermally stable DNA polymerase to remain heat resistant during the process. Originally, the DNA copying enzyme was used in vitro by Kary Mullis. The separation of both the strands of DNA takes place at a very high temperature of 96 °C; thus, in the original experiment, the DNA polymerase isolated from *E. coli* did not sustain, and therefore, it was later substituted with *Taq* polymerase. With time, a great deal of modifications has been made to the original PCR process as it was a very inefficient, time-consuming process.

Types of PCR

(a) **Quantitative PCR techniques:** It is also termed as qRT-PCR or real-time PCR as it provides information about the amount of DNA in real time, i.e., at a particular time, how much amount of DNA has been amplified?
(b) **Qualitative PCR techniques:** It works on the same principle as other PCR techniques except that it identifies the presence or absence of the product of the target DNA segment. It is an excellent technique to be performed in cloning or to detect a pathogenic gene. It is the only technique to identify if an individual has ever been infected with a specific pathogen before [34].
(c) **Conventional PCR:** The basic principle is same as the normal PCR. The primers bind to both the DNA strands and prevent the replication to move further [35].
(d) **Multiplex PCR:** This type of PCR distinguishes between varied pathogens in the same sample and differently recognizes sequences with exons and introns in the target DNA segment. The primer designing is unlike the other PCR techniques as they are meant to adhere to more than one target DNA sequence. Consequently, the length of the sequence must be of different size to obtain accurate targeted sequences with minimal time consumption [36].
(e) **Nested-semi-nested PCR:** In this type of PCR, two separate primers are used to bind to the same sequence. The second set which is the complementary set is shorter than the first set, i.e., amplified sequence. The contamination in the products is reduced by using nested PCR which arise due to the amplification of nonspecific primer binding sites [37].
(f) **Colony PCR:** This type of PCR screens the plasmids containing the desired DNA insert [38]. The bacterial or yeast colonies are grown on the agarose growth media and selectively picked to be added to the PCR master mix or in autoclaved water. Subsequently, PCR is performed to identify the presence of target DNA segment.

(g) **Quantitative/semiquantitative real-time PCR:** Here, to determine the quantity of amplification in a given sample, probes and fluorescent dyes like SYBR Green Master Mix are incorporated in the process [39]. In semiquantitative PCR, the amplification of c-DNA is done followed by gel electrophoresis and then quantified by densitometry. The limitation of this technique is that it does not provide the real time value of the amplification as compared to qRT-PCR.

(h) **Standard PCR:** This is the general PCR technique which detects the presence of leucocytes DNA/heamo compounds [40].

Application These are the following applications of PCR:

- **In tuberculosis diagnosis:** PCR is very helpful in the detection of *M. tuberculosis*. Since the studies have used different techniques to detect products along with the number and types of samples used, using PCR in the diagnosis of TB for different clinical specimen is not clear [41]. This makes it difficult to do the comparison between the reported sensitivities and specificities.
- **In Human Genome Project:** The Human Genome Project was initiated to determine the complete sequence of nucleotide bases for each functional unit in the genome. Most of the mapping techniques used in HGP relied on PCR. About five billion nucleotide bases are present in a somatic cell of humans. Considering 1500 nucleotides to be present in a gene, then even 30,000 functional units of DNA would also make only 1% of the total DNA per cell [42].
- **Analysis of disease genes:** PCR is an exceptional technique for the detection of pathogenic genes. Since PCR amplifies a small amount of segment to such large extent, it makes it extremely easy for determining a disease gene [43]. For an instance, a small amount of fetal cells obtained from amniotic fluid is adequate to identify the existence of any diseased genes. Furthermore, modified techniques of PCR including ligase chain reaction (LCR) and PCR-15 single-strand conformational polymorphisms (PCR-SSCP) analysis are competent in determining point mutations. For very little samples, the DNA sequences are read through reverse transcription PCR or RT-PCR.
- **Genetic analysis of organism:** To understand an organism at their molecular level, their genetic makeup needs to be identified and analyzed. In the past, various techniques such as VNTR- and RFLP-based methods have been developed to determine an individual uniquely. However, PCR has achieved this aim due to its uncomplicated and easy procedure. In addition, another PCR-based technique used for genomic DNA amplification performed with the help of short, nonspecific primers is random amplified polymorphic DNA (RAPD) analysis [44].

22.3 Electroencephalogram (EEG)

Introduction Encephalography has undergone a great progress during the course of more than 100 years of its history. Richard Caton, an English physician in 1875, has for the first time discovered the presence of electrical currents in the brain. Electroencephalogram (EEG) of brains of animals such as monkeys and rabbits were observed for the very first time by Caton. Years later, in 1924 an ordinary radio equipment was used by a German neurologist named Hans Berger, so that amplified electrical activity of the brain can be measured over the scalp. He has further depicted the weak electrical current on a paper strip and proclaimed that even without exposing the brain by opening the human skull, these currents can be recorded. Moreover, he gave the term "electroencephalogram" for the first time to describe the electric potential in the human brain. He was also precise about his suggestion that when there is a change in the general status of the subject, e.g., relaxation to alertness, the brain activity changes in a regular and detectable way. Later the term "alpha rhythm" was introduced in 1934 signifying steady oscillations around 10 to 12 Hz by Adrian and Matthews, who have for the first time reported in their paper about the concept of "human brain waves" [45]. The electrical activity generated by the brain structures can be efficiently read over the scalp by medical imaging technique, known as electroencephalography. An alternating type of electrical activity recorded over the surface of the scalp using metal electrodes and conductive media is known as EEG. Electrocorticogram is the term given when EEG is measured from the cortical surface directly, and when it is measured by depth probes, it is termed as electrogram. This article deals with the EEG which has been recorded from the head surface. Thus, electroencephalography is a process that can be applied without any risk or limitations to the patients, normal adults, and children repeatedly in a noninvasive way.

Principle Local current flow is generated upon the activation of neurons (brain cells). The current flow produced during the excitation of the dendrites of numerous pyramidal neurons of the cerebral cortex is the one that is mostly measured by EEG. The electrical dipoles created between the soma, i.e., the body of neurons and branches of neurons, i.e., the apical dendrites, are due to the differences of electrical potentials, produced by summation of postsynaptic graded potentials from pyramidal cells. The direction of electrical current in the brain is governed by the membrane potential, and it mostly constitutes the ions such as sodium, potassium, calcium, and chloride ions which are pumped by the channels present in neuronal membrane [46]. Involvement of various types of neurotransmitters along with variation in the types of synapses makes the detailed microscopic picture more sophisticated. Also, the electrical activity can be recorded from the head surface only when it is produced by the large number of active neurons. The current moves through the skin, skull, and various other layers in between electrode and neuronal layers. Amplification of weak electrical signals recorded with the help of scalp electrodes is showed on the paper or saved in the computer [47]. EEG is considered as the powerful technique because of its ability to depict both the usual and unusual electrical activities of the brain and plays a prominent role in the field of neurology and clinical neurophysiology.

Brain Waves

From peak to peak, the amplitude value lies between 0.5 and 100 μV which is further amplified before plotting.

The categorization of brain waves has been done as follows:

- Delta wave – 0.5–4 Hz.
- Theta wave – 4–8 Hz.
- Alpha wave – 8–13 Hz.
- Beta wave – > 13 Hz.

The normal rhythm is denoted by the alpha rhythm which is better observed in the posterior and occipital region. This refers to the situation of relaxation and closing of eyes. Whereas, under normal condition, beta waves are dominant when the eyes are opened; delta waves, on the other hand, are related to the condition of deep sleep.

EEG Recording Techniques

The following components are required for the encephalographic measurement:

- Electrodes.
- Amplifiers with filters.
- A/D converter.
- Recording devices.

Application

Clinical EEG is done by using scalp electrodes and the total time taken is about 20–30 minutes. It is done under the following circumstances:

- For distinguishing epileptic seizures from other diseases such as syncope (fainting), psychogenic non-epileptic seizures, migraine variants, and subcortical movement disorders.
- When primary psychiatric syndromes such as catatonia has to be differentiated from "organic" encephalopathy or from delirium.
- To be used for detection of brain death.
- To do the prognosis of the conditions such as when the patients is in coma.
- To decide whether the anti-epileptic medication has to be stopped or not.

When the patients are having a seizure, routine EEG might not act as sufficient tool for diagnosis. Those patients are then admitted to the hospital, and the EEG along with audio recording and time-synchronized video is done for several days or weeks. The an ictal recording of an epileptic is done in between the period of two seizures to observe whether the problem is due to the epileptic seizure or not and from which part of the brain the seizure activity is arising.

Usually epilepsy monitoring is typically done:

- To distinguish the different types of spells such as syncope (fainting), migraine variants, psychogenic non-epileptic seizures, and subcortical movement disorders from epileptic seizures.
- To characterize treatment for seizures.
- To design a proper work-up for seizure surgery after localizing the region of the brain from where the seizure has originated.

Moreover, certain procedures are also monitored using EEG such as:

- For determining the depth of anesthesia.
- In carotid endarterectomy for indirect indication of cerebral perfusion.
- For monitoring the effect of amobarbital during the Wada test.

In order to monitor the function of the brain, EEG can also be used in intensive care units:

- For monitoring nonconvulsive seizures/nonconvulsive status epilepticus.
- For treating the refractory seizures or increased intracranial pressure in medically induced coma patients and to observe the effect of sedative/anesthesia.
- To monitor conditions of secondary brain damage, e.g., subarachnoid hemorrhage.

22.4 Magnetic Resonance Imaging (MRI)

Introduction In 1952, Bloch and Purcell were acknowledged with Nobel Prize for Physics because they have discovered nuclear magnetic resonance (NMR) phenomenon, which they both have described experimentally in the year 1946 [46]. Over the time, the rapid evolution of the technique has taken place which has further introduced wide-bore superconducting magnets (approximately 30 years ago) useful for clinical applications. In the year 1980 in Nottingham and Aberdeen, the clinical magnetic resonance images were produced for the first time, and meanwhile the magnetic resonance imaging (MRI) has become a widely accessible and influential clinical tool [47, 48].

Principle The combined usage of radio waves, magnetic field, and particular character of hydrogen atom present in all tissues of body is specifically dependent on the principles of nuclear magnetic resonance (NMR). The nuclei of H atoms are found to spin in random directions like tops. During MRI, the orientation of nuclear spins of all the hydrogen atoms occurs in a single direction. Precession is produced due to change in the angular momentum produced by the torque. In a uniform magnetic field, the precession of protons with the same mass takes place at the same frequency, termed as Larmor frequency, given by the formula $f = \gamma B$, where:

f = frequency of rotation.
γ = Larmor constant.
B = magnetic field strength.

The value of γ for a proton is given as 42.58 million of cycle per Tesla (T), i.e., the precession of proton will take place at a rate of 42.58 million cycle per second in 1 Tesla. The proton will tend to precess with B1 if the second field B1 is applied perpendicular with B0, and the strength of B1 will determine the rotation. While the proton gets returned to their original lower energy configuration in Z axis when the field B1 is removed and MRI measures this relaxation. The processed frequency is referred to as radio signal or RF signals because it is in radio range. The energy transfer takes place by the excitation of proton from higher energy transverse plane to lower energy longitudinal plane and requires collision between the protons. Hence, the energy required for collision will be more for free water because it has lesser intermolecular interaction than that of bound water which relaxes faster. The material characteristic determines the factor T1. The time required to reach 63% magnetization is denoted by T1, and the magnetic recovery depends on it. The value of T1 for the human body ranges in between 0.1 and 4.0 sec. The value of T1 decreases by IV contrast like Gd-DTPA (gadolinium-diethylenetriamine pentaacetate). This value is higher for bounded water in comparison with free water. If the tissue samples have different T1 value portions, they will relax differently, e.g., the portion having the lesser T1 value will relax faster. And in tissue which has higher T1 value, its MR signal will produce darker image, e.g., the spinal fluid appears darker than vertebrae in MRI. Fourier transformation of the radio signals or MRI signals occurs which splits into curves, and each curve have a unique frequency that helps in generating the picture by the computer [49, 50].

Fast magnetic resonance imaging The quality of patient care can be improved by enhancement of MRI speed. Development in the MRI technique has taken place at a faster rate and has offered other MRI applications such as cardiac imaging, diffusion imaging, functional MRI, and interventional MRI. These new applications have encouraged the MRI engineers and scientists to develop faster imaging methods. Moreover, the improvements in MRI hardware have provided the possibility of improving the imaging speed. The strength of the main magnetic field has also increased from 0.2 T to 1.5 T and 3 T systems which has also made the tool more popular.

Spiral Imaging It is also referred to as conventional regridding and was firstly defined by Meyer et al., and Jackson et al. later extended the technique. In imaging applications, like in coronary artery imaging, the spiral trajectories are utilized as they have greater advantage over simpler Cartesian raster method because they can traverse the *k*-space more efficiently [51, 52].

Parallel Imaging To improve the imaging speed, the multiple receiver coils were started to be used during the late 1980s and early 1990s [53, 54]. However, imaging using multiple receiver coils side by side was possible only after the simultaneous acquisition of spatial harmonics (SMASH) technique [55], and the sensitivity encoding (SENSE) technique was developed [55].

Image Evaluation The quality of image plays a prominent role in medical imaging as it helps the physicians to examine, diagnose, plan the therapy for, and apply and assess it [56]. Evaluation of the MR image quality appears to be an important issue in MR research. A very upfront way is used by the researchers which involves the simple display of the images from phantom or anatomy and specifies the quality of image as to how good or bad it is. This might also involve the display of control image for comparison [57]. Specific methods are applied for specific applications and tasks. Some researchers display the profile for spiral imaging applications along a straight line in order to compare its image quality, while others simulate the point spread function (PSF) to display the effects of reconstructions [58].

Application

The patient undergoing an MRI examination of the brain is generally placed in supine position. Various brain-related disorders can be diagnosed with the help of brain MRI including neurodegenerative conditions such as Alzheimer's disease, brain defect by birth, tumors and infection of the brain, mild cognitive impairment, and bleeding in the brain. For detection of different brain disorders, an appropriate imaging plane is decided, i.e., sagittal plane for multiple sclerosis and sagittal or coronal sections for lesion within the corpus callosum. An ordered combination of RF and gradient pulses determines the standard sequences designed to acquire the data in order to form the image which are FLAIR, Fat-Sat T2, FLAIR + Gd, SWI, PD/T2/DwI/ADC, STIR TOF MRA, etc. The marker of axonal loss and neuronal destruction, black holes, or persistent TI hypointense lesion in multiple sclerosis can be imaged and diagnosed by MRI.

- **In maternal diagnosis:** Fetal imaging has become easier with the technical development of MRI. Due to fetal motion, it was difficult to visualize the fetal anatomy especially the brain, thorax, abdomen, and pelvis which has now become easier with the use of MRI. The fetus of as early as 18 weeks can be efficiently visualized using the technique. Apart from neurological studies, MRI is also often used to examine the abnormalities in other organs like lungs, liver and chest of the fetus, urinogenital structure and the neck masses. Maternal diagnosis is also done using MRI that helps in evaluating the functionality and condition of placenta, tumors, volume of amniotic fluid, and maternal cerebral blood flow changes. After ultrasound MRI is the mostly used tool because of its complexity, availability, and cost [59, 60].

- **Full-body MRI:** Detection of skeletal metastasis, i.e., a major orthopedic complication of failed cancer treatment, can be done by implicating the whole body. This procedure serves as an alternative to skeletal scintigraphy, in which phosphates or diphosphates labeled with technetium-99 m are used. Abundance of proton in the matrix forms the foundation of MRI. Lesions in the spine and pelvis can also be detected by whole-body MRI, and it is also able to evaluate the soft tissue organs simultaneously. The patient in which the tumor has spread all over the body can be accessed using this technique. It can serve as a diagnostic tool for the patients with polymyositis and can be helpful in fat measurement [61, 62].

22.5 Computed Tomography Scan (CT Scan)

Introduction When the modern cross-sectional imaging was not developed, the diagnosis of neurological diseases, such as stroke, through radiology was a crucial task. Merely indirect signs of processes involved in the disease were depicted by the methods including pneumoencephalography, contrast angiography, and myelography which were associated with inherent risks. The introduction of computed tomography (CT) for clinical applications in diagnostic imaging in the late 1970s, made the field of medical engineering more empowering. An in-depth understanding of the anatomy and pathological processes in the human body using this cross-sectional imaging technique has assured the possibilities to overcome various limitations of plain radiographs. After the X-rays were developed in 1895, the discovery of CT is considered as the greatest innovation in the field of radiology. For the discovery, G.N. Hounsfield and A.M. Cormack in 1979 were acknowledged and were further rewarded with the Nobel Prize in medicine. The evolution of CT technology in the 1970s has taken place from the "mere concept" to a "technological marvel" and further from a single-detector "CAT" (computerized axial tomography) scanners to spiral scanners which has advanced to the state-of-the-art MDCT (multidetector CT) and dual-source scanners in the 1990s [63–65].

Types of CT Scanner

(a) **Single-slice axial CT scanners:** To move toward the higher level for utilizing tomography for diagnosis, the advent of computers was needed which took place during the 1960s. The basic theory forms the foundation of a CT image which is comparable to undeveloped tomography. And multiple passes of X-ray emitter over the region of interest are done on the patients instead of a single (or complex) pass. A single X-ray emitter and a solitary detector element of fixed longitudinal (slice) width were used to acquire the data sets from the most primitive, first-generation machines. After every acquisition of the data ray, the next ray was obtained only when the equipment experienced a translation. In order to acquire other data rays set from the same slice, the procedure was repeated a number of times across the body surface, and then a slender rotation of the equipment was done. Afterward, the whole table was then forwarded for the another slice, and all of the information for the slice was obtained [66].

(b) **Single-slice axial CT scanners:** In late 1980s, an innovative phase in CT data attainment was marked by the introduction of the slip ring technology. Electrical power is now directly conducted by the insulated brushes from the gantry housing to the emitter/detector system instead of moving the patient in order to unwind the equipment. This allows to continuously acquire a helix of data when the patient easily moves under the scanner as the amount of dead time gets minimized. The axial decreased values for every pixel are delivered by processing the data "ribbon." The presentation for every different slice has notably improved slightly. And the set of data is greatly enhanced for the overall exam. A volumetric helix is now obtained in the place of thick and occasionally non-contiguous image slices, permitting easier and faster reformatting. The scan time is also reduced as there is no need of moving the patient between the slices [67]. The new helical technology has introduced a new adjustable parameter, known as pitch. There is no motion in the table while the image is being developed, and the thickness of slice or collimation in fixed quantity makes the conservative scanner acquisition and display comparatively straightforward. These relationships on the other hand are not maintained during volumetric acquisition. The data sets are more complex when acquired from these machines as compared to that obtained from the single-slice scanners. The quality of image for multislice scanners is not directly related to pitch, while for single-slice machines, it is directly related [64].

(c) **Multislice CT scanning:** Another innovation in CT technology has occurred in the late 1990s. Four or more rows of detectors in the longitudinal (z) direction including helical scanners were introduced. This has allowed to collect the information from numerous overlying helical ribbons at the same interval. Separate data channels (helixes), up to 64, can be found on current machines. This has increased the coverage in the longitudinal direction at any given period. Furthermore, CT tube cooling constraints have been largely eliminated on further advancement of hardware technology. These multislice or multidetector CT scanners can be utilized to scan the body's larger section in a lesser time period [68]. The complexity of data sets assimilated from these machines is more than that obtained from the single-slice scanners. Image quality too is not directly related to pitch in the case of multislice scanners. Certain pitches are preferable over others, depending on the reconstruction technique used to get the best image quality [69]. Scan time is also decreased because of the multichannel capabilities and hardware improvements. Every visualized slice can at the moment be attained within a second as the slices are reconstructed from multiple data channels which can be implicated for physiological fast-scanning applications. Perhaps, during iodinated contrast administration, the brain perfusion in cerebral ischemia can be assessed by repetitive scanning of the parenchyma [70].

Applications

A wide range of structures can be examined by CT which gives a detailed description and makes the technique advantageous over conventional radiography or ultrasonography. CT is selected for the cases according to the need of patients or based on different factors such as cost-efficiency, convenience, and accessibility to other imaging facilities. The technique is quite expensive than ultrasonography but cheaper than that of magnetic resonance imaging (MRI).

- **Diagnosis of the brain:** The presence of high contrast because of bone and air-containing structures makes CT the well-suited technique for head examination. It also eliminates the drawbacks of superimposition which affect radiography because of its ability to produce cross-sectional images. Hence, the animals which suffer from head trauma are assessed using CT. When access to MRI is limited, or the person cannot manage to pay for the technique, CT serves as a suitable alternative. CT provides less anatomic details than MRI but is still capable of detecting hydrocephalus, intracranial hemorrhage [71], and intracranial masses [72]. Scanning of the brain using CT is usually done by injecting contrast medium intravenously. In normal condition, the blood-brain barrier generally stops the contrast medium to reach to the tissues of neurons. But when the blood-brain barrier is damaged by inflammatory or neoplastic lesions, the contrast medium gets accumulated within the lesions, thus making the damage of tissue detectable.
- **Diagnosis of the thorax:** A helical CT scan technique is generally favored for the diagnosis of the thorax as it minimizes the scan time and further decreases the chances of movement artifacts subsequent from breathing of patient. Except for the heart which appears blurry in CT scan, the technique can be used for imaging all the thoracic structures. CT images of the lungs are often useful as they reveal severe lesions (e.g., metastases) which were not noticeable on survey radiographs. It also enables us to do detailed assessment of the lesion that can provide useful information which can help in surgical planning. Thoracic CT is generally done when there are cases of suspected pulmonary metastasis [73], thoracic mass, and nontraumatic ("spontaneous") pneumothorax [74].
- **Detection of musculoskeletal structure abnormality:** As CT uses a multidirectional X-ray beam, the drawback of the superimposition is eliminated by the production of cross-sectional images which allows it to be the tool more sensitive over the conventional one. The conditions which are difficult to be detected by conventional radiography can be precisely detected by using CT such as examination of the canine elbow, particularly in dogs with suspected fragmented coronoid process [74] or incomplete ossification of the humeral condyle [75]. The occurrence of a condylar fracture can be explained using CT which enables us to find the marks of inadequate ossification of the humeral condyle. Moreover, the technique allows us to go for prophylactic treatment in animals that are at a risk for fracture. The animals suspected with the risk of intervertebral disk prolapse, neoplasia, or fracture can be examined by doing the CT of the spine. The guidance provided by CT can be used to obtain the biopsies of the lesion [76].

22.6 Conclusion

Disease diagnosis is equally essential as the treatment since it provides information crucial to explain the symptoms appearing along with the histopathological findings to present a more insightful and accurate etiology of the disease toward appropriate and effective treatment options. Clinical laboratories increasingly rely on latest and routinely improving disease diagnostic techniques. These techniques have been advancing according to the challenges faced during the diagnosis and are a huge milestone in the field of medicine. A number of pathophysiological processes that were not clearly detectable beforehand are now accessible via these diagnostic techniques. Conclusively, these diagnostic techniques provide evidences from efficient early-stage detection to final confirmations about an ailment without causing any damage to the tissue and a hope to design techniques in the future with paramount accuracy and minimum discomfort to the patient.

References

1. Coons AH, Creech HJ, Jones RN (1941) Immunological properties of an antibody containing a fluorescent group. Proc Soc Exp Biol Med 47(2):200–202
2. Van Weemen B, Schuurs A (1971) Immunoassay using antigen—enzyme conjugates. FEBS Lett 15(3):232–236
3. Yalow RS, Berson SA (1960) Immunoassay of endogenous plasma insulin in man. J Clin Invest 39(7):1157–1175
4. Leinikki PO, Shekarchi I, Tzan N, Madden DL, Sever JL (1979) Evaluation of enzyme-linked immunosorbent assay (ELISA) for mumps virus antibodies. Proc Soc Exp Biol Med 160 (3):363–367
5. Ukkonen P, Penttinen K, Granström ML (1981) Mumps-specific immunoglobulin M and G antibodies in natural mumps infection as measured by enzyme-linked immunosorbent assay. J Med Virol 8(2):131–142
6. O'Kennedy R, Byrne M, O'Fagain C, Berns G (1990) Experimental section: a review of enzyme-immunoassay and a description of a competitive enzyme-linked immunosorbent assay for the detection of immunoglobulin concentrations. Biochem Educ 18(3):136–140
7. Kato K, Hamaguchi Y, Okawa S, Ishikawa E, Kobayashi K, Katunuma N (1977) Use of rabbit antibody IgG bound onto plain and aminoalkylsilyl glass surface for the enzyme-linked sandwich immunoassay. J Biochem 82(1):261–266
8. Nandi S, Maity S, Bhunia SC, Saha MK (2014) Comparative assessment of commercial ELISA kits for detection of HIV in India. BMC Res Notes 7(1):436
9. Tarigan S, Indriani R, Durr PA, Ignjatovic J (2015) Characterization of the M2e antibody response following highly pathogenic H5N1 avian influenza virus infection and reliability of M2e ELISA for identifying infected among vaccinated chickens. Avian Pathol 44(4):259–268
10. Hunsperger EA, Yoksan S, Buchy P, Nguyen VC, Sekaran SD, Enria DA, Vazquez S, Cartozian E, Pelegrino JL, Artsob H (2014) Evaluation of commercially available diagnostic tests for the detection of dengue virus NS1 antigen and anti-dengue virus IgM antibody. PLoS Negl Trop Dis 8(10):e3171
11. Welch RJ, Chang G-JJ, Litwin CM (2014) Comparison of a commercial dengue IgM capture ELISA with dengue antigen focus reduction microneutralization test and the centers for disease control dengue IgM capture-ELISA. J Virol Methods 195:247–249

12. Hosseini S, Azari P, Farahmand E, Gan SN, Rothan HA, Yusof R, Koole LH, Djordjevic I, Ibrahim F (2015) Polymethacrylate coated electrospun PHB fibers: an exquisite outlook for fabrication of paper-based biosensors. Biosens Bioelectron 69:257–264
13. Schieffelin J, Moses LM, Shaffer J, Goba A, Grant DS (2016) Clinical validation trial of a diagnostic for Ebola Zaire antigen detection: design rationale and challenges to implementation. Clin Trials 13(1):66–72
14. Aria L, Acosta ME, Guillen Y, Rojas A, Meza T, Infanzón B (2016) ELISA Chagas test IICS V. 1 evaluation for the diagnosis of Chagas disease. Memorias del Instituto de Investigaciones en Ciencias de la Salud 14(3):7–13
15. Lauricella MA, Maidana CG, Frias VF, Romagosa CM, Negri V, Benedetti R, Sinagra AJ, Luna C, Tartaglino L, Laucella S (2016) An rK28-based immunoenzymatic assay for the diagnosis of canine visceral leishmaniasis in latin America. Am J Trop Med Hyg 95(1):92–98
16. Hinckley AF, Connally NP, Meek JI, Johnson BJ, Kemperman MM, Feldman KA, White JL, Mead PS (2014) Lyme disease testing by large commercial laboratories in the United States. Clin Infect Dis 59(5):676–681
17. Prince HE, Lapé-Nixon M, Givens TS, Bradshaw T, Nowicki MJ (2017) Elimination of falsely reactive results in a commercially-available West Nile virus IgM capture enzyme-linked immunosorbent assay by heterophilic antibody blocking reagents. J Immunol Methods 444:24–28
18. Chin AR, Fong MY, Somlo G, Wu J, Swiderski P, Wu X, Wang SE (2016) Cross-kingdom inhibition of breast cancer growth by plant miR159. Cell Res 26(2):217
19. Scholler N, Crawford M, Sato A, Drescher CW, O'Briant KC, Kiviat N, Anderson GL, Urban N (2006) Bead-based ELISA for validation of ovarian cancer early detection markers. Clin Cancer Res 12(7):2117–2124
20. Wingeier M, La Marca-Ghaemmaghami P, Zimmermann R, Ehlert U (2017) Is salivary estriol detectable in very early pregnancy? J Mater Neonatal Med 30(2):228–232
21. Chard T (1992) Pregnancy tests: a review. Hum Reprod 7(5):701–710
22. Karen A, De Sousa NM, Beckers J-F, Bajcsy ÁC, Tibold J, Mádl I, Szenci O (2015) Comparison of a commercial bovine pregnancy-associated glycoprotein ELISA test and a pregnancy-associated glycoprotein radiomimmunoassay test for early pregnancy diagnosis in dairy cattle. Anim Reprod Sci 159:31–37
23. El-Bali M, Zaglool DA, Khodari YA, Al-Harthi SA (2016) Appraisal of prenatal anti-toxoplasma gondii (IgG+ IgM)-IHA/IgM-ELISA screening in single samples via IgG avidity test. J Egypt Soc Parasitol 46(1):201–208
24. Makunyane L, Moodley J, Titus M (2017) HIV transmission in twin pregnancy: maternal and perinatal outcomes. South Afr J Infect Dis 32(2):54–56
25. Pizza M, Scarlato V, Masignani V, Giuliani MM, Arico B, Comanducci M, Jennings GT, Baldi L, Bartolini E, Capecchi B (2000) Identification of vaccine candidates against serogroup B meningococcus by whole-genome sequencing. Science 287(5459):1816–1820
26. Miura K, Orcutt AC, Muratova OV, Miller LH, Saul A, Long CA (2008) Development and characterization of a standardized ELISA including a reference serum on each plate to detect antibodies induced by experimental malaria vaccines. Vaccine 26(2):193–200
27. Voronin Y, Zinszner H, Karg C, Brooks K, Coombs R, Hural J, Holt R, Fast P, Allen M, Allen M (2015) HIV vaccine-induced sero-reactivity: a challenge for trial participants, researchers, and physicians. Vaccine 33(10):1243
28. Smalley C, Erasmus JH, Chesson CB, Beasley DW (2016) Status of research and development of vaccines for chikungunya. Vaccine 34(26):2976–2981
29. Mullis KB, Faloona FA (1987) Specific synthesis of DNA in vitro via a polymerase-catalyzed chain reaction. Methods Enzymol *155*:335–350
30. Mullis KB (1990) The unusual origin of the polymerase chain reaction. Sci Am 262(4):56–65
31. Olson M, Hood L, Cantor C, Botstein D (1989) A common language for physical mapping of the human genome. Science 245(4925):1434–1435

32. Kubista M, Andrade JM, Bengtsson M, Forootan A, Jonák J, Lind K, Sindelka R, Sjöback R, Sjögreen B, Strömbom L (2006) The real-time polymerase chain reaction. Mol Asp Med 27 (2–3):95–125
33. Morillo JM, Lau L, Sanz M, Herrera D, Silva A (2003) Quantitative real-time PCR based on single copy gene sequence for detection of Actinobacillus actinomycetemcomitans and Porphyromonas gingivalis. J Periodontal Res 38(5):518–524
34. Garibyan L, Avashia N (2013) Research techniques made simple: polymerase chain reaction (PCR). J Invest Dermatol 133(3):e6
35. Ke D, Ménard C, Picard FJ, Boissinot M, Ouellette M, Roy PH, Bergeron MG (2000) Development of conventional and real-time PCR assays for the rapid detection of group B streptococci. Clin Chem 46(3):324–331
36. Brookes AJ (2012) High muliplex nucleic acid amplification. In: Google Patents
37. Bharathi MJ, Rameshkumar G, Ramakrishnan R, Venugopal Reddy YC, Shivkumar C, Ramesh S (2013) Comparative evaluation of uniplex, nested, semi-nested, multiplex and nested multiplex PCR methods in the identification of microbial etiology of clinically suspected infectious endophthalmitis. Curr Eye Res 38(5):550–562
38. Plourde-Owobi L, Seguin D, Baudin M-A, Moste C, Rokbi B (2005) Molecular characterization of Clostridium tetani strains by pulsed-field gel electrophoresis and colony PCR. Appl Environ Microbiol 71(9):5604–5606
39. Ferre F (1992) Quantitative or semi-quantitative PCR: reality versus myth. Genome Res 2(1): 1–9
40. Marshall OJ (2004) PerlPrimer: cross-platform, graphical primer design for standard, bisulphite and real-time PCR. Bioinformatics 20(15):2471–2472
41. Pahwa R, Hedau S, Jain S, Jain N, Arora VM, Kumar N, Das BC (2005) Assessment of possible tuberculous lymphadenopathy by PCR compared to non-molecular methods. J Med Microbiol 54(9):873–878
42. Consortium EP (2007) Identification and analysis of functional elements in 1% of the human genome by the ENCODE pilot project. Nature 447(7146):799
43. Van Dongen J, Macintyre E, Gabert J, Delabesse E, Rossi V, Saglio G, Gottardi E, Rambaldi A, Dotti G, Griesinger F (1999) Standardized RT-PCR analysis of fusion gene transcripts from chromosome aberrations in acute leukemia for detection of minimal residual disease. Leukemia 13(12):1901
44. Williams JG, Kubelik AR, Livak KJ, Rafalski JA, Tingey SV (1990) DNA polymorphisms amplified by arbitrary primers are useful as genetic markers. Nucleic Acids Res 18(22):6531–6535
45. Bronzino JD(1995) Principles of electroencephalography. In: The biomedical engineering handbook, 1
46. Purcell EM, Torrey HC, Pound RV (1946) Resonance absorption by nuclear magnetic moments in a solid. Phys Rev 69(1–2):37
47. Hawkes R, Holland G, Moore W, Worthington B (1980) Nuclear magnetic resonance (NMR) tomography of the brain: a preliminary clinical assessment with demonstration of pathology. J Comput Assist Tomogr 4(5):577–586
48. Smith F, Hutchison J, Mallard J, Johnson G, Redpath TW, Selbie R, Reid A, Smith C (1981) Oesophageal carcinoma demonstrated by whole-body nuclear magnetic resonance imaging. Br Med J (Clin Res Ed) 282(6263):510–512
49. De Wilde J, Rivers A, Price D (2005) A review of the current use of magnetic resonance imaging in pregnancy and safety implications for the fetus. Prog Biophys Mol Biol 87(2–3): 335–353
50. Walker RE, Eustace SJ (2001) Whole-body magnetic resonance imaging: techniques, clinical indications, and future applications. In: Seminars in musculoskeletal radiology: 2001: Copyright© 2001 by Thieme Medical Publishers, Inc., 333 Seventh Avenue, New . . .; 2001: 005–020

51. Meyer CH, Hu BS, Nishimura DG, Macovski A (1992) Fast spiral coronary artery imaging. Magn Reson Med 28(2):202–213
52. Jackson JI, Meyer CH, Nishimura DG, Macovski A (1991) Selection of a convolution function for Fourier inversion using gridding (computerised tomography application). IEEE Trans Med Imaging 10(3):473–478
53. Hutchinson M, Raff U (1988) Fast MRI data acquisition using multiple detectors. Magn Reson Med 6(1):87–91
54. Kelton JR (1989) An algorithm for rapid image acquisition using multiple receiver coils. In: Proceeding of the ISMRM, Amsterdam, 1989, 1172
55. Ra JB (1991) Fast imaging method using multiple receiver coils with subencoding data set. In: Proceedings of the SMRM 10th Annual Meeting, San Francisco
56. Xue P, Thomas CW, Gilmore GC, Wilson DL (1998) An adaptive reference/test paradigm: application to pulsed fluoroscopy perception. Behav Res Methods Instrum Comput 30(2): 332–348
57. Van de Walle R, Barrett HH, Myers KJ, Aitbach M, Desplanques B, Gmitro AF, Cornelis J, Lemahieu I (2000) Reconstruction of MR images from data acquired on a general nonregular grid by pseudoinverse calculation. IEEE Trans Med Imaging 19(12):1160–1167
58. Ahunbay E, Pipe JG (2000) Rapid method for deblurring spiral MR images. Magnetic Resonance in Medicine: An Official Journal of the International Society for Magnetic Resonance in Medicine 44(3):491–494
59. McKenna D, Meehan C, Alhajeri A, Regan M, O'Keeffe D (2007) The use of MRI to demonstrate small bowel obstruction during pregnancy. Br J Radiol 80(949):e11–e14
60. Bulas D, Egloff A (2013) Benefits and risks of MRI in pregnancy. In: Seminars in perinatology, Elsevier, pp 301–304
61. Schick F (2005) Whole-body MRI at high field: technical limits and clinical potential. Eur Radiol 15(5):946–959
62. Lauenstein TC, Freudenberg LS, Goehde SC, Ruehm SG, Goyen M, Bosk S, Debatin JF, Barkhausen J (2002) Whole-body MRI using a rolling table platform for the detection of bone metastases. Eur Radiol 12(8):2091–2099
63. Goldman LW (2007) Principles of CT and CT technology. J Nucl Med Technol 35(3):115–128
64. Hsieh J 2009 Computed tomography: principles, design, artifacts, and recent advances. In: 2009: SPIE Bellingham, WA
65. Hounsfield G (1976) Historical notes on computerized axial tomography. J Can Assoc Radiol 27(3):135–142
66. Curry TS, Dowdey JE, Murry RC: Christensen's physics of diagnostic radiology: Lippincott Williams & Wilkins; 1990
67. Fishman EK, Jeffrey RB: Spiral CT: principles, techniques, and clinical applications: raven Pr; 1995
68. Kalra MK, Maher MM, D'souza R, Saini S (2004) Multidetector computed tomography technology: current status and emerging developments. J Comput Assist Tomogr 28:S2–S6
69. Sabarudin A, Subramaniam C, Sun Z (2014) Cerebral CT angiography and CT perfusion in acute stroke detection: a systematic review of diagnostic value. Quant Imaging Med Surg 4(4): 282
70. Hunter GJ, Silvennoinen HM, Hamberg LM, Koroshetz WJ, Buonanno FS, Schwamm LH, Rordorf GA, Gonzalez RG (2003) Whole-brain CT perfusion measurement of perfused cerebral blood volume in acute ischemic stroke: probability curve for regional infarction. Radiology 227 (3):725–730
71. HOPKINS AL, WHEELER SJ (1991) Subdural hematoma in a dog. Vet Surg 20(6):413–417
72. Turrel J, Fike J, LeCouteur R, Higgins R (1986) Computed tomographic characteristics of primary brain tumors in 50 dogs. J Am Vet Med Assoc 188(8):851–856
73. Waters DJ, Coakley FV, Cohen MD, Davis MM, Karmazyn B, Gonin R, Hanna MP, Knapp DW, Heifetz SA (1998) The detection of pulmonary metastases by helical CT: a clinicopathologic study in dogs. J Comput Assist Tomogr 22(2):235–240

74. Yoon J, Feeney DA, Cronk DE, Anderson KL, Ziegler LE (2004) Computed tomographic evaluation of canine and feline mediastinal masses in 14 patients. Vet Radiol Ultrasound 45(6): 542–546
75. Carpenter L, Schwarz P, Lowry J, Park R, Steyn P (1993) Comparison of radiologic imaging techniques for diagnosis of fragmented medial coronoid process of the cubital joint in dogs. J Am Vet Med Assoc 203(1):78–83
76. Macari M, Bini EJ, Xue X, Milano A, Katz SS, Resnick D, Chandarana H, Krinsky G, Klingenbeck K, Marshall CH (2002) Colorectal neoplasms: prospective comparison of thin-section low-dose multi–detector row CT colonography and conventional colonoscopy for detection. Radiology 224(2):383–392

Part IX

Neuro Disability and Neurorehabilitation

Molecular Insights into the Pathophysiology of Neurological Disorders 23

Arpita Devi

Abstract

Disabilities are limitations or restrictions to oneself. It may be mental or physical. Physical disability is limitation to one's physical functions or mobility. Physical disabilities can be acquired at any time – before birth, after birth, during adulthood or old age. There can be various reasons why one becomes physically disabled. The reasons may be genetic, diseases or accidents. The physical disabilities are often related to our nervous system. Damage to neurons often leads to nervous system disorders which are seen in the form of physical or mental disability. But sometimes deformities of limbs or damage to locomotory tissues also results in physical disability. Microorganism-borne diseases such as polio also cause deformities of the limbs. Some forms of physical disability can be prevented. Polio can be prevented by vaccination, and care during pregnancy and childbirth can prevent prenatal disabilities. Genetic physical disabilities as well as age-related disabilities, on the other hand, are difficult to prevent. However, through physiotherapy and communication therapy, the patients are able to live an independent life.

Keywords

Spina Bifida · Muscular Dystrophy · Poliomyelitis · Diabetes · Cerebral Palsy · Stroke · Epilepsy · Alzheimer's Disease · Parkinson's Disease

23.1 Introduction

Physical disability is the inability to perform physical activity. Physical disability can be congenital or acquired. Congenital disability is acquired before birth due to genetic reasons or accidents during pregnancy. Acquired physical disability is the

A. Devi (✉)
Department of Molecular Biology and Biotechnology, Tezpur University, Tezpur, Assam, India

S. Paul (ed.), *Application of Biomedical Engineering in Neuroscience*,
https://doi.org/10.1007/978-981-13-7142-4_23

disability acquired after birth due to accidents or infections such as polio (https://hwa.org.sg/general-information-on-physical-disabilities/). Based on the time at which the disability has occurred, physical disability is of three types:

(a) Prenatal: It is acquired during pregnancy due to accidents met by the mother or exposure to some diseases or substances.
(b) Perinatal: It is acquired between some weeks before birth to about one month after birth. This can be due to premature delivery, brain injury during birth, etc.
(c) Postnatal: It is gained after birth due to accidents, obesity, diabetes, infections, etc.

23.1.1 Categories Under Physical Disability (https://hwa.org.sg/general-information-on-physical-disabilities/)

The two major categories are:

1. Musculoskeletal Disability: It is the lack of ability to perform the activities that require movements of the body parts. The reasons for this are deformation of muscular or skeletal system, diseases or degeneration of locomotory tissues. The disabilities in this category are:
   (a) Loss or deformity of limbs: Limbs can be lost or damaged due to disease or accidents.
   (b) Osteogenesis imperfecta: It is due to defective development of connective tissue. The bones get softened that leads to bending or twisting of the skeletal system, and the patient experiences deformed posture.
   (c) Muscular dystrophy: It is a hereditary disorder. There is progressive loss of strength of the muscles around the neck, shoulders and hips.
2. Neuro-musculo Disability: It is a disability arising due to disorder or degeneration of the nervous system. The patient is unable to control the movements of affected body parts. The disabilities in this category are:
   (a) Cerebral palsy: It is a group of disorders which causes impairment of motor function due to damage of the brain. The person suffers from movement and coordination problems, intellectual, visual and speech problems, etc.
   (b) Spina bifida: It occurs due to defective development of the spinal canal. The spinal cord is protected by a closed bony encasement. The disease arises when the protective bony encasement is incompletely closed. There is sensory loss which leads to paralysis of the lower part of the body.
   (c) Poliomyelitis: It is caused by poliovirus affecting the lower motor-neuron system. The person has weakness in the body and both legs.
   (d) Stroke: It is caused by haemorrhage inside the brain. There is a sudden numbness or muscle weakness in one side of the body due to loss of sensory motor functioning.
   (e) Spinal cord injury: It is of two types – paraplegia, where there is total or partial loss of function of sensory or motor neurons in the lower part of the

body and lower limbs, and tetraplegia, where there is total or partial loss of function of sensory and motor neurons of the whole body including all four limbs.

23.2 Diseases Leading to Physical Disability

23.2.1 Spina Bifida

The neural tube of the embryo develops into the spinal cord and brain [1]. The process of formation of the neural tube is called neurulation [1]. Primary neurulation is a developmental process that involves shaping, folding and midline fusion of the neural plate. Later primary neurulation forms the brain and the region of the first five sacral vertebrae [1]. In human, primary neurulation is completed by about four weeks post-conception. Neural folding does not occur in secondary neurulation, but the caudal portion of the spinal cord is formed by secondary neurulation [1]. Defect in primary neurulation causes a condition called spina bifida (meningomyelocele) [2]. It occurs due to failure of fusion in the caudal region of the neural tube [2]. Animal model studies showed that spina bifida can result from disturbances in cell adhesion or alterations in neural plate shaping or bending that prevent apposition of the neural folds [3]. As the neural tube of the embryo is open, it is exposed to the amniotic fluid environment [2]. The bifid neuroepithelium undergoes normal neuronal differentiation initially. However, with time the exposed spinal cord turns haemorrhagic, and neurons die due to toxicity of amniotic fluid [4]. Axonal connections as well as the functions are lost [4]. Hence, the pathophysiology of spina bifida is a two-step process: failure in neural tube closure and neurodegeneration of the foetus in uterus.

There are many causes of spina bifida. Less intake of natural folate, or folic acid or vitamin B9, before and during beginning of pregnancy, often increases risk of spina bifida in the foetus [5, 6]. Folic acid is known to participate in two biochemical pathways, nucleic acid synthesis and methylation reactions that control expression of genes. 5,10-methylenetetrahydrofolate reductase is an enzyme encoded by MTHFR gene [5]. This enzyme produces 5-methyltetrahydrofolate, which converts homocysteine into methionine [5]. The MTHFR C677T variant, where cytosine is replaced by thymine at codon 677 in either maternal or foetal gene, is a trigger for spina bifida [5]. Disturbance in folate metabolism results in increased homocysteine concentration [5, 7]. Homocysteine caused malformation of the neural tube in some animal models [7]. Vitamin B12 might also be associated with neural tube defects [8]. Pre-gestational diabetes of the mother may also cause spina bifida in the foetus [9]. Studies from mouse models suggest that glucose has a direct teratogenic effect [10]. Excess glucose might alter the expression of genes involved in embryonic development [11]. If the foetus is exposed to drugs such as valproic acid or carbamazepine alone, or in combination with each other or other anticonvulsants in the uterus, it increases the risk of spina bifida [12]. Mechanism of action of valproic acid and carbamazepine in the development of spina bifida is not known. Free radicals formed as by-products of drug metabolism or their effects on folate metabolism might damage the developing embryonic tissues [12]. Fumonisins (a class of myocotoxins) can produce neural tube defects in animal models [13]. Thus, there may be an association between maternal exposure to fumonisins and increased risk of spina bifida in foetuses (Fig. 23.1).

23.2.2 Muscular Dystrophy

Muscular dystrophies are a group of hereditary diseases. It causes progressive muscle weakness due to intrinsic biochemical defects of muscle. The most common human muscular dystrophy is the X-linked recessive Duchenne muscular dystrophy (DMD) and Becker muscular dystrophy (BMD) [14]. With the discovery of dystrophin and its interacting partners, the mystery surrounding muscular dystrophy has begun to unfold. Absence of one of these proteins could change the integrity of the muscle fibres [15]. The absence of dystrophin causes change in the membrane structure by delocalisation of the dystrophin-interacting proteins from the membrane [15].

Calcium homeostasis is important in muscle function. Calcium accumulation of hypercontracted fibres in biopsies of DMD patients has led to the possibility of the role of calcium in the pathophysiology of DMD [15]. Increased influx of Ca^{2+} through a dystrophin-deficient membrane has been found [16]. This influx may occur mostly through voltage-independent calcium channel [16]. Submembranous concentration of Ca^{2+} may increase abnormally, but confirmation is needed at physiologic values of membrane potential. Continuous increase in calcium concentration in cytosol leads to activation of various proteases, mainly calpains, which may result in the destruction of membrane proteins which, in turn, will facilitate calcium entry [17]. Eventually, excessive calcium ion may lead to cell death by triggering necrosis or apoptosis [18].

Also it has been found that muscles from DMD patients have high levels of products of lipid peroxidation as well as high levels of enzymes responsive to oxidative stress [19]. Both the evidence suggests that oxidative stress occurs in dystrophic muscle. It is not yet proved that oxidative damage of sarcolemma promotes dystrophinopathy, although dystrophin-deficient muscle cells are found to be more susceptible to oxidative stress *in vitro*. Rando and his team showed that dystrophin-deficient myotubes are killed more easily upon exposure to oxidants [20]. However, other metabolic stresses did not easily kill the dystrophin-deficient myotubes [20]. The question of dystrophin deficiency and defects in generation of free radical in dystrophic muscle can be answered through neuronal nitric oxide synthase (nNOS) [21]. nNOS is displaced from the cell membrane in dystrophin-deficient muscle, and its concentration is reduced to less than 20% of normal levels. Even mRNA level of nNOS is also decreased. As a result, the production of nitric oxide (NO) is drastically reduced in dystrophin-deficient muscles [21]. These

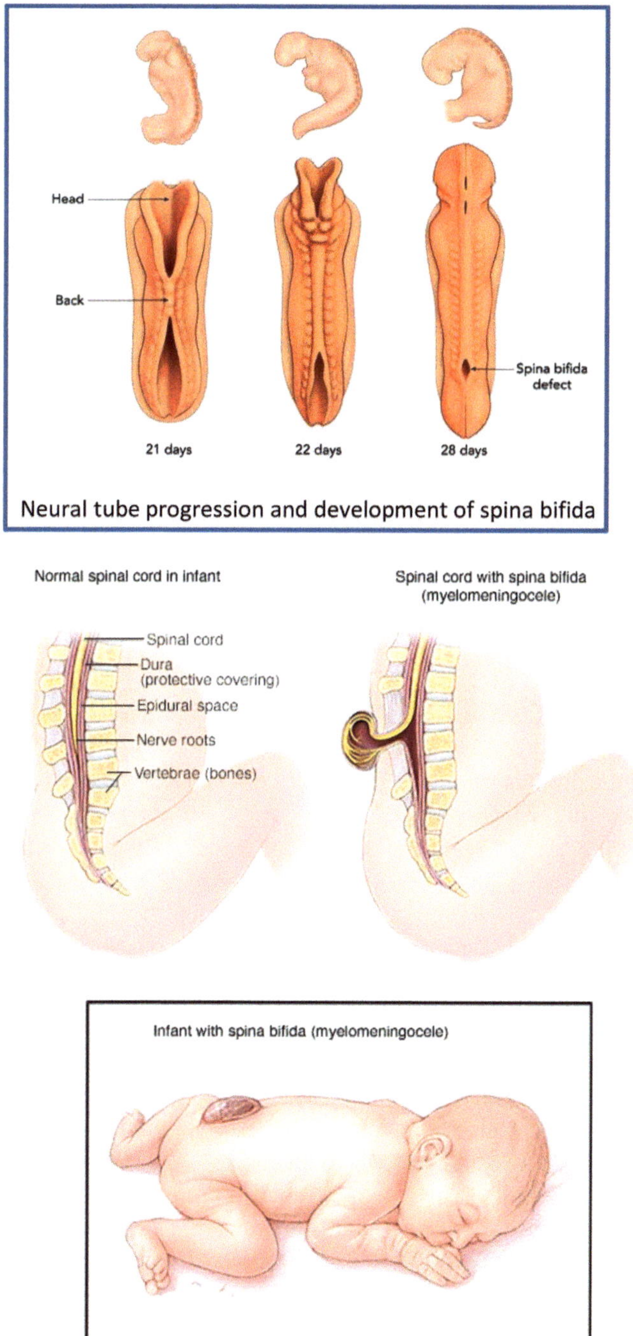

Fig. 23.1 Development of spina bifida. (Image source: Mayo foundation for medical education and research and https://www.childrensmn.org)

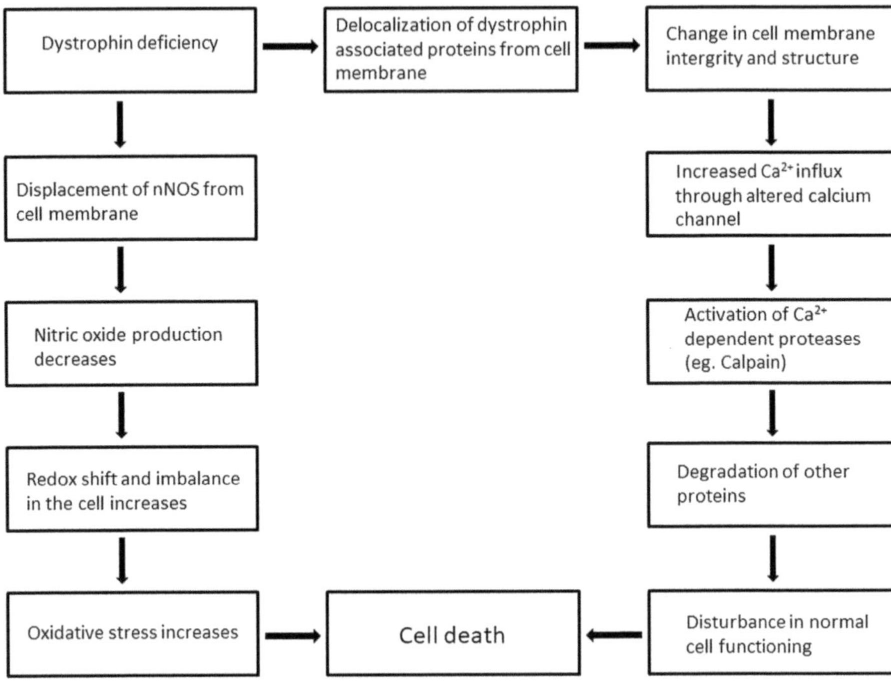

Fig. 23.2 Cellular events leading to cell death in case of dystrophin deficiency

disruptions in the production of NO in muscle could cause major redox environment shifts in muscle (Fig. 23.2).

23.2.3 Poliomyelitis

Poliomyelitis is a well-known, highly infectious viral disease. The virus belongs to the Picornaviridae family (https://www.who.int/biologicals/areas/vaccines/poliomyelitis/en/). This disease causes crippling deformities in the patients. Poliovirus was isolated in 1909 [22]. In 1981, the genomic structure of the virus and its pathogenesis was elucidated [23, 24]. Salk and Sabin gave two vaccines, inactivated polio vaccine (IPV) and oral polio vaccine (OPV), in the years 1955 and 1961, respectively (https://www.who.int/biologicals/areas/vaccines/poliomyelitis/en/). America was declared polio free by the World Health Organization (WHO) in 1994. Western Pacific Region was declared polio free in 2000 and European Region in 2002. In 2013, only Nigeria, Pakistan and Afghanistan remained polio endemic (https://www.who.int/en/news-room/fact-sheets/detail/poliomyelitis).

The poliovirus enters through the oropharynx and multiplies in the tonsils and lymph nodes of the neck. Then, it invades the Peyer patches and small intestine [25]. The incubation time ranges from 2 to 35 days. It is also said that, sometimes, the virus may invade the tonsils through the bloodstream. After 3 to 5 days, the virus is excreted through the stool and can also be obtained from the throat swabs of infected patients [25]. There can be total absence of symptoms during this period. Respiratory tract infection, gastroenteritis and influenza-like symptoms can occur. The presence of virus in the blood may subside due to the circulating antibodies or the virus' entry to the central nervous system (CNS). The virus has high affinity for cellular receptor CD155 (Cluster of Differentiation 155). CD155 helps in cell attachment of the virus and entry of viral genome [26]. The anterior horn cells of the spinal cord are extensively damaged. This causes limb paralysis [25]. Then, the virus spreads to the posterior horn cells of the spinal cord and motor neuron of the thalamus and hypothalamus of the brain. In bulbar form of poliomyelitis, the brainstem is damaged, which may be fatal [25]. There is accumulation of polymorphonuclear neutrophils, plasma cells and microglia in the infected cells. Infected cells then get ingested by macrophages causing degeneration of axons. Muscular atrophy also occurs, leading to flaccid paralysis [25]. Respiratory paralysis may even cause death (Fig. 23.3).

23.2.4 Diabetes

Diabetes is a well-known complication these days. Diabetes itself is not a disability, but secondary diabetic complications may lead to physical disability. Secondary diabetic complications include cataract, diabetic neuropathy, retinopathy and nephropathy. Cataract and diabetic neuropathy can be considered physical disabilities as the vision as well as the nervous system of the patient is compromised. During diabetes, the blood sugar level remains continuously high. Blood sugar can easily diffuse into the neurons and nephrons of the kidney, retina and lens of the eye. As a result, the glucose level inside those cells increases. As the concentration of glucose reaches 30 mM, a NADPH-dependent cytosolic oxidoreductase, aldose reductase (AR), gets activated. AR converts glucose into sorbitol. Sorbitol cannot diffuse through the cell membrane and remains inside the cell. Thus, there occurs an osmotic imbalance. As a result, more and more water is taken inside. This causes cataract [27]. Also, during conversion of glucose to sorbitol, NADPH is used up in

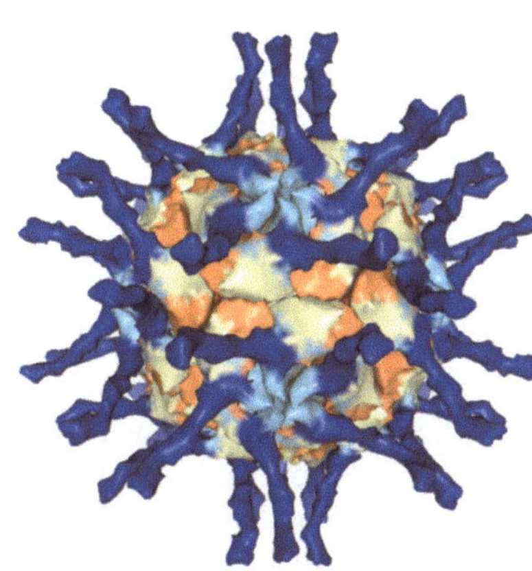

Fig. 23.3 Electron microscopy image of poliovirus type 1 (strain Mahoney) at 15 Å resolution. PDB ID: 1NN8

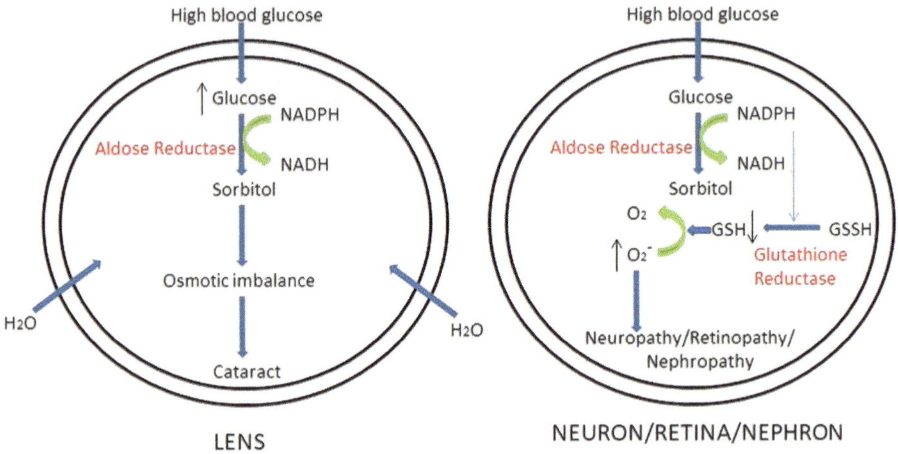

Fig. 23.4 Progression of secondary diabetic complications

the cell. NADPH is important for conversion of oxidised glutathione (GSSH) to free glutathione (GSH). GSH is an antioxidant. Due to less NADPH, there is less GSH in the cells. Thus, the cells are more damaged by free radicals. When extremely damaged by free radicals, cells even die, causing retinopathy, neuropathy and nephropathy [27]. Thus, due to cataract and retinopathy, the vision is compromised, and due to neuropathy, the nervous system is weakened leading to many other complications (Fig. 23.4).

Degradation of skeletal muscles due to diabetes again leads to physical disability [28]. The exact mechanism behind this is not known, but diabetic neuropathy is considered as one of the causes. Fat infiltration in muscle cells is another cause of muscle quality degradation [29].

23.2.5 Cerebral Palsy

Cerebral palsy (CP) is a collection of nonprogressive disorders of posture, coordination and motor skills, causing muscle movement limitation. It occurs in early childhood. It is of many types [30]:

(a) Spastic cerebral palsy: It is characterised by stiff muscles and exaggerated reflexes. It is the most common form of CP occurring in almost 70% of people with CP.
(b) Ataxic cerebral palsy: It is characterised by disorganised voluntary muscle movement observed in approximately 5–10% cases of CP.
(c) Athetoid cerebral palsy or dyskinetic cerebral palsy: It is characterised by abnormal movement in arms and legs.
(d) Mixed cerebral palsy: It has symptoms of all the forms of CP, i.e. athetoid, ataxic and spastic CP appearing simultaneously.

Many factors cause cerebral palsy such as preterm birth, lack of oxygen to the brain during pregnancy and delivery, infections during pregnancy such as German measles and herpes simplex, jaundice of the infant, head injury during pregnancy, brain infection such as encephalitis and meningitis, etc. In preterm infants, spastic paralysis is the most prevalent pathological condition. It is associated with intraparenchymal haemorrhage, diffuse white matter injury and periventricular cavitary lesions. Corticospinal tract is disrupted, and motor disorders develop by influencing the motor neurons of the brain and spinal cord. The cerebellum and basal ganglia also influence the final muscle tone in CP [31].

Main factors for cell death in hypoxic, ischemic and inflammatory conditions are production of pro-inflammatory cytokines, deprivation of growth factors, oxidative stress, extracellular matrix modification, release of glutamate and activating excitotoxicity. In preterm newborns or during pregnancy, all these factors cause primary defects in the process of myelination and gliosis and result in thalamic degeneration with maldevelopment in the secondary cortex and the thalamus. First, microglial cells respond to any kind of injury. They mediate neurotoxicity by expressing both glutamatergic receptors and adenosine receptors which function in inflammatory response. Oligodendrocyte progenitor cells express α-amino-3-hydroxy-5-methyl-4-isoxazolepropionic acid (AMPA) receptor on their soma and the *N*-methyl-D-aspartate (NMDA) receptor on their processes in response to oxidative stress and glutamate excitotoxicity. During pregnancy, the cortical connections of the foetal cerebral cortex and cortical neurons depend on distribution of extracellular matrix and axonal guidance molecules such as netrins, slits, ephrins, semaphorins, cell adhesion molecules, etc. Axonal guidance molecules are produced within the subplate zone and periventricular crossroads of growing axons. Disturbances in these mechanisms in case of preterm infants cause alteration of the development of cortical connections. Excitotoxicity may be due to lack of reuptake of excitatory neurotransmitter, glutamate. Due to excitotoxicity, more and more Ca^{2+} ions are taken in via NMDA receptor. The increase in Ca^{2+} results in free radical production. Free radicals cause a number of effects such as cell membrane damage and mitochondrial dysfunction and direct the cell towards apoptosis. Apoptosis is triggered by upregulation of caspase, BAX (Bcl-2-associated X protein), FAS death receptor, calpain and PARP [31].

23.2.6 Stroke

Stroke is a neurological disease which results from poor blood flow to the brain. There are two types of stroke: ischemic stroke and haemorrhagic stroke [32]. In ischemic stroke, blood supply to the brain may be obstructed due to (a) thrombosis (blood clot in the blood vessels), (b) embolism (embolus in the blood vessel), (c) systemic hypoperfusion (decrease in blood supply due to some conditions like shock) and (d) cerebral venous sinus thrombosis (presence of blood clot in the dural venous sinuses). Haemorrhagic stroke are of two types: (a) intracerebral haemorrhage (bleeding in the brain) and (b) subarachnoid haemorrhage (bleeding between arachnoid mater and pia mater). Brain tissue is fully dependent on glucose and oxygen for functioning. The result of ischemia is reduced blood supply to the brain which results in reduced oxygen and glucose supply to the brain. If ischemia continues, it will result in a stroke [32].

Other than depletion of blood supply, many acute and chronic infective conditions are also known to precede stroke. Most often, these are of bacterial origin. *Chlamydia pneumoniae* is a bacterium known to cause pneumonia. They affect adherence of macrophages or monocytes to endothelial cells and vascular endothelium and thus have a role in atherogenesis [33]. This microorganism has been found in atherosclerotic plaques of coronary artery and middle cerebral artery, and chlamydial lipopolysaccharide has been found in serum immune complexes of patients who have suffered from stroke [34]. Other microorganisms associated with ischemic stroke are *Helicobacter pylori* and herpes virus [34].

23.2.7 Epilepsy

Epilepsy is a nervous system disorder characterised by repeated and sudden interruptions in normal activity of the brain. These interruptions are called epileptic seizures. Various researchers are convinced that the hyperexcitability of neurons is due to the imbalance between two excitatory and inhibitory neurotransmitters, glutamate and γ-aminobutyric acid (GABA), respectively. Glutamate (an amino acid) is an important excitatory neurotransmitter in the brain. It depolarises neurons and generates excitatory postsynaptic potentials. Glutamate receptors, namely, α-amino-3-hydroxy-5-methyl-4-isoxazolepropionic acid (AMPA), *N*-methyl-D-aspartic acid (NMDA), kainite and G-protein-coupled glutamate receptors, are upregulated. As a result, signal transmission via glutamate increases manifold. This phenomenon can be seen as interictal spike on an electroencephalogram known as paroxysmal depolarising shift. Paroxysmal depolarising shift is nothing but depolarisation due to a huge excitatory synaptic potential. It is dependent on activation of AMPA and NMDA receptors [35]. On the other hand, GABA is an important inhibitory neurotransmitter that hyperpolarises neurons and generates inhibitory presynaptic potentials. GABA receptors are of two types: $GABA_A$ and $GABA_B$ receptors. $GABA_A$ receptors increase influx of chloride and rapidly mediate inhibitory presynaptic potentials. G-protein-coupled $GABA_B$ receptors increase potassium conductance and decrease the calcium influx and mediate slow inhibitory presynaptic potential. Reduction in synthesis and release of inhibitory GABA and reduction in GABA receptors may increase the probability of generating excitatory postsynaptic potentials and therefore induce epileptic seizures [35] (Fig. 23.5).

Mutation in genes of ion channels is another cause of epileptic syndrome [36, 37]. Mutation in genes results in synthesis of defective protein or no synthesis of protein at all. With the advancement in the field of genetics and molecular biology, it was possible for the researchers to show that hyperexcitability of neurons is a result of mutations in genes encoding ion channel proteins. Neuronal excitation is a normal process. Normally, cation channels are responsible for neuronal

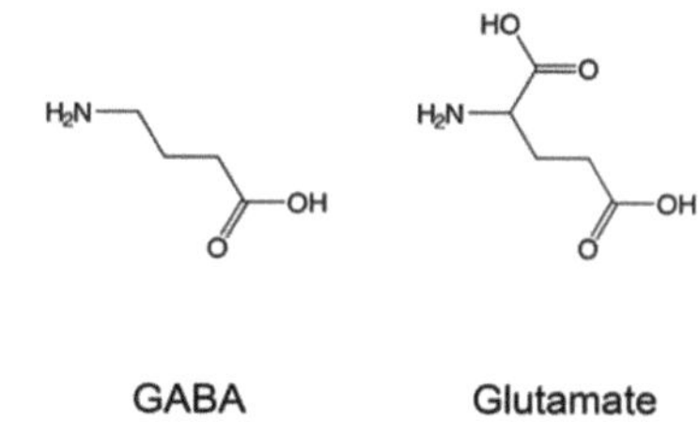

Fig. 23.5 Chemical structures of neurotransmitters, GABA and glutamate

excitability, and anion channels are responsible for reducing the neuronal excitatory process. Cationic and anionic channels maintain the influx and outflux of ions. Any imbalance of ion charges due to non-functioning of either anion or cation channel can induce epileptic seizures [36, 37]. Mutations in genes expressing ion channels (K^+, Na^+, Cl^-, Ca^{2+}) and receptors of acetylcholine and GABA have also been reported in onset of epilepsy [37].

23.2.8 Alzheimer's Disease

Alzheimer's disease (AD) is a neurodegenerative disorder of irreversible memory loss and cognitive function. Different researchers have put forward different hypotheses on the progression of this disease. Among all the hypotheses, the highly accepted one is the amyloid beta (Aβ) hypothesis. According to this hypothesis, the amyloid precursor protein (APP) is abnormally cleaved by β- and γ-secretases. Normally APP is cleaved by α- and β-secretase and releases a fragment which may be involved in apoptotic process. But due to abnormal cleavage, Aβ peptide fragments are formed, and they spontaneously aggregate into an insoluble mass and are deposited as plaques. Aβ peptides are of varying lengths, but 40-residue and 42-residue peptides are more common [38]. It has been shown in various reports that Aβ42 oligomers induce oxidative stress and tau hyperphosphorylation and have harmful effects on nerve synapses and mitochondria. Aβ42 plaques also attract microglia [39]. Pro-inflammatory cytokines are produced and released in response to microglial activation. The released cytokines, in turn, stimulate other astrocyte-neuron to produce more Aβ42 peptides [40]. Aβ peptide aggregates may be the key factor for the neuronal and vascular degeneration in the brain.

Another hypothesis is the tau hypothesis. According to this hypothesis, hyperphosphorylated tau protein pairs with another tau protein. They form neurofi brillary tangle-like structure inside nerve cell bodies. Due to these tangles, the microtubules disintegrate, and the cytoskeleton of the neuron is destroyed. Thus, the neuronal communication gets affected first which ultimately results in cell death [41] (Fig. 23.6).

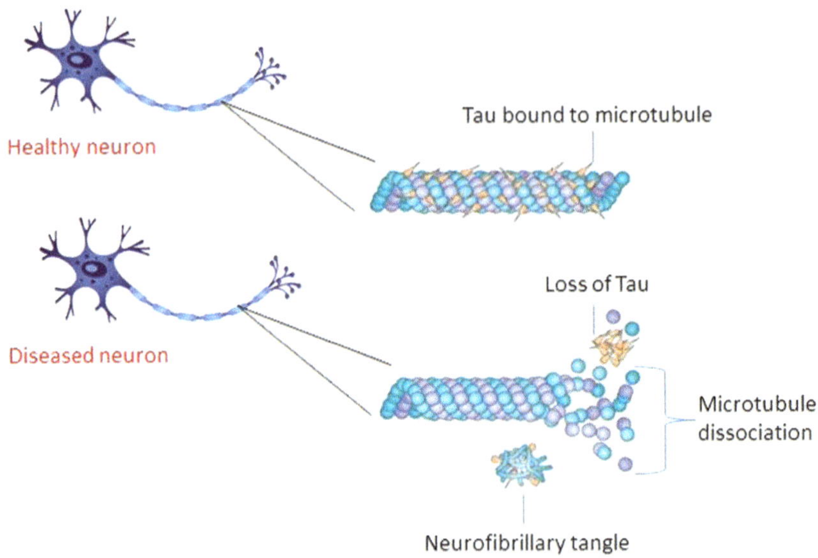

Fig. 23.6 Stabilisation of microtubules by Tau. Hyperphosphorylated Tau begins to form neurofibrillary tangles, and microtubules start to dissociate

23.2.9 Parkinson's Disease

It is a well-known neurodegenerative disease that mainly affects the brain. The parts of the brain that are affected include the midbrain, brainstem, cerebral cortex and olfactory tubercle and some parts of peripheral nervous system [42]. The major cause of neuronal death was oligomerisation or aggregation of a protein named alpha-synuclein. Being insoluble, α-synuclein aggregates to form Lewy bodies in neurons. Lewy bodies directly do not cause cell death, but it leads to other effects that cause cell death [43]. But, formation of Lewy bodies is a pathological marker for development of Parkinson's disease. Another mechanism of neuronal death is disruption in autophagic pathway [44]. Autophagy is a mechanism by which damaged cell components are broken down and recycled for use. Autophagy is a cell survival mechanism; disruption in this mechanism can lead to adverse effects like cell death. During autophagy, autophagosomes are formed which are supposed to clear aggregated protein. However, in neurodegenerative diseases, the ratio of protein aggregates formation and clearance is disturbed. With more and more aggregates accumulating in the cell, normal functions of the cells are disturbed and the cell dies [44] (Fig. 23.7).

Dysregulated mitochondrial function is another cause of neuronal death in Parkinson's disease [45]. Mitochondria are the energy-producing organelle of the cell. Any alteration in mitochondrial functioning leads to inhibited energy production which ultimately results in cell death. Two proteins, PINK1 (PTEN-induced kinase 1) and Parkin (E3 ubiquitin ligase), drive autophagy of mitochondria, known

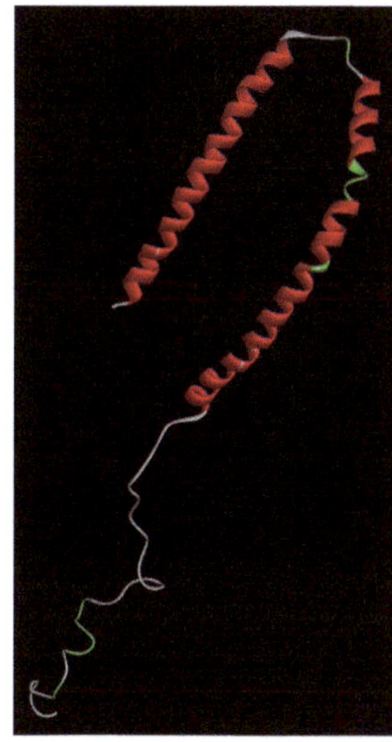

Fig. 23.7 X-ray crystallographic structure of α-synuclein. PDB ID: 1XQ8

as mitophagy. PINK1 is generally transported into the mitochondrion, but when mitochondria are damaged, PINK1 accumulates on the surface. Parkin is recruited by accumulated PINK1. Parkin starts the breakdown of dysfunctional mitochondria. In Parkinson's disease, genes encoding PINK1 and Parkin are mutated, thus forming misfolded proteins. Mitophagy is not possible with misfolded PINK1 and Parkin, causing morphological and functional changes in mitochondria and eventually cell death [45, 46]. Mitochondrial DNA (mtDNA) mutations also increase with age. Thus, the probability of neuronal death due to dysfunctional mitochondria also increases with death.

ROS (reactive oxygen species) are highly reactive chemical moieties containing oxygen. They are toxic to the DNA, proteins and other organelles of the cell. Mitochondria are one of the sites where ROS is generated. Mitochondria also have the ability to remove ROS; thus, they maintain the ratio of ROS generation and removal. With increase in age, mitochondrion loses the ability to remove ROS. Thus, there is an increased production of ROS. High level of ROS in a cell is also one of the reasons of neuronal death leading to Parkinson's disease [46] (Fig. 23.8).

Another major mechanism for brain cell death is the blood-brain barrier (BBB) breakdown. There are three types of cells which regulate the blood-brain barrier, namely, endothelial cells, pericytes and astrocytes. The blood-brain barrier is necessary in regulating the inflow and outflow of molecules in and out of the brain. In neurodegenerative diseases, blood-brain barrier breakdown does not occur everywhere but in specific regions of the brain, such as the substantia nigra in case of Parkinson's disease and the hippocampus in case of Alzheimer's disease [47]. Protein aggregates, which are released into the blood or cytokines from neuroinflammation, may interfere with cell receptors present on endothelial cells, pericytes and astrocytes and alter their function. Mainly, vascular endothelial growth factor (VEGF) receptors are deregulated in neurodegenerative diseases. Vascular

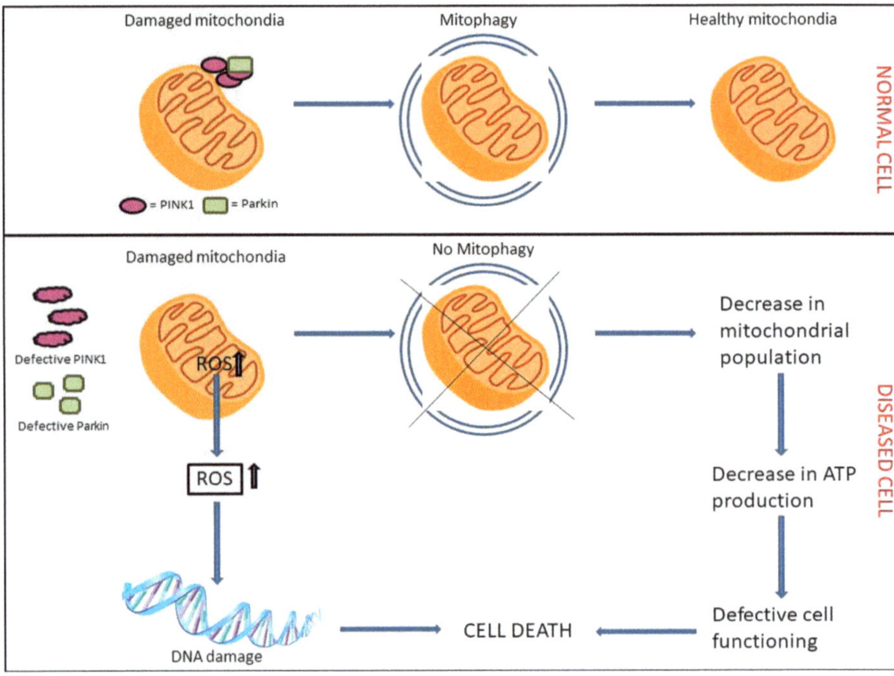

Fig. 23.8 Cell death in absence of mitophagy

endothelial growth factor (VEGF) interacts with VEGF receptor and signals for cell proliferation. But the VEGF receptors are damaged in Parkinson's disease and Alzheimer's disease. Thus, the cell stops growing, and therefore, new capillary formation via angiogenesis does not occur. Damage of cell receptor also affects the ability of cells to adhere to one another through adherens junctions. With lack of new capillary formation, the existing capillaries break down, and cells begin to dissociate from each other. This leads to the breakdown of gap junctions of endothelial cells. Gap junctions are important because they prevent large, unwanted, toxic or harmful molecules from entering. As gap junctions break down, plasma proteins as well as cytokines, protein aggregates, etc. enter into the extracellular matrix of the brain. This process is also called vascular leakiness, where blood and blood proteins enter (leaks) into the brain due to capillary degeneration. This imbalance is a major threat to the brain, and the brain cells are bound to alter their function and shift towards apoptosis.

23.3 Treatment for Physical Disabilities

Treatment of conditions leading to physical disabilities depends on the cause, type and stage of the disease. When treatment is not possible, there is always the need to learn how to survive with the disabilities. Rehabilitation programmes are done to make the patients confident in their current situation. It is necessary to teach the patients how to manage their condition independently. In the case of children, it is very important to make them feel loved all the time for their mental health. Physiotherapy or physical exercise is another way that has been found effective in many cases. The exercises are planned according to the patient's need. Communication therapy is another way by which the deaf, dumb and blind patients are taught to communicate with others. Vaccination is another way to prevent the microorganism-borne diseases that lead to physical disabilities.

References

1. Gilbert SF, Barresi MJF (2016) Developmental biology, 11th edn. Oxford University Press, New York
2. Greene ND, Copp AJ (2014) Neural tube defects. Annu Rev Neurosci 37:221–242
3. Mohd-Zin SW, Marwan AI, Abou Chaar MK, Ahmad-Annuar A, Abdul-Aziz NM (2017) Spina bifida: pathogenesis, mechanisms, and genes in mice and humans. Scientifica 2017:1–26
4. Copp AJ, Adzick NS, Chitty LS, Fletcher JM, Holmbeck GN, Shaw GM (2015) Spina bifida. Nat Re Dis Primer 1:15007
5. Greenberg JA, Bell SJ, Guan Y, Yu YH (2011) Folic acid supplementation and pregnancy: more than just neural tube defect prevention. Rev Obstet Gynecol 4(2):52
6. Safi J, Joyeux L, Chalouhi GE (2012) Periconceptional folate deficiency and implications in neural tube defects. J Pregnancy 2012:1–9
7. Imbard A, Benoist JF, Blom H (2013) Neural tube defects, folic acid and methylation. Int J Environ Res Public Health 10(9):4352–4389
8. Molloy AM, Kirke PN, Troendle JF, Burke H, Sutton M, Brody LC et al (2009) Maternal vitamin B_{12} status and risk of neural tube defects in a population with high neural tube defect prevalence and no folic acid fortification. Pediatrics 123(3):917–923
9. Chen CP (2005) Congenital malformations associated with maternal diabetes. Taiwanese J Obstetrics Gynecol 44(1):1–7
10. Zabihi S, Loeken MR (2010) Understanding diabetic teratogenesis: where are we now and where are we going? Birth Defects Res Part A 88(10):779–790
11. Davidson CM, Northrup H, King TM, Fletcher JM, Townsend I, Tyerman GH, Au KS (2008) Genes in glucose metabolism and association with spina bifida. Reprod Sci 15(1):51–58
12. Koren G, Nava-Ocampo AA, Moretti ME, Sussman R, Nulman I (2006) Major malformations with valproic acid. Can Fam Physician 52(4):441–442
13. Missmer SA, Suarez L, Felkner M, Wang E, Merrill AH Jr, Rothman KJ, Hendricks KA (2005) Exposure to fumonisins and the occurrence of neural tube defects along the Texas–Mexico border. Environ Health Perspect 114(2):237–241
14. Bellayou H, Hamzi K, Rafai MA, Karkouri M, Slassi I, Azeddoug H, Nadifi S (2009) Duchenne and Becker muscular dystrophy: contribution of a molecular and immunohistochemical analysis in diagnosis in Morocco. Biomed Res Int 2009:325210
15. Deconinck N, Dan B (2007) Pathophysiology of duchenne muscular dystrophy: current hypotheses. Pediatr Neurol 36(1):1–7
16. Dudley RW, Danialou G, Govindaraju K, Lands L, Eidelman DE, Petrof BJ (2006) Sarcolemmal damage in dystrophin deficiency is modulated by synergistic interactions between mechanical and oxidative/nitrosative stresses. Am J Pathol 168(4):1276–1287
17. Vanderklish PW, Bahr BA (2000) The pathogenic activation of calpain: a marker and mediator of cellular toxicity and disease states. Int J Exp Pathol 81(5):323–339
18. Orrenius S, Nicotera P (1994) The calcium ion and cell death. J Neural Transm Suppl 43:1–11

19. Choi MH, Ow JR, Yang ND, Taneja R (2016) Oxidative stress-mediated skeletal muscle degeneration: molecules, mechanisms, and therapies. Oxidative Med Cell Longev 2016:1–13
20. Rando TA, Disatnik MH, Yu Y, Franco A (1998) Muscle cells from mdx mice have an increased susceptibility to oxidative stress. Neuromuscul Disord 8(1):14–21
21. Chang WJ, Iannaccone ST, Lau KS, Masters BS, McCabe TJ, McMillan K et al (1996) Neuronal nitric oxide synthase and dystrophin-deficient muscular dystrophy. Proc Natl Acad Sci 93(17):9142–9147
22. Paul JR (1971) A history of poliomyelitis. In: A history of poliomyelitis. Yale University Press, New Haven/London
23. Racaniello VR, Baltimore D (1981) Molecular cloning of poliovirus cDNA and determination of the complete nucleotide sequence of the viral genome. Proc Natl Acad Sci 78(8):4887–4891
24. Kitamura N, Semler BL, Rothberg PG, Larsen GR, Adler CJ, Dorner AJ, Emini EA, Hanecak R, Lee JJ, van der Werf S, Anderson CW (1981) Primary structure, gene organization and polypeptide expression of poliovirus RNA. Nature 291(5816):547
25. Mehndiratta MM, Mehndiratta P, Pande R (2014) Poliomyelitis: historical facts, epidemiology, and current challenges in eradication. The Neurohospitalist 4(4):223–229
26. Lange R, Peng X, Wimmer E, Lipp M, Bernhardt G (2001) The poliovirus receptor CD155 mediates cell-to-matrix contacts by specifically binding to vitronectin. Virology 285 (2):218–227
27. Devi A, Reddy AB, Yadav UC (2018) Aldose reductase inhibitors in the functional foods: regulation of diabetic complications. In: Functional food and human health. Springer, Singapore, pp 555–574
28. Perry BD, Caldow MK, Brennan-Speranza TC, Sbaraglia M, Jerums G, Garnham A, Wong C, Levinger P, Asrar Ul Haq M, Hare DL, Price SR et al (2016) Muscle atrophy in patients with Type 2 Diabetes Mellitus: roles of inflammatory pathways, physical activity and exercise. Exerc Immunol Rev 22:94–109
29. Kalyani RR, Corriere M, Ferrucci L (2014) Age-related and disease-related muscle loss: the effect of diabetes, obesity, and other diseases. Lancet Diabet Endocrinol 2(10):819
30. O'Shea TM (2008) Diagnosis, treatment, and prevention of cerebral palsy. Clin Obstet Gynecol 51(4):816–828
31. Marret S, Vanhulle C, Laquerriere A (2013) Chapter 16: Pathophysiology of cerebral palsy. In: Handbook of clinical neurology. Elsevier, Amsterdam, p 111
32. Donnan GA, Fisher M, Macleod M, Davis SM (2008) Stroke. Lancet 371(9624):1612–1623
33. Chen J, Zhu M, Ma G, Zhao Z, Sun Z (2013) Chlamydia pneumoniae infection and cerebrovascular disease: a systematic review and meta-analysis. BMC Neurol 13:183. https://doi.org/10.1186/1471-2377-13-183
34. Heuschmann PU, Neureiter D, Gesslein M, Craiovan B, Maass M, Faller G et al (2001) Association between infection with Helicobacter pylori and Chlamydia pneumoniae and risk of ischemic stroke subtypes: results from a population-based case-control study. Stroke 32 (10):2253–2258
35. Guerriero RM, Giza CC, Rotenberg A (2015) Glutamate and GABA imbalance following traumatic brain injury. Curr Neurol Neurosci Rep 15(5):27
36. Escayg A, Goldin AL (2010) Sodium channel SCN1A and epilepsy: mutations and mechanisms. Epilepsia 51(9):1650–1658
37. Lerche H, Shah M, Beck H, Noebels J, Johnston D, Vincent A (2012) Ion channels in genetic and acquired forms of epilepsy. J Physiol 591(4):753–764
38. Hamley IW (2012) The amyloid beta peptide: a chemist's perspective. Role in Alzheimer's and fibrillization. Chem Rev 112(10):5147–5192
39. Barton S (2006) Microglia give amyloid plaques the brush off. Nat Rev Neurosci 7:254–255
40. Wang WY, Tan MS, Yu JT, Tan L (2015) Role of pro-inflammatory cytokines released from microglia in Alzheimer's disease. Ann Transl Med 3(10):136
41. Chun W, Johnson GV (2007) The role of tau phosphorylation and cleavage in neuronal cell death. Front Biosci 12:733–756

42. Galvan A, Wichmann T (2008) Pathophysiology of parkinsonism. Clin Neurophysiol 119 (7):1459–1474
43. Yasuda T, Nakata Y, Mochizuki H (2012) α-Synuclein and neuronal cell death. Mol Neurobiol 47(2):466–483
44. Nixon RA, Yang DS (2012) Autophagy and neuronal cell death in neurological disorders. Cold Spring Harb Perspect Biol 4(10):a008839. https://doi.org/10.1101/cshperspect.a008839
45. Reeve AK, Grady JP, Cosgrave EM, Bennison E, Chen C, Hepplewhite PD, Morris CM (2018) Mitochondrial dysfunction within the synapses of substantia nigra neurons in Parkinson's disease. npj Parkinson's Dis 4(1):9
46. Hwang O (2013) Role of oxidative stress in Parkinson's disease. Exp Neurobiol 22(1):11–17
47. Desai BS, Monahan AJ, Carvey PM, Hendey B (2007) Blood–brain barrier pathology in Alzheimer's and Parkinson's disease: implications for drug therapy. Cell Transplant 16 (3):285–299

Smart Rehabilitation for Neuro-Disability: A Review

24

Sateesh Reddy Avutu, Sudip Paul, and Dinesh Bhatia

Abstract

Neurological disorders are increasing globally due to various factors such as change in lifestyle patterns as well as personal and professional stress. Scientists belonging to various research and development organizations from a variety of disciplines are working together to invent new rehabilitation devices. Emerging trends in electronics and information technology-enabled devices provide a platform for scientists to engage in innovation to enhance the independence of patients with neurological disorders. This chapter summarizes the use of advanced technology to create smart rehabilitation devices such as wearable devices, non-invasive stimulators, brain-computer interface (BCI), and neuro-robotic systems. By the end of this chapter, the reader will have acquired knowledge about various advanced smart neuro-rehabilitation devices related to several neurological disorders. The current challenges faced by neuro-rehabilitation scientists are discussed at the end of the chapter.

Keywords

Neuro-rehabilitation · Wearable devices · Neural stimulators · Neuro-robotic systems

24.1 Introduction

The methodology used to assist a person to achieve psychological, physical, emotional, and vocational potential and engage in social activities so that dignity, self-respect, and a good quality of life can be maintained is called *rehabilitation*. Neuro-rehabilitation aims to assist a person to recover from a nervous system injury in order to minimize negative effects resulting from the injury. Rehabilitation should start

S. R. Avutu · S. Paul (✉) · D. Bhatia
Department of Biomedical Engineering, North-Eastern Hill University, Shillong, Meghalaya, India

S. Paul (ed.), *Application of Biomedical Engineering in Neuroscience*,
https://doi.org/10.1007/978-981-13-7142-4_24

with contacting the patient and end with enhancing independence, irrespective of the disability. Any rehabilitation device or methodology is designed to improve physical independence and provide psychological support, occupational integration, social integration, and mobility. Stroke, brain injury, spinal cord injury, Parkinson's disease, Alzheimer's disease, and multiple sclerosis are common neurological disorders. The tremendous advances in the field of electronics and information technology-enabled services in the past two decades have led the revolution in the medical devices' manufacturing industry. In view of product segmentation, neuro-rehabilitative devices are classified as wearable devices, non-invasive stimulators, brain-computer interface, and neuro-robotic systems.

Wearable devices are used to recognize or record physical activity. Non-invasive stimulators are used to stimulate various body parts such as bone, the spinal cord, nerves, and the vagus nerve. Brain-computer interface deals with movement control of any rehabilitation device through biological signals such as electromyography (EMG), electroencephalogram (EEG), and electrooculography (EOG). The neuro-robotic system can be used for motor control, action selection, memory and perception.

24.2 Wearable Devices

Wearable devices are particularly useful for pain management, physiological signal tracking, glucose monitoring, sleep monitoring, and brain activity of individuals in the home. A wearable device provides timely access to rehabilitation, which is a key factor for neurological disorder patients. These devices are classified based on their proximity to the human body, appearance, function, and other parameters. Physiological and movement-sensing data collection hardware, the communication setup to transfer the data to a remote center, as well as data analytical tools to extract the relevant information, are the main building blocks of any wearable medical device. A graphic representation of a wearable sensor-based health-monitoring system is shown in Fig. 24.1.

24.2.1 Physiological Signal-Tracking Devices

The continuous sensing and recording of physiological parameters helps people to track their metabolic status. Emerging technology in electronics, microfluidics, electrochemical biosensors, and artificial intelligence algorithms facilitate the process for manufacturers who build wearable devices to generate real-time medical data compatible with the Internet of Things. The design characteristics of wearable devices vary with respect to their application. Wearable devices that can track biochemical and electrophysiological signals can be categorized as skin-mounted or implantable, as shown in Fig. 24.2. Skin-mounted devices should be designed in such a way that there is compatibility between the mechanical properties of the

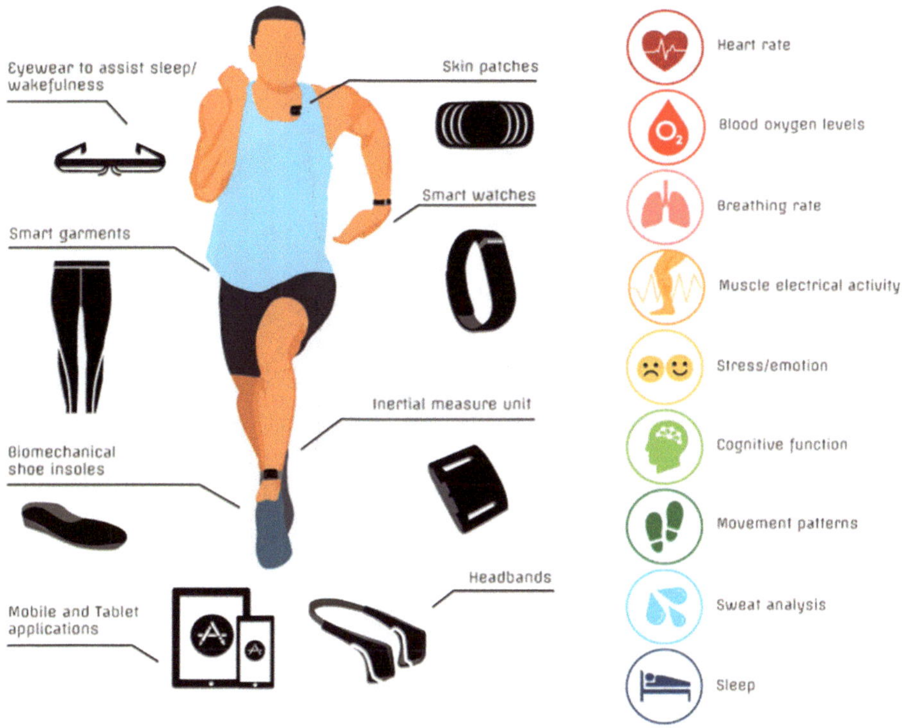

Fig. 24.1 Graphic representation of a wearable sensor-based health-monitoring system [1]

device surface and the skin. Heikenfeld et al. aptly explained the historical development of a number of variable sensors [3].

The smart electronic shoe has the ability to define the gait of a person. A wide range of health and disease conditions can be predicted through walking style. Bjoern M. Eskofier et al. did a comprehensive review of various medical aspects of smart shoes and digital health applications [4]. Smart socks can be used to measure foot pressure and walking distance. The user can view the results via a mobile app on a smart phone [5]. Tsung-Chien Lu et al. have explained the various health applications and historical developments in smart watch technology [6]. The emerging trends in 3D printing technology have led to the invention of the digital tattoo. Flexible materials are used to make thin and "invisible" sensors interwoven with the tattoo. Crystal M. Glassy et al. aptly explained current trends and limitations of the digital tattoo [7].

24.2.2 Brain Activity and Sleep-Monitoring Devices

The EEG can be used to record and monitor brain activities such as sleep patterns, cognition, gait, and gaze analysis. Various devices used for brain activity monitoring

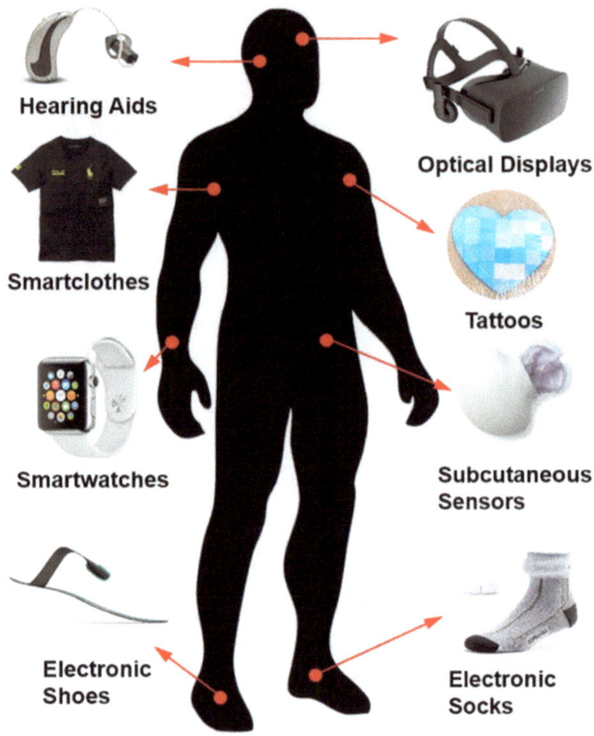

Fig. 24.2 Wearable devices to track the biochemical and electrophysiological signal [2]

and the outcomes of these devices are shown in Table. 24.1. KwangSuk Park et al. discussed the role of smart technologies for sleep monitoring at home, as well as parameters representing the quality of sleep [9]. Depending on the nature of the neurological disorder, continuous EEG monitoring is often required. Yuichi Kubota et al. gave a comprehensive review of various methodologies utilized for continuous EEG monitoring in ICU [10].

24.2.3 Smart Garments

Emerging technologies in chemical engineering, material science, and sensors have led to the invention of smart garments containing interlaced biosensors to capture a variety of electrophysiological signals. Smart garments are the basic building blocks for the Internet of Things-based healthcare-monitoring systems. Tiago et al. have provided information about IoT-based garments, as well as the past, present, and future of smart wearables and clothing. Mulatier et al. [11] explained historical developments where electronic components are integrated with textiles, as shown in Fig. 24.3 [12].

Table 24.1 Various devices used for brain activity monitoring showing outcome of the device [8]

Device Nature	Relevant Device Manufactures	Outcome of Device
Portable EEG headband	MUSE (InteraXon Inc. Toronto, Canada), Emotiv EPOC (Emotiv Inc., Sydney, Australia)	Event-related brain potential
Wrist-worn actigraphy device	Actiwatch2 (Philips Respironics, Murrysville, PA), Motionwatch 8 (CamNtech, Cambridge, UK)	Sleep quality and quantity and circadian rhythms
Peripheral arterial tone wrist/ hand-worn device	WatchPAT (Itamar Medical, Caesarea, Israel)	Sleep architecture
Forehead-worn sleep monitor	Sleep Profiler (Advanced BrainMonitoring, Carlsbad, CA)	EEG, EOG, EMG, ECG
Non-contact sleep sensor	Beddit 3 Sleep Monitoring System (Apple, Cupertino, CA)	Sleep quality and quantity and circadian rhythms
Wearable gait monitor	McRoberts (The Hague, Netherlands)	Detailed gait measurements
Gait monitoring insole	Moticon insole (Moticon GmbH, Munich, Germany)	Detailed gait measurements
Cognitive function	Project EVO (Akili Interactive Labs, Boston, MA)	Multitasking
Mobile eye-tracking solutions	SensoMotoric Instruments (SMI, Teltow, Germany)	Eye-tracking endpoints

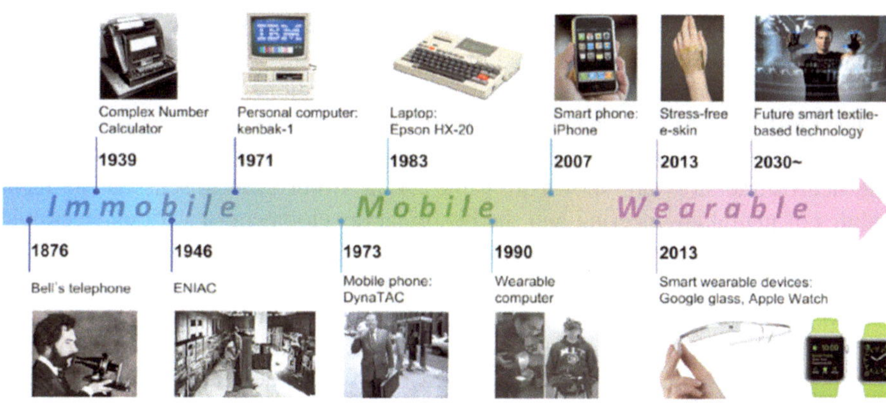

Fig. 24.3 Integration levels of smart clothing [12]

24.3 Neurostimulators

The purposeful modulation of the nervous system's activity using either invasive or non-invasive methods is called *neurostimulation*. Micro-electrodes are needed for this invasive procedure. Non-invasive stimulation can be done through transcranial

Table 24.2 Commercially and experimentally available neurostimulators [13]

Device	Sensor	Sensed variable
Cochlear implant[a]	Microphone	Air pressure variations
Foot-drop stimulator (external)[a]	Under-heel switch or force sensor or tilt sensor	Under-heel pressure or leg tilt
Deep brain stimulator implant[a]	EMG and accelerometers: experimental	Tremor: experimental
Grasp-release stimulators[a]	Push-button or earpiece	Taps, tooth clicks, head nods
StimRouter implant[a]	Earpiece	Tooth clicks
Visual cortex implants	Video	Visual field
Retinal implants	Video, eye position sensor	Visual field, eye position
Somatosensory cortex implants	Force or strain sensors	Pressure or movement
Spinal cord epidural stimulator implants (spatiotemporal)	Video motion capture, cortical electrode arrays	Limb position or cortical commands
Intra spinal micro stimulation (limb movements)	External goniometers	Limb position
Intra spinal micro stimulation (bladder controls)	Bladder and sphincter pressures	Urethral pressure sensor
Intra spinal micro stimulation (respiration)	EMG	Genioglossus EMG

([a]commercial, experimental)

magnetic stimulation or transcranial electric stimulation; each procedure has its own advantages and disadvantages. Based on the nature of the disorder, stimulation can be done on the brain, deep brain, spinal cord, heart, or bone.

The most successful implantable neurostimulator is the cochlear implant used for severe sensorineural hearing-loss patients. Commercially and experimentally available neurostimulators are shown in Table 24.2. Arthur et al. discussed the main features of neural engineering and the historical development of various electronic devices related to neuro-rehabilitation [13]. Shepherd et al. discussed the development of neurostimulators and gave a brief summary of preclinical safety and value issues. The animal models used to evaluate the safety and efficiency of neurostimulators are thoroughly explained in research [14]. Neurostimulation can be classified as spinal cord stimulation (SCS), peripheral nerve stimulation, and intracranial stimulation. Intracranial stimulation is also classified as deep brain stimulation and motor cortex stimulation.

Spinal cord stimulation (SCS) uses electrical stimulation through leads placed in the spinal dorsal epidural space. The basic mechanisms involved in conventional SCS are shown in Fig. 24.4. The activation of inhibitory interneurons in the dorsal horns can be achieved by antidromic DC afferent stimulation. The orthodromic impulses in the DCs activate the neurons in locus coeruleus (LC) in the brainstem

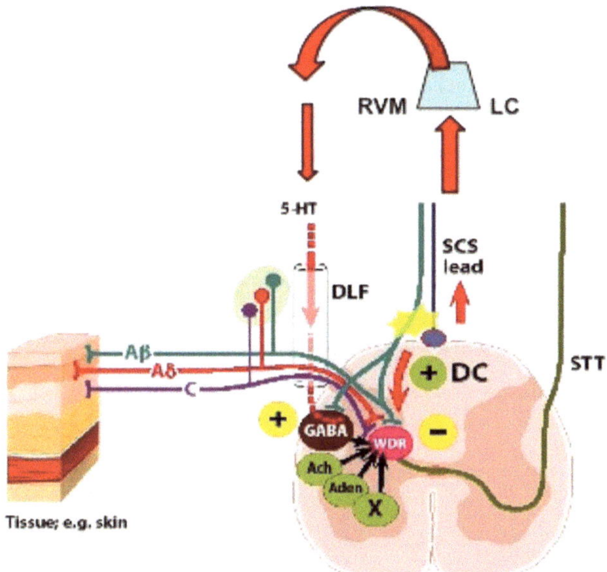

Fig. 24.4 Basic mechanisms involved in conventional SCS

and the neurons in the RVM, leading to descending inhibition and further strengthening the segmental inhibitory mechanisms. Bengt Linderoth et al. briefly explained the hypothetical mechanisms and action of conventional and novel spinal stimulation algorithms [15].

Peripheral nerve stimulation (PNS) is commonly used to treat chronic pain. The peripheral nerves are located beyond the brain and spinal cord. Electrical pulses like mild tingles are used to stimulate the nerves. The person is able to control the stimulation by adjusting the corresponding parameters. PNS is used for specific clinical indications such as complex regional pain syndromes, back pain, migraine headaches, and urinary bladder incontinence.

Deep brain stimulation (DBS) has, for several decades, been used to treat intractable pain. More recently, these methods have been used to treat involuntary movements associated with multiple sclerosis, dystonia, and Parkinson's disease. The principle of DBS is similar to that of the cardiac pacemaker. Dysfunctional circuits are used to send continuous electrical signals to specific target areas of the brain to stop the impulses that cause the neurological dysfunction. The DBS system consists of three components: thin electrodes implanted in the brain, connecting leads for the internal pulse generator (IPG), and a neurostimulator. The advantages and disadvantages of DBS are shown in Table 24.3.

Jonathan R Ellenbogen and Keyoumars Ashkan explained common deep brain stimulation targets corresponding to various disorders [16]. A graphic summary of common neuromodulator devices is shown in Fig. 24.5.

Table 24.3 Advantages and disadvantages of DBS

Advantages	Disadvantages
Stimulation can be applied to both sides of the brain for control of symptoms affecting both sides of the body	Temporary pain and allergic reaction at the implantation site
The effects are reversible	Temporary tingling in the face or limbs
The potential side effects can be minimized by adjusting the stimulation parameters, and efficiency will improve over time	Slight paralysis, speech, and vision problems, as well as loss of balance
It provides continuous symptom control	Dizziness and concentration difficulties
Patients have the flexibility to undergo other treatment options such as stem cell or gene therapy	Reduced coordination and a shock sensation

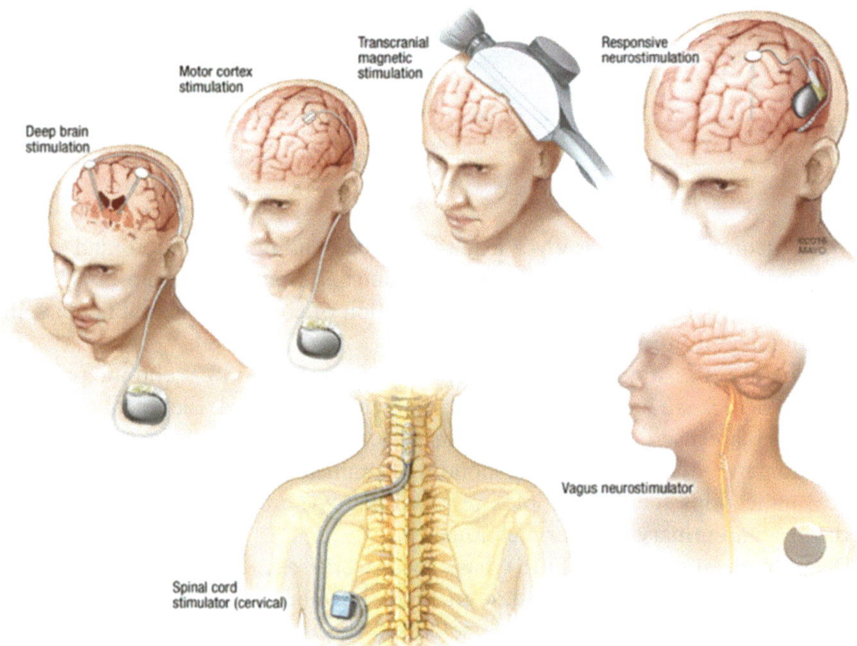

Fig. 24.5 Graphic summary of common neuromodulator devices

Motor cortex stimulation (MCS) is much easier than DBS. In this method, the electrodes are placed on the motor cortex surface, whereas in DBS, the electrodes are placed in the brain, as shown in Fig. 24.5. The stimulation process is performed in two steps. At step one, the patient is required to undergo functional MRI to map the brain's motor cortex. At step two, the neurosurgeon will estimate the electrode placement over the targeted area of the brain, based on the precise roadmap of the brain. C. Michael Honey et al. discussed the differences between future research challenges regarding DBS and MCS in the design of invasive neurostimulators [17].

24.4 Brain Computer Interface

Brain-computer interface (BCI) is a neural control interface used to interact with computers through brain activity that is measured by electroencephalography. Any standard BCI system can be built with four major parts: signal acquisition, signal processing, feature extraction, and classification. The main application of BCI for rehabilitation engineering is to control the movement of any device such as a prosthesis, orthosis, wheelchair, or robot. Sometimes the output of the BCI is used to direct electrical stimulation of the brain or muscle. Based on the placement of the electrodes, BCIs can be classified as invasive or noninvasive. In a noninvasive BCI system, the electrodes are mounted on the scalp or external surface of the head. Signals like slow cortical potentials, sensorimotor rhythms, P 300 event-related potential, steady-state visual-evoked potential, error-related negative-evoked potentials, blood oxygenation levels, and cerebral oxygenation changes can be measured. Electrodes or multi-electrode grids are used for signal acquisition, such as local field potentials, single unit activity, and multiunit activity. Calcium channel permeability and electro-cortico graphic oscillations can be acquired from the electrodes placed at the cortical surface. The general framework for a BCI system is shown in Fig. 24.6. Ujwal Chaudhary et al. explained the various applications of BCI systems in rehabilitation engineering [18].

Reza Abiri et al. gave a comprehensive review of EEG-based brain-computer interface paradigms. The advantages and disadvantages of current BCI paradigms – from a variety of perspectives – and various EEG decoding algorithms with classification methodologies were evaluated. Potential problems with the current systems as well as possible solutions were identified [19]. The compression between different modes of signal acquisition is shown in Table 24.4.

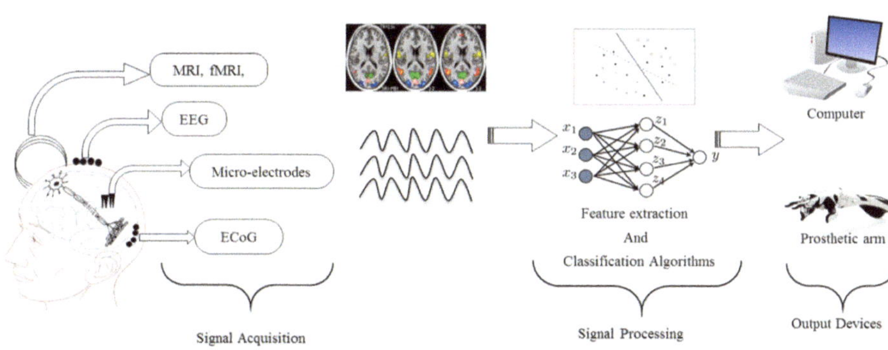

Fig. 24.6 The general framework of a BCI system [20]

Table 24.4 Comparison of different signal acquisition methods

Highly invasive	Non-invasive	Partially invasive
Very short-term recording	Long-term recording	Short-term recording
Risk of infection and inflammation	No risk of infection	Risk of infection
Higher spatial resolution	Poor spatial resolution	High spatial resolution
Medical training and assistance required	Medical training and assistance not required	Medical training and assistance required
Difficult to monitor	Easy to monitor	Difficult to monitor
Costly	Cost efficient	Costly

Table 24.5 The various classifiers and their features [22]

Name of classifier	Features
LDA	Widely used in BMI-based systems as well as BMI systems based on EAs. Dimensionality is a problem when data with low numbers of observations is encountered. Lack of efficiency when noise and outliers are encountered. Generalizable as a nonlinear classifier using kernel methods
SVM	Linear and stable. Robustness against overtraining. Generalizable as a nonlinear classifier using kernel methods. Lack of efficiency when noise and outliers are encountered
LVQ	Nonlinear and unstable, less frequently used in BMI-based systems
PNN	Less frequently used in BMI-based systems. Robustness against noise and outliers. High computation. Nonlinear and unstable
RBF-NN	Nonlinear and stable. Performance dependent on the number of selected centers in the hidden layer
MLP	Nonlinear and unstable. Performance dependent on the number of hidden layers as well as the number of neurons in the hidden layers. Widely used in BMI-based systems as well as BMI systems based on EAs
NB	Less frequently used in BMI-based systems. Acceptable speed when observations of a large dimension are encountered. Simplicity and low computation. Nonlinear and stable
k-NN	Inefficiency when observations of a large dimension are encountered. Varying performance based on data type. Nonlinear and stable
LR	Less frequently used in BMI-based systems. Nonlinear and unstable
Boosting	Lack of efficiency when noise and outliers are encountered. Non-linear
Bagging	Acceptable performance when unstable classifiers are used as base classifiers. Robustness against noise and outliers. Acceptable generalizability

Rajesh PN Raogave gave a comprehensive review of recent progress in simultaneous decoding and encoding for closed-loop control and plasticity induction, and introduced a unifying framework for developing brain co-processors based on deep learning and neural networks. This framework helps to achieve desired behaviors such as targeted augmentation of brain function [21]. FarajollahTahernezhad-Javazm et al. gave a comprehensive review of theoretical and experimental evolutionary methods for EEG-based BMI systems [22–23] (Table 24.5).

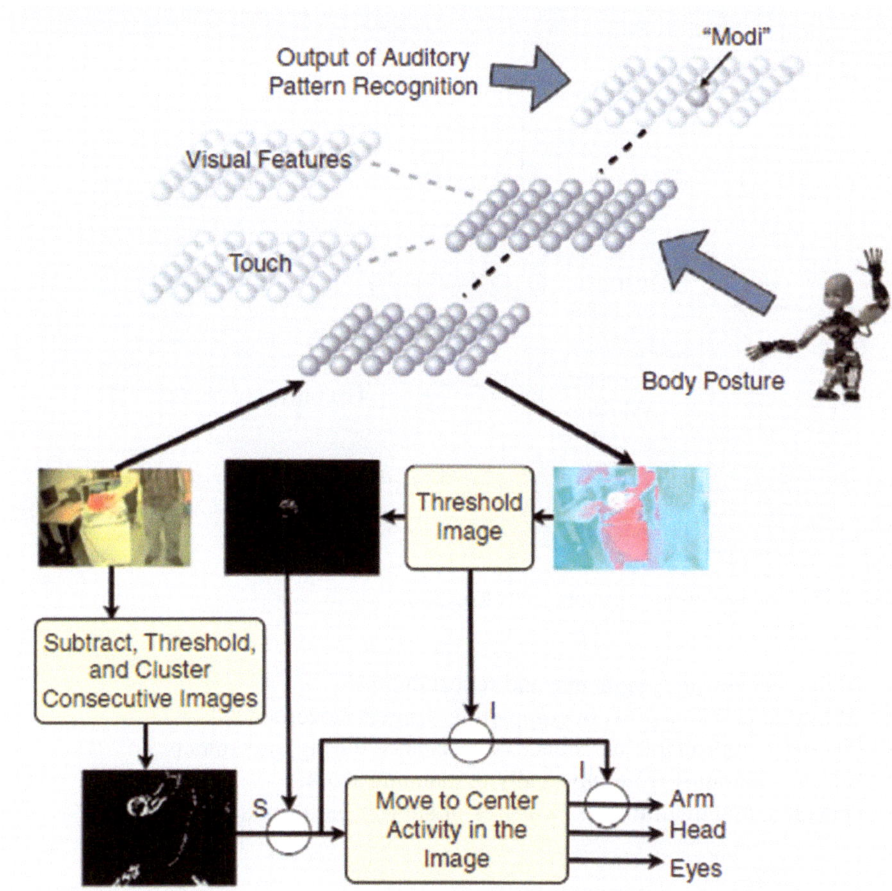

Fig. 24.7 The cognitive architecture for developmental language-learning experiments

24.5 Neuro-Robotic Systems

A robot is an electromechanical device endowed with artificial intelligence that has the ability to take decisions autonomously or partially autonomously as a result of programming. A neuro-robotic system belongs to interdisciplinary research and makes use of neuro-science, artificial engineering, and robotics. Locomotion and motor control, learning and memory systems, and sensory perception are the sub-classes of neuro-robotic models. Emerging technology development in electronics and information technology has led to innovations, with many commercial neuro-prosthetic, robotics, and human-robot interaction applications. The cognitive architecture for the developmental language-learning experiments is shown in Fig. 24.7. The ideal features [24] of the neuro robot are shown below:

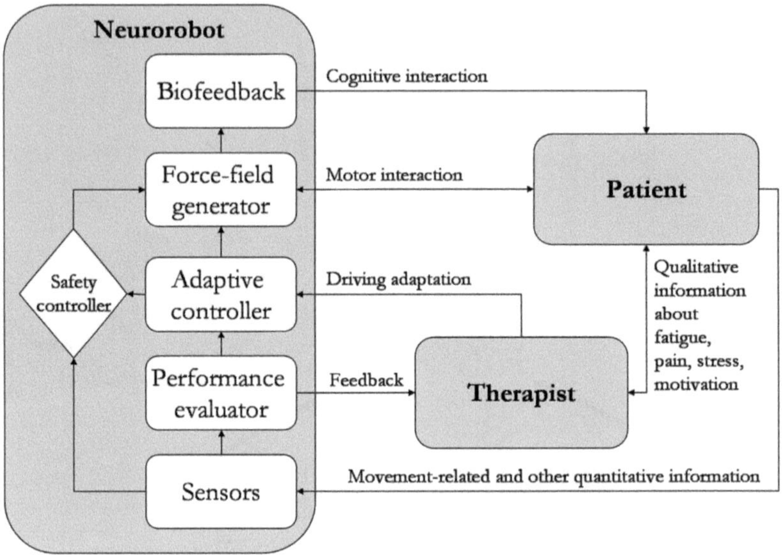

Fig. 24.8 The ideal patient-therapist robot loop [24]

- High mechanical compliance and repeatability.
- Ability to assist patients in completing desired movements.
- Precisely controllable assistance or resistance during movements.
- Relieves therapist from physically demanding work.
- Objective and quantifiable measures of subject performance.

Ethical rules for robot programmers who utilize artificial intelligence are based on three laws of robotics. Marco Iosa et al. accurately explained the laws for neuro-rehabilitation robots [24]. The ideal patient therapist robot loop is shown in Fig. 24.8.

- A robot should not allow a human being to harm or injure another human being.
- A robot must obey the orders given by the operator. An order conflicting with the first law is an exception.
- A robot must defend its own existence as long as such protection does not contradict the above two laws [24].

BCI is an emerging technology that acts as an alternative communication system between the brain and an output device. It provides an augmentative communication between the brain and the output device by creating a muscle-free channel [26]. It is useful in creating location and mobility rehabilitation for people suffering from various neurological disorders (Fig. 24.9).

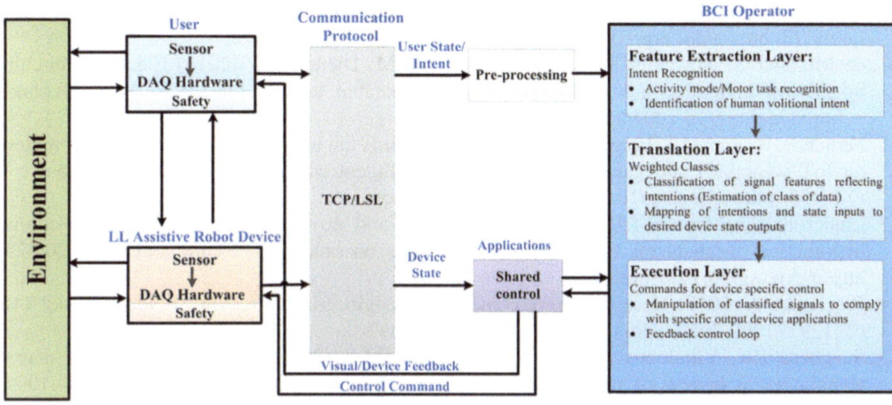

Fig. 24.9 The general framework for BCI-controlled neuro-assistive device control [25]

24.6 Conclusion

This chapter summarizes the use of advanced technology to make smart rehabilitation devices. The emerging trends in the design of non-invasive stimulators, brain computer interface (BCI), and neuro-robotic systems are explained.

References

1. Eskofier B, Lee S, Baron M, Simon A, Martindale C, Gabner H, Klucken J (2017) An Overview of Smart Shoes in the Internet of Health Things: Gait and Mobility Assessment in Health Promotion and Disease Monitoring. Appl Sci 7(10):986. https://doi.org/10.3390/app7100986
2. Gonçalves C, Ferreira da Silva A, Gomes J, Simoes R (2018) Wearable E-textile technologies: a review on sensors, actuators and control elements. Inventions 3(1):14. https://doi.org/10.3390/inventions3010014
3. Heikenfeld J, Jajack A, Rogers J, Gutruf P, Tian L, Pan T et al (2018) Wearable sensors: Modalities, challenges, and prospects. Lab Chip R Soc Chem 18:217. https://doi.org/10.1039/c7lc00914c
4. Peake JM, Kerr G, Sullivan JP (2018) A critical review of consumer wearables, mobile applications, and equipment for providing biofeedback, monitoring stress, and sleep in physically active populations. Front Physiol 9. https://doi.org/10.3389/fphys.2018.00743
5. Yetisen AK, Martinez-Hurtado JL, Ünal B, Khademhosseini A, Butt H (2018) Wearables in medicine. Adv Mater 30. https://doi.org/10.1002/adma.201706910
6. Byrom B, McCarthy M, Schueler P, Muehlhausen W (2018) Brain monitoring devices in neuroscience clinical research: The potential of remote monitoring using sensors, wearables, and mobile devices. Clin Pharmacol Ther 104(1):59–71. https://doi.org/10.1002/cpt.1077
7. Glassy CM, Glassy MS, Aldasouqi S (2012) Tattooing: medical uses and problems. Cleve Clin J Med 79:761. https://doi.org/10.3949/ccjm.79a.12016
8. Kubota Y, Nakamoto H, Egawa S, Kawamata T (2018) Continuous EEG monitoring in ICU. J Intensive Care 6:39. https://doi.org/10.1186/s40560-018-0310-z
9. Lu T-C, Fu C-M, Ma M, Fang C-C, Turner A (2016) Healthcare Applications of Smart Watches. Appl Clin Inform 07(03):850–869. https://doi.org/10.4338/aci-2016-03-r-0042

10. Park KS, Choi SH (2019) Smart technologies toward sleep monitoring at home. Biomed Eng Lett 9:73. https://doi.org/10.1007/s13534-018-0091-2
11. de Mulatier S, Nasreldin M, Delattre R, Ramuz M, Djenizian T (2018) Electronic circuits integration in textiles for data processing in wearable technologies. Adv Mater Technol 3. https://doi.org/10.1002/admt.201700320
12. Fernandez-Caramés T, Fraga-Lamas P (2018) Towards the internet-of-smart-clothing: a review on IoT wearables and garments for creating intelligent connected E-textiles. Electronics 7 (12):405. https://doi.org/10.3390/electronics7120405
13. Linderoth B, Foreman RD (2017) Conventional and novel spinal stimulation algorithms: hypothetical mechanisms of action and comments on outcomes. Neuromodulation 20:525. https://doi.org/10.1111/ner.12624
14. Prochazka A (2017) Neurophysiology and neural engineering: a review. J Neurophysiol 118 (2):1292–1309. https://doi.org/10.1152/jn.00149.2017
15. Shepherd RK, Villalobos J, Burns O, Nayagam DAX (2018) The development of neural stimulators: a review of preclinical safety and efficacy studies. J Neural Eng 15:041004. https://doi.org/10.1088/1741-2552/aac43c
16. Abiri R, Borhani S, Sellers EW, Jiang Y, Zhao X (2019) A comprehensive review of EEG-based brain-computer interface paradigms. J Neural Eng 16:011001. https://doi.org/10.1088/1741-2552/aaf12e
17. Chaudhary U, Birbaumer N, Ramos-Murguialday A (2016) Brain-computer interfaces for communication and rehabilitation. Nat Rev Neurol 12:513
18. Ellenbogen JR, Ashkan K (2018) Neurosurgery of epilepsy, movement disorders and pain. Surgery (United Kingdom) 36:655. https://doi.org/10.1016/j.mpsur.2018.09.002
19. Honey CM, Tronnier VM, Honey CR (2016) Deep brain stimulation versus motor cortex stimulation for neuropathic pain: a minireview of the literature and proposal for future research. Comput Struct Biotechnol J 14:234. https://doi.org/10.1016/j.csbj.2016.06.003
20. Bablani A, Edla DR, Tripathi D, Cheruku R (2019) Survey on brain-computer interface. ACM Comput Surv 52(1):1–32. https://doi.org/10.1145/3297713
21. Cangelosi A, Invitto S (2017) Human-robot interaction and neuroprosthetics: a review of new technologies. IEEE Consumer Electron Mag 6:24. https://doi.org/10.1109/MCE.2016.2614423
22. Iosa M, Morone G, Cherubini A, Paolucci S (2016) The three laws of neurorobotics: a review on what neurorehabilitation robots should do for patients and clinicians. J Med Biol Eng 36:1. https://doi.org/10.1007/s40846-016-0115-2
23. Rao RP (2019) Towards neural co-processors for the brain: combining decoding and encoding in brain–computer interfaces. Curr Opin Neurobiol 55:142. https://doi.org/10.1016/j.conb.2019.03.008
24. Tahernezhad-Javazm F, Azimirad V, Shoaran M (2018) A review and experimental study on the application of classifiers and evolutionary algorithms in EEG-based brain-machine interface systems. J Neural Eng 15:021007. https://doi.org/10.1088/1741-2552/aa8063
25. Deng W, Papavasileiou I, Qiao Z, Zhang W, Lam KY, Han S (2018) Advances in automation technologies for lower extremity neurorehabilitation: a review and future challenges. IEEE Rev Biomed Eng 11:289–305. https://doi.org/10.1109/RBME.2018.2830805
26. Tariq M, Trivailo PM, Simic M (2018) EEG-based BCI control schemes for lower-limb assistive-robots. Front Hum Neurosci 12(August):312. https://doi.org/10.3389/fnhum.2018.00312

MIX
Papier aus verantwortungsvollen Quellen
Paper from responsible sources
FSC® C105338

If you have any concerns about our products,
you can contact us on
ProductSafety@springernature.com

In case Publisher is established outside the EU,
the EU authorized representative is:
Springer Nature Customer Service Center GmbH
Europaplatz 3, 69115 Heidelberg, Germany

Printed by Libri Plureos GmbH
in Hamburg, Germany